A HISTORY OF
TECHNOLOGY

A HISTORY OF
TECHNOLOGY

EDITED BY
TREVOR I. WILLIAMS

VOLUME VI
THE TWENTIETH CENTURY
c. 1900 TO *c.* 1950
PART I

CLARENDON PRESS · OXFORD
1978

Oxford University Press, Walton Street, Oxford OX2 6DP

OXFORD   LONDON   GLASGOW
NEW YORK   TORONTO   MELBOURNE   WELLINGTON
IBADAN   NAIROBI   DAR ES SALAAM   LUSAKA   CAPE TOWN
KUALA LUMPUR   SINGAPORE   JAKARTA   HONG KONG   TOKYO
DELHI   BOMBAY   CALCUTTA   MADRAS   KARACHI

© OXFORD UNIVERSITY PRESS 1978

British Library Cataloguing in Publication Data

A history of technology.
  Vol. VI
  I. Williams, Trevor Illtyd
  609   T15   78–40539

  ISBN 0–19–858151–3

Typeset by Gloucester Typesetting Co. Ltd.
Printed in Great Britain
by Fletcher & Son Ltd., Norwich

PREFACE

THE five previous volumes of this *History* were published over the period 1954–8. In the preface to the fifth volume I joined with my then co-editors—Charles Singer, E. J. Holmyard, and A. R. Hall—in listing 'a number of reasons (why) it would be impracticable to extend its range to include the twentieth century'. Readers of these two further volumes are, therefore, entitled to inquire what has happened to cause this conclusion to be revised.

One reason then advanced was that the amount of additional space required would be excessive. This was certainly true at the time: the subvention made available to bring the original work to a conclusion was exhausted and it did not then appear that the demand for such a substantial work in this field would be sufficient to make it self-sustaining. In the event, this judgement proved wrong. The demand for the original edition has proved several times greater than even modest optimism had suggested at the outset. Additionally, there have been foreign editions in Italian and Japanese, as well as a large printing in the U.S.A. by the Library of Science Book Club. All this indicated that an extension of the work would be both welcome and feasible, even though a survey of the ground to be covered indicated that two further volumes, not one, would be necessary to treat the first half of this century at the same sort of level as the five earlier volumes had reviewed the history of technology from the dawn of civilization to 1900.

Another reason advanced for originally making 1900 the *terminus ad quem* was the acknowledged difficulty of treating comparatively recent events in a way that distinguishes what is historically significant from what is not. To an extent, the mere passage of time has diminished this difficulty. Had we tried, at that time, to bring the story up to 1950 we should, at the end, have been treating not history but current affairs. Now, in the late 1970s, we have at least the benefit of the lapse of a quarter of a century since the end of our new period in which to make some sort of assessment. Not much, perhaps, but historians of other aspects of human activity have latterly shown a growing readiness to record and assess comparatively recent events and this may give confidence to historians of technology of their ability to do likewise.

The third reason that led us to pause at the beginning of this century was that 'it would . . . be impossible to present the recent development of technology in the relatively non-technical manner adopted in these present volumes.' This remains true, although we overestimated the difficulties in respect of a number of fields. Again the passage of time has to some extent changed the situation. The success of technology in the twentieth century, and in particular its dramatic influence on the course of the Second World War, left governments in no doubts about the potential contribution of science to improving the material prosperity of mankind—with all that that implied for social improvement—and its immense strategic importance. This was reflected in a great expansion of school and university education: broadly based but with emphasis on numeracy and an understanding of basic scientific principles. This movement has now been operative long enough for it to have produced a whole new generation of readers capable of understanding a degree of technical exposition that their predecessors would have found difficult. Moreover, the interests of our readers have significantly changed. For the original volumes it sufficed, very largely, to concentrate on what things were made and how they were made. Today, however, the obvious impact of economic, social, and political factors on the development of technology has engendered a growing interest in these aspects that cannot be ignored. While they are individually complex, and their interplay more so, they are subjects of common interest and debate and the difficulties of expounding them to a general readership are of a different kind, and more surmountable, than those presented by the purely scientific aspects.

How far these non-technological factors are relevant is, of course, a matter of debate. I would not accept the extreme view that the ultimate historical work is one in which every facet is considered roughly in proportion to its supposed significance. Some degree of compartmentalization seems to me inescapable: the separate stocks of knowledge thus established can then be drawn on to provide a variety of broader—but not exhaustive—syntheses. Equally, however, I do not support a narrow interpretation that tries to divorce the history of technology altogether from external circumstances. It is an inescapable fact that the history of technology is often profoundly affected by extraneous events. No one would question, for example, that the history of the development of atomic energy would have been totally different but for the Second World War; indeed, it would probably scarcely have featured in these volumes at all. The history of technology influences, and is influenced by, events in the world at large. Surely the significance of events—the reasons that led up to them and the consequences that followed—is as interesting and important as the events themselves? This applies with all the greater force if one believes that the lessons of the past can be a guide to the future. To emphasize this, these volumes are prefaced by a short general historical review of world history, to remind the reader of the general background to the more specialized chapters that follow.

These, then, were the guiding principles, but to translate them into a workable plan proved a long and difficult undertaking. Naively, one tends to cherish the belief, against past experience, that ultimately there will emerge— as an unpromising sediment suddenly changes in the test-tube to a mass of brilliant crystals—a scheme in which every subject to be considered finds a logical place, duplication is avoided, and there are no loose ends. The reality is very different: no plan is ideal, though some are demonstrably better than others. Thus certain topics can be considered under various headings. Fertilizers, for example, could equally appropriately be placed under the headings of their manufacture by chemical industry or their use in agriculture. With polymers, too, we must consider both their manufacture as a raw material and their subsequent conversion into clothing, electrical fittings, domestic equipment, and paints. Should one attempt to consider all these different aspects together, or should they be divided among chapters on the chemical, agricultural, textile, electrical, and paint industries? Again, duplication is in theory to be avoided, if only on the ground that it wastes valuable space. Yet in practice it cannot be avoided altogether without destroying the unity of individual chapters. Finally, one cannot ignore the limitations as well as the strengths of contributors, and the allocation of subjects has had to take this into account.

Again, there is the matter of length. There is a case for allocating space on the basis of, say, the economic importance of the technology concerned. Yet the technology of some of the basic industries—coal, glass, and ceramics, for example—remained relatively simple compared with that of, say, the chemical or the new electronic industries. I tried, therefore, to allocate space in accordance with the needs of each subject for its adequate treatment as well as of its importance, as they appeared before a single contribution had been received.

Finally, there is the matter of internal discrepancies. Ideally, of course, there should be no inconsistencies over simple matters of fact—dates, initials, place-names, etc—and it is the Editor's task to eliminate these. But there is a second level of inconsistency which is less easily dealt with. For their various purposes different authors derive statistical information, for example, from different sources; these do not necessarily agree among themselves. There is a further level of inconsistency, much more difficult to assess, that arises when there are clashes about matters of opinion. Such may arise, for example, in assessing the root causes of technological innovation or the influence of government control. Generally, this is symptomatic of the fact that the subject is developing and there is still plenty of room for argument. In such cases, it is no part of the editorial function to demand a single interpretation: the reader must be free to draw his own conclusions.

Having regard to all these constraints, it has been necessary to adopt a pragmatic approach. While I hope that no major subject has been omitted, I make no apology for a limited amount of repetition, nor for the fact that a few topics may seem to be considered a little out of context, nor that the opinions of some contributors may be at variance with those of others. The plan of such work can be argued interminably but if a start is to be made at all a firm decision must be taken and adhered to. Like their predecessors, these volumes make no claim to be a definitive history. The intention has been to provide a general conspectus which may serve as a basis for more specialized studies.

From the point of view of these new volumes, the post-war burgeoning of science and technology has not, it must be admitted, been wholly helpful. Twenty-five years ago, it seemed that science and technology could not go wrong, that the possibilities for improving material prosperity—the rock upon which social progress must be founded—were almost unlimited. In the event, the heady euphoria of those days has not proved justified. In spite of—or, as some would hold, because of—technological progress the world is still full of strife. An uneasy peace has been maintained, in the sense that another world war has not engulfed us, but the universal brotherhood of man seems as remote as ever. Recession and inflation are, at this moment, almost worldwide afflictions. In such circumstances, provision for the present volumes of the *History* has necessarily been more austere than that for its predecessors. Editorial staff, secretarial help, funds to provide special illustrations, and all those aids forthcoming in more prosperous days have been lacking. However, by way of compensation for this manifest disadvantage I have been fortunate in having had much informal advice from friends and colleagues as the project took shape.

As indicated above, I have thought it desirable to lay emphasis on economic, social, and political factors. All contributors have been asked to keep these in mind, but in addition a number of chapters dealing specifically with such subjects have been included. It is abundantly clear that the history of technology is not determined simply by man's ability to do things. Innovation is dependent on the existence of a favourable social milieu; on the availability of capital and the readiness of those who control it to deploy it in a particular direction; on the existence of an appropriately educated public; and so on. Nor can we divorce from the history of technology the role of management and the unions. In our own time the Manhattan Project and the Moon landings represent the peak of technological achievement, but it is fair to say that they were a triumph as much for the managers of these immensely complex operations as for the scientists and engineers acting in a strictly technical capacity.

Subject to this change of emphasis, these volumes are an extension of their predecessors. While they provide in themselves a history of technology in the twentieth century, almost all aspects of this have their roots in the practice of earlier days; there is, therefore, extensive cross-referencing to the earlier volumes. While the general remit to contributors was that they should confine themselves to the period 1900–50, this has been treated fairly elastically. Some subjects, such as the computer, appear here for the first time, and it would have been illogical to ignore nineteenth-century work which was fundamental to modern developments. At the other end of the scale, the reader would surely have been disappointed if there had been no mention of space flight or of the beginning of atomic power. Generally, the precise period covered by each chapter has been determined by the need to avoid, as far as possible, too abrupt a beginning or leaving the end hanging in the air.

While the conception and planning of such a work as this must necessarily be the responsibility of the Editor its execution is dependent on the contributors. The number of professional historians of technology is rather small and the interests of the majority lie mainly in periods earlier than the present century. It became clear at an early stage that a great deal of reliance would have to be placed on contributors who were not primarily historians but had nevertheless taken a serious interest in the recent history of their own particular fields. To all such contributors I am particularly grateful, because the preparation of their chapters to the exacting requirements of the work as a whole has for them been exceptionally demanding. When all have been so helpful, it may seem invidious to mention any particular contributor by name. I must, nevertheless, acknowledge the exceptional help I received from Lord Hinton in connection with the section on atomic energy. The preparation of this was beset with difficulties, not least the sad death of Robert Spence while he was engaged in preparing the section on the chemical aspects. Apart from dispelling great editorial anxieties, the book has benefited by having much more than was originally planned from the pen of one who can fairly be said to have himself contributed very significantly to shaping the history of atomic energy.

As for the Chinese philosopher every quality has its anti-quality, as the physicist now deals in matter and anti-matter, so the Editor of any substantial work must reckon with the non-contributor as well as the contributor. Statistically, it is almost inevitable that a few contributors will fail altogether to deliver the goods they have contracted for. Happily, the non-contributor has been a relatively minor problem with the present work, though some gaps have had to be bridged on an *ad hoc* basis. Where topics were essential it proved possible to incorporate them within other chapters; I am greatly indebted to the contributors concerned for their understanding of my problems.

Since this is an international work, contributors were asked to use metric units as far as practicable. It was not felt necessary, however, to make conversions with the same rigour as would be necessary, for example, with a textbook. For example, the trouble and expense of redrawing graphs and diagrams or preparing entirely new values for tables was, in most cases, regarded as unnecessary. For many purposes, too, it is relative, rather than absolute, values that matter, so the actual units used may be immaterial.

The units of mass and length are at least strictly related, so that any reader wishing to convert from one system to another in a particular context can do so without very much trouble, but another problem of units is much less easily and precisely resolved. This is the problem of monetary units. Trade figures are often expressed not in terms of weight or volume, but of monetary value, using the currency of the country concerned. Here we are on very treacherous ground indeed, for over any considerable historical period not only do the rates of exchange vary considerably but there are often different rates for different kinds of transaction. For example, one standard reference work informs us that in 1942 Spanish output of electrical appliances was valued at 500m. pesetas (*Encyclopaedia Britannica*, Vol. 21 p. 146D, 1947), and that in 1937 the value of German exports of similar goods was 312m. marks (*Ibid*, Vol. 10, p. 251). To establish any realistic relationship between such figures is very difficult. During the first half of this century the number of U.S. dollars to the pound sterling fell from 4·86 to 2·80. Over the same period the number of Japanese yen to the pound varied from 10 to 1010. In many cases, currencies were revalued and new units introduced. To readers who want to thread their way through these perplexities I would commend R. L. Bidwell's *Currency conversion tables* (Collings, London, 1970), which covers the period 1870–1970.

The original *History* was made possible by the enlightened patronage of Imperial Chemical Industries, who provided the very substantial subvention necessary to commission the text and illustrations and work them up into a form suitable for publication. For these two additional volumes there has been no such subvention—nor, indeed, was it sought—but it is proper to acknowledge that they would not have been compiled but for the existence of their predecessors. It has been a pleasure to have collaborated again with the Oxford University Press, and I particularly acknowledge the assistance I have had on editorial matters and in connection with illustrations. Finally, I have special pleasure in having Dr T. K. Derry among my contributors and encouragers. Although we have for many years been geographically separated by the North Sea, and within the present work 56 chapters intervene between his historical introduction and my own final conclusions, I have the happiest recollection of the period when we collaborated closely on our *Short history of technology*. He taught me a lot about the techniques of the professional historian and I believe that in the process he gained a useful insight into the ways and attitudes of industry.

TREVOR I. WILLIAMS

Oxford, May 1977

CONTENTS

VOLUME VI

VOLUME VII

ILLUSTRATIONS

I

THE SETTING IN WORLD HISTORY

T. K. DERRY

1. THE WORLD IN 1900

AT the dawn of the twentieth century the condition of the civilized world was primitive in terms of today's technologies, but in some other important respects it was not altogether unenviable. Looking back from the political, social, and moral perplexities of our own era across the chasm created by two world wars, we see the benefits of political stability and a widely held belief in the idea and ideals of human progress. Although democracy had as yet made few inroads upon the authority of the property-owning classes, methods of government were becoming more equitable and humane, while the standard of life for the masses was on the up-grade in the rapidly expanding industrial regions and even in some of their rural hinterlands. Liberal opinion watched with intense satisfaction the introduction of more systematic poor relief, of easier access to primary education, and of some form of representative institutions in almost every major state except Russia—and even there the 1890s had witnessed the growth of factory industries, raising new hopes among the many peasant families for whom there was no land available. Moreover, all Europe was influenced by contact with the free and prosperous communities which had grown up in the United States during the era of immigration and land settlement which had followed the civil war of 1861–5.

There had been no war between great powers for thirty years and no general war since the fall of Napoleon, but Europe contained a danger zone in the Balkan Peninsula, where various Christian peoples sought to escape from Turkish rule. The situation was complicated by a struggle for influence in the Balkans between the two neighbouring great powers, Russia and Austria-Hungary. The latter was a mosaic of nationalities precariously held together by the Habsburg monarchy, but its rulers were allied with Germany, the preponderant power of the continent. In the 1890s, however, a rival camp began to take shape through the alliance of Russia with France, where the loss of Alsace-Lorraine to Germany in the war of 1870 had not been forgotten. Hence an increased competition in armaments, which the Tsar vainly attempted to check by calling an international conference at The Hague (1899). This led to the establishment of the Hague Court, to which disputes between states could be referred for arbitration, but the principal peace-keeping agency continued to be the Concert of Europe—a system of informal negotiation among the great powers, which had enabled them in recent years to accomplish even such a difficult task as the partition of Africa without ever coming to blows among themselves.

Warfare at this time meant chiefly limited expeditions in Africa and other 'colonial territories', the inhabitants of which could offer little resistance to the white man's superior military technology. The period was nevertheless marked by three small wars which gain significance in retrospect. In 1894-5 the Japanese, who had begun the modernization of their island state less than thirty years before, won an easy victory over the huge, obsolescent Chinese Empire: Russia, Germany, and France united to compel them to relinquish most of their gains, but Japanese troops were included in the international force which put down the anti-foreign movement of the Boxers in 1900, after which all China seemed ripe for division into spheres of influence. In 1898, the Americans had briefly fought Spain over Cuba, one result being to strengthen their interest in the Far East through the occupation of the Philippine Islands. And in 1899 the British became involved in a long war with the Dutch Afrikaners in South Africa, who were eventually beaten but who never became fully reconciled to membership of what seemed then to be an invincible empire.

Imperialism had provided the great powers with worldwide markets and sources of cheap raw materials as a solid basis for their prosperity, while the growth of their trade with one another made them mutually dependent; at the turn of the century two-thirds of the trade of the whole world was in European hands. A large increase in the supply of gold encouraged a trade boom, during which the gold standard spread to every major state except China and capital moved more freely than ever before in search of profitable openings. Britain's accumulated wealth and long commercial experience gave her a clear lead in the provision of financial services of all kinds; her mercantile marine was unchallenged; her textiles were in worldwide demand; and her retention of the free-trade principles which other powers had abandoned provided her people with cheap food and other advantages. But Britain was no longer the privileged pioneer of modern industry—only one important link in a chain of industrial communities which stretched across Europe from Glasgow to Turin, from Warsaw to Barcelona.

France, Italy, Sweden, and Switzerland each now held the lead in some major manufactures, but it was the unification of Germany in 1871 which had set up Britain's most serious European rival. By 1900 their superior discipline, organization, frugality, and appreciation of scientific factors had put the Germans slightly ahead in iron and steel production and very far ahead of the British in the new chemical and electrical industries. The steel production of the United States by this time equalled that of Britain and Germany combined, but during the period of the 'moving frontier' American interest in foreign trade had been concerned chiefly with agricultural exports, manufactures rising slowly to one-third of the total. The huge domestic resources of their own continent made the American people in any case less dependent than others upon competing for foreign markets; their policy in China, for instance, advocated the 'open door' of equal commercial opportunities.

Although very few countries in Europe had such completely democratic institutions as the United States, the franchise was now sufficiently widespread for liberal opinion everywhere to base its hopes for the future upon the voice of the people. It was true that in Austria-Hungary, the Balkans, and some other discontented areas what the people were demanding was big frontier changes to conform with the principle of nationality, while throughout the continent conscript armies were maintained for the contingency of war. But would not the new century so enrich the life of the individual that democracy must spell peace?

The advance of technology was gathering speed: it rendered armaments every year more lethal and destructive, but it also extended the average length of life and improved the average standard and mode of living. This hope of continuing progress through international collaboration in the arts of peace found notable expression in December 1901, when the five Nobel prizes were awarded for the first time 'to those who had conferred the greatest benefit on mankind' in the fields of physics, chemistry, physiology or medicine, idealistic literature, and 'work to promote fraternity between nations'.

II. THE GROWTH OF INTERNATIONAL TENSION, 1901–14

In spite of tension developing between nations, these years did not altogether disappoint the hopes of the world's liberals. World trade flourished, and although rising prices brought about a fall in real wages in many countries in 1909–12, the workers were protected to some extent by widespread imitation of the social insurance laws which had originated in Germany as a safeguard against the growth of socialism. Social democratic parties, backed by trade unions, were nevertheless becoming a widely accepted phenomenon of political life, and parliamentary institutions of a rudimentary kind came into existence in the Russian empire and even in Turkey. Manhood suffrage extended as far as the Austrian half of the Habsburg monarchy, while in two of the smallest European nations (Finland and Norway) the liberal trend even brought about the enfranchisement of women.

In 1907 a second international conference at The Hague had many additional members, including the Latin American states, and was backed by the dynamic American President, Theodore Roosevelt. Yet the Americans showed no desire to be drawn into the efforts of the Concert of Europe to preserve peace through the maintenance of the balance of power; they themselves had employed high-handed methods to secure the territory on which to build the Panama Canal, but even before its opening in 1914 they had shown that their rapidly expanding navy was intended primarily to assert their rights against Japan in the Far East and their claims against turbulent neighbours in the Americas. Internal developments in any case absorbed their attention. These included the rapid exploitation of new inventions, from the combine harvester to the model 'T' Ford motor-car; the struggle to preserve natural resources from the worst abuses of the Trusts; and the absorption into an increasingly urbanized American society of a record number of European immigrants. In 1901–10 these were arriving at the rate of more than 813 000 a year, 70 per cent of them from the less easily assimilated peoples of southern and eastern Europe.

The second Hague conference was as unsuccessful as its predecessor in its attempt to relieve the mounting pressure of competitive armaments. This had become all the more dangerous because Britain was no longer an uncommitted observer of the two rival camps in Europe, able to use its full influence against aggressive tendencies on either side. This fundamental change in British foreign policy came about through the establishment in 1902–7 of links with Japan, with France, and with Russia.

The growth of the British Empire had caused a bitter quarrel with the French regarding the control of Egypt and the Sudan, conflicts with Russia in Central Asia, and disagreements with Germany over annexations in tropical Africa. Their unchallenged naval supremacy had long enabled the British to treat this latent hostility with unconcern, but from 1900 onwards the Germans were building a fleet which would eventually be bigger than the French or Russian fleet and might join with them in an effective challenge. The German terms for an alliance having proved too high, the British government secured alternative help in the Far East by a treaty with Japan; the latter power then checked the growth of Russian influence in China by a war in which Russia's French allies were prevented from intervening because in that case Britain was pledged to intervene on the other side. The treaty was later extended to cover the position in which either party was at war in China or the British in India against even one power.

Meanwhile, in 1904 the British government had taken the still more momentous step of substituting an *entente cordiale* for the many long-standing disagreements with France; in particular, the French recognized Britain's existing control of Egypt in return for British recognition in advance of the control which France planned to establish over the sultanate of Morocco, adjoining their colony of Algeria. In 1905, however, when Japanese victories in the Far East had temporarily destroyed Russia's value as an ally for France, the Germans tested the strength of her new 'understanding' with Britain by ostentatiously championing Moroccan independence. Aware that in a one-front war their country would be quickly overrun by the German army, the French agreed that the Moroccan question should be referred to an international conference, where a compromise was agreed upon. In 1911 Germany challenged this settlement in turn, but her opposition was finally bought off by a cession of French colonial territory in west Africa. During the first crisis the French were given moral support from Britain in the shape of military conversations as to the help which might possibly be given, but even this gesture was concealed from all except four members of the Cabinet. In the second crisis, help could have taken the form of an expeditionary force, for which preparations had been made, and British opinion was aroused by public statements to the effect that France must not be left to negotiate alone with Germany where vital British interests were affected.

In the intervening years Anglo-German relations had deteriorated. The British were inclined to resent the dominant position of the new Germany on the continent, especially in matters of trade, a feeling which became articulate as soon as the growth of the German navy was seen to endanger the safety of the island and its imperial communications. The competition was rendered more acute as the result of a British technological triumph in 1906–7, when the first all-big-gun battleship was built, with turbine engines (partly oil-fuelled) which would enable it to pin down any enemy for destruction at a range beyond risk of torpedo attack by surface vessels. Keeping the design of H.M.S. *Dreadnought* a secret as long as possible, Britain secured a temporary paramountcy over all other navies; and when the Germans began to design their own dreadnoughts, they were further handicapped because it was necessary to widen the Kiel Canal, so that it was not until the summer of 1914 that their new ships could be moved freely between the Baltic and North Seas. In the long run, however, a development which rendered Britain's big lead in older ships almost valueless stimulated German competition; by 1914 the British lead in dreadnoughts had been reduced to about 50 per cent, an advantage which submarine and minefield might put at risk.

A change in Anglo–Russian relations was one of the by-products of the catastrophic defeats suffered by Russian arms in their war with the Japanese; their navy was completely outfought and their armies never overcame the handicap of dependence upon the single-tracked and incomplete trans-Siberian railway. By the peace which President Roosevelt negotiated for them at Portsmouth, New Hampshire, in August 1905, the Russians ceded some of their more recent acquisitions in the Far East to the Japanese and left it open for them to annex Korea and penetrate the markets of southern Manchuria. The loss of prestige almost completed the ruin of the Tsardom, which in the preceding January had rashly suppressed a peaceful workers' demonstration in St Petersburg by measures which cost at least 130 lives.

In October 1905 industrial unrest in Russia took the more formidable shape of a general strike, to which the Tsar capitulated by the formal concession to his people of full civil liberties and an elective Duma controlling both legislation and the administrative machinery. However, the collaboration between moderate constitutionalists and the several doctrinally divided socialist parties, such as the Bolsheviks—who had organized workers' Soviets in St Petersburg and Moscow—did not survive the return of the armies from the Far East to restore order. By 1907 Russian nationalism rather than any form of liberalism dominated the Duma, which supported the repression of the non-Russian peoples of the Empire, such as the Poles, the Finns, and especially the Jews. But the peasantry were freed by law from the restraints of the communal village system, so that a prosperous class of yeomen farmers or *kulaks* came into existence and Siberia was for the first time colonized extensively with the help of the new railway. These developments in turn stimulated the growth of industry—mining, metallurgy, oil, textiles; in 1890–1910 the consumption *per capita* of cotton cloth doubled.

The Russian economy had long been dependent upon French loans, but in 1906 the Tsar's government received significant help also through the City of London. In the following year the understanding with France grew into the 'triple entente', as Britain and Russia abandoned old quarrels on the borders of India and agreed to share the exploitation of Persia, the Russians in the north and the British in the south-east, where they secured oil for the dreadnoughts. This *rapprochement* was regarded with reserve on both sides, because the Russian ruling class found the German social outlook more congenial while British Radicals deplored connivance at a regime of repression. Yet the connection remained, and helped to determine the alignment of the European states in the fateful summer of 1914.

In its capacity as the greatest of Slav powers, the empire of the tsars regarded Serbia, which forms the nucleus of modern Yugoslavia, as its natural satellite. Austria-Hungary, too, included several Slav peoples among its subjects, though their political influence was greatly restricted, and had also been authorized by the Concert of Europe to administer the Turkish province of Bosnia, whose inhabitants were likewise Slavs. In 1908, when a revolution set up a less despotic form of government in Turkey, Austria-Hungary seized its chance to annex Bosnia outright, after proposing that the Russians should receive compensation—which was in the end denied them. The announcement of full support by Germany for her Austrian ally then compelled Russia, which had not yet recovered from its defeat in the Far East, to accept the rebuff. The Turkish revolution also led on to the two Balkan Wars of 1912–13, in which almost all that remained of Turkey in Europe was divided up among the small states of the peninsula. Russia's Serbian satellite was the principal victor, but was denied the Adriatic coastline to which her armies had advanced; the Concert of Europe intervened—for what proved to be the last time—to adjust matters, so that the Great Powers escaped involvement in a minor conflict.

Although these last years before Armageddon were marked by special military expenditure in Germany, France, and Russia, while Britain too perfected her expeditionary-force arrangements, two moves were made to render economic rivalries less dangerous. In the early months of 1914 Britain and Germany reached agreement over a hypothetical partition of the Portuguese colonies in Africa, and Britain and France together came to terms with the Germans over their project for a Baghdad Railway, which would have carried German influence from Constantinople, where it was already very strong, into the oil-bearing lands of the Middle East. But on 28 June the bitter political conflict over the Balkans was unexpectedly reopened by the assassination of the heir to the Austro-Hungarian thrones in Serajevo, the capital of their newly annexed province, at the hands of a Bosnian Serb. The Austro-Hungarian government, knowing that the advance of Slav nationalism imperilled the existence of the composite Habsburg realms, was alone in deliberately risking a general war in order to destroy Serbia; but the German Emperor and government at the outset gave their ally *carte blanche*. Nearly four weeks passed before the Serbs were confronted with an ultimatum, to which they replied submissively, offering to refer outstanding matters to arbitration by the Hague Court; by then it was too late for Germany to restrain her ally—and the Powers accordingly stumbled like sleep-walkers into war.

An immediate partial mobilization of Austrian forces against Serbia occasioned the first military preparations of the Russians as allies of the Serbs and of the Germans as allies of the Austrians. At that stage the views of military experts began to outweigh those of the politicians, because a general mobilization had become an immensely complicated technical process, designed to bring huge conscript armies to the frontiers in a given order of battle, accompanied by their munitions and all other supplies. The Russians, conscious of the inadequacy of their railway network and other handicaps, were the first to mobilize on a general basis, but the other continental powers quickly followed suit. The British navy undertook to safeguard the Channel coast of France, if necessary, the French fleet having for the past two years been concentrated by mutual agreement in the Mediterranean. What carried the British people unitedly into the war, however, was not the imperfectly comprehended *entente cordiale* but the German plan of operations; this required the deliberate violation of the neutrality of Belgium, which had been guaranteed by both Britain and Prussia since 1839.

III. THE FIRST WORLD WAR AND POST-WAR RECONSTRUCTION, 1914–25

In the first month of the war the German military machine, of which Europe had long stood in awe, nearly achieved decisive success. By the end of August the German armies had swept across eastern Belgium and were approaching Paris, having driven back the French and the small British Expeditionary Force, which risked encirclement, while the German defence had repulsed the main French offensive with very heavy losses in Lorraine. An unexpectedly rapid penetration by the Russian armies across the German frontier in the east had also been turned back, again with heavy loss to the attacking forces. In early September, however, some hesitation by the German commanders in carrying out the last stages in the advance into northern France enabled the French and British to make a successful stand at the River Marne in front of Paris, after which both sides extended their flank rapidly northwards. The war of movement then came to an end on the western front, as both sides constructed a continuous trench-line from the Swiss frontier to the sea. On the eastern front the war never became in the same degree immobile; by November the Germans were within 35 miles of Warsaw and the Russians were already feeling the shortage of munitions which was to hamper their whole war effort. But their numbers exceeded those of their German and Austro-Hungarian opponents put together, and an early penetration into Austrian Galicia had shown that the Russian morale was generally superior to that of the soldiers of the Habsburg monarchy, more than half of them drawn from disaffected nationalities.

The long-distance naval blockade of the German coast, set up at the outbreak of the war, could take effect only gradually: American insistence on neutral rights enabled the Germans to import through Dutch and Scandinavian territory, and their own technological skill produced valuable substitute materials. In the early months of 1915 the Allies tried to take advantage of Turkey's adhesion to the Central Powers by forcing their way to Constantinople, first by naval action and, when the big ships were halted by minefields, by the disastrous landings on the Gallipoli peninsula. These efforts helped indeed to bring Italy into the alliance; but their eventual failure stiffened Turkish resistance in the lands of the Middle East, prevented the Allies from attacking Austria-Hungary through the Balkans, and above all destroyed the only hope of supplying the Russians with sufficient arms. In the campaigns of 1915 and 1916 the Tsar's forces still won big victories at the southern end of their front, so that Romania joined the Allies—only to have her oil-wells and cornfields overrun by the enemy—but farther north the German armies made deep inroads into Russian territory. Meanwhile the western front remained almost stationary in spite of a whole series of sanguinary battles, such as Verdun and The Somme: machine-guns and barbed wire broke the force of even the most determined surprise attacks, whereas recourse to bombardment by heavy artillery gave the enemy time to strengthen his defences in the rear. At sea the only major encounter likewise showed the strenth of defensive tactics, since the German battle fleet made good its escape under cover of night after inflicting twice as heavy losses as it received; although the Battle of Jutland had no effect on the British naval blockade, our proved inferiority in shells, and in some aspects of ship design, engendered caution.

The war might have ended in a compromise, such as was canvassed in influential quarters on both sides in the later months of 1916, had not two events opened up new perspectives. In February 1917 the Germans began unrestricted submarine warfare against all merchant shipping which might approach the British Isles. They had attempted this policy two years earlier, but abandoned it then on account of the American outcry against its inhumanity and illegality; this time they counted on forcing a British surrender before a hypothetical American intervention in the war could be effective—and they were not very wide of the mark. When the Americans declared war in April, submarine sinkings had reached a height which caused 1 November to be stated in naval circles as representing the limit for British endurance. Help came promptly from the American navy and shipyards, as well as through an increase in the flow of the munitions which the Allies had been purchasing on the other side of the Atlantic since the outbreak of the war. But what ended the immediate danger was the systematic organization of convoys—which the British Admiralty had stubbornly opposed as a merely defensive policy—and to a less extent the employment of new technical devices for underwater detection and destruction. In the meantime, however, a political change began to open up a second chance for the Germans to win the war on land before any large number of American soldiers could be trained and deployed in Europe.

From March 1917 onwards, Russia was in the throes of revolution. The gross mismanagement of the widespread war effort, especially in the sphere of supply, caused widespread disaffection, so that the food riots in the capital led to the replacement of the discredited tsardom by a provisional government, which was formed by the Duma with the nominal support of the Petrograd Workers' and Soldiers' Soviet. While a sweeping programme of domestic reform awaited the meeting of a constituent assembly, to be elected by manhood suffrage, the new administration remained loyal to the Allied cause and even launched a short-lived offensive against the Austrians. But Russia's peasant armies were now interested primarily in returning home to share in a redistribution of their native soil, so the spontaneously formed Soviets of peasants, workers, and soldiers followed the lead given by Lenin and Trotsky with their promise of 'peace, land, and reform'. The Bolshevik revolution which they organized on 7 November 1917 was the work of a very small minority, whose dismissal of the constituent assembly after a single meeting marked the end of Russia's brief experience of western forms of representative government. A political revolution of a familiar kind had given place to something wholly new—the seizure of economic power by a small but resolute minority. While in the long term this was to transform the political and economic institutions of half the world, its main immediate impact was on the military situation. Between December 1917 and March 1918 the new masters of Russia negotiated the Peace of Brest-Litovsk, which not only opened border regions from Finland to the Ukraine for German exploitation, but also released large German armies for use elsewhere.

On the western front, 1917 had been marked by two unprofitable Allied offensives, of which the first reduced the French armies to mutiny and the second, which began with the first large-scale use of Britain's new weapon, the tank, failed to exploit the surprise and degenerated into almost purposeless carnage in the Flanders mud. In addition, British and French troops had been sent to Italy, where German support had enabled the Austrians to sweep down from the Alps. In 1918 the transfer of veteran forces from the eastern front gave the Germans the means of launching a series of offensives from March onwards, which put both the Channel Ports and Paris in acute danger. However, the Allies at long last established unity of command and, with the help of the newly arrived American reinforcements, were each time able to hold out by a narrow margin until the exhaustion of the German supplies ended the chance of a complete break-through. This was the final turning-point. In July the Allies began to counterattack all along the line, and within a month the German generals were advising their government that the war was lost. Nevertheless, the discipline and morale of their troops were such that resistance in both Turkey and Austria-Hungary had collapsed completely before the Germans laid down their arms on 11 November, at which date the battle line was still a long way from their own frontiers.

The war which shattered four empires—Russia, Germany, Austria-Hungary, and Asiatic Turkey—had cost the lives of at least 10m. combatants: the Russians and the Germans had lost most, the French more than a million, the British from their larger population a little less than a million, the forces of the United States 115000. Civilians had suffered heavily from food-deficiency diseases, induced chiefly by the Allied naval blockade, but air bombing—which developed rapidly in the later war years—had not been used extensively against purely civil targets. Four years of heavy fighting had indeed done great damage in parts of Belgium and northern France, but in 1919 the treaty of Versailles provided that the Germans were to pay unspecified sums in reparations, legally grounded on a forced acknowledgement of war guilt.

Although the peace settlement had other punitive features, such as the confiscation of every German colonial possession, the permanent disarmament of Germany was envisaged as a step towards a general scaling-down of armed forces, while the redistribution of enemy territory in Europe met the most urgent claims of nationality, notably by the revival of Poland and the setting-up of the new composite states of Czechoslovakia and Yugoslavia. Another reason why the reconstruction of Europe seemed to presage peace was the institution of completely democratic systems of government in Germany and, in principle, in every other region (except Russia) where they had not existed before the war. Moreover, the edifice was crowned by a League of Nations, designed to organize the civilized world for peaceful cooperation and to compel the submission of any aggressor state by the united imposition of economic or (in the last resort) military sanctions. In spite of the failure of its main designer, President Woodrow Wilson, to bring in the United States, the League ideal became an article of faith with European liberals.

In 1919 Communism in its Bolshevik form showed signs of spreading from Russia into the defeated states of Central Europe, which partly accounts for the support given by the victorious Allies to the White (i.e. counter-revolutionary) Russian forces attacking the Bolshevik basis of power in Moscow and Petrograd. Such a venture was also tempting because of the availability of Allied troops and supplies which had been sent to Russia before the armistice to try to re-create an eastern front against the Germans. The main advance of the White generals was from Siberia along the railway, but they also marched in from the south and north-west. They failed by a narrow margin: politically, the Whites were too reactionary to engage the sympathy of the peasants; strategically, they were worsted by Trotsky, who conducted the campaign in such a way that the Red Army derived full profit from operating on interior lines. One important result of this troubled period was that the Poles forced the cession of border provinces containing about 4m. Russians. A second was a long-term legacy of mistrust felt by the Russians for all the powers which had fomented the civil war. The most important immediate result, however, was that a serious famine and other extreme hardships caused Lenin to relax Communist principles in favour of a more moderate 'new economic policy'. Russia was still treated as a pariah, but the Communist propaganda organization, known as the Third International, was no longer regarded as a serious threat by any western state except perhaps Italy, where it provided one of the excuses for the seizure of power in November 1922 by Benito Mussolini as the head of the anti-Communist *fascisti*.

In the same year the changing balance of power in the world as a whole found expression in the Washington Naval Treaty, which restricted capital ships in the ratio of the U.S.A. and Britain 5, Japan 3, France and Italy 1·75 apiece. Not only was Britain no longer able to compete for naval supremacy against the wealth of America, but in the Far East it would require their full cooperation to meet a Japanese challenge. This fact found further recognition in ancillary treaties for asserting the independence of the Chinese Republic (founded in 1912), of whose internal troubles the Japanese had taken advantage during the war, and for cancelling the Anglo–Japanese alliance. This last measure was influenced by the wishes of the British Dominions, which now asserted a separate interest in world affairs as a sequel to their full participation in the war and the peace conference.

One intractable problem remaining from the latter was that of the reparations payments, which the Allies hoped might offset their huge debt to America for munitions but on which the Germans almost immediately defaulted. The British were ready to accept any settlement which might encourage world trade, but in 1923 the French tried to extort payment by occupying the Ruhr, the only effect being a runaway inflation which ruined the solid German middle class but not the big industrialists. However, in the following year a workable system of reparations payments was set up through the intervention of America, and this in turn led to the Locarno Treaties, under which Britain (but not the Dominions) guaranteed the Franco-German and Belgo-German frontiers along with Italy and the three border states concerned. Although there was no similar protection for the frontier between Germany and Poland—including a much-resented 'corridor' which provided Polish access to the sea—it was widely believed that the German people had voluntarily accepted the post-war settlement of Europe. Accordingly, in 1925 the peoples of the industrial heartlands of the Continent were encouraged to regard their diminished share in world trade, together with the reduction in their overseas investments and their heavy indebtedness to the United States, as handicaps which the war had imposed but which a reconstructed Europe would now be able to overcome. In spite of significant unemployment in traditional export industries, such as textiles, engineering, and shipbuilding, Britain followed the example of Sweden in returning to the gold standard at the pre-war rate of exchange with the American dollar.

IV. THE END OF THE LIBERAL ERA

The later 1920s were a period of renewed hope, in which men and women of liberal sympathies could believe that the civilized world was moving, slowly but once again certainly, towards the rule of enlightened and peace-loving parliamentary democracies. The enormous prosperity of the United States offered an alluring example of democratic achievement, and the overflow of American wealth provided the loans for industrial enterprise in Europe, especially in Germany. The mass-production of motor-cars, electrically operated domestic appliances, and especially radio sets offered new amenities; by 1929 the capacity for steel output of the four main European centres averaged one-third above the pre-war amount. Britain, where the rise in steel output was only one-quarter, continued indeed to suffer severely from unemployment, and the General Strike of 1926 marked the fact that its coal-mining shared with the other heavy industries in a loss of foreign markets which the enhanced value of the pound had accentuated. Yet in a five-year period (1925–9 inclusive) during which the value of European trade rose by 22 per cent and world trade by 19 per cent, it seemed natural to look rather to the profits which might result from the resumption by the City of London of its position as the world financial centre. Moreover, the fact that the strike had entailed no serious disorders and that it had been preceded and followed by periods of minority Labour government under a smoothly functioning parliamentary system seemed to justify the belief in universal suffrage as a panacea, at least when safeguarded by the rule of law and habits of compromise.

The League of Nations had its brief heyday after the admission of Germany in 1926, its annual Assembly serving as a forum of world opinion. Although the United States remained outside, hopes of collective security were strengthened by the Kellogg Pact, an undertaking sponsored by the American Secretary of State, in which almost every government renounced 'the resort to war as an instrument of national policy'. In 1925 the League had appointed a Preparatory Commission which was to pave the way to an eventual world disarmament conference, and in 1930 America, Britain, and Japan agreed to extend the existing naval limitations to ships of smaller categories. The world appeared to be settling down. The Imperial Conference of 1926, for instance, had a far-reaching significance, since its adoption of the doctrine that Britain and the fully autonomous Dominions (which now included southern Ireland) remained 'united by a common allegiance to the Crown', suggested that there was a form of imperial unity which might still apply when the Indian Empire and the colonies in Asia and Africa likewise achieved full self-government, towards which they were gradually advancing. Even China seemed to be recovering from its long civil wars, when the nationalists of the Kuomintang set to work to establish a single government in 1927 after the expulsion of Communist Russian advisers.

This last event reflected a momentous change inside Soviet Russia, where the death of Lenin in 1924 had been followed by a struggle for power between Stalin and Trotsky, from which the former emerged triumphant with a policy of building up economic strength by a drastic collectivization of agriculture and concentration upon heavy industry. In 1928 the first Five-Year Plan spelt death to the dispossessed *kulaks* and inaugurated a decade of convulsive social changes, during which Soviet foreign policy remained on the defensive. Such events as the cynically staged treason trials of Stalin's opponents (including the army commanders) and rumours of the mass purges, in which untold thousands of obscurer victims likewise perished, made it difficult to credit the U.S.S.R. with the new strength which came from an industrial growth of 12–15 per cent per annum.

In October 1929 the record boom in America gave place to a stock market collapse which was likewise of record dimensions. Its effects spread quickly to central and western Europe, where the calling-in of private loans from American sources destroyed the basis of recent prosperity, especially in Germany; when the flow of reparations ceased in consequence, the former Allies defaulted on their war debts, greatly to the detriment of their relations with the United States. There was general consternation over the mounting totals of unemployed—some twelve millions in the U.S.A., six in Germany, and three in Britain. The British government abandoned its traditional free-trade regime, but this did not check the decline in world trade; in the United States a similar self-interest dictated a blank refusal to cooperate with a World Economic Conference, which tried in 1933 to deal with the currency chaos which had followed the general abandonment of the gold standard. The economic disaster reversed the liberal trend in politics as well. Countries such as Poland and Yugoslavia, where national and social cleavages had long made the position of parliamentary institutions precarious, began to conform to the pattern of the one-party, fascist state which had been established step by step in Italy. Even in France, to which the economic crisis came later than elsewhere, reactionary organizations were by 1933 a threat to the survival of the Republic, while in Britain and the United States the electorate gave the executive an unusually free hand to grapple with the crisis. At Westminster the 'National Government', whose request for a so-called 'doctor's mandate' was endorsed by a huge majority in 1931, was not seriously challenged until the outbreak of war, although unemployment was only half cured. In Washington, President Franklin D. Roosevelt's New Deal policies aroused bitter opposition from affected interests, but they were also more thoroughgoing than the planning of the British economy; he was re-elected with an increased majority in 1936 and a slightly smaller one in 1940.

Neither the American nor the British government was disposed to take up the challenge in the Far East when the Japanese—from motives which included the collapse of their raw silk and textile exports in the general economic crisis—conquered Manchuria by a military advance which brought their armies south to the vicinity of the Great Wall and even Peking. Since the League of Nations was, therefore, powerless to intervene in this case of utterly unprovoked aggression, the principle of collective security began to be discredited. In 1937 the Japanese launched a second attack, which in two years overran the coastal regions as far as Canton, though part of the hinterland was still held by the Kuomintang and part by the little-regarded Communist militia of Mao Tse-Tung, while a Japanese advance against Outer Mongolia had been sharply repelled by the Russians. The eyes of the western world, however, were still directed elsewhere.

In Germany, a succession of ineffective Cabinets had failed to relieve the masses of the scourge of unemployment or the middle class of a haunting fear of the recurrence of the inflation which had been their ruin in 1923—the year when a certain Adolf Hitler had vainly attempted to seize power in Munich. Now unemployment and insecurity enabled this demagogue of genius to build up his National Socialist Party to a dominant position, not merely in street riots, but in the Reichstag. Summoned to the chancellorship by the President of the Republic in January 1933, he quickly found pretexts for outlawing the Communists and silencing the other opposition parties, whilst attracting nationwide support by the lure of uniforms, parades, and slogans, and above all by his own hypnotic oratory with its incitement to antisemitism and other irrational hatreds. Words were accompanied by deeds: the activities of his private army and elite guards (the S.A. and S.S.) were reinforced by the secret state police or *Gestapo*, and before the outbreak of war 18 concentration camps were already in operation. In June 1934 the S.A. leaders and other potential opponents were murdered outright by Hitler's orders; two months later the death of the President enabled him to become head of state. This change was supported by a 90 per cent majority in a plebiscite and by an oath of personal loyalty from the officers of the small professional army—the last independent factor in German politics—who had been conciliated by the timely action against the S.A.'s alleged pretensions.

The project of closely unifying and then vastly expanding the territory inhabited by the 'German race' had been outlined in *Mein Kampf*, the book which Hitler wrote after the Munich fiasco, and in his first year of office the advertised programme prompted Germany's withdrawal both from the League of Nations and from the recently inaugurated World Disarmament Conference. The democratic powers accepted his explanations, which were invariably plausible and pacific, albeit the sequel was his open defiance of the treaty of Versailles by the reintroduction of conscription and the revival of the German air force, announced early in 1935. The Soviet government, on the other hand, knowing that Hitler's grand expansion of Germany was to be at

Russia's expense, joined the League of Nations and sought security through a defence pact with France.

For about four years, however, Hitler was preoccupied with rearmament, a process which helped also to solve Germany's pressing unemployment problem. In the meantime, he derived an unexpected advantage from the Italian invasion of Ethiopia, when the League was finally discredited by the unsuccessful application of economic sanctions; oil being excluded, these served only to drive Mussolini into the arms of his fellow-dictator. Accordingly, in March 1936 Hitler seized his chance to send troops back into the Rhineland, which had been demilitarized in 1919 to handicap any future German invasion of France. Since this grave act also violated the Locarno pacts, it destroyed Hitler's claim that his policies aimed only at redressing the inequalities of the Versailles settlement. But neither the French nor the British people was psychologically prepared for prompt military action, in face of which (as Hitler later admitted) the Germans 'would have had to withdraw with their tails between their legs'. From the summer of 1936 onwards the dictators were further united by their support for Franco's rebellion in Spain, which was abetted by Italian troops and German aircraft; British and French support for the government side proved wholly ineffective and was not concerted with Soviet Russia, which sent in munitions—and political advisers.

In 1935 the British government had alienated much French opinion by negotiating a treaty which allowed the Germans to build up to one-third of British naval strength; but this was also the year of Britain's first major rearmament programme. This had no effective counterpart in France, where the internal strife between reactionary groups and the left-wing *Front Populaire* (which included the powerful Communist party) hampered all industrial activities. In both countries it was widely believed that they were defenceless against German bomber attack, though in 1935 the newly formed *Luftwaffe* did not yet exceed the R.A.F. in numbers; a fortunate chance that year saw the invention of radar, which was so greatly to strengthen the defence of Britain in her time of peril. This was soon to come, for by November 1937 at latest (as surviving records show) Hitler gave the order for forcing the pace to his generals, who after the Rhineland episode could no longer challenge his strategic judgment. By 1945 Germany would lose the advantage of her armaments drive, because her potential enemies would then have technically superior armaments in mass production, which it would take time as well as inventive skill to overtake. To obtain the coveted *Lebensraum* in eastern Europe, the Reich must quickly be built up by annexations and other means to a position of complete security in all other parts of the continent. The chance of American intervention was in any case discounted: public opinion there had not favoured support for the League of Nations in the event of oil sanctions against Italian aggression in 1935, and since then the neutrality laws had imposed a complete ban on any commercial or financial relations which might involve the United States in a repetition of the events of 1914–18, when it had allegedly been inveigled into Europe's war.

The absorption of Austria into the Reich in February 1938 was followed by that of the Sudeten districts of Czechoslovakia in September, when Britain and France at the Munich Conference accepted that this, too, was an annexation of territory inhabited by Germans and did not justify a war. In March 1939, however, when the rump of Czechoslovakia was seized likewise, an Anglo–French guarantee to Poland, followed by similar offers to states in south-east Europe, barred the way to further German advances. By this time Britain was spending a larger proportion of G.N.P. on her arms programme than Germany, and she even introduced conscription at this juncture to strengthen the morale of the French, who were disinclined to look further than the defence of the Maginot Line. But neither power could seriously oppose a German invasion of Poland without military support from Soviet Russia, of whom they had fought shy during the Munich crisis, for which they now negotiated with a dilatoriness expressive of extreme distrust for their prospective Communist partner. The distrust was mutual, but it was not until 19 August that Stalin finally decided that his advantage lay in signing a non-aggression pact with Germany, which offered Russia at least a respite from German attack and the chance of acquiring Polish and other buffer territory which would strengthen its defences. Since Hitler no longer had any reason to compromise over his Polish claims, the pact induced him to embark forthwith upon a war which he expected the western allies to abandon as useless, once Poland had been obliterated from the map.

V. THE SECOND WORLD WAR

The war which Hitler had begun as a limited war for Poland gained momentum by its success, so that by the summer of 1941 he was master of virtually the entire European continent—up to the frontiers of his hated Soviet ally. His most conspicuous advantage was the skill with which the German generals employed their lead in technologically advanced armaments, such as tanks, bomber aircraft, and mechanized infantry transport, to overwhelm opposition by their superior mobility. The *Blitzkrieg*, or 'lightning war', required the use of fully trained armoured formations to break through the enemy line on a narrow front and penetrate at high speed into the rear areas, while dive-bombing aircraft gave close support to the advance and also added to the panic among civilian refugees. Its first success was the overrunning of Poland in the first month of the war. The second was the break-through in the Ardennes in May 1940, which nearly resulted in the cornering of the British Expeditionary Force at Dunkirk and led on to the surrender of France while the Maginot Line was still intact. The third success came in the spring of 1941, when Yugoslovia and Greece were overwhelmed and 373 000 prisoners taken at the cost of 5000 Germans killed or wounded, while a British army was again driven off the mainland and even from the island of Crete.

At the outset the psychological advantage likewise rested with the Germans, after a six-year period in which the youth of the nation had been schooled for total war in military and paramilitary organizations and the entire population had been exposed to a fanatical propaganda on behalf of a dynamic leader who seemed always able to get the better of the 'corrupt democracies'. For their opponents, on the other hand, the winter of 1939–40 was the period of the 'phoney war', when the partitioning of Poland between Germany and Russia deprived them of any immediate objective and the Russian attack on Finland —whose heroic resistance received immense publicity—led to thoughts of a diversionary campaign in the north, discounting the risk of Russian inter-vention. This atmosphere of unreality helps to explain the unreadiness for all-out resistance shown by the four small neutral democracies of Denmark, Norway, Holland, and Belgium, when the Germans sprang on them in April and May 1940. That the French people as a whole were also psychologically unready to withstand the shock of the German attack may be imputed largely to the bitter class conflict between the parties of the left, which had triumphed briefly in the *Front Populaire*, and those reactionary groups which were to rule unoccupied southern France with Hitler's approval for more than two years after the armistice of June 1940. How the British people as a whole would have reacted to an invasion in that deadly summer of 1940 cannot be determined, for the morale which carried them through the night bombing of London and other industrial centres in the following winter was streng-thened in the meantime. On the one hand, the premiership of Winston Churchill, which began on 10 May, had built up a sense of unity, confidence, and purpose by the cumulative effect of his oratory, showmanship, and sense of history. On the other hand, the Battle of Britain, in which the *Luftwaffe* failed to win control of the air for the projected invasion, had given the British a victory of which every newspaper reader could keep the score. The courage and self-sacrifice of the R.A.F. pilots set an inspiring example, but the successful operation of the radar screen and the new eight-gun fighter aircraft also showed the initiated that, given time, British military technology might draw ahead of the enemy.

The economic advantage, however, was that which Germany retained longest. The effects of the British naval blockade were reduced from the outset of the war by the workings of the German–Soviet pact; at the end of 1939 the Germans annexed the industrial districts of western Poland; and after the campaigns of 1940 and 1941 they were free to exploit the resources of virtually the entire continent west of the Soviet borders for their war machine—except in so far as resistance movements inspired from Britain could prevent adherence to the 'European New Order'. The British in their beleaguered island received, indeed, an increasing quantity of supplies from across the Atlantic: at first they had to fetch and pay cash; in the summer of 1940, 50 much-needed destroyers were sent in exchange for the lease of British bases in the Caribbean; and at last the uniquely generous Lend–Lease Act of March 1941 removed any financial restrictions upon American assistance. But from the summer of 1940 the German blockade of Britain was based on possession of the entire coastline from the North Cape to the Bay of Biscay, with aircraft supplementing the depredations of submarines, and in the first half of 1941 the sinkings continued to increase. Britain's sense of isolation was mitigated by the support of the Commonwealth and Empire (except Eire): Canadians helped to garrison the island itself, while the forces of the other three Dominions and India enabled Egypt to be used as a base for the swift conquest of the Italian colonies in North Africa, after Mussolini had judged the collapse of France to be the right moment for joining the winning side. But even in this theatre the arrival of German armoured units, to strengthen what was left of the Italian armies, and the loss of Crete might have led to the fall of Egypt and the Middle East; Persia was now the world's fourth-largest source of oil, and the Iraqi supply was pipe-lined to the Mediterranean.

In less than six months (22 June–11 December 1941) the war in western Europe and the Mediterranean basin, in which Germany held most of the winning cards, became a world war which she was bound eventually to lose. Hitler had decided before the end of the previous year to strike his next main blow against the U.S.S.R., in the belief that the *Lebensraum* of his dreams was now attainable because its demoralized government and people could not withstand the well-tried methods of the *Blitzkrieg*. The Japanese warlords likewise believed that a 'Co-prosperity sphere of south-east Asia' was within their grasp if they acted boldly enough; by the summer of 1941 they had added Indo-China to their area of control on the mainland, and they did not lose belief in their ability to outface the Americans, even when the latter replied with an embargo on oil and scrap-iron which would make war ultimately inevitable. The German invasion of Russia in June was accordingly followed by the Japanese attack on the American Pacific fleet at Pearl Harbor, Hawaii, in December, whereupon Hitler recklessly declared war on America, a country of which he had little knowledge, but whose unneutral support for the British he naturally resented. The fact that the two wars were thus joined immediately in one enabled Churchill—who had been prompt to welcome Soviet Russia as Britain's ally—to obtain for the next two years a disproportionate influence on Anglo–American global strategy, the stronger partner having agreed that the defeat of Germany must have priority over that of Japan.

Down to the late autumn of 1942 aggression triumphed, nevertheless, on a scale unparalleled in the annals of mankind. Although the German armoured thrusts into Russia failed by a narrow margin to capture Moscow and Leningrad before the first winter, when the appalling cold enabled the defence to rally, in the next summer they swept on to the edge of the Caucasus with its oilfields and to the lower Volga valley, where in September they laid siege to Stalingrad. The Japanese successes were still more rapidly achieved. The initial disablement of the American Pacific fleet by surprise attack from aircraft carriers was followed immediately by air strikes which prepared the way for the invasion of the Philippines and by landings in the north of the Malay Peninsula, where two British capital ships were likewise destroyed from the air. Two months later an overland advance brought about the humiliating British capitulation at Singapore, while their complete ascendancy at sea enabled the Japanese to seize the wealth of the Dutch East Indies and to gain control of a vast perimeter of islands, from the Andamans to northern New Guinea, the Solomons, and the Gilbert and Marshall groups, with an outpost in the western Aleutians. For a short period both India and northern Australia were endangered, and as the surrender of Hong Kong had been capped by the expulsion of the British forces from Burma, the fate of China—though now accepted as the ally of the western powers—appeared to be sealed.

In 1942, however, the Americans won a decisive victory over the enemy carrier fleet off Midway Island, which checked their forward thrusts. But the Japanese still offered a desperate resistance in New Guinea and the neighbouring archipelagos, so that about one and a half years passed before sufficient resources were available in the Pacific theatre to justify a general onslaught on the island perimeter. For the Americans first built up their military potential for Europe, where a slow but rewarding strategy of indirect approach was adopted in accordance with the British assessment of the hazards of landing in the face of fully prepared German defences.

In early November 1942, the British army in Egypt, employing large numbers of a superior quality of tank, won a decisive victory at El Alamein and Anglo–American forces landed in French North Africa. Although southern France was seized by the Germans, the Allies succeeded in clearing the North African coast from both ends. Next July they crossed to Sicily, whereupon the Italian government disowned Mussolini and surrendered, but the German defence of the mainland was so resolute that the Allies did not enter Rome until 4 June 1944. In the same period the Russians had transformed the situation on the vast eastern front. In February 1943 they took 200000 prisoners in the capitulation which ended Hitler's reckless attempts to gain 'the city of Stalin'. In July, they blocked the last major German offensive in one week, and then attacked sector by sector with superior numbers and unprecedented concentrations of fire-power, until by the early summer of 1944 Russian soil was almost freed from the invader and the line of battle reached into Romania, Poland, and Finland.

The western allies had assisted the Russian recovery—to an extent that Stalin never acknowledged—with a supply of trucks, planes, and tanks, which had been sent in from the outset in spite of grave losses at sea. By the summer of 1943 the Anglo–American supply position as a whole was no longer in peril, since the combination of American mass-production of shipping with new British technical devices for the protection of convoys was at last winning the Battle of the Atlantic. The Americans also cooperated in attempts to lower civilian morale and the output of the enemy's war industries by stepping up the air attacks on Germany, to which Britain had devoted much of her limited resources while she had no other means of taking the offensive; but until 1944 their impact was much below expectation. Then, however, the almost continuous Allied sorties began to cripple the German arms manufactures, while also securing control of the air over the intended zone of invasion in France; without that control the second front, long demanded by the Russians, could not have been effectively set up.

Landings on the Normandy beaches began at dawn on 6 June 1944. In the first three weeks the Allied position was consolidated with a determination which built on far-sighted technological and logistical preparations. The Germans attempted a distraction at this juncture with the pilotless aircraft sent against London, but their far more lethal jet-propelled rockets were fortunately not ready until September, by which time the Allies had swept across France into the Low Countries. Since these operations in the west were synchronized with Russian advances all along the line, the end was no longer in doubt, though even as late as December Hitler hoped that his final counter-attack in the Ardennes might induce the western allies to make a separate peace. Instead, they halted at an agreed line on the Elbe, whilst the Russians smashed their way into Berlin and figured as the liberators of central and south-eastern Europe.

In May 1945 the war in the Far East still presented a double problem, even though the situation had been transformed in the preceding 18 months by the two-pronged advance of American amphibious forces across the Pacific, leading to the recapture of the Philippines and the establishment of bomber bases within reach of Japan, while the British reconquest of Burma had also brought succour to the Chinese. One problem was created by the fanatical bravery with which the Japanese had defended every inch of ground: how was their homeland to be subjugated without incurring disproportionate losses? The other problem was created by the Russian determination to exploit their victories in the interest of Communism, already shown in the case of Poland: what if the participation of Russian armies in the conquest of Japan, due to follow three months after the end of the war in Europe, should have similar results?

Both problems were solved by recourse to the annihilating new weapon, on which British and Canadian scientists had worked throughout the war but which from 1942 onwards was being prepared in the U.S.A. from its ampler resources. After a single trial explosion atomic bombs were dropped on Hiroshima and Nagasaki; the second drop was preceded by the Russian declaration of war, but was immediately followed by the Japanese surrender. Russia regained rather more than she had lost in the Far Eastern war of 1904–5, but the future of Japan itself lay securely in American hands. On a global basis too, America had reached a pinnacle of power. The U.S.S.R. had been devastated and drained of men by four years of murderous battles; Britain's resources were so completely exhausted that in the last decisive year her forces in every field of conflict had necessarily played second string to the Americans. They, on the other hand, had sustained fewer casualties than the British from a population three times as large, had suffered no weakening of their economy through the war effort, and had newly demonstrated their possession of a new and ineluctable weapon.

VI. THE WORLD IN THE 1950S

The loss of life in the Second World War was about three times as heavy as in the First; approximately one-third of its victims were civilians, among whom the air-raid casualties spring most rapidly to mind, but they were few in comparison with the 5·7m. Jews done to death in German concentration camps; and while the total losses for the United Kingdom and the United States were respectively 338 000 and 298 000 lives, those of the U.S.S.R. were of the order of 18 millions. The destruction of property and exhaustion of physical resources, extending from Europe into the Far East, were likewise many times greater than before. Consequently, the most important aspect of the immediate post-war situation was not the setting-up of the new United Nations, nor the punishment of Europe's chief tormentors by the Nuremberg Tribunal, nor even the virtual partitioning of Europe—including a greatly diminished Germany—between the Communist east and the capitalist west; what mattered most was the attitude of America to its possession of the atomic know-how demonstrated with such deadly effect at Hiroshima. The monopoly was jealously guarded even from the British, who had contributed very largely to the common achievement, and all atomic information was withheld from the Russians, whose desperate economic position was believed by wishful thinking in America to impose an almost insuperable handicap to their acquirement of the bomb by independent effort.

This fact helps to explain the speed with which the grudging cooperation of wartime passed into covert hostility and finally into a Cold War which often threatened to become hot. Less than two years after the end of the war in Europe the Truman Doctrine offered support to 'all free peoples resisting attempted subjugation', and was followed up by the massive Marshall Aid Programme which rescued the west European economies from a collapse that might have proved fatal to their freedom. By 1949 the expansion of Communist control into Czechoslovakia and an attempt to oust their western allies from Berlin—which survived an eight-months Russian blockade by means of a huge air-lift—had led to the cordoning-off of eastern Europe through the NATO defence pact. But in August of the same year Russia's first atomic explosion marked the approaching end of the era in which the influence of America's wealth and avowed championship of free institutions was reinforced by possession of a technical monopoly of supreme importance. Three years later, when the Americans produced the enormously more powerful thermo-nuclear bomb, the Russians were only 10 months behind them, and at the end of the decade, when the first earth satellites were put into orbit by the Russians, their technologists appeared also to lead in the development of the I.C.B.M. for global war.

The eventful 1950s were marked by three other major changes in world affairs. The first was the emergence of the Chinese People's Republic; after

Mao-tse Tung's forces had expelled the Kuomintang from the mainland, this became by degrees a third world power, able to thwart the aims of the Americans (which had U.N. backing) in the Korean War of 1950–3 and subsequently to measure itself against the U.S.S.R. as the authoritative source of Communist theory and practice for the hitherto uncommitted peoples of the Far East and other regions. The second change was the break-up of the European colonial empires, which began in Asia from the time of the Japanese withdrawal and which spread rapidly through Africa as well, when the Anglo–French débâcle at Suez in 1956 showed that the former imperial powers could no longer protect their interests. In some ways this break-up was a triumph for liberal principles; the independence of Israel, for example, might be viewed as an act of atonement, and the Republic of India emerged hopefully from British tutelage as by far the most populous state that had ever attempted the practice of democracy. Yet by 1960, when the original fifty members of the United Nations had doubled, the proportion of stable governments and harmonious communities which it represented was alarmingly reduced. The third change was in western Europe, which by 1952 had celebrated its emergence from the acute post-war difficulties by establishing the Coal and Steel Community; in 10 years the steel output of the six member states doubled without any increase in labour, and they were led in the meantime to unite more closely in the Common Market. By the 1970s the United Kingdom was to accept the politico-economic lesson of the new situation, namely that the future of the off-shore islanders lay with cultivating their continental connections rather than in trying to maintain a 'special relation' with America or the direction of the Commonwealth—ambitions which every year became less realizable as Britain's relative strength declined.

In 1963 the G.N.P. of western Europe reached two-and-a-half times its pre-war level, and even in Britain (where inflation was already a recurrent problem) the masses had 'never had it so good'. If the employment of nuclear energy as a power source for industry made slower progress than was expected, ocean-bed supplies of gas and oil promised a windfall. Automation and the use of electronic equipment of all kinds transformed the execution of industrial processes; chemical research provided an endless variety of synthetic materials; air travel by the jet-propelled planes which had been one of the last of the great wartime innovations became a holiday adventure for many; and the television set was only one of a host of amenities and labour-saving devices that became commonplace in the rebuilt and refurnished homes of the west European peoples. Where America had led the way, Japan also followed, while Russia's achievements in the nuclear field suggested that in consumer goods, too, she might soon draw level.

Nevertheless, when the first Russian cosmonauts were put into orbit in 1961, they gazed down on a world which is clearly a single unit with limited physical resources. The technology which had already accomplished so much —and which was soon to bring Americans to the Moon—faces a still more intractable problem than those of space travel. The spread of medical knowledge, the use of multifarious new drugs, and higher standards of hygiene have lowered the global death-rate (especially infant mortality), while psychological and other factors delay the adoption by many peoples of the birth-control techniques long practised in the industrial countries of the west. In spite of the world wars, estimated global population had risen in the first half of the century from 1610 to 2509 millions; by 1970 the total advanced to some 3650 millions, of whom at least one-half suffered in some degree from malnutrition. With the help of U.N. agencies such as F.A.O., W.H.O., and UNESCO, technologists have stimulated food production among formerly backward peoples by improving the strains of seed and breeds of cattle, by introducing better equipment for their fields, forests, and fisheries, and by the adaptation of fertilizers, weed-killers, and pesticides to every variety of soil and climate. Even so, after three-quarters of a century of unparalleled technical advances, it was still uncertain whether Man's management of his now fairly well defined total resources would enable him to escape eventual catastrophe from the tendency for numbers to increase faster than the means of subsistence, enunciated by the sombre genius of Thomas Malthus when the great industrial changes were in their infancy.

BIBLIOGRAPHY

BARRACLOUGH, G. *An introduction to contemporary history*. Watts, London (1964).

MOWAT, C. L. (ed.) *The New Cambridge Modern History*, Vol. 12 (2nd edn.). *The shifting balance of world forces 1896–1945*. Cambridge University Press (1968).

LANDES, D. S. *The unbound Prometheus. Technological change and industrial development in western Europe from 1750 to the present*, chapters 5–8. Cambridge University Press (1969).

THOMSON, D. *World history from 1914 to 1961* (2nd edn.) Oxford University Press (1963).

GATZKE, H. W. *The present in perspective. A look at the world since 1945* (3rd edn.) John Murray, London (1966).

DERRY, T. K. and JARMAN, T. L. *The European world 1870–1975* Bell, London (1977).

2

THE SOURCES OF INNOVATION

DAVID SAWERS

I. THE CHARACTERISTICS OF INNOVATION

THE first half of the twentieth century saw the sources of innovation begin to change. Industry began to found research and development establishments, to provide an internal source of innovations, while governments began to spend more lavishly on research and development in the quest for ever more effective weapons, and, more recently, to develop products which industry would not develop on its own. These changes supplemented the sources of innovation which had existed in the nineteenth century—primarily the independent inventor and the entrepeneur–industrialist—but they did not supplant them. The sources of innovation thus became more diverse: the industrial R and D establishment enlarged the class of salaried engineers and scientists seeking opportunities for technical change, while the independent inventor and small-firm entrepeneur became less numerous, but continued to exist as a competing source of innovations.

Technological innovation is not only a commercial activity, as the growing military concern with it has emphasized; its objectives are most often commercial, but it can equally well take place in non-commercial activities as diverse as medical care and warfare. For this chapter, I shall define it as the application of novel technology, or a novel application of existing technology, to some practical purpose. It embraces the whole process of technical change, whether in industry or not, and covers minor technical improvements as well as radical changes in design or the introduction of major new products like the aeroplane or the motor-car. Innovation is thus characterized by variety, and the various forms are likely to be influenced by different factors: the distinction between major and minor innovation within industry is particularly relevant. Minor improvements are more often introduced by people inside a company than are radical innovations, and they provide much of the activity of industrial research laboratories. Military innovations will be more

heavily influenced by the desire for improved performance than by consideration of cost than will their industrial counterparts. The growing importance
of military innovation—and of government support for related civil technologies—in this century seems to have increased the inclination to put
performance first among some engineers and scientists.

Discussion of the influences on innovation is complicated by its dual nature
as both a technical and a commercial (or military, or social) process. In the
most familiar instance of industrial innovation it is therefore subject to influences from the technical side—what is or might be technically possible;
and from the market side—what is or might be commercially wanted.
These two sides can be distinguished as representing what technology can
supply, in the form of inventions, and what the market demands. Successful
innovations represent a combination of these two factors: what is technically
possible and what is commercially demanded at the price at which it can be
supplied. Neither factor can be predicted at all accurately, which produces
the uncertainty that surrounds any innovation. The relative importance of
the two factors in producing the successful innovations has been argued long
and inconclusively, for the evident reason that their roles cannot easily be
disentangled. Their relative importance can also be expected to vary from
case to case: the technical side will be more important in areas where the
technology is changing fast, so that new technical possibilities are frequently
appearing.

The speed of technical progress in an industry will partly be determined by
factors which firms cannot control: whether there is rapid progress in any
scientific research relevant to its activities; whether the technology itself has
scope for rapid improvement; or whether outside inventors produce ideas
which change the technology. The first case is not a common one; discoveries
in basic research are rarely directly applicable, and generally provide background knowledge which aids technical progress, rather than being the prime
cause of this progress. Science and technology obey differing laws, with
science seeking knowledge of the universe and technology that which is useful;
their progress has often been at different speeds, they have often progressed
fastest in different places, and the characteristics of the successful scientist
and the successful technologist are generally distinct. There is rarely a causal
chain from basic research to applied research, development, and application.
Where there is a link, it is unusual for the time-interval between a scientific
discovery and its application to be short, as it was with nuclear fission. The
importance of science will thus vary from industry to industry and from time

to time; but it is generally a secondary, permissive factor rather than the determining one in setting the pace of technical change.

The scope for changing the technology itself usually does most to determine the rate of technical change. The maturity of the technology influences the scope for improving it; technologies often go through life cycles, in which they start out from an invention—or, less often, a scientific discovery—on which a growing edifice of further inventions and minor improvements is built. Over time, progress will tend to become more difficult and more expensive to achieve, though a further major invention may alter the situation and reduce the costs of technical progress again. Aircraft engines provide one example; by the 1940s the development of piston engines was becoming increasingly expensive, and engines were becoming increasingly complex. The introduction of the jet engine, invented by people who saw a need for an engine that would be more effective than the piston engine at high altitudes and speeds, provided the opportunity to accelerate technical progress [1]. By the 1970s, improving the performance of the jet engine had become more difficult and expensive, and engines had become more complex; if progress is to accelerate, some new break-through will have to be made.

Outside inventors remain a significant force in stimulating technical progress by injecting basically novel ideas into a technology—as they did with the jet engine (p. 819). Other examples include the catalytic cracking process for petroleum, where an outsider introduced this new process to the oil industry when thermal cracking methods seemed to be reaching limits to their development; looms for weaving cotton and wool, where the first radical change was introduced by an independent inventor and an engineering firm that was new to the business; and typesetting machinery, where photosetting methods were introduced by outsiders some seventy or eighty years after the last major change in methods had been made.

In all these examples where technical possibility has taken the leading role in bringing about an innovation, the commercial demand of course had to exist as well: it was something which could not easily be forecast in many cases, especially where a radical innovation was involved, but something which was essential if the invention was to become an innovation. It seems broadly true that technical possibility will count for more in producing major innovations than in producing minor ones: major innovations generally involve a radical change in technology. Minor changes are more likely to be produced by modifying what is already known than by extending knowledge; the results of this work, and the demand for it, are more easily forecast than those for major changes. Or the customer may say that some improvement in performance would be desirable, and the supplier be able to meet this desire by minor changes. But if the customer asks for a silent jet engine, a cure for cancer, or a car engine with a thermal efficiency of 100 per cent, he will not get it. Some demands cannot be met, at whatever price.

Another distinction among industrial innovations is that between product and process innovation. This distinction is sometimes dependent on the producer concerned: one firm's process innovation may be another firm's new product. But there is a real distinction between innovations in the method of producing an article and changes in the product itself, or the introduction of a completely new product. An innovation in production methods may involve changes in the production hardware used, or changes in the ways in which available hardware is used; in chemicals, it may mean introducing a new production process, again with new hardware, or changes in the ways in which an existing process is operated. It may be internal to the company making the innovation, though a product innovation, by definition, cannot be; and assessing the market demand is thus less relevant. The market may be within the firm. Outsiders are less likely to propose a process innovation than a product innovation, because they will have less knowledge of how a process can be improved. Process innovations are therefore more likely to be developed within the firm concerned.

Process innovation will be especially important in those activities where a relatively standardized product is made: steel, for example, or basic chemicals like ethylene or sulphuric acid. There may well be some scope for improving the quality of the product, perhaps by changes in production methods; but the main emphasis in such activities will be on reducing production costs, because goods will sell at a standard price and profits will thus depend on costs. The major recent innovations in the steel industry, such as oxygen steelmaking and continuous casting, have been primarily aimed at reducing costs; process innovations have also been particularly important, and numerous, in the chemical industry.

Uncertainty is the characteristic common to all forms of innovation. The degree of uncertainty about the outcome depends on the degree of technical novelty in the innovation, so that a minor improvement can be expected fairly confidently to work, and the effect on sales or costs will be reasonably easy to predict. Uncertainty about the market will probably be greater than uncertainty about achieving technical objectives. Major innovations are a different story; the risk of technical failure will be high, as will the risk that development costs will substantially exceed the estimates. Studies of both military and civil projects show that the greater the degree of technical advance sought in a project, the greater is the likely overrun on the development costs. Uncertainty about commercial success, if technical success can be achieved, will be increased by the difficulty of forecasting demand for a novel product and by the long period that may elapse between the decision to try to develop the new product and its arrival on the market. Competing products may appear in this period, destroying or reducing the expected markets; alternative technical routes to achieving the same objectives may appear, within or outside the organization launching the innovation; or tastes in the market may change, reducing—or increasing—demand. Flexibility to adjust to changing technical and commercial circumstances is thus the most necessary characteristic of the successful innovating organization.

Technical innovation is thus a highly diverse activity. It is most familiar in industry, where its dual nature as both a technical and a commercial activity confuses attempts to detect what determines its rate, to say nothing of increasing the difficulty of successfully managing innovation. It can apply to processes and the ways in which they are used, as well as to products; and it can result from ideas arising outside industry or within it. The ultimate test of success is a commercial one, but the prerequisite for commercial success is technical success. Engineers, scientists, and businessmen are thus all engaged in this activity; and their effective collaboration, as well as their individual competence, is needed for success.

II. THE MOTIVATION OF INDUSTRIAL INNOVATORS

The motives of industrial innovators are rendered complex by the combination of commercial and technical factors involved. Commercial success and financial profit, for the company or the individual, must be a major, if not the main objective of the industrial innovator; but it would simplify unjustifiably to suggest that it is the only one. Few people can be solely motivated by the desire for profit, and the appeal of the task itself seems especially strong where technical novelty and technical problems are involved. Some innovators seem to be motivated as much by the desire to produce technical improvements for their own sake as by the desire to market a profitable new product. The fascination of the task helps to explain the dedication of the innovator, and the effort that is often put into launching an innovation. 'Resistance to change' is an overworked phrase, but those who wish to do something different have to persuade others that the change is worth while and worth risking their own or, more often, their company's money on attempting. Where technological innovation is involved, they have to make the new product or process work reliably, and at the desired cost. These tasks of persuasion and of technical development are taxing ones, which would not be undertaken successfully, and perhaps not at all, unless the innovators were strongly motivated to succeed. The innovator must therefore believe that his task is worth doing, and this belief will usually imply that he thinks that it will produce some 'better' product or process, as well as one that will be a commercial success.

At the end of his exhaustive study of the major innovations in petroleum refining processes in the first half of this century, J. L. Enos concluded that there was an economic motive in each case—'to achieve better results in each case and so to increase profits' [2]. But the personal goals of the innovators were difficult to determine, except that financial gain was not the most important one. Summing up, Enos concluded: 'We can only hypothesize as to what really motivated these innovators; from our evidence it appears to be a combination of curiosity and creativeness. They were driven to satisfy an observed need. They sought recognition and respect resulting from technical achievement and financial profit. These appear to be more important characteristics than acquisitiveness.' [3]

Innovators in the aircraft industry during the first thirty years of this century similarly seem to have been motivated by a mixture of economic and more creative urges. A designer like J. K. Northrop seems to have been at least as much driven by a desire to make aircraft more efficient as to make money when he pioneered well streamlined aircraft with the Lockheed Vega in 1927, and developed the all-metal structure that was to make the Douglas DC-2 and DC-3 so durable. After helping to design these aircraft, he went on to devote himself to an ultimately abortive attempt to develop an all-wing aircraft, because it seemed more efficient than the conventional combination of wing and fuselage. Alexander Lippisch, the pioneer of delta-winged aircraft, was similarly driven by a desire for higher performance and greater efficiency from aircraft, developing his ideas with gliders and light aircraft in the 1930s. The earlier pioneers of the aircraft industry such as L. Blériot, H. Farman, the Breguet brothers, H. Junkers, and A. Fokker were mostly businessmen as well as engineers before they started to work on aeroplanes; they devoted much of the money they had already made, much energy, and sometimes their lives to the hazardous and, initially, unrewarding activity of launching a new industry on a fascinated world that could not provide any large market—until the First World War created one. Their motive, again, was far more than a desire for profit.

The greater organization of innovation that came with the industrial research laboratory and the growth of managerial control within firms does not seem to have changed the motives of the innovators, or eliminated the need for individuals to champion ideas and take decisions. Greater anonymity is a general characteristic of the large organizations, but individuals become more evidently important to firms when major innovations are taking place. A few men were the key figures in the innovations in the American electronics industry in the 1950s and 1960s, and their moves affected the prosperity of companies. Studies of innovations, even relatively minor ones, in the 1960s showed the importance of individuals to success, especially that of a champion for the project among senior managers in the company [4]. The motivation of the technical staff seems no different from that of earlier innovators, as described by Enos; they have less opportunity for financial gain than their predecessors, unless their services are being competed for by several firms, but their motives remain to earn respect from technical achievement and financial profit.

One example from International Business Machines is worth quoting. The disc memory unit, the heart of the modern computer, was developed in one of the I.B.M.'s laboratories as a bootleg project. Management said that the project should be dropped because of budget difficulties, but a handful of men in the research laboratory carried on with its development because they believed in it [5]. Managers now may be less profit-orientated than their predecessors were; but innovation is a means to recognition for managers, as well as for engineers and scientists, and recognition can bring promotion.

The motivation of the company as a whole to innovate is that a new product gives a temporary monopoly, whether or not patent protection is available, and this monopoly provides an opportunity to expand sales and to increase profit margins. A new process will similarly be monopolized by the company for a time—unless it chooses to sell licences to the know-how or patents—and will give the company an advantage in production costs or product quality, allowing it to increase profit margins or reduce prices and increase sales. The extent of the monopoly will depend on the novelty of the innovation; a minor improvement may be quickly copied by competitors, or matched by a different improvement; a major innovation may take longer to copy, unless competitors were working on the same lines before it was launched. Boeing, for example, gained an advantage of only a year over Douglas in selling jet airliners, because Douglas had been preparing to launch one themselves before Boeing took the decision to launch the 707; but this lead in time still helped Boeing to gain the lead in sales.

Companies can, successfully, follow differing policies towards innovation. Some will choose to be first with new products, bearing the risk of innovation to gain the profits of leadership; others will choose to be second, eschewing the risks and profits in the hope of a more stable existence. The choice will be influenced by the attitudes and abilities of the companies' staffs; people who are interested and skilled in the technology are more likely to be successful innovators than those who are not. The policy which proves more successful will also depend on the circumstances of the industry, especially the scope for, and cost of, technical change; but no one policy can be described as 'right' for all companies in all industries. Reluctance to innovate may reflect a company's recognition of its own limitations, or a sound judgement of the road to commercial success in its industry. Innovation is one element in a company's commercial policy, and cannot be considered in isolation from all the other factors influencing it.

III. THE SOURCES OF INVENTIONS

The supply of ideas for technical change, of inventions, has already been mentioned as an important factor in determining the rate of innovation. The establishment of industrial research laboratories introduced a new source of inventions in the first half of this century, but independent inventors remained a major source of inventions—especially of the more radical ones, it would appear. Out of one sample of 70 inventions, mostly from the first half of this century, J. Jewkes, D. Sawers, and R. Stillerman found that more than half had come from independent inventors [6]. These were: air-conditioning; air-cushion vehicles; automatic transmission; Bakelite; the ball-point pen; catalytic cracking of petroleum; 'Cellophane'; chromium plating; Cinerama; the cotton picker; the cyclotron; domestic gas refrigeration; electric precipitation; the electron microscope; the gyro-compass; the hardening of liquid fats; the helicopter; insulin; the jet engine; Kodachrome; magnetic recording; the Moulton bicycle; penicillin; photo-typesetting; the 'Polaroid' Land camera; power steering; quick freezing; radio; Rhesus haemolytic disease treatment; the safety razor; the self-winding wrist watch; streptomycin; the Sulzer loom; the synthetic light polarizer; titanium; the Wankel engine; xerography; the zip fastener.

The inventions in the sample which were attributed to industrial corporations were: acrylic fibres; 'Cellophane' tape; Chlordane, Aldrin, and Dieldrin; continuous hot strip rolling; crease-resisting fabrics; DDT; the diesel-electric locomotive; Duco lacquers; float glass; fluorescent lighting; Freon refrigerants; methyl methacrylate polymers; modern artificial lighting; Neoprene; Nylon and Perlon; oxygen steelmaking; polyethylene; semi-synthetic penicillins; silicones; synthetic detergents; television; Terylene; tetraethyl lead; the transistor.

Some other studies, covering smaller samples, have reached broadly similar conclusions. D. Hamberg, in a study of 27 inventions produced between 1946 and 1955 [7], found that 12 had come from independent inventors. Studies of some industries, such as that by Enos of petroleum refining, have also found that many important inventions came from outsiders. In other industries, such as plastics, studies have shown [8] that industrial research laboratories have made the major contribution.

Generalization is dangerous where the evidence is inconclusive, but the cases quoted do include a number where radical inventions have been produced by outsiders, even in industries which undertook research themselves. Catalytic cracking of petroleum, 'Cellophane', the electron microscope, insulin, the jet engine, Kodachrome, penicillin, photo-typesetting, radio, streptomycin, the Sulzer loom, titanium, the Wankel engine, and xerography are all inventions which industrial research laboratories might have been expected to produce, and which in most cases have led to further research within such laboratories to exploit the break-through that they represented. Inventions produced within industry include such major advances as crease-resisting fabrics, DDT, float glass, Nylon and Perlon, polyethylene, synthetic detergents, television, and the transistor. But even the most research-intensive of companies is far from self-sufficient in inventions; of the 25 most important product and process innovations introduced by du Pont between 1920 and 1949, 10 were based on inventions by du Pont's staff. They were responsible for 5 out of 7 process innovations, and 5 out of 18 new products, with a share in the invention of one other [9].

The contribution of organized industrial research seems to have been in increasing the amount of minor innovation that is undertaken, rather than in increasing the amount of major innovation or in making industry self-sufficient in inventions. Studies of the research and development done by industry show that the majority of projects are relatively short term, being expected to recoup their costs within 5 years [10], and that they involve relatively little technical risk: the average probability of technical completion of a project was better than 50:50 in a sample studied by E. Mansfield [11], while, in part of the sample, the majority of projects which did not achieve their technical objectives had been abandoned for commercial reasons. Thus only about a quarter of projects appeared to have been technical failures. Firms have apparently concluded that the best return from their investment in R and D lies in the relatively safe, short-term project that will bring quick returns. If this is generally true, it helps to explain why independent inventors have continued to contribute a significant share of major inventions.

IV. INNOVATION AND INDUSTRIAL STRUCTURE

The relationship that may exist between the structure of an industry—whether it is competitive, oligopolistic, or monopolistic—and the propensity of firms in that industry to innovate has fascinated economists for much of this century. The evidence remains inconclusive, for several reasons; the most influential is that the structure can be only one of several factors influencing the rate of innovation, the technical possibilities for innovation being one of the major factors (as discussed in Section I). But the situation is even more complex, for the structure of an industry may well be influenced by the nature of the technology; where, for example, the technology requires heavy expenditure to launch new products, the size of firm that is successful is likely to be large and the industry is therefore likely to be oligopolistic or even monopolistic. Conversely, the technology may be influenced by the structure, for if structure does affect the propensity to innovate, firms in those industries which have a structure that encourages innovation will spend more on R and D, and so accelerate the development of the technology.

The value of economists' studies is further limited by the difficulty in measuring concentration relevantly, when large firms often operate in several industries; and of relating competitiveness, as observed by the businessman, to any measures of concentration that may be used. Though no definitive answer can be given to the question 'How much does industrial structure determine the propensity to innovate?' much suggestive evidence is available.

The broad but tentative conclusions from this evidence are that competition provides an incentive to innovate, and that in many industries the largest firms are not the most effective innovators, or the heaviest spenders on research and development. The main qualification needed to these statements is that the situation varies from industry to industry, and within an industry it may vary over time, because the nature of the technology has its influence.

This qualification relates more to the effects of size than of competition; the effectiveness of the largest firms will be affected by the cost of innovation. Where the cost is high, as in the aircraft or computer industries, the largest firms may be the most effective innovators simply because they are the only firms which can afford to launch a major new product. But high costs of innovation, determined largely by the technology, will change over time, as they have done with aircraft and computers.

One might expect that the effect of competition, even among a few companies, would be to encourage firms to seek better products or cheaper production processes than their rivals possess, so that they can gain the temporary monopoly that seems to be the main incentive to innovate. Monopolists might be expected to have less incentive to innovate because the potential benefits from innovation would be smaller when they already possess a monopoly. Evidence to support the first view comes from a number of sources. Competition between the producers and operators of civil aircraft in the U.S.A. in the 1930s seems to have stimulated the rapid technical progress that took place then. In Europe, where competition was absent, technical progress was largely absent as well. More recently, the existence of de Havilland in Britain and Boeing in the United States, companies anxious to break into the airliner market, accelerated the introduction of the jet airliners [12]. In the electronics industry, the development of semiconductor devices seems to have been accelerated by the fierce competition that existed, especially in the U.S.A.; it is difficult to believe that progress would have been so rapid if, say, Bell Laboratories had been able to maintain rigid control over the market through patents. Competition again seems to have stimulated innovation in the pharmaceutical industry in the last thirty years. The rate of invention and innovation seems to have been stimulated in the aluminium industry in the U.S.A. by the change from a monopolistic structure before 1941 to an oligopolistic structure after the war. The two new entrants to the industry, Kaiser and Reynolds, each generated as many inventions as the previous monopolist, Alcoa [13]. Competition was fierce in the early life of the computer, with I.B.M. entering the market relatively late when it found that computers were taking business from its existing office machines [14].

Statistical comparisons between the rate of innovation in concentrated and unconcentrated industries do not support the belief that concentrated industries are more innovative, though their results are hedged about with qualifications. The rate of innovation cannot be directly measured, statistically; it is difficult to introduce adequately the effects of differing technical possibilities for innovation into the calculations; and the concentration index usually employed—the proportion of the industry's sales attributable to the four, eight, or twenty largest firms—does not allow for potential competitors with new products, or for the weak relationship between the statistical definition of an industry and product markets as seen by a firm.

W. R. Maclaurin avoided some of these difficulties, but introduced others, by comparing his judgement of the innovative performance of thirteen U.S. industries between 1925 and 1950 with his judgement of the degree of monopolization in 1950. He found the ranking of industries under the two headings did not coincide [15]. The obvious weakness of this study was the dependence on judgement to measure innovative performance, and monopoly power; but other studies using different measures have reached broadly similar conclusions. F. M. Scherer used the number of patents as an indicator of innovative performance, and the four-firm concentration index as an indication of concentration, and found no evidence that innovative performance increased with higher concentration [16]. G. J. Stigler used the fall in labour inputs per unit of output as a measure of technical progress between 1899 and 1937, and found progress had been most rapid in industries where concentration had declined, and least rapid where concentration had remained high throughout the period [17]. When the study was extended to cover 1939–64 by B. T. Allen, however, he found no significant differences between the performance of industries with differing degrees of concentration [18]. Neither the number of patents nor the growth of labour productivity is a good indicator of technical progress, but the fact they each produce broadly similar results, and ones similar to Maclaurin's judgemental study, suggests that the results have some validity.

Other studies have used expenditure on R and D or employment of technical staff as a measure of innovative performance; as they represent only inputs to innovation, they are even less satisfactory indicators than patents or the growth of labour productivity. The results suggest that research intensity rises as industrial concentration rises, but that maximum research intensity is reached when the four biggest firms hold about half the market. At higher levels of concentration, research intensity seems to fall. Thus the oligopolistic industries, where the market is dominated by a few companies, seem to be those likely to spend most on research; but there is no evidence that this expenditure is matched by higher innovative output in such industries.

Causal links between concentration, rate of innovation, and research intensity have not been established by these statistical studies. Such a linkage is most plausibly provided by the concept of the technological life cycle that was described in Section 1; the technology relevant to parts of the aircraft, chemical, electrical, electronic, oil-refining, and scientific instrument industries seems to have made technical change relatively easy for much of the present century. The main reason why it was made easy was that the technology was at a fluid, fast-growing stage. In the aircraft industry, progress was rapid before 1914, because simple improvements to aerodynamic and structural efficiency were made very cheaply; rapid again in the early 1930s, with the introduction of more sophisticated improvements to aerodynamics, the introduction of metal structures, and more efficient engines; then slower until the introduction of the jet engine and the swept-back wing in the 1940s and 1950s. The cost of launching a new aircraft, at 1976 prices, may have been roughly £50000 to £100000 before 1914, rising to £2–3m. in the early 1930s, and £300m. in the 1950s; the increase partly reflects the rising size of the aeroplane, from the two-seater before 1914, to the 14-seater in the early 1930s, to the 140-seater of the 1950s. But the increasing cost of making progress seems the main factor, as the technology of aircraft became more mature, and the easier ways of improving efficiency were used up. Designers therefore found that they were getting closer to the limits of what was technically possible at the time, which increased the need for testing, and the risk of failure. Higher safety standards, in this case, also made more thorough testing necessary, and helped to push up the cost of launching a new product. The magnitude of the cost reflects the amount of testing needed to achieve high performance safely.

Much the same phenomenon can be seen in other industries; new antibiotics could be discovered more easily in the 1940s and 1950s than in the 1960s or 1970s; the early electronic computers could be developed far more cheaply than the current models; the development cost of the Burton thermal cracking process in the 1900s was perhaps a thirtieth of that of the Houdry catalytic cracking process in the 1930s. If the cost of launching a new product or process is high, the number of firms that can survive in an industry will be small; if the cost rises sharply, the number will fall. High launching costs mean that few producers can earn a profit in a given market, while high and rising costs make it likely that some firms will go out of business if their products fail. Concentration in the industry will therefore rise, as has happened most noticeably in the aircraft, computer, and electrical industries.

A rapid rate of technical change, producing pressure on firms to spend heavily on R and D, has less effect on concentration than rising and high costs of launching each new product. In the former situation, firms can limit their total annual expenditure by limiting their product range; though firms with wide product ranges seem better able to exploit the results of R and D expenditure than smaller firms, there are also gains to efficiency from specialization. Relatively low launching costs for each product also make entry to the industry less risky and easier; new firms have successfully entered the electronic and pharmaceutical industries since 1945, but not the aircraft industry.

The nature of the technology can thus raise the size of firm that can exploit it effectively. Large firms are therefore better equipped to innovate than small ones in some industries, especially those where technical progress remains relatively rapid and important for commercial success, but where the technology is becoming mature. The advantage, contrary to the belief of J. K. Galbraith [19], comes not from any general suitability of large firms for innovation, but from the characteristics of certain branches of technology at certain times—which happen to be recent times. Large firms may indeed have some positive disadvantages for launching radical innovations: size creates a need for organization, and organization is upset by change; innovation requires decisions taken in uncertainty, and such decisions are best taken by one or a few people who know whatever information is available. But in large organizations the leaders know and understand less of what their staff are doing than in small ones. The motivation for success may also be weaker in a big organization, where personal contacts will be more distant and financial involvement in the project is less likely.

Small organizations are more efficient than large ones as developers because information flows more quickly, decisions therefore can be better informed and made more quickly, and staff may have a financial commitment as shareholders. Large firms have tried to reproduce the conditions of small firms when handling radical innovations, by establishing subsidiaries to manage them; but the results do not seem to have been wholly successful [20], perhaps because large firms are not staffed by entrepreneurs, or perhaps because those in charge of the subsidiaries look over their shoulders at their superiors in the parent firm. The need for large financial resources to launch innovations thus produces a managerial problem that stretches the ability of top management; American companies seem able to cope most effectively, as their leadership in the aircraft and nuclear industries helps to demonstrate.

Big firms therefore dominate technological innovation in certain industries because of the nature of the technology, and despite the handicaps which size imposes on them. Statistical studies either omit the industries where technology seems to have made large firms most necessary, or cover all industries; the results are not, therefore, wholly relevant to this topic. Mansfield studied the relationship between size of firm and output of important innovations in the American coal, petroleum, pharmaceutical, and steel industries. The biggest companies had produced a higher proportion of innovations than of output in the coal and petroleum industries, but a smaller proportion in the pharmaceutical and steel industries. The most innovative size of firm was the sixth largest in the coal and petroleum industries; the tenth or twelfth largest in the pharmaceutical industry; and among the smallest firms in the steel industry [21]. In all these industries, the cost of launching an important innovation seems to be within the financial means of most companies, and even relatively small firms are much larger than the usual definition of a small firm (less than 200 employees is often taken as the boundary in Britain).

Small firms, however defined, play a varied role as innovators. Their importance seems to depend again on the nature of the technology more than anything, so that small firms can make a significant contribution to innovation where costs are low. In Britain, small firms have, since 1945, contributed a larger share of important innovations than of output in the scientific instrument industry, parts of the engineering industry, and the paper industry. They also made a significant contribution to innovation in the electronics, carpet, textile, leather, timber and furniture, and construction industries. They made little or no contribution to innovation in the aerospace, chemical, motor vehicle, steel, glass, cement, shipbuilding, and aluminium industries [22]—in many of which no producer of the final product will have fewer than 200 employees. Taking all industries, small firms contributed 10 per cent of innovations but 21 per cent of output. In the sample of inventions studied in *The sources of invention*, firms which were relatively small for their industry seem to have launched a significantly smaller proportion of these inventions than did large firms, though many of the big firms had relied on ideas from outsiders.

New firms are the most extreme example of the small firm, and their role as innovators comes in much the same situations: where technology is flexible, and costs are low. But they are especially important where the technology is new. New companies contributed most to the launching of new products like the aeroplane, the motor-car, and radio—even if, now the technology is mature, small firms contribute little to innovation in these industries. If there is no existing product that resembles the new one, and if a new product has been invented outside a firm, a new firm is better equipped than existing firms to launch the product. Its staff will have less to unlearn about technical and commercial policies, and can more easily adopt new ones appropriate to the product; new production methods may have to be devised, and the lack of an inheritance will make it easier to do so; the commitment of management to the new product will be stronger.

In the first quarter of this century, a new firm for a radically new product was the norm; more recently, it has become abnormal. Apart from the aeroplane and radio, Bakelite, the gyro-compass, the safety razor, the self-winding wrist watch, and the zip fastener were early twentieth-century innovations launched by new firms. In the last fifty years, the ball-point pen, photo-typesetting, and the Moulton bicycle are the only clear cases from the sample of inventions in *The sources of invention* where new firms have helped to launch innovations. Small firms which were new to the industry launched some 30 per cent of the major innovations in semiconductors in the U.S.A. from 1951 to 1968 [23], but new firms have been unimportant in this industry in Europe. One reason for this apparent decline of the new firm may be a decline in the number of radical innovations which they are most suited to launching; another is the growing stability of industrial structure produced by the increasing preponderance of the large public company; another is the growing desire among such companies to diversify their product ranges. Yet another is the growing difficulty of raising risk capital with which to launch a new company making a new product, caused by high taxation of personal incomes and wealth, especially in Britain. The first and the last may be the main reasons, judging from the role that new companies have played in the U.S.A. with radical innovations like semiconductors.

Evidence about the relationship between technology and industrial structure is far from definitive; but it seems to point towards technology having more causal influence. The nature of the technology appears to influence the nature of the industrial structure more than the structure influences the technology. Economists who have argued that large firms in concentrated industries are effective instruments for accelerating innovation seem to have mistaken the direction of causation.

V. THE INFLUENCE OF GOVERNMENTS

Governmental interest in innovation during the first half of this century was mainly limited to weapons. But governments began to show an interest in the progress of industrial innovation, especially where it could strengthen a country's military potential. Governments had most influence on commercial innovation through the operation of the patent system, with its protection for the innovator—most valuable for the independent inventor and small or new firm. The patent system may therefore have been more influential earlier in the century than in the last forty or fifty years.

Two world wars effectively strengthened military interest in technology, and the willingness of governments to spend public money upon it. But the influence of changes in technology itself, produced without immediate military uses, was also significant; if there had been no aeroplanes, rockets, or nuclear fission, governments might well have spent less than they did on weapons. Presented with these inventions, at varying stages in their development, they found them sufficiently valuable as potential weapons to deserve lavish support. Aeroplanes existed without military finance for nearly a decade, and the effect of the military expenditure from 1914 to 1918 was to accelerate development along lines already established, rather than to produce improvements in basic efficiency. The latter came in the 1920s and early 1930s, more for commercial than for military purposes though helped by background research done in government establishments, and the government subsidies to air transport itself. The military aircraft in the war of 1939–45 owed more to commercially than to militarily sponsored innovations. Military support accelerated innovation more noticeably with aero engines than with aeroplanes themselves, especially through the backing given to the jet engine—an invention that had not been produced through government-funded research. Rockets were taken from enthusiasts who saw them as means of interplanetary travel, but who could not afford the costs of development, and were turned, with military funds, into weapons which could then be developed into vehicles for space travel. Nuclear fission was taken into military custody at an even earlier stage, virtually straight from the laboratory discovery; and the resulting development programme set a precedent for extravagant, and sometimes fast, government-financed development programmes that has continued to this day. It also provided the basis for nuclear generation of electricity, which may yet prove of great economic value.

The main contribution of the military to technology has been money. Without this money, atomic and hydrogen bombs and rockets to carry them would not have been built; and the benefits of nuclear energy, space travel, and jet-propelled airliners would have come later or not at all. Other technologies, such as radio, electronics, metallurgy, explosives, and naval architecture gained some military money, and development was no doubt accelerated by its availability; but the economic benefit from most of the accelerated development seems to have been small, especially if allowances are made for what was lost by diverting many engineers and scientists to military work. The total volume of R and D, and of the number of people so employed, would have been much smaller since about 1940 if military expenditure on technology had not grown; but the people employed in this way would have done something, if not other R and D work, if they had not been employed with military money. The benefits from this alternative activity were lost because of the military R and D, so whatever gains to the economy there were from it have to be set against this immeasurable loss. The gains from military R and D were not a free gift to the economy.

Direct government support for non-military innovation was on so small a scale for most of the period up to 1950 that its influence on the pace of innovation was bound to be slight. Most of the government support before 1950 was in the form of background research, done by government laboratories for the benefit of industries which were of some military, strategic, or social value: aviation, chemistry, radio, medicine, and agriculture all gained some support. The work that most obviously influenced innovation was the research of the American National Advisory Council for Aeronautics and of the research centre at the University of Göttingen in Germany. Other government research may have raised technical standards, especially in industries like agriculture where each firm was too small to do research itself, and supplied information that was helpful to innovators; but any beneficial influence on the cost or level of innovation is, inevitably, more or less unobservable.

The short-term monopoly given by patents has been used by governments to encourage innovation for centuries, and they remained the government measure that most influenced innovation up to 1950, but their effects have only recently been examined methodically. Under present conditions, and taking only their effect on the decisions of firms, it seems that they do little to change industrial behaviour outside the pharmaceutical industry. In other industries, too much knowledge about how a product is made or a process is operated is needed for imitation to be easy; so the extra protection of the patent is not vital to preserving a monopoly for some time. In pharmaceuticals, this protection is valuable because imitation is easy, and the existence of patent protection may well have encouraged research and innovation in this industry [24].

The effects of patents on the actions of an independent inventor or small firm are very different from those on a large company. An independent inventor would find it difficult to get his invention developed and marketed without the legal protection of the patent, unless he could finance the development and marketing himself. Patents secure him against the theft of his ideas by those he may approach as business partners, and increase the potential profit that his financial backers hope to gain. Small firms, which cannot exploit a market quickly and may need to raise external finance to launch an innovation, similarly gain from patent protection. All innovations from independent inventors have not been dependent on patent protection, however; in the aircraft industry, patents were never important. Though the Wright brothers sold their patents, the purchasers lost on the deal, and patents had little effect on the development of the aeroplane in the pioneering days; the pioneers easily invented ways around the Wright patents. Patents have been relevant only where a clearly definable part of an aeroplane, like a type of wing flap, has been invented. Most of the pioneers were also able to finance their work themselves, so that they had less need of patent protection as a means of raising capital. The design of an aeroplane is in any case not patentable, so that the important gains in efficiency through changing design were not affected by patents. Radio was a case where the reverse applied, and patents were probably essential to financing the development work of Marconi and the other pioneers. They helped Marconi, for example, to raise funds from the public to finance his experiments.

Patents were not universally essential, therefore, to the launching of important innovations by independent inventors; but they were probably essential in many cases. Without patent protection, it is likely that fewer innovations would have been launched which were based on the work of independent inventors, and that the rate of innovation would have been lower. The difference would probably have been greatest in the first quarter of the century, when the majority of important inventors seem to have originated with independent inventors; the increased contribution of corporate research would have reduced the effect in the last fifty years, but the continuing importance of the independent inventor suggests that patent protection still increases the rate of innovation.

VI. CONCLUSIONS

Innovation is a process that is imperfectly understood, and the factors which have influenced its rate and direction in this century cannot therefore be clearly established. But my judgement on the available evidence is that the increasing organization of research and of industrial management has modified

rather than transformed the factors which had influenced innovation before 1900: the businessman's desire 'for profit and growth, combined with
the engineers' and the scientists' desire to improve, have remained the underlying influences on the rate of innovation. Rivalry, actual or potential, between
firms also remained a major influence, despite the growing concentration of
industry. Big companies became more influential, especially in industries
where the nature of the technology made the cost of innovation high; but
their dominance in these industries has resulted from the nature of the
technology, not because there are any general economies of scale in innovation. External factors, such as wars and the governmental interest in military
technology that they produced, did not again revolutionize the situation.
More money was spent on R and D, so that more people were engaged in it;
but the benefits of military R and D for civil purposes seem to have been
small. Evolution rather than revolution characterized the changes which we
have seen in this century.

REFERENCES

[1] SCHLAIFER, ROBERT, and HERON, S. D. *Development of aircraft engines and fuels*. Division
    of Research, Graduate School of Business Administration, Harvard University. Cam-
    bridge, Mass. (1950).
[2] ENOS, J. L. *Petroleum progress and profits*, p. 225. M.I.T. Press, Cambridge, Mass. (1962).
[3] ENOS, J. L. *op. cit.* [2], p. 229.
[4] LANGRISH, J. *et al. Wealth from knowledge*. Macmillan, London (1972).
[5] *Economic concentration: Hearings before the Sub-committee on Anti-trust and Monopoly of
    the Committee on the Judiciary, U.S. Senate, Eighty-ninth Congress*. Part 3, p. 1207.
    Evidence of D. Schon.
[6] JEWKES, J., SAWERS, D., and STILLERMAN, R. *The sources of invention*. (2nd edn.) Macmillan,
    London (1968).
[7] HAMBERG, D. *Research and development*. Random House, New York (1966).
[8] FREEMAN, C. The plastics industry. *National Institute Economic Review*, November 1963.
[9] MUELLER, W. F. The origins of the basic inventions underlying du Pont's major product
    and process innovations, 1920–1950. In NELSON, R. R. (ed.) *The rate and direction of
    inventive activity*. Princeton University Press, Princeton, N.J. (1962).
[10] NELSON, R. R., PECK, M. J., and KALACHEK, E. D. *Technology, economic growth and public
    policy*. The Brookings Institution, Washington D.C. (1967).
[11] MANSFIELD, E., RAPOPORT, J., SCHNEE, J., WAGNER, S., and HAMBURGER, M. *Research and
    innovation in the modern corporation*. Norton, New York (1971).
[12] MILLER, R. and SAWERS, D. *The technical development of modern aviation*. Routledge and
    Kegan Paul, London (1968).
[13] PECK, M. J. *Competition in the aluminium industry, 1945–1958*, pp. 201–4. Harvard Univer-
    sity Press, Cambridge, Mass. (1961).
[14] JEWKES, J., SAWERS, D., and STILLERMAN, R. *op. cit.* [6].
[15] MACLAURIN, W. R. Technological progress in some American industries. *American Econo-
    mic Review*, **44**, 178 (1954).

[16] SCHERER, F. M. Firm size, market structure, opportunity and the output of patented inventions. *American Economic Review*, **55**, 1097 (1965).

[17] STIGLER, G. J. Industrial organization and economic progress. In *The State of the social sciences* (ed. L. D. WHITE). University of Chicago Press, Chicago (1956).

[18] ALLEN, B. T. Concentration and economic progress. *American Economic Review*, **59**, 386 (1969).

[19] GALBRAITH, J. K. *American capitalism*. Hamish Hamilton, London (1956).

[20] CAUDLE, P. G. The management of innovation and the maintenance of productive efficiency. *In* BOWE, C. (ed.) *Industrial efficiency and the role of government*, H.M.S.O., London (1977).

[21] MANSFIELD, E. *Industrial research and technological innovation—an econometric analysis*. Norton, New York, for the Cowles Foundation for Research in Economics at Yale University (1968).

[22] FREEMAN, C. *The economics of industrial innovation*. Penguin, Harmondsworth (1974).

[23] TILTON, J. E. *International diffusion of technology: The case of semi-conductors*. The Brookings Institution, Washington D.C. (1971).

[24] TAYLOR, C. T., and SILBERSTON, Z. A. *The economic impact of the patent system: A study of the British experience*. Cambridge University Press (1973).

BIBLIOGRAPHY

Books

ALLEN, J. A. *Studies in innovation in the steel and chemical industries*. Manchester University Press, Manchester (1967).

BRIGHT, A. A. *The electric lamp industry*. Macmillan, New York (1949).

BROOKS, P. W. *The modern airliner*. Putnam, London (1961).

ENOS, J. L. *Petroleum progress and profits*. M.I.T. Press, Cambridge, Mass. (1962).

FREEMAN, C. *The economics of industrial innovation*. Penguin, Harmondsworth (1974).

GIBBS-SMITH, C. H. *The invention of the aeroplane, 1799–1909*. Faber and Faber, London (1966).

JEWKES, J., SAWERS, D., and STILLERMAN, R. *The sources of invention*. (2nd edn.). Macmillan, London (1968).

MACLAURIN, W. R. *Invention and innovation in the radio industry*. Macmillan, New York (1949).

MANSFIELD, E. *The economics of technological change*. Norton, New York (1968).

MILLER, R., and SAWERS, D. *The technical development of modern aviation*. Routledge and Kegan Paul, London (1968).

NELSON, R. R. (ed.) *The rate and direction of inventive activity*. Princeton University Press, Princeton, N.J. (1962).

SCHLAIFER, R., and HERON, S. D. *Development of aircraft engines and fuels*. Division of Research, Graduate School of Business Administration, Harvard University, Cambridge, Mass. (1950).

STURMEY, S. G. *The economic development of radio*. Duckworth, London (1958).

Articles

FREEMAN, C., YOUNG, A., and FULLER, J. K. The plastics industry: A comparative study of research and innovation. *National Institute Economic Review*, November 1963.

——, HARLOW, C. J., and FULLER, J. K. Research and development in electronic capital goods. *National Institute Economic Review*, November 1965.

KAMIEN, M. I., and SCHWARTZ, N. L. Market structure and innovation: A survey. *Journal of Economic Literature*, March 1975.

3

THE ECONOMICS OF TECHNOLOGICAL
DEVELOPMENT

F. R. BRADBURY

NEW inventions do not in themselves suffice to bring about technological progress; the availability of capital to exploit them has always been a major consideration. As the twentieth century advanced and technology became increasingly complex, the sums involved increased enormously and economic factors became correspondingly more critical. This chapter discusses in general terms the economics of technological change, the financing of innovation, and the growth of corporations in the first half of the twentieth century. Technological change is so varied in its scope and nature —as these collected volumes bear witness—that much important detail is lost in general statements about financial and economic aspects. Because of this, the chapter is designed to display some of the important factors influencing, and being influenced by, technological change and to give a sketch of interacting and dynamic complexity in which technology and technological change is but one thread—albeit an important one.

I. THE GENERAL PICTURE

The period covered by these new volumes, the first half of the twentieth century, was one of very great technological development. The calendar of innovation provided by T. K. Derry and Trevor I. Williams in their *Short history of technology* [1] makes this very clear. A synoptic world view of technological developments during the first half of this century shows not only that there were great advances but that these were very unevenly distributed among countries, with technological leadership moving from one country to another—at one time Britain, at another Germany, and in our period largely the United States of America, with Japan showing a remarkable rate of development in more recent times.

Moreover, the industries of the world advance unevenly: at one time coal, steel, and railways, at another electrical industries, at another polymers and plastics, at another electronics and communications; the leaders of one period may be the laggards of the next. The determinants of technological advance at any particular time or place are complexities we cannot hope to dispel here; the sources of innovation are considered elsewhere (Ch. 2). But we may note the light which J. Schumpeter and J. Schmookler threw on this matter. For Schumpeter, invention—the uncovering of potentially useful technology—was an externality, something which industrial innovators absorbed from outside their system of work and used, events occurring outside industry and outside its control but exploited by it [2]. Schmookler, by contrast, argued that periods of technological development and investment called up the inventions which were appropriate to industrial demands and opportunities [3]. On this view, economic and social forces impose the pattern of new technological developments, perhaps by selecting from the many candidate inventions those which were perceived to match needs of the time and rejecting or neglecting those which were not so seen. The Schmookler view is supported in his writings by much empirical information on patents and industrial investment, and it goes some way to explaining how industrial development and innovation is clumped in particular nations at particular times and not evenly spread the world over: in short, rapid and growing technological development and investment begets further technological development.

This theme, although poorly illuminated by writings in the literature, is one of very great importance. If we may take the liberty of stepping forward from our period into the 1970s we find much controversy in international forums such as O.E.C.D. and the U.N. on problems of technological development of less developed countries (L.D.C.s). The governments of L.D.C.s the world over are striving to induce technological development in their countries by importing technology. If the problem of how technology breeds technology were better understood, the tasks and strategies of L.D.C. governments might be greatly eased. W. Parker, writing in 1961, made a comment on the history of technological development which put the question provocatively. He wrote:

The rate of flow of Western thought, popular taste and technology has been strongly affected by the development of these techniques of social communication. But the direction in which culture traits and knowledge have flowed over this communication net is a more puzzling and profound question. What does knowledge consist of, and why has it moved out from Europe, rather than in from Asia and Africa? Why is it that the things the Europeans have brought have been the good, the penetrating things, and why have local cultures proved so brittle before their advance? These questions cannot be answered by reference to the culturally conditioned elements in human character. Western ideas have on occasion been spread by force, just as Islam spread in the seventh and eighth centuries. Before the advent of mass communication techniques, the most common way was through the conversion of a small group at the top of the existing social order. Given this sort of revolutionary overthrow of the old society, the education of a new generation may be sufficient to alter radically its character. But the spread of Western ideas is not due wholly to a combination of power and brainwashing; one must postulate instead some universal characteristics of human nature to which scientific thought and its accompanying technology, and even the curious combination of values placed on goods, enterprise and the saving of human labor, make an irresistible appeal. [4]

It is germane to ask why so much value is set on technological development; why, today, is 'technology transfer' such a strong political issue in the U.N.? The answer to this is generally agreed by economists; the level of technology determines the consumption of factors of production, labour, capital, and materials, and improvements in technology make production more efficient and are therefore economically desirable. A substantial part of this chapter will be devoted to explaining the technological–economic interactions which drive forward the adoption of improved technology. The value put upon technological development is its usefulness in improving the way in which resources are used to satisfy man's needs. The economist assumes that the innovations chosen, given a wide range of choice, are progressive, the term being used in the sense of improved efficiency of conversion of resources.

The effects of technological development on industrial organization will also be explored. We have noted how technology begets technology; is it true that technological developments cause concentration of productive resources, that is, give rise to large firms and corporations? We are fortunate in having detailed studies of factors affecting industrial concentration, principally in the work of Joe S. Bain. We will treat this in detail but here we may note that technological development is a factor favouring concentration of production; there are, however, other forces, towards and against, and it is certainly not true that technological advances lead inevitably to concentration of production.

One aspect of concentration which is much discussed in the context of technological development is the provision of resources to finance major innovations. We shall present figures to show the very heavy investments which major developments such as catalytic cracking of oil to petrochemical products, or nuclear energy, demand. It is argued, by J. K. Galbraith, for example, that only large corporations can amass the liquidity of resources required to sustain such large negative cash flows during the time of major innovations. It is undeniably the case that such ventures are beyond the means of small firms and, indeed, often beyond the big ones too. This leads to government financing of major technological developments where the resource demands are beyond the limits of private industry. When innovations of this magnitude are taken out of the realm of normal market forces decisions on investment are political ones and the sifting of efficient from inefficient use of resources is taken out of the market. Investment appraisal is replaced by cost–benefit analysis, and the results may make good politics and poor economics. This phenomenon has become most marked in the years after 1950 and we shall not therefore pursue it further.

The final general point to be made on economics of technological development is on the subject of technology itself. Technology tends to be regarded, perhaps especially so by readers of a history of technology, as a collection of machines or plants or productive systems. In reality, it is much more than this. Technology is the way of doing, the total operation of the productive process involving people as well as hardware. Because technology includes people and their way of working (Ch. 5), it is not sensible to regard it as a prime mover or causative agent in historical development. While it is possible to identify technological landmarks or milestones and to describe any current production technology—say, the manufacture of nylon shirts—such a technology is not a single entity but the combination of many: the machines, the materials, the people, and the organization of work. Therefore, when we write of the economics of technological change, we are really trying to describe economic aspects of changing patterns of work with particular reference to the degree of effectiveness attained in converting inputs to products. It is easily seen that technology in the sense described is very sensitive to many factors, social as well as economic and political, and one is perhaps attempting, in an essay on economics of technological change, the interesting but dubiously useful task of tracing the path of one thread in the warp of an elaborately woven fabric. However, as other contributors seek to unravel other threads, we may hope that, collectively, some coherent pattern will become apparent.

II. VARIETY AND SCOPE OF TECHNOLOGY

The variety of technological change to be encompassed may be illustrated by Table 3.1, due to J. Enos [5] and reproduced in E. Mansfield [6], which lists forty-six important product and process innovations and the estimated time interval between invention and innovation. We shall come later to discuss the two terms *invention* and *innovation*; at this point we are interested simply in the variety displayed by the innovations, in gestation time, and in complexity, from radar to the homely zipper.

TABLE 3.1

Estimated time interval between invention and innovation, selected industries

Invention	Interval (years)	Invention	Interval (years)
Distillation of hydrocarbons with heat and pressure (Burton)	24	Steam engine (Watt)	11
		Ball-point pen	6
Distillation of gas oil with heat and pressure (Burton)	3	DDT	3
		Electric precipitation	25
Continuous cracking (Holmes-Manley)	11	Freon refrigerants	1
		Gyro-compass	56
Continuous cracking (Dubbs)	13	Hardening of fats	8
'Clean circulation' (Dubbs)	3	Jet engine	14
Tube and Tank process	13	Turbo-jet engine	10
Cross process	5	Long playing record	3
Houdry catalytic cracking	9	Magnetic recording	5
Fluid catalytic cracking	13	Plexiglas, lucite	3
Catalytic cracking (moving bed)	8	Nylon	11
Gas lift for catalyst pellets	13	Cotton picker	53
Safety razor	9	Crease-resistant fabrics	14
Fluorescent lamp	79	Power steering	6
Television	22	Radar	13
Wireless telegraph	8	Self-winding watch	6
Wireless telephone	8	Shell molding	3
Triode vacuum tube	7	Streptomycin	5
Radio (oscillator)	8	Terylene, Dacron	12
Spinning jenny	5	Titanium reduction	7
Spinning machine (water frame)	6	Xerography	13
Spinning mule	4	Zipper	27
Steam engine (Newcomen)	6		

Sources: J. Enos [5] and E. Mansfield [6]

Another way of looking at the problem of variety in technological change is in terms of technological novelty and origin. M. Gibbons and R. D. Johnston [7] have analysed a random selection of significant U.K. product innovations. Their classification showed only 18 per cent of these to be new and to involve technical change by U.K. firms. This points to the international character of technology and suggests that much of what is new in a particular location may embody little technical novelty.

One may attempt to group technological changes by their technical nature. C. Freeman [8] writes of innovations in products, processes, energy, and materials; we might add another category, communications, which would fall into his category of multi-component systems. Economic classification of technological change might distinguish disembodied from capital-embodied innovations (indicating whether or not the new technology, to be applied, has to be embodied in new plant and equipment), those with high ratio of development to production costs, those which have high entry costs, and so on.

One confusion to be alert to is that of identifying technologies with industries. We may define an industry as a group of establishments engaged primarily in the same or closely related types of business, but within any industry there are many technologies. In the textile industry, for example, there are employed at least six of the major product and process innovations listed in Table 3.1, to say nothing of a limitless number of others—chemical, mechanical, electronic, etc. Nonetheless, much of the information we have on economics and financing of technological development is, quite naturally, available on an industry basis, and it is to industries, all with their own mix of technologies, that we must turn in the consideration of technology and the growth of corporations.

III. INDUSTRIAL STRUCTURE

During the first half of the twentieth century—and, indeed, prior to this period—there were very great changes in the concentration of industry in industrialized countries. M. A. Utton for example illustrates concentration of industry in the U.K. [9]. Using the work of R. Evely and I. M. D. Little [10], Utton summarized concentration indices of 220 industries into the 16 broad categories shown in Table 3.2. The relative concentrations of the groups are displayed as average levels of concentration of employment and of net output. The levels of concentration used in Utton's table are derived from industry concentration ratios (per cent of total employment or of net output by the three or four largest firms) by grouping (into the 16 broad categories) and weighting the concentration ratios of individual industries by the numbers employed in that industry to give the average employment concentration, or by the net output of that industry to give the average net output concentration. 'Generally speaking', wrote Utton, 'those sectors of industry associated with very large capital requirements appear amongst the most heavily concentrated, such as chemical and allied trades, electrical engineering and electrical goods, vehicles, and iron and steel and non-ferrous metals. On the other hand, sectors usually associated with smaller firms tended to rank low in terms of concentration, e.g. cotton, woollen and worsted, clothing and footwear, and building and contracting.'

TABLE 3.2

Concentration of employment and net output by industry groups

Broad industrial category	Number of industries	Average level of concentration	
		Employment (%)	Net output (%)
Chemical and allied trades	16	51	46
Electrical engineering and electrical goods	8	48	46
Vehicles	8	41	44
Iron and steel and non-ferrous metals	15	39	40
Drink and tobacco	7	36	42
Mining and quarrying and mining products	21	35	41
Shipbuilding and non-electrical engineering	20	31	32
Food	18	30	35
Other metal industries	27	29	32
Other textiles	18	27	36
Paper and printing	10	21	24
Cotton	3	21	18
Other manufacturing and service trades	23	20	23
Woollen and worsted	6	18	18
Clothing and footwear	17	14	12
Building, contracting and civil engineering	3	12	11
Total	220	29	33

Sources: Evely and Little [10] and Utton [9]

Utton compared changes in concentration over time in industry in the U.K. and the U.S.A. He observed that up to 1950 trends in the two countries were similar, upwards, with something of a reverse during and immediately after the Second World War. Since about 1951, however, market concentration seems to have increased noticeably in U.K. manufacturing industry and to differ in this from the U.S.A., where it has remained fairly stable.

There is an extensive literature on industrial concentration, the important source book being Joe S. Bain's *Industrial organization*. To illustrate the relation between technology and industrial concentration we take Bain's analysis of the American scene. Bain finds it convenient to identify three phases, 1870 to 1905; 1905 to 1935; and 1935 to the early 1960s. The picture which emerges is of rapid concentration of manufacturing industries up to 1905, an even more marked concentration of the service industries in 1905 to 1930, followed by a less obvious increase in the most recent period. (It will be noted that Utton's analysis shows U.K. concentration advancing more rapidly than in the U.S.A. in the post-1950 period.)

The forces acting for and against concentration of industry are identified by Bain as follows:

   (i) Technological considerations. Technological developments—such as the engineering of continuous processes in the chemical industry—encourage concentration to make it possible to take advantage of the resultant economies of scale.

  (ii) Sales promotion. Effective exploitation of sales promotion methods needs large markets and this drives towards larger industrial groups. Bain comments that this drive may out-pace the technological one, leading to a degree of concentration beyond what is needed simply to exploit technological opportunities.

 (iii) The urge to control markets and reduce competition—'monopolistic considerations' in Bain's terminology.

 (iv) The urge to set up barriers to new entrants to the business area by strategic acquisitions, such as key raw material supplies.

  (v) Financial operations, which we might today dub conglomerate formation or asset stripping.

 (vi) Opposing forces operating against concentrations include antitrust legislation; the desire to retain sovereignty over business operations; and market growth (which aids would-be entrants).

The relative strengths of these forces will differ from period to period and from country to country.

Before we go on to see how the events of Bain's three periods are analysed in terms of the foregoing factors, we should turn aside to assess the magnitude of the resources needed to exploit some technological developments of the study period.

IV. FINANCING OF INNOVATION

Freeman analyses the estimated expenditures of time and money in developing new cracking processes in the petrochemical industry. Table 3.3 is drawn from the work of Enos [11]. It will be observed that some of the processes shown there are included in the 46 of Table 3.1.

The sums involved are very large because of the long periods which elapse perforce between early development and successful exploitation. The story of catalytical cracking as related by Freeman reveals the astronomical costs of a major technological development [8]. E. J. Houdry, a wealthy French engineer, spent $3m. of his own fortune but two further contributions of $4m. each from Sun Oil and Socony Vacuum were needed for the successful development of a process. Houdry's success stimulated a concerted effort by

TABLE 3.3

Estimated expenditure of time and money in developing new cracking processes

Process	Development of new process		Major improvements to new process		Total	
	Time	Estimated cost ($000)	Time	Estimated cost ($000)	Time	Estimated cost ($000)
Burton	1909–13	92	1914–17	144	1909–17	236
Dubbs	1917–22	6 000	1923–31	> 1 000	1909–31	> 7 000
'Tube and tank'	1918–23	600	1924–31	2 612	1913–31	3 487
Houdry	1925–36	11 000	1937–42	n.a.	1923–42	> 11 000
Fluid	1938–41	15 000	1942–52	> 15 000	1928–52	> 30 000
TCC and Houdri-flow	1935–43	1 150	1944–50	3 850	1935–50	5 000

Sources: Enos [11] and Freeman [8]

TABLE 3.4

Influence of capacity on production cost of ethylene, 1963

Production costs (£000)	Capacity and output (000 tons)		
	50	100	300
Capital investment			
Battery limits plant	2 200	3 100	5 400
Off-site facilities	650	900	1 600
Total (excluding working capital)	2 850	4 000	7 000
Current costs per year			
Net feedstock	250	500	1 500
Chemicals	50	100	300
Utilities	300	600	1 800
Operating labour and supervision	50	50	50
Maintenance at 4 per cent of battery limits plant cost	90	125	215
Overheads at 4 per cent of battery limits plant cost	90	125	215
Depreciation	250	355	620
Total	1 080	1 855	4 700
Current costs (£) per ton of ethylene	21·6	18·6	15·7

Source: Freeman [8]

a group called Catalytic Research Associates, including Kellogg, I. G. Farben, Indiana Standard, Jersey Standard, Shell, B.P., Texaco, and O.U.P. who eventually, after an expenditure of over $30m., developed a successful improved (fluid bed) catalytic cracking process.

The cracking story illustrates not only the vast sums required in technological development, but also the way in which many companies may come together to tackle the immense development problems. It should be noted that Catalytic Research Associates was not a business merger of the participating companies but a co-operative research and development undertaking. This is an example of how research and development (R and D) costs may be shared and the fruits exploited, partly by royalties to the consortium (Jersey Standard's share of royalties was over $30m., received 14 years after the successful development) and partly by exploitation of scale economies the new technology made possible. The first Dubbs unit had a capacity of 500 barrels per day (1925) but by 1956 fluid bed crackers had capacities of 100000 barrels per day. The effects of scale economies on production costs is well illustrated by Table 3.4, from which it may be seen that production costs fall dramatically with scale but demand very large capital investments. The attraction of such economies and the need for large capital resources to achieve them are some of the forces tending to concentrate industry into fewer larger units, as described by Bain.

V. CHANGES IN RATE OF CONCENTRATION OF INDUSTRY WITH TIME

Let us return to Bain's analysis of the three periods—1870 to 1905; 1905 to 1935; and 1935 onwards—in the U.S. scene. First two fallacies are rejected: (1) that concentration of industry continuously increases; (2) that concentration is adequately explained by effects of technology and distribution scale effects.

In Bain's first period there were great technological changes: in America, Remington's typewriter, Bell's telephone, Edison's phonograph, Edison's carbon filament lamp, Carnegie's first big steel furnace, Niagara Falls hydroelectric installation, the Eastman Kodak camera, Northrop's automatic loom, the first main-line electric train, Ford's motor car, and Orville Wright's powered flight; for a chronological table of these and other major innovations of the period, see Derry and Williams [1]. Mechanical devices and mass production methods greatly enlarged the minimal competitive size of plants. Excess capacity was to some degree avoided by expansion through mergers—

a device for expanding plant size without building excess capacity. There was a strong drive to achieve market control, fed by the excess capacity which was created by expansion of individual plants. Other technological developments, such as growth of railways, also increased competition by opening up bigger market areas and further encouraged tendency to monopoly responses.

In the period 1905–35 the rapid expansion of manufacturing industry size eased, partly because the scale demands had been largely met, partly because restraining legislation, notably the Sherman Act, was being enforced more firmly than in the previous period. In contrast to manufacturing, however, service industries underwent much concentration. As in the earlier period of concentration of manufacturing, technological developments—electric power generation and distribution for example—were an active incentive towards concentration. Bain records that utilities concentration went further than manufacturing concentration had done, a factor being that utilities industries were regulated by public authorities and these are not greatly influenced by anti-trust legislation.

In the period 1935 onwards there seems to have been in America a roughly balanced situation, legal constraints and growing markets holding in check the pro-concentration forces.

Summarizing this section, we observe that technological development is a force driving industry to merger and concentration but that counter-forces restrain this to the extent that at the end of our period the American scene is in a roughly balanced state where the degree of industrial concentration showed little further change. We observe that there is a balance of forces involved and whether concentration occurs or not depends not only on the state of technological development but of market growth, anti-monopoly legislation, and other factors which will have different strengths at different parts of the globe at different periods of history. In this connection we recall Utton's observation that in the post-1950 period U.K. concentration proceeded more rapidly than it did in the U.S.A.

The significance of market growth as a force counter to concentration should not be underestimated. If because of lack of technological development an industry suffers declining output or supply fails to grow, costs may be expected to rise and firms decline with consequent increase in concentration (S. E. Boyle [12]). If we accept this argument, it follows that technological development favours concentration of industry both in its progress, by providing opportunities to exploit scale economies, and in its failure to progress, by inducing a decline in business in which firms may merge or disappear. This apparent illogicality may be resolved when we turn to consider the mechanism by which moribund technologies are displaced.

A counterforce to those tending towards concentration of industrial under-takings is economic nationalism. It is of much interest to observe that econo-mic nationalism shaped to a degree the great British chemical enterprise of our period, Imperical Chemical Industries Ltd. We are fortunate in having two volumes of the history of that company, commissioned by I.C.I. and written by W. J. Reader [13]. Reader's analysis of the formation of I.C.I. in 1926, the famous transatlantic negotiation and agreement of Harry McGowan and Alfred Mond, is an illuminating case-study of concentration and exempli-fies well the main factors promoting concentration as identified by Bain. I.C.I. came into existence in 1926 by merger between Brunner Mond and Co. Ltd., Nobel Industries Ltd., the United Alkali Co. Ltd., and the British Dyestuffs Corporation Ltd. The demands of technology for growth in capital investment were certainly one factor leading to the merger. Nobel Industries had large resources in cash and securities but was short of investment oppor-tunities; whereas Brunner Mond had the ammonia project at Billingham which was short of the resources necessary for expansion. The merger's out-come included the application of Nobel resources to fertilizer plant invest-ment.

The merger was also defensive in character, and the process of merging was catalysed by the formation of a German heavyweight, I. G. Farbenin-dustrie A.G. in the autumn of 1925. This culminating step in a continuing process of concentrating the German chemical industry produced a complex large by American standards and massive by British. Moreover, the German pattern of business was one of aggressive overseas sales. Reader's account of the negotiations in which I.C.I. and I.G. attempted to exchange technical information and to share markets contains a phrase which illustrates vividly Bain's main theses: '[I.C.I.] accepted it (the idea of a profit pool), taking as a guiding principle "the avoidance of economic wastage, of duplication of plant, of capital expenditure, and of competition between the parties"—a perfect short statement of the rationalizers' creed, to which Mond so heartily subscribed.'

The formation of I.C.I. was also motivated by grand and sometimes euphoric expressions of an imperialist style of thinking by its two founders, McGowan and Mond. Reader quotes Mond as follows: 'as becomes more apparent day by day, the trend of all modern industry is towards greater units, greater co-ordination for the more effective use of resources . . . But this process . . . is leading towards a further series of economic consequences. One of the main consequences is the creation of inter-relations among industries which must seriously affect the economic policies of nations.' McGowan was more succinct and, if possible, even more ambitious. The formation of I.C.I., he told a representative of du Pont in December 1926, 'is only the first step in a comprehensive scheme . . . to rationalize (the) chemical manufacture of the world.' Together they wrote in November 1926 to the President of the Board of Trade: 'We are Imperial in aspect and Imperial in name' and frequently stated their aim of keeping the development of the chemical industry in British territory in British hands.

Turning to the counterforce of economic nationalism referred to above we find that the merger from the beginning is aware of the importance of overseas manufacture. Reader writes as follows:

The other main aspect (of the merger) was overseas manufacture. Brunner Mond had never put up works overseas and the main manufacturing companies, overseas, which came into I.C.I. were those founded by Nobels to make explosives in three of the main mining countries of the world: Canada, South Africa, and Australia. They represented a response to economic nationalism in the Dominions, each anxious to encourage the growth of home industries to displace imports, and there were local partners in each. Moreover, although each was in origin an explosives company, they were all developing into diversified enterprises roughly on the model of I.C.I. at home, so as to take, in the growing chemical industry of each Dominion, a place as dominating as that of I.C.I. in the chemical industry of Great Britain. They were the main instruments of I.C.I.'s development policy overseas, because it was recognized that in the future, even in imperial territories, British imports would increasingly be kept out in favour of local manufacture and that the only way British interests could be developed overseas would be through manufacturing enterprises set up jointly with local investors to employ local labour.

The interest of this comment is particularly in relation to the historical timing. Whereas today (see below) the demand for local manufacture, and input of technology rather than its products, is a dominant political theme and something that has received a name—technology transfer—it must be seen as very advanced thinking by the founders of I.C.I. in 1926.

For an economist's comment on economic nationalism in the current technological-economic scene the reader is referred to A. Lindbeck [14].

This chapter is directed towards economics and technological development and not to political factors, which are discussed elsewhere. However, it must be observed that the structure of industry, which we have been considering in economic terms, and especially economic nationalism mentioned above, is profoundly affected by war and war preparations. The Reader history of I.C.I. and L. F. Haber's *Chemical industry 1900–1930* [15] are valuable source books for a study of this aspect of technological development. There is little doubt that national emergencies force the pace of innovation. The point is well made by Reader:

The growth points in IG were all in materials useful in war, particularly if blockade were to cut off supplies of natural materials such as rubber, wool, cotton, or oil. The point was noticed, and in England a good deal of ignorant scorn was poured on German ersatz materials, which were taken as evidence of strategic weakness in the German economy. What was less readily comprehended—even, perhaps, in some high quarters of I.C.I.—was that these materials were characteristic of the most advanced practice of the contemporary chemical industry, especially in the organic field, and their development foreshadowed the lines along which the industry was to advance towards the world of the nineteen-fifties and sixties.

An important facet of the economics of technological development is the rise and decline of British industrial strength. The rise is prior to our period but the decline spreads through it. Many articles and opinions on the reasons for the decline of Britain's technological industrial position have been produced. We shall refer to the analysis by J. H. Dunning and C. J. Thomas [16]. These authors looked at the performance of British industry in 1914 and found the following points of weakness relative to competitive industry elsewhere.

(1) Britain's former products, in which she had excelled, were now being supplied by her customers: while her technological ability in the newer products and processes lagged behind those of Germany and the U.S.A.
(2) The economy was over-specialized, in output and markets—coal mining, iron and and steel, and textiles being dominant.
(3) Technology lagged behind in Britain even in staple industries.
(4) British industries' failure to exploit inventions as fast as the competition.
(5) Addiction by Britain to *laissez-faire* and unrestricted competition, whereas the U.S.A. and Germany preferred concentration of production. (British industry concentrated, too, at a later period.)
(6) Legislative entanglements of businessmen by government, e.g. in patent law.
(7) Its highly prosperous state deprived enterprise of the drive relative to competition.
(8) Export of capital to the Dominions and elsewhere at the expense of domestic capital formation.

The last point raises again the recurrent theme of economic nationalism and transfer of technology. Although the founders of I.C.I. were keen to export British chemical technology to the Dominions, there are lobbies which strive today to restrain technology transfer from developed countries on the grounds that this weakens the domestic economy. There is, however, no consensus that technology transfer is harmful to the transferror nation.

VI. R AND D EXPENDITURE AND INDUSTRIAL GROWTH

We have seen above what great resources are needed to sustain major technological developments through the invention/innovation process to profitable exploitation. This factor itself may act as a determinant of viable firm size. Freeman shows how in an industry in process of rapid technological development a minimal size of R and D resource is necessary for survival [17]. In its absence the firm will not be able to respond quickly enough to competitive developments and may, as a result, be put out of business.

It is estimated by Freeman that the R and D effort required to nurture a machine-tool control system is £300 000 to £600 000. A small scientific computer may absorb £1m. to £2m. in R and D costs. If the R and D costs are spread over three years, and assuming R and D costs to be 2 per cent of sales, the firm innovating the machine-tool control device demands a minimal sales volume of £5m. to £10m. and the computer innovator £17m. to £33m. By a further step in the argument, £2·5m. of sales is equated with a work force of 1500, leading to the conclusion that the machine-tool control R and D will not be supported by a firm employing less than 3000 and the computer R and D by not less than 10 000 [18].

The ratio of R and D costs to sales—the most commonly used measure of R and D activity levels in industry—is set in the passage above at 2 per cent. The relation of R and D expenditure to sales differs very greatly among industries. Mansfield [6], for example, quoted in Table 3·5, estimates the ratio of R and D costs to sales for the year 1955 to vary from 5·92 per cent for drugs to as little as 0·75 per cent for steel.

The variations must reflect different expectations from research and development expenditures in different industries and this in turn must depend to a large extent on the rate of technical change recently experienced in the industry. Freeman [19] puts the matter this way:

Broadly speaking, fundamental discoveries in science and technology dictate the areas in which research and development is likely to be fruitful; the development of new products and techniques in these areas stimulates demand and hence provides the basis for growth of output there. Added to this, research will tend to be concentrated in the producer goods industries since it is essentially through the improvement of producer goods—machines and materials—that progress is achieved. Further, in any field, a firm or industry with a big expansion in demand for its products, and with constant pressure for products which show an improved performance is more likely than others to undertake research work.

Freeman goes on from this to advocate provision in Britain for more research expenditure to stimulate more rapid industrial growth.

Expenditure on R and D by industry in the expectation that this would lead to new products and new methods of production through invention and innovation was an observable feature of industrial and government resource allocation during the period. The amount so spent differed widely among industries and with time. We have information on R and D expenditure as a percentage of turnover for a number of industries in the United States of America for the end of the review period, and some projections forward to later years, thanks to the investigations of Mansfield [6]. Table 3.5 is a small section of a larger one presented by Mansfield.

TABLE 3.5

R and D expenditure as a percentage of sales

Industry	1945 R and D expenditure (% of sales)	Standard deviation	1955 R and D expenditure (% of sales)	Standard deviation
Chemical	2·38	0·92	3·32	0·91
Petroleum	0·558	0·24	0·686	0·238
Drug	3·58	1·58	5·92	1·26
Steel	0·302	0·096	0·75	0·395
Glass	1·11	0·45	2·26	0·47

Source: Mansfield [6]

Mansfield drew three conclusions from the data he assembled. First, that R and D expenditure increased enormously over the period; second, that there is not only great variation in expenditure on R and D among industries, but also that there is great variation within industries among firms (the standard deviation being high); third, that variation in expenditure decreased during the period (the standard deviation increasing less rapidly than the mean).

In another section of the same work, Mansfield developed a production function relating output of the firm to labour and capital inputs and included also terms for R and D expenditure, the annual rate of 'depreciation' of an investment in R and D, the rate of technological change that would occur if R and D expenditure by the firm were to cease, and elasticity of output in relation to accumulated past R and D expenditures.

The argument is too long and complex to be reproduced here but the results are of great interest, even though it is necessary to reiterate Mansfield's warning to take them with considerable caution. Despite this, the work is a rewarding study and does fulfil at the very least Mansfield's own hopes: 'Having cursed the dark loudly and publicly, my purpose here is to stimulate discussion of the problem [of estimating the returns from industrial research and development] and to light a few candles, limited though their power may be' [6].

Another approach to the rewards for R and D investment is made by Freeman. Having stated that the fruits of research, in the form of new products and new techniques of production, are an essential part of economic progress [19], Freeman presents the relation between R and D expenditure and industrial growth in the U.S.A. and the U.K. The paper was written in 1962 and refers to growth between 1935 and 1958 and R and D expenditure (as a ratio of ultimate output) in 1958. A single line fits well the scatter of points, both the American and British figures behaving similarly, showing strong positive correlation between growth and R and D expenditure in both countries. Freeman's analysis also reveals unevenness of expenditure on R and D among industries; nine-tenths of total industrial R and D expenditure in his sample went to capital goods and chemicals, despite the fact that these same industries employed less than half the total industrial labour force.

The subject of 'how much' R and D in industry would be incomplete without reference to an important contribution to the subject by R. R. Nelson [20]. In an article entitled 'The simple economics of basic research' written in 1959 (under the shadow of *Sputnik*) Nelson attempted to estimate the net social value of basic research—by which term he designates scientific research having only loosely defined goals, in contrast to the problem-solving goals of applied scientific research. The amount spent on basic research in 1953 in the U.S.A. he estimated at $435 m. or 8 per cent of total U.S. expenditure in that year on research and development. He saw the fruits of science as the ability to predict facts about a phenomenon without, or prior to, experiment. This lowers the expected costs of invention and of innovation. Nelson then argued that the nature of the uncertainties involved in the inventive process is such as to make applied research relatively unlikely to result in significant breakthroughs, save by accident:

for, if significant breakthroughs are needed before a particular practical problem can be solved, the expected costs of achieving this breakthrough by a direct research effort are likely to be extremely high; hence applied research on the problem will not be undertaken, and invention will not be attempted. It is basic research, not applied research, from which significant advances have usually resulted. It is seriously to be doubted whether X-ray analysis would ever have been discovered by any group of scientists who, at the turn of the century, decided to find a means for examining the inner organs of the body or the inner structure of metal castings. Radio communication was impossible prior to the work of Maxwell and Hertz. Maxwell's work was directed toward explaining and elaborating the work of Faraday. Hertz built his equipment to test empirically some implications of Maxwell's equations. Marconi's practical invention was a simple adaptation of the Hertzian equipment. It seems most unlikely that a group of scientists in the mid-nineteenth century, attempting to develop a better method of long-range communication, would have developed Maxwell's equations and radio or anything nearly so good.

Since basic research adds to knowledge which is socially valuable and since firms funding basic research can appropriate only a proportion of the social value they thereby generate, Nelson argued that 'private profit opportunities alone are not likely to draw as large a quantity of resources into basic research as is socially desirable'. This conclusion is music to the ear of the fund-starved academic curiosity-oriented researcher today, but we have to remember that we are here considering the 1950s, when the costs of the social fruits of basic research were not seriously questioned, as they are today.

VII. INVENTION AND INNOVATION

Whatever the effect of technological developments on industrial concentration as a whole, there is well-documented evidence to show that technical change—innovation—has become, during the first half of this century, increasingly the fruit of corporate research. There is some danger of confusion if the words 'invention' and 'innovation' are loosely used. The confusion is confounded by a change of emphasis during the period under review. Schumpeter is rightly credited with a clear formulation of the activities embraced by the terms. Invention is the concept, an idea or perception of useful new technology, be this a product, a process, a structure, or a design. Innovation is much further down the line, the realization of the new technology in the hands of its user. The innovation process may be long and will include the development phases of the R and D activity. Schumpeter saw invention as something extraneous to industry, the innovation process as an internal function making use of inventions external to it. During the period under review this picture of invention as something external to industry has changed with the inventive process becoming internalized as part of the R and D process of industrial laboratories and design offices.

This change has been described well by Freeman [8]. Referring to J. Jewkes's work, Freeman wrote:

The brief account has stressed the shift from the inventor-entrepreneur of the nine-teenth century towards large-scale corporate R and D. This differs sharply from the interpretation given by Jewkes *et al.* They minimise the differences between the nine-teenth and twentieth centuries and generally belittle the contribution of corporate professional R and D.

The underestimation of the contribution of corporate effort to technological development stems from a concern with invention to the neglect of the vastly greater resource-consuming activity of innovation.

It is this aspect of resource consumption in technological development that engages our attention in this chapter on economic aspects. At the end of the period we are considering, in 1956, the world's first commercial nuclear power plant came on stream at Calder Hall. Freeman describes this as the culmina-tion of earlier tendencies towards large-scale professional R and D, charac-teristic of the twentieth century. Innovation on this scale demands government support if it is to proceed. Table 3.6, taken from Freeman, illustrates the dimensions of development costs for nuclear power.

TABLE 3.6

Development costs estimates (£m) of nuclear power plants (1971)

Reactor type	Actual	Projected additional
Gas-cooled, Magnox mark 1	20	
Gas-cooled, A G R mark 2	114	
High-temperature reactor, mark 3	25	32
Steam-generating heavy water reactor	78	16
Fast-breeder reactor	205	124
Total	442	172

Source: Freeman [8].

These sums are much greater than those needed for chemical process develop-ment. Freeman makes this comment:

The need for such massive public investment arose because nuclear-engineering pro-cesses carry to an extreme degree all the tendencies which have been discussed in relation to chemical and oil refinery processes. The very heavy costs and long gestation period arose from the extraordinary complexity of the design problems, involving new materials, instruments, components and equipment of all kinds to satisfy the exacting requirements and safety standards of the new technology. At every stage intimate collaboration was necessary between nuclear engineers and scientific research teams investigating fundamental problems, so that large R and D groups were essential. The Harwell Laboratory in the 1960s was spending about £5 million per annum on the background research necessary to support the Authority's other design and development activities. Altogether the Atomic Energy Authority was spending about £50 million *per annum* on civil research and development over the last fifteen years, and similar sums have been spent by public authorities in several industrialised countries.

VIII. THE DYNAMICS OF TECHNOLOGICAL CHANGE

The process by which new technology displaces old is without doubt the most interesting aspect of the economics of technological change. In what follows we shall draw heavily on the work of W. E. G. Salter who, in a brilliant monograph in the late 1950s, described the dynamics of the system in which technological change occurred. In this work Salter developed a model, which we shall try to describe, and then went on to put it to the test of matching it against available empirical data for the period 1924 to 1950. We shall present some of this evidence too, since it serves not only to test Salter's model but also to provide an interesting picture of what technological changes were significant in the period under review.

The engine of change, if we may use that expression, involves the interaction of inputs and outputs of the productive activity with market forces and the progressive disturbance of these relations by innovation and technological change. Basic to it all is the economists' notion of a production function, variously written, but in Salter's terminology having this form:

$$O = f_n (a, b, c \ldots)$$
$$O = f_{n+1} (a, b, c \ldots)$$
$$\ldots \ldots \ldots \ldots \ldots \ldots$$
$$O = f_{n+t} (a, b, c \ldots)$$

where O is output, a, b, c are inputs of factors of production, and $n, n+1 \ldots n+t$ are consecutive time periods. The impact of new knowledge, of invention and innovation, is to lead to new production functions, each of which is superior to its predecessors in the sense that less of one or more of the factors of production is required to produce a given output.

Fig. 3.1, taken from Salter, shows graphically the effect of progressive changes in the factor f. For ease of presentation, only two factors of production, labour, and capital, are considered in Fig. 3.1. The vertical axis measures unit labour requirements for a given output and the horizontal axis measures unit capital requirements. Successive curves move inwards towards the origin, illustrating how new technical knowledge offers new techniques and improved

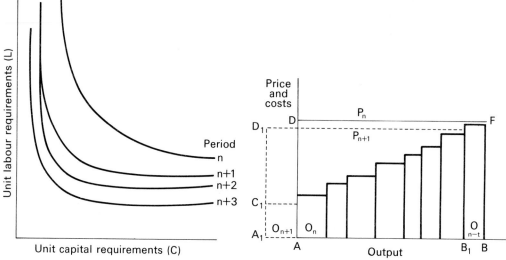

FIG. 3.1. Changing production functions.

FIG. 3.2. The overthrow of existing technology by new.

productivity. The actual amounts of labour and capital consumed to make the given output will depend, of course, on their relative prices. The best-practice technique at each date takes account of both technical and economic conditions and is the technique which yields minimum costs in terms of the production function and relative prices which hold at each date. (By reference to the original work the reader will see that the point of contact of the tangent of slope equal to ratio of capital and labour prices with the appropriate production function indicates the current best-practice technique.) Salter writes:

. . . successive best-practice techniques differ in two respects which, although difficult to distinguish in practice, are analytically quite distinct. First, today's best-practice techniques differ from yesterday's in that they make use of the new knowledge acquired between today and yesterday. This is represented by the shift in the production function; new knowledge by easing technical restraints has opened up a superior range of production-possibilities. Secondly, changed relative prices have altered the technique which is economically appropriate.

Either influence is sufficient to result in a continuous flow of new techniques of production. If knowledge were constant and labour became dearer relative to real investment, best-practice techniques would become increasingly mechanised in the sense that the input of labour would decrease and that of investment increase. This transaction would take place in two ways. First, businessmen would find it profitable to employ in new uses machines and methods already developed by engineers and machine-makers; for example, the extension of industrial-type hoists and conveyor systems to building operations. Secondly, engineers and machine-makers would be under pressure to apply the existing fund of knowledge to the design of methods which were labour-saving and investment-using. Such methods would be new only in the sense they were new designs; they would involve little or no new knowledge but would be largely a rearrangement of existing technological knowledge into forms appropriate to the new factor prices. For example, the small tractors designed for use in horticulture are much more a new design to suit changing factor prices than the result of significant advances in technological knowledge. On the other hand, if knowledge alone were changing, there would also be a flow of new best-practice techniques. These would not necessarily be more or less mechanised than their predecessors. Changes in the extent of mechanisation would depend on whether the knowledge was of a character which tended to save labour more than investment or vice versa.

The next step in the development of the model is to introduce the different response of producers to recovery of variable and fixed costs. Because, in the famous dictum of F. R. Jevons, 'bygones are bygones' in capital costs, technologies of varying degrees of obsolescence can and do coexist in the overall production process. We illustrate this by a second figure taken from Salter's work (Fig. 3.2).

All technology is, of course, outmoded and to a varying degree obsolescent whenever it is embodied in hardware. But the capital cost of its installation having been incurred, this cost becomes a 'bygone' and the plant may be operated to make a positive contribution to cash flow so long as the realized prices more than cover the variable or recurrent costs. In the diagram the vertical axis represents costs and prices and the horizontal axis volume of output by different technologies, the most recently added being to the left, the oldest to the right. All technologies along AB are viable so long as the price of product stands at AD, above the variable costs (represented by the height of the blocks) of all plants. An investment in new technology A_1A, of variable costs A_1C_1 and capital costs C_1D_1, however, makes plant B_1B non-viable, unless the owners of A_1A continue to sell at the old price AD. If they do so they will be earning greater-than-normal profits on the investment (capital cost, C_1D_1, is assumed to include normal profit). The greater-than-normal profits will attract new investment, depressing the price to A_1D_1 and causing B_1B to be scrapped. It should be noted that capacity A_1A is shown to be greater than that which it displaces. This represents an increase in output which is just sufficient in relation to demand conditions to reduce price by an amount equal to the total cost saving of the new best-practice technique. The industry is now in a position of momentary equilibrium where no further existing plants will be scrapped or new plants constructed.

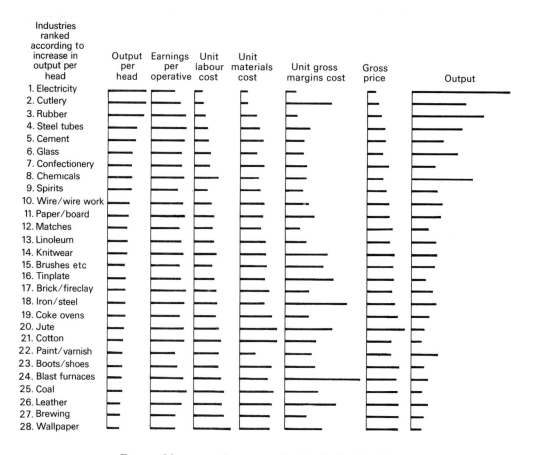

FIG. 3.3. Movement of output per head and related variables.

The 'engine' of technological change, as so far described, is a gross simplification. Salter identified five problems which his model in its simple form assumes away: the effect of semi-independent machines within a plant; resale value; technical change leading to improved products; the effect of non-technical factors, and the effect of non-competitive market conditions. These factors cannot be considered within the present compass, but the interested reader will find them separately discussed in Salter's book.

The pay-off from technological development emerging from the process described by Salter in his model is strikingly demonstrated by Enos in the work referred to earlier [11]. He compared the costs of a unit of gasolene transport performance (100 ton-miles), manufactured by the Burton process

(1914) and the Fluid process (1955). The 1955 technology showed savings over the 1914 under all heads: raw material, energy, labour, maintenance, catalyst, royalty, depreciation, taxes, and insurance, the net effect being $0·263 for the Fluid process compared with $1·471 for the Burton, or an average increase in overall productivity of 4·3 per cent per year.

We may perhaps sum up the Salter model by referring to his survey of productivity movements in British industry in the period 1924 to 1950; that is, roughly the second half of the period we are reviewing.

Twenty-eight industries were selected, on the basis of the quality of the information available, ranging from coal-mining to cutlery and brushes and brooms. Productivity movements over the period for each industry were compared with other relevant variables—prices, wages, costs, output, and employment. There was a wide diversity in the performances of the 28 industries over the 27-year period, as indicated by the various statistics. Five industries—electricity, rubber, chemicals, cutlery, and steel tubes—more than quadrupled output; six showed absolute declines—brewing, tinplate, cotton, coal-mining, wallpaper, and jute. Employment figures are much less varied, but there are large variations in output per head, six industries more than doubling output per head and five showed increases of less than 25 per cent. The great dispersion of output and output per head is in contrast to the small dispersion in earnings movement.

By correlation analysis to show what association there might be between the collected variables, Salter was able to arrive at some important conclusions. Fig. 3.3, reproduced from Salter, shows very plainly what the patterns of changes were in the 28 industries through the 27-year period. In Salter's words:

This empirical analysis suggests that uneven rates of productivity growth are closely associated with the main features of the inter-industry pattern of growth. Industries which have achieved substantial increases in output per head have, in general, been successful in other aspects: their costs have risen the least, the relative prices of their products has fallen, output has expanded greatly, and in most cases employment has increased by more than the average. On the other hand, industries with small increases in output per head are generally declining industries—at least in relative terms. Their costs and selling prices have risen the most, output has increased much less than average (or even fallen), and increases in employment are below average . . . The diagram brings out the parallel variation in many of these variables, and shows that most of the important changes in industrial structure—as far as they are revealed by this sample—are associated with differential movements of labour productivity.

The question that arises from this analysis is what mechanisms are at work to bring about such great differences in productivity and what is the basis of the relationships uncovered? After exploring and dismissing a number of possible explanations—such as an improvement over the years in the personal efficiency of labour—Salter concludes that there are two mainsprings of the varied pattern of changes in productivity: economies of sale and increases in knowledge leading to improved techniques of production.

In an addendum to Salter's book, W. B. Reddaway compared the 1924–50 rankings with the 1954–63 rankings of 23 industries for which information was available for the two periods. Whereas in growth of output the rankings of industries are broadly similar for the two periods, in output per head the rankings for the two periods showed zero correlation. Reddaway comments: 'There are—to say the very least—far more exceptions than there were in the case of output to any generalization that the previous period's star performers remain stars, and the laggards remain laggards.'

The unevenness of productivity performance among industries through different periods is not incompatible with Salter's designation of increases in knowledge of techniques of production as a major determinant of productivity improvements, since the science relevant to innovation in products and processes is itself characteristically uneven in its advances. Chemicals, plastics, and, in an earlier era, electricity, are examples of science-based major technological growing-points. The age and rate of growth of investment in an industry may also interact with increases in knowledge and technological innovation as described by Freeman in the passage quoted on p. 62.

Schmookler develops a related theme at length [3]. He presented data on patents relating to an industry, and investment, showing a very strong positive correlation between them. One of Schmookler's figures, for example, plots railway passenger car output and patents annually from 1871 to 1949. Both variables move in marked harmony over both the long and the short run. Output generally leads patents at the troughs of the curves, with leads and lags roughly even at the peaks. Many other instances of this kind of relationship are presented by Schmookler, who goes on to argue that the correlation is too close to allow the deduction that fluctuations in invention cause those in output—which is determined by many factors, including the state of the economy and population growth. The underlying connection, Schmookler argues, is the hard work and money required to bring innovations to fruition, and that these resources will not be forthcoming without commensurate prospective rewards in income or prestige; and those in turn will be favourably affected by high volume of sales of the product or process the innovation seeks to improve. The conclusion is that, to a degree, demands of growing industrial development call up the inventions and innovations they require.

In an economic report to the President of the United States of America [21] in 1962 there is a passage which effectively links some of the ideas discussed in previous paragraphs. It reads:

Technological knowledge sets limits on the productivity of labor and capital. As the frontiers of technology are pushed ahead, industrial practice and productivity follow, sometimes pressing close on the best that is known, sometimes lagging behind, with the gap varying from industry to industry and from firm to firm. A stimulus to economic growth can come either from increasing the rate at which the frontiers are advancing or from bringing the technology actually in use closer to the frontiers. The advance of technological knowledge depends on the amount and effectiveness of the human and material resources devoted to research and development.

Whatever the causes of uneven technological development may be, it emerges strongly from the Salter analysis that technological improvements tend to save all factors of production—labour, capital, and materials—and so give rise to general reduction in costs. It should be kept in mind that when writing 'technological improvement' in describing its effect on productivity, we use the term in something of a blunderbuss sense to embrace many things which affect production. All economists make this reservation. Let us quote one of them, Mansfield [6]:

Because technological change is measured by its effects, and its effects are measured by the growth of output unexplained by other factors, it is impossible to sort out technological change from the effects of whatever inputs are not included explicitly in the analysis. The customary measures are plagued by the difficult problems, both theoretical and practical, in evaluating entirely new products.

The dependency of consumption of factors of production on technological levels, noted by many other economists, is fundamental and explains why technological advance is equated with 'progress' in the economist's sense. It need hardly be added that technological innovation can by no means be generally equated with 'progress' once we invoke the social and ethical connotations of the word. Staying with the economists' restricted use of the term we may note that R. M. Solow, making an analysis of productivity of American industry over a period close to that of Salter's (1909–49) concluded that gross output per man-hour doubled over the interval, with $87\frac{1}{2}$ per cent of the increase attributable to technical change and the remaining $12\frac{1}{2}$ per cent to increased use of capital [22]. It should be added that Solow, like Mansfield and others, uses the term 'technical change' as a portmanteau word to include any kind of shift in the production function: 'showdowns, speed-ups, improvements in the education of the labour force, and all sorts of things . . .' Although it is evidently the most important factor in productivity, the inability to measure technical change and its all-embracing nature give rise to the curious situation in which it is regarded by economic analysts as a 'residual' factor.

IX. TECHNOLOGY AND THE STRUCTURE OF THE ECONOMY

An important element of the model of technological obsolescence is its extension to the structure of the economy. In our discussion so far we have focused on the mechanisms by which obsolescent productive units are displaced by technologically and economically superior methods. But, as Salter makes clear, the process of displacement does not stop at within-industry replacements; by the same workings of the 'engine' capital may move to production of different goods and displace one type of product by another. The effect is illustrated by employment and output figures for U.K. electricity, cutlery, rubber, chemicals, and steel tubes, in Table 3.7 from Salter.

TABLE 3.7

Relative importance of certain industries

Percentage of total sample represented by these industries with respect to	1924 (%)	1950 (%)
Employment	7·8	20·2
Volume of output (at 1935 prices)	10·5	42·6
Value of output (at current prices)	10·9	24·5

Source: W. E. G. Salter. *Productivity and technical change,* 1969

These industries were relatively unimportant in 1924, but were dominant 27 years later. Technologically progressive industries squeeze other, less progressive, ones in two ways: (*a*) by providing acceptable substitute products—nylon for cotton, plastics for steel, cement for bricks and timber—and (*b*) by raising the costs of the factors of production as the demand on these grows with expansion of the markets for the new goods. Technologically backward industries find themselves under pressure; few technological opportunities lead to a performance which compares unfavourably with the average of the economy as a whole and, with rising costs and prices, the backward industries decline.

We may observe that equating technological change and innovation with 'progress' becomes even more suspect when we see the engine of increased profit opportunity driving the economy into new products down a 'best technology' slope. In the decades following the close of the period under review we have moved a long way towards applying social sanctions to the operation of this engine, by a process now styled 'technology assessment'.

It is perhaps appropriate to conclude this chapter with a further reference to that doyen of industrial organization studies, Joe S. Bain. Here is how he deals with technical progress:

> Since we are unable to distinguish good from bad performance in the dimension of technological progress, we are unable to establish empirically the conditions of market structure which might favor good progress. An alternative is to theorize about the effects of market structure on progressiveness. Unfortunately, the indications of such theorizing are so inconclusive as to be almost useless. We therefore drop the matter of technological progressiveness, and turn to other matters.

It should perhaps be added that in the passage quoted, Bain was not calling into question the beneficial effects of improved production techniques. The difficulty which caused him to turn aside from the question of technological progressiveness was the daunting prospect of trying to evaluate how well an industry has exploited the opportunities for invention and innovation open to it. With this view we sympathize and take it as the key with which to close this chapter: the complex factors comprising the determinants of technological change. Bain writes:

> Reliance on a criterion of 'gross progress' can lead to nonsensical conclusions. Application of this criterion, for example, would find that the petroleum refining industry showed 'good' progress in the last forty years, with a steady stream of technological developments. The flour-milling industry, however, would be rated as evidencing 'poor' progress because over the same period firms in that industry were able to find or introduce only minor improvements in a technique which was fully developed by the early years of this century. The comparison is unfair, and the conclusions are meaningless, because both of the wide difference in the intrinsic complexity of the manufacturing process involved, and of the wide difference in the ages of the two industries.

And there we leave this fascinating fabric of interacting factors styled economic aspects of technological development.

REFERENCES

[1] DERRY, T. K., and WILLIAMS, TREVOR I. *A short history of technology*. Clarendon Press, Oxford (1960).

[2] SCHUMPETER, J. *The theory of economic development*. Oxford University Press, London (1961).

[3] SCHMOOKLER, J. *Journal of Economic History*, March 1962, p. 1; reprinted in N. ROSENBERG (see Bibliography).

[4] PARKER, W. *Economic development and cultural change*, October 1961, p. 1; reprinted in N. ROSENBERG (see Bibliography).

[5] ENOS, J. L. *The rate and direction of inventive activity*. Princeton University Press, New Jersey (1962).

[6] MANSFIELD, E. *Industrial research and technological innovation*. Norton New York (1968).

[7] GIBBONS, M., and JOHNSTON, R. D. *The interaction of science and technology*. Department of Liberal Studies in Science, University of Manchester (1972).

[8] FREEMAN, C. *The economics of industrial innovation*. Penguin, Harmondsworth (1974).

[9] UTTON, M. A. *Industrial concentration*. Penguin, Harmondsworth (1970).

[10] EVELY, R., and LITTLE, I. M. D. *Concentration of British industry*. Cambridge University Press (1960).

[11] ENOS, J. L. *Petroleum progress and profits*. M.I.T. Press, Cambridge, Mass. (1962).

[12] BOYLE, S. E. *Industrial organization*. Holt Rinehart and Winston, Inc., New York (1972).

[13] READER, W. J. *Imperial Chemical Industries—A history*, Vol. 2, Oxford University Press (1975).

[14] LINDBECK, A. *Kyklos*, **28,** 23 (1975).

[15] HABER, L. F. *The chemical industry 1900–1930*. Clarendon Press, Oxford (1971).

[16] DUNNING, J. H., and THOMAS, C. J. *British industry*. Hutchinson, London (1961).

[17] FREEMAN, C., HARLOW, C. J., and FULLER, J. K. *National Institute Economic Review*, **34** (1965).

[18] TOWNSEND, H. *Scale, innovation, merger and monopoly*. Pergamon Press, Oxford (1968).

[19] FREEMAN, C. *National Institute Economic Review*, Nos. 20, 21 (1962).

[20] NELSON, R. R. *Journal of Political Economy*, **297** (1959).

[21] UNITED STATES PRESIDENT'S COUNCIL OF ECONOMIC ADVISERS. *Economic Report of the President*, January 1962, p. 123.

[22] SOLOW, R. M. *Review of Economics and Statistics*, **39,** 312 (1957); reprinted in N. ROSENBERG (see Bibliography).

BIBLIOGRAPHY

BAIN, JOE. S. *Industrial organization*. Wiley, New York (1968).

ROSENBERG, N. *The economics of technological change*. Penguin, Harmondsworth (1971).

SALTER, W. E. G. *Productivity and technical change*. Cambridge University Press (1969).

4

MANAGEMENT

GLENN PORTER

THE notion of management as a science or a technology is one associated primarily with the present century. As Sidney Pollard has noted, the widespread study and public discussion of management did not occur 'in effect until the twentieth century' [1]. Like most allegedly practical fields of knowledge, management as a body of ideas arose only gradually and in response to widely-felt needs. Not until relatively recently in man's past did the number and variety of highly organized activities—mostly in the business system—reach a sufficiently high level to generate the systematic production of general ideas about how to order those activities effectively and efficiently.

If, among countless competing explications, we take Henri Fayol's classic formulation as a working definition, then what managers do is plan, organize, command, coordinate, and control human activities in an effort to reach some particular goal [2]. Many activities other than business obviously require managerial talent: the elaborate undertakings of state and church in ancient times, for example, certainly did so. An important body of ideas and practices arose dealing with the problems of administration in general, and with public administration in particular [3]. In this chapter, however, 'management' will refer primarily to business administration, because the figure of the manager has come to be associated most closely with business enterprise. Further, managers normally are thought of as creatures and agents of industrialization. Since the characteristic organizational form of the urban, industrial West today is the large, complex, bureaucratic organization, it is hardly surprising that management is fundamentally and distinctively a twentieth-century phenomenon. It had, however, many important roots in the more distant past, especially in the latter half of the nineteenth century.

The two business institutions that first created the problems that management is supposed to solve were the factory and the railway. In the nineteenth century, large numbers of firms faced similar problems for the first time in administering highly detailed, complicated, and interrelated economic activities on an ongoing basis. For a long time, individual owner-managers struggled with their problems in an *ad hoc*, learning-by-doing fashion, usually assuming that few challenges were common to such seemingly different enterprises as cotton mills, railways, and iron foundries. When one of them encountered what would now be thought of as a managerial problem, he could normally draw on no rationalized, codified body of ideas embodied in such forms as books or university courses. Instead, most early managers of factories had to rely on themselves or on the limited experience of their associates, As more and more industrial units proliferated, however, and as their leaders encountered new and increasingly complex challenges, some persons began to think and write about the common aspects of those challenges.

From the very beginning of the study of management, the topic has been very often restricted to the area of decisions concerning how the enterprise was to do what it had set out to do, not with the fundamental choices governing which products or functions the firm was to engage in. Those latter, basic decisions, which define the purpose and identity of an enterprise, may be thought of collectively as strategic, or long-run decision-making, or as *entrepreneurial* decisions. Tactical, or short-run choices, on the other hand, concern the best ways of implementing the organization's basic strategies. The study of management in the past most often meant the analysis of tactical, not strategic, planning and decision-making, though that has changed to some extent in recent decades.

Once factories appeared and spread, the challenges of managing these institutions began to attract growing attention. Among the first problems addressed was one that remains an important part of one branch of management thought—the productive benefits of the division of labour. By breaking the task of manufacturing into small, distinct stages and having workers specialize in a single stage, output could be increased above the level that was possible when a skilled worker handled several or all phases of the manufacture of some item. Adam Smith's famous pin factory passages in *The Wealth of Nations* (1776) constitute the best-known early example of the elaboration of those principles. The next great early work appeared approximately half a century later, after the factory and mechanized production in general had advanced to new heights in Britain, continental Europe, and the U.S.A. Charles Babbage's *On the Economy of Machinery and Manufactures* (1832) extended Smith's ideas about the division of labour by advocating the objective collection of information on the nature and performance of work, by discussing cost accounting, incentive systems, and what he called 'domestic arrangement, or the interior economy of factories,' as well as by offering many intriguing notions about political economy and the present state and future prospects for manufacturing in Europe.

Such milestones as Babbage's commentary, brilliant though it was, remained isolated examples of individual analyses, pieces that did not add up to anything that could be called a science or a technology [4]. Several other sets of ideas appeared by the end of the nineteenth century, however, that collectively lifted management to the loftier status of something that could be generalized and taught.

An important but often slighted element in that process was the coming of the railroad, which presented new and tremendously difficult challenges to business leaders [5]. A wide range of intricate and interrelated activities had to be co-ordinated over vast spaces when railroads began to extend over hundreds of kilometres. Such puzzles as the efficient movement of traffic; the intelligent utilization of motive power and rolling stock; the avoidance of accidents; and the handling of cash by large numbers of scattered employees, were among those with which the pioneers of railroading struggled. The problems eventually called forth a well-defined set of managerial practices, and the new ideas came not from interested theoreticians like Babbage but from the managers who were forced to deal with unprecedented problems. A generation of railroaders— especially Daniel C. McCallum (general superintendent of one of the first American trunk lines of the 1850s, the New York and Erie); Albert Fink (president of the Louisville and Nashville); and J. Edgar Thomson (head of the 'standard railroad of the world', the Pennsylvania)—worked out and wrote about managerial structures and practices that would be taken up later by the giant industrial firms that arose at the end of the nineteenth and the beginning of the twentieth century. A century ago, the managers of the railroads had come to employ a very significant portion of the body of techniques, structures, and principles that make up modern management. These included: the recognition of general bureaucratic principles of clear lines of authority, responsibility, and communications; the importance of feedback in decision-making; distinctions between line and staff personnel; the need to separate strategic and operational (or tactical) decision-making; the creation of departments defined by function; the use of both the traditional centralized managerial structure and the new, decentralized one with relatively autonomous divisions responsible for implementing the firm's long-run strategies; the elaboration of cost accounting systems dealing with both fixed and variable costs and embodying the principle of 'control through statistics'; the rational evaluation of managerial performance; and the use of the holding company as an organizational device. Some of these ideas—especially those flowing from the distinction between strategic (or entrepreneurial) and operational (or tactical) decisions—went beyond the idea of management as the mere pursuit of internal efficiency. Like the railroads' path-breaking ventures into new, oligopolistic forms of competition, they pointed the way to a business world vastly different from the classical competitive environment assumed by Adam Smith and Charles Babbage.

Although the managerial creativity and sophistication of the railroaders far exceeded that of those studying the internal efficiency of the factory, the latter received (and still do receive) much more attention from the public and from scholars. The first significant body of thought (after Babbage) on the administration of the factory appeared in the last three decades of the nineteenth century, when a group of American engineers and manufacturers including Henry R. Towne, Henry Metcalfe, and Frederick A. Halsey led a movement now known as systematic management [6]. This was a consequence of 'the growth of the plant and the increasing complexity of manufacturing operations', plus 'the advent of trained engineers who rapidly assumed managerial as well as technical responsibilities' [7]. Publishing their ideas in engineering journals such as the *American Machinist*, the *Transactions of the American Society of Mechanical Engineers*, and *Engineering Magazine*, these men addressed what they felt was a widespread lack of system and method in manufacturing, and the resulting loss of managerial control over the internal processes of production. They proposed as solutions improved methods of production and inventory control (identifying and recording data on the materials, tools, and man-hours involved in each stage of production), better coordination of output with demand patterns, and development of cost accounting for the various steps in manufacturing (including the identification of overhead costs). Partly in response to growing troubles with labour, beginning in the 1880s they also turned their attention to new ways of paying workers, adding various forms of profit-sharing to the two established patterns of payment, per-day wages or payment by the piece (or volume produced). All these changes had the effect of reducing the control over the work process that had been exercised by workers, by foremen, and by various lower-level managers, all at the expense of higher management, which became at once more aware of, and in charge of, the details of the manufacturing process.

The systematic management movement had many proponents, and a large number of different administrative systems were suggested by the authors of the literature. Joseph Litterer, the chief student of systematic management, concluded that there were, however, two common factors. One was 'a careful definition of duties and responsibilities coupled with standardized ways of performing these duties,' and the other was 'a specific way of gathering, handling, analyzing, and transmitting information' [8]. The managerial systems worked out by these engineers, Litterer argued, 'represent what can be called the beginnings of a management technology' [9].

I. TAYLORISM

By far the best known aspect of this phenomenon, of course, was the so-called scientific management movement, closely identified with a single man: Frederick W. Taylor (1856–1915). Taylor was many things in his career— apprentice, machinist, foreman, shop superintendent, engineer, inventor of major advances in metal working (especially in high-speed tool steel), management consultant, and publicist. He is most famous for his development and promotion of 'the Taylor system' of shop management, a school of administrative thought that caught the imagination not only of engineers and businessmen but of people in many other walks of life, such as government and education [10]. Scientific management took on the trappings of a kind of secular religion; Taylor was the messiah, and his followers who spread the word were (and still are) commonly referred to as 'disciples'.

Intellectually, Taylorism owed a great deal to the systematic management group, for scientific management basically was no more than elaboration of the principles of the older movement. Over a period of several decades, working primarily at Midvale Steel Company, Manufacturing Investment Company (a paper manufacturer), and Bethlehem Steel, Taylor worked out his particular system. It began to attract wide attention, beginning with his delivery of a speech before a meeting of the American Society of Mechanical Engineers in 1895.

The system was based on the idea that workers turned out much less than they could, and that if the requirements of each job and the abilities of each worker could be 'scientifically' determined, and if workers received proper pay incentives for producing at their capacity, productivity, profits, and wages could all be substantially improved. This would result in the end of the practice of 'soldiering' (the deliberate restriction of their output by workers) and would erase forever the ancient conflict between owner (or manager) and worker, thus solving what Taylor referred to as 'the labor problem'.

The first essential was to learn what constituted a 'fair day's work', or the standard daily output level by a 'first-class worker' using the most efficient methods and tools for carrying out the task. Time study, using the stopwatch (which became the symbol for Taylorism), could tell what that standard was by measuring how long it took an able worker to complete a job under optimum conditions. This was the key element that Taylor felt made his system scientific; he believed that the setting of standard productivity levels could be determined objectively. Another important element of an allegedly scientific nature was the fitting of 'first-class' workers to appropriate jobs. Management's task was not only to determine and provide the optimum working procedures to permit the production of a high standard output, but also to discover for what job or jobs the individual worker was best suited. This implied testing an employee to see if his abilities enabled him to be, for example, a first-class machinist or a first-class coal shoveller. (Taylor believed that every worker was a first-class one for *some* job; otherwise he was simply lazy.) Other parts of the Taylor system included pay schemes to reward workers who met or exceeded the standard output levels; the creation of 'functional foremen' who would specialize in the various tasks that had traditionally been combined in a multi-purpose foreman; and a planning department to control and coordinate the production process.

Many of the contributions of both systematic and scientific management, it should be noted, amounted to 'reinventing the wheel'. Managers of the railroads had already faced and solved a number of similar problems, especially in the areas of coordinating the flow of output to match demand, cost accounting, and standard systems for recording and communicating data, as well as the creation of specialized staff to monitor information flows. The railroaders' work, however, appears to have had less effect than might be expected on the engineers who concentrated on factory administration. Analyses of the work of the major participants in systematic and scientific management seems to indicate relatively little awareness of the writings of railroad men like McCallum, Fink, and Thomson (though Frederick Taylor did specifically acknowledge a debt to railroad accounting). The managerial ideas of the railroad pioneers did not, of course, often grace the public prints; instead they appeared in annual reports and other internal documents, and they generally dealt with much bigger problems than shop management.

The engineers' somewhat surprising failure to make full use of a potentially helpful body of thought is perhaps most interesting for the light it sheds on the relatively primitive state of society's mechanisms for gathering and diffusing managerial knowledge before the twentieth century. That situation was improved by the appearance of more and more organizations of engineers and managers (such as the American Society of Mechanical Engineers) and by the gradual creation of a substantial number of professional business schools at the close of the nineteenth and the opening of the twentieth century.

II. BUSINESS EDUCATION

The United States led the way in this latter development, though it soon spread to other industrial nations. Until the closing portion of the nineteenth century, explicitly commercial or business education found little place in American secondary or higher education. When the subject was taught in such schools or in vocational schools, the curricula included such subjects as penmanship, 'business arithmetic', composition and grammar, typing, and shorthand.

The first attempt to make business education analogous to professional training for medicine or the law came at the University of Pennsylvania. There in 1881 businessman Joseph Wharton endowed the Wharton School of Finance and Commerce. It offered a two-year undergraduate programme, which was expanded to four years in 1895. Others eventually appeared: in 1898 business schools were begun at the University of Chicago and at the University of California; New York University's came in 1900, and others soon followed. The first graduate schools of business appeared at Dartmouth College in 1900 (the Tuck School) and at Harvard in 1908 [11]. These university programmes offered courses in finance, transportation (primarily railroad management), accounting, business law, insurance, and the emerging body of administrative systems produced by the systematic and scientific management movements.

These shifts in education paralleled the proliferation of magazines and journals devoted to various aspects of business management, and ideas from the scientific management group began to appear even in specialized trade journals. The cumulative effects of these changes was to improve society's ability to disseminate new ideas about administration, as more and more institutions came into existence, specializing in the task of learning and teaching about new developments in the administration of business enterprise. The influence of Taylorism in the universities and the journals was strong, and Taylor's followers and imitators—such as Carl Barth, C. Bertrand Thompson, Frank and Lillian Gilbreth, H. L. Gantt, Horace K. Hathaway, Harrington Emerson, and Morris L. Cooke—refined and extended his ideas, carrying the movement's influence across America and around the world.

Many of those followers and imitators increasingly found themselves wrestling with what was perhaps the weakest aspect of Taylorism: its failure to deliver on its promise to deal effectively with 'the labor problem', a promise that had made it so attractive to businessmen [12]. A storm of public criticism arose, focusing on the dehumanizing tendency of Taylorism, its apparent goal of reducing the worker to the status of a machine by placing all the control over precisely how and when jobs were to be done in the hands of management. The worker was supposed, in the Taylor scheme, to be motivated purely by a desire for economic gain; nothing else mattered. The proponents of the next major developments in administrative thought addressed themselves to this shortcoming in the engineer's concept of management.

In contrast to the focus on the details of the actual work process, or industrial engineering, there arose several other, quite different sets of concerns, all focused on the worker and not on the work. The first sign of this shifting concern came in the form of 'welfare work'. This was a paternalistic movement 'based on the belief that voluntary efforts by employers to improve the lot of the workman encouraged self-betterment, loyalty, and cooperation—that they inspired the employee to become a better person and a better worker' [13]. This was, of course, an attitude almost as old as the factory itself, but in the late nineteenth and early twentieth centuries it reappeared in a peculiarly strong form. Rejecting the view that workers cared only for pay, reformers like the Filenes in Boston and John H. Patterson of the National Cash Register Company led a movement for such amenities as profit-sharing plans, physical improvements in the work environment, educational opportunities, and desirable housing for workers. 'Welfare secretaries' were hired by some companies, and this eventually led to the establishment of personnel departments. All this was designed both to do right and to do well, that is, to impove the workers' lives as well as raising productivity and profits; incentives other than pay were thought to be the key.

III. INDUSTRIAL PSYCHOLOGY

It was, however, far from a 'scientific' approach, and management analysts began to search for approaches to the study of the worker that seemed as objective as Taylorism. One important example of such approaches came from a group of German-trained psychologists, among whom the most prominent were Hugo Munsterberg and Walter Dill Scott. Experimental psychology was applied to industry via aptitude and intelligence testing. The premise was that different personalities were suited to different types of work, and that one could identify the relevant psychological qualities associated with various jobs, identify the personality traits of individuals, and then match up the right persons and tasks. Industrial psychology quickly spread to other uses, such as the testing of recruits and draftees for the military.

Beginning in the 1920s, the industrial psychologists encountered heavy assaults from another 'people-oriented' group of would-be management scientists—the industrial sociologists. The greatest figure among that assemblage was Australia's Elton Mayo, and the most significant forum was what came to be known as the Hawthorne experiments. The experiments arose in the context of studies of the effects of monotony and fatigue on worker productivity. At the Western Electric Company's Hawthorne Works in Chicago, a study was conducted to measure the effects of various levels and kinds of lighting on the output of workers. The results showed no significant difference between the workers with good lighting and the control group with the prevailing standard illumination in the plant: the productivity of both the observed groups rose. This led to further investigations, in which Mayo participated, and which he subsequently interpreted so effectively that he became the major figure in industrial sociology. In brief, the entire body of Hawthorne experiments suggested that worker output was not merely a function of having an optimum production system in the factory, or even of placing psychologically well-suited persons in such factories. It also depended on the interpersonal relations of workers in the factory and on the state of their personal lives off the job. Group dynamics and the overall psychological well-being of workers, it seemed, were very important factors in productivity. The approach of Mayo, his associates, and his intellectual heirs to management has been to explore the full implications of the perception that the worker has an extremely complicated set of needs and desires in addition to those that can be satisfied by monetary rewards. Proponents of this branch of management thought have tried to 'humanize' the work environment, which, they felt, had been dehumanized by the Taylor approach.

Many important additions to the modern body of knowledge about management had very little indeed to do either with administrative systems for organizing efficient manufacturing plants or with the various 'people-oriented' ideas about management. These fall largely into one of two categories: advances in organizational or managerial structure on the one hand, and the innovation of new competitive strategies on the other. Neither sort of change ever led to anything comparable to the cult that grew up around scientific management, but they have fundamentally altered the nature of business in the market economies of the West and Japan and have been of greater consequence for twentieth-century management than was the Taylor phenomenon.

IV. THE ADVENT OF BIG BUSINESS

As was pointed out earlier, the railroads played a pathbreaking role in creating systems of business management very different from those useful in the small, single unit, single product, single function enterprises of the late eighteenth and early nineteenth centuries. The organizational forms, administrative methods, and competitive strategies they were forced to invent spread to many firms in manufacturing as big 'business' appeared in that sector [14].

In many of the new, technologically advanced industries that were to form the backbone of twentieth century economies—such as petroleum, steel, chemicals, electricals, machinery, autos, and rubber—as well as in mass retailing and transportation, the kind of traditional firm envisaged by Smith and Babbage was supplanted by a new institution. The leading historian of that institution, Alfred D. Chandler, Jr, terms it 'the multi-unit enterprise'. Chandler's definition is the best: 'Such enterprises have consolidated under the control of a single office many different units, usually carrying on a number of different activities. Individually each of these units—factories; sales, purchasing, and financial offices; mines; plantations; laboratories; and even shipping lines—could be and often have been operated as independent business enterprises. United, they formed a single administrative network through which flowed orders, reports, information, and, of course, goods and services [15].

This phenomenon differed fundamentally from the older, traditional firm, and these differences made for revolutionary changes in management. Modern, giant enterprises represented vastly larger pools of capital, and much more of the capital required was now fixed rather than working capital. Further, the owners of these new institutions were no longer the same persons as the managers of the enterprises [16]. Ownership often became dispersed among many people because most of the new, giant corporations ultimately required more capital than an individual or a few associated individuals could provide, compelling the use of publicly sold shares in most big businesses. Further, the number of managers necessary to run the huge institutions far exceeded those that could be supplied in the traditional manner of utilizing the owning family's relatives as administrators. Management quickly came to be professional rather than personal, as the rising institutions in the field of business education indicated. The number of managers proliferated as the multi-unit enterprise spread, creating for the first time a real managerial class, far more numerous than the previous generation of factory managers, and engaged for the most part and for the first time in the direction of other managers.

The distinguishing characteristic of the novel institution, however, was that it represented the unification under a single office of what in the more distant past would have remained separate businesses, that it was multi-unit. This commonly meant that the firm not only had more than one manufacturing plant, but that it included units that did not manufacture but performed other functions. These could include purchasing, wholesaling, retailing, finance, transportation, research and development, and the extraction of raw materials. The combination of several of these functional activities within a single firm came to be known as vertical integration. Some businesses integrated into raw materials because of their need to assure a large, steady flow of inputs into their production processes. Integration into marketing, however, was much more common. Enterprises dealing in a commodity that did not fit readily into existing marketing channels were compelled to undertake that function; others moved into marketing because increasingly concentrated markets made it profitable to do so rather than to continue utilizing the traditional networks of independent middlemen [17]. Vertical integration often allowed firms to insulate themselves from market fluctuations to a greater degree than had been possible before the coming of large-scale enterprise, especially if they could also succeed in identifying their products or services in the minds of consumers through brand-name advertising. It also permitted a much better co-ordination of the flow of production and contributed to the superior competitive position of many large firms. Naturally, vertical integration also added to the top managers' burdens, because they now had to worry about administering a multi-functional operation.

V. DIVERSIFICATION

The constant pressure to keep the resources that the large firm embodied employed as fully as possible was perhaps the major engine of the growth of

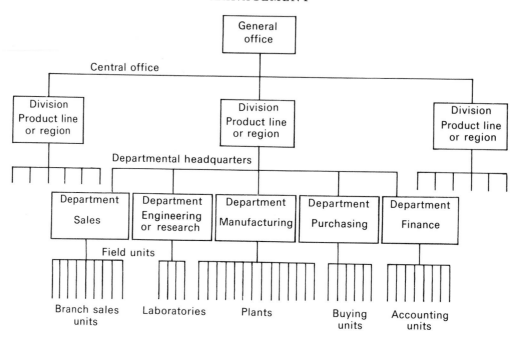

FIG. 4.1. The general structure of the multi-unit enterprise.

such enterprises once they came into existence [18]. One of the main avenues that modern businesses have followed in consequence of that pressure has been to diversify the line of goods or services they produce. Giant firms often expanded into new activities because they had either a scientific or technical base of expertise that was transferable to a different product line (for example, a base in chemicals or electronics) or because they had some functional expertise (usually marketing) that was similarly applicable to handling different goods or services. This ability to add several wholly new products or services had not characterized most single-plant manufacturers of the nineteenth century. It was a new and powerful weapon in management's arsenal. In the years since the Second World War, the appearance of conglomerates—financial amalgamations of wholly unrelated companies not produced by transferring some existing expertise within the firm—has carried the idea of diversification to its furthest point. Like vertical integration, however, diversification brought with it a price: substantially increased complexity for the manager. Diversification often put unbearable strains on the organizational structure of corporations and led to the innovation of a general structure that has come to characterize the multi-unit enterprise in the United States,

Europe, and Japan (see Fig. 4.1). In an extension of the idea first developed in the middle of the last century by railroad men like McCallum and Thomson, relatively autonomous divisions are organized along either product or geographical lines, with functional departments within each division. This organizational structure has permitted the expansion of the enterprise beyond the wildest imaginings of the engineers who gave the general study of management its origins.

The managerial developments implicit in the multi-unit enterprise arose in their full-blown form in the U.S.A., Japan, and Britain. Slowly in the years before the Second World War, and very rapidly afterwards, they spread across continental Europe. Not only is it 'historically a new institution', but it 'has become an engine of modern economic development' [19]. What the factory was to the nineteenth century, the multi-unit enterprise has been to the twentieth. The factory was the cradle in which modern management was born, and the multi-unit business is the environment in which it has taken its place among the major technologies of the present era.

In recent decades the study of management has pursued a number of new directions, and this has been reflected in the curricula at business schools. The older Tayloresque emphasis on the internal administration of factories persists in the study of production, but modern business is a vastly more complex phenomenon than the one-product, one-factory world assumed by classical economists and the analysts of first systematic, then scientific management. The fields of finance and marketing and the areas of international and general management have eclipsed production as the most intricate and pressing challenges for twentieth-century executives. The modern firm, as we have seen, is often a giant, multi-plant, highly diversified organization competing in various industries and markets around the world.

The older definitions of management as the internal administration of production have become increasingly irrelevant in an environment in which managers can no longer be expected to apply simple, mechanical decision rules about price and output in the fashion assumed by the theory of the firm. Market forces are often only one of the many sets of variables that interact in the process of managerial decision-making [20].

Managerial analysts have responded to the tremendously complex tasks of executives by making more and more use of other disciplines, especially psychology, sociology, economics, mathematics, statistics, and organization theory [21]. Increasingly, computer technology, probability and game theory, and simulation have become important elements in modern management studies. These changes constitute attempts to recognize and deal with the difficult challenges of managing the giant, complex organizations that mark the twentieth century.

From its beginnings, management as a science or technology represented a response to changing, widely perceived problems. The definition of what falls within the purview of the analysis of management has steadily expanded as the limitations both of classical economics and of older assumptions about human behaviour became ever more apparent. Those limitations became clear primarily because the manager's tasks grew in number and in kind, and this process seems likely to continue. The resulting improvements in our understanding of management have contributed to our ability to improve the level of material existence in technically advanced societies. They represent a significant part of the 'software' of twentieth-century technology.

REFERENCES

[1] POLLARD, S. *The genesis of modern management: a study of the Industrial Revolution in Great Britain*, p. 296. Penguin, Harmondsworth (1968).

[2] FAYOL (1841–1925) published his ideas in the *Bulletin de la Société de l'Industrie Minérale* in 1916 under the title Administration industrielle et générale: prévoyance, organisation, commandement, coordination, contrôle.

[3] KAUFMAN, H. Administration, in *International encyclopedia of the social sciences*. Crowell, Collier, and Macmillan, New York (1968).

[4] POLLARD, S. *Op.cit.* [1], pp. 295–6.

[5] CHANDLER, A. D., Jr. (ed.) *The railroads: the nation's first big business*. Harcourt, Brace, and World, New York (1965).

[6] LITTERER, J. A. *Business History Review*, **37**, 369–91 (1963).

[7] NELSON, D. *Managers and workers: origins of the New Factory System in the United States 1880–1920*. University of Wisconsin Press, Madison (1975), p. 48.

[8] LITTERER, J. A. *Op.cit.* [6], p. 389.

[9] LITTERER, J. A. *Op.cit.* [6], p. 388.

[10] HABER, S. *Efficiency and uplift: scientific management in the Progressive Era, 1890–1920*. University of Chicago Press (1964).

[11] BAUGHMAN, J. P. Management, in *Dictionary of American economic history*. Charles Scribner's Sons, New York (In press).

[12] NELSON, D. *Op.cit.* [7].

[13] NELSON, D. *Op.cit.* [7], p. 101.

[14] PORTER, G. *The rise of Big Business, 1860–1910*. Crowell, New York (1973).

[15] The quotation is from one of the most recent of Chandler's several formulations on this subject, his essay entitled The multi-unit enterprise: a historical and international comparative analysis and summary in WILLIAMSON, H. E. (ed.) *Evolution of international management structures*. University of Delaware Press, Newark (1975).

[16] The classic statement of this point is BERLE, A. A., JR., and MEANS, G. C. *The modern corporation and private property*. Macmillan, New York (1932).

[17] PORTER, G., and LIVESAY, H. C. *Merchants and manufacturers: studies in the changing structure of nineteenth-century marketing.* Johns Hopkins Press, Baltimore (1971).

[18] PENROSE, E. *The theory of the growth of the firm.* Blackwell, Oxford (1966).

[19] CHANDLER, A. D., JR., *Op.cit.* [15], p. 227.

[20] CYERT, R. M., and MARCH, J. G. *A behavioral theory of the firm.* Prentice-Hall, Englewood Cliffs, N.J. (1963).

[21] See CYERT, R. M. Business Administration, in *International encyclopedia of the social sciences.* Crowell, Collier, and Macmillan, New York (1968).

BIBLIOGRAPHY

AITKEN, H. G. J. *Taylorism at Watertown Arsenal: scientific management in action, 1908–1915.* Harvard University Press, Cambridge, Mass. (1960).

BARITZ, L. *The servants of power: a history of the use of social science in American industry.* Wesleyan University Press, Middletown, Conn. (1960).

BARNARD, C. *The functions of the executive.* Harvard University Press, Cambridge, Mass. (1938).

BAUGHMAN, J. P. (ed.) *The history of American management: selections from the* Business History Review. Prentice-Hall, Englewood Cliffs, N.J. (1969).

——. Management, in *Dictionary of American economic history.* Charles Scribners' Sons, New York (In press).

CHANDLER, A. D., JR. (ed.) *The railroads: The nation's first big business.* Harcourt, Brace, and World, New York (1962).

——. *Strategy and structure: chapters in the history of the industrial enterprise.* M.I.T. Press, Cambridge, Mass. (1962).

COPLEY, F. B. *Frederick Winslow Taylor, father of scientific management,* 2 vols. Harper, New York (1923).

CYERT, R. M., and MARCH, J. G. *A behavioral theory of the firm.* Prentice-Hall, Englewood Cliffs, N.J. (1963).

DALE, E. *The great organizers.* McGraw-Hill, New York (1960).

GEORGE, C. S., JR. *The history of management thought.* (2nd edn.) Prentice-Hall, Englewood Cliffs, N.J. (1972).

HABER, S. *Efficiency and uplift: scientific management in the Progressive Era, 1890–1920.* University of Chicago Press (1964).

KAKAR, S. *Frederick Taylor: a study in personality and innovation.* M.I.T. Press, Cambridge, Mass. (1970).

MAYO, E. *The human problems of an industrial civilization.* Macmillan, New York (1933).

MERRILL, H. F. (ed.) *Classics in management.* American Management Association, New York (1960).

NELSON, D. *Managers and workers: origins of the New Factory System in the United States 1880–1920.* University of Wisconsin Press, Madison (1975).

PENROSE, E. *The theory of the growth of the firm.* Blackwell, Oxford (1966).

POLLARD, S. *The genesis of modern management.* Penguin, Harmondsworth (1968).

PORTER, G. *The rise of big business, 1860–1910.* Crowell, New York (1973).

PUGH, D. S., HICKSON, D. J., and HININGS, C. R. *Writers on organizations.* (2nd edn.) Penguin, Harmondsworth (1971).

ROETHLISBERGER, F. J., and DICKSON, W. J. *Management and the worker.* Harvard University Press, Cambridge, Mass. (1949).

SIMON, H. A. *Administrative behavior* (2nd edn.) Macmillan, New York (1957).

SLOAN, A. P., JR. *My years with General Motors.* Doubleday, New York (1964).

TAYLOR, F. W. *Scientific management.* Harper, New York (1947).

WILLIAMSON, H. F. (ed.) *Evolution of international management structures.* University of Delaware Press, Newark (1975).

5

TRADE UNIONS

HAROLD POLLINS

I. INTRODUCTION

THE development of technology cannot sensibly be studied without a consideration of the role of the people who are employed to make, maintain, and operate machines and equipment. This chapter examines one of many possible aspects, namely the relationship between trade unions and technological change in the main industrial countries.

Associations of workpeople formed to protect and further their interests are a normal feature of industrial society. Their name, 'trade unions' or 'labour unions', correctly implies that they arise from and focus their attention on their members' employment. We expect them to be concerned with remuneration for work and the host of considerations that come under the heading of 'terms and conditions': hours, promotion, safety, training, entry to the trade, and many others. But they do not confine themselves to the workplace and unions interpret their role widely so as to incorporate the furtherance of their members' interests in society generally. They are often associated with political parties which, in some countries, took the initiative in establishing them as instruments to change the social and economic system. Others, such as those in North America and Japan, have been non-ideological in the sense of consciously working within and accepting the existing pattern of social, economic, and political relationships.

The history of trade unionism normally describes the development of permanent organizations, but it would be improper to ignore the early history of labour movements when attempts, usually short-lived and unsuccessful, were made by groups of workers to make an impact on their working lives. Whether or not they had political aims they were regarded as subversive, and they were indeed usually illegal. Towards the end of the nineteenth century, however, unions were legalized in many countries—Japan and Russia being notable exceptions—but this did not mean that either governments or employers readily accepted them. Twentieth-century trade union history continued to be a story of struggle.

However widely or narrowly they interpret their purpose, unions at the very least aim to reduce the unilateral power and authority of management. Workpeople are at the receiving end of decisions made by others, especially those who employ them. Unions react defensively and challenge the authority of employers who, even if benevolently disposed, are not likely to accept willingly such usurpation. Even positive suggestions by workers or unions to improve productive techniques have often been rejected by managers and employers who regarded these matters as their province. Some employers came to recognize the existence of unions—although this acceptance has varied considerably over time and from country to country—because the law forced them to do so, or because they had little alternative in the face of union pressure, or because they saw that unions not only expressed the differences that existed between employers and workers but institutionalized the conflict and so helped to manage and resolve it. In general, though, union pressure to reduce managerial prerogatives has seldom been enthusiastically welcomed by upholders of ideas of traditional authority structures.

The impact of technological change on workers and their unions has varied greatly, as has its reception: the positive acceptance by some groups has been balanced by the defensive and hostile suspicion of others. Technological innovation has not necessarily been to the workers' disadvantage. While some skills were eliminated or reduced, others expanded and new ones were created. Machinery often reduced heavy physical labour. It could lead to higher wages. It might displace labour, but it might create new jobs. Beneficial or not, technical advance was quite clearly accompanied by new health hazards for workers. Power-driven machinery with moving parts was more likely to lead to accidents at work. Deeper coal-mines could contain dangerous gases and be liable to explosion and flooding, and coal-cutting machines produced more of the dust which caused chest and lung diseases. The use of siliceous abrasives in the sand-blasting of castings produced silicosis. The chemical industry is characterized by two particular hazards: the possibility of disaster by explosion and fire; and disease caused by specific response to contact with particular substances. These were not all novelties of the modern era: as early as the seventeenth century cases were recorded of workers poisoned through constant handling of lead (Ch. 23, p. 592). But the expansion of the chemical industry and the greater use of newer materials in the twentieth century produced fresh hazards. The carcinogenic activity of β-naphthylamine, used as an intermediary in dyestuffs manufacture until its dangerous properties came to light, is a classic example. In many cases, as in that just cited, the knowledge of a health risk developed only many years after workers came to use such materials; and unions, which were naturally concerned with these matters, were no more perceptive than others in their awareness of possible dangers. Once recognized, the remedy often lay in legislation to prohibit or restrict the use of certain materials, or—more generally, to comprehend the use of machinery as well as materials—to impose a legal obligation on employers to safeguard workers in their employ. The relevant statutes often encompassed, in addition, measures affecting the physical environment of the workplace (ventilation, cleanliness), as well as hours of work.

Indeed, pressure for legislation on industrial relations matters is one of several methods used by unions to pursue their aims. In some countries, especially in western Europe, the role of law has been extensive and many 'terms and conditions' have been determined in that way. Where the law has been less evident, as in Britain and the U.S.A., more emphasis has been placed on negotiation with employers. Nevertheless this method—collective bargaining—has been a characteristic technique everywhere, although its content has varied from country to country and indeed from industry to industry. A third method is that of unilateral decisions by unions, laying down their members' terms of employment without consulting or bargaining with employers. Its effectiveness depends very much on the discipline and loyalty of the members; it was used, for instance, by unions representing skilled workers which could enforce their policies because their members could not easily be replaced. Only a small proportion of the labour force had this degree of power, and increasingly the majority of industrial workers came to work machines which required little all-round skill. They and their unions could seldom control the supply of labour and for them collective bargaining was the more appropriate method, or obtaining remedies through legislation, or by putting pressure on governments.

Despite the fact that unions in all countries have broadly similar characteristics, their structures, methods, and policies have varied greatly. We would expect that each country's social, political, and economic conditions would markedly influence the development of its unions; and the British trade union movement, the first to develop, did not provide the model for others to follow. Few other countries went along the road of one central (if weak) Trades Union Congress, of minimal State intervention, and an absence of divisions along lines of politics or religion. Unions separately organized as Socialist, Liberal, Christian, or, after the First World War, Communist were (and are) common features of many European countries. Whereas most unions aim to build a sound organization based on dues-paying members, in France and Italy these prosaic matters have been eschewed as has, until recently, collective bargaining, the unions being rather vehicles of political protest in the traditions of their countries. In France, for example, mass demonstrations from time to time have forced governments to act to improve wages and conditions. In the western democracies the unions, however much they have become reformist and integrated into society—notably after the Second World War—remained independent of government. In Soviet Russia during the inter-war years they developed as state agencies with the prime task of encouraging higher output.

For much of the first half of the twentieth century the unions in most countries were comparatively weak. They seldom organized even one-half of the industrial labour force, and in a number of countries hardly existed. We can virtually ignore their history in Japan before 1945, or in Russia before 1917. Under Fascist or Nazi regimes they were abolished or hamstrung: in Italy, 1924–43; in Germany, 1933–45; and in the countries occupied by Germany during the Second World War. This does not mean that workers even in the most difficult circumstances were not able to protest in one form or another, but obviously unions as such could not do a great deal. Furthermore, the most detailed studies of union industrial activities refer to Britain and America (plus some from other countries); to a great extent the published histories of continental European unions concentrate on their political activities and the fissiparous conflicts between the separate and opposing union groups. The available information does not always allow us to identify and isolate, in these cases, the policies pursued by unions towards technical change.

II. THE BEGINNING OF THE CENTURY

The 1890s conveniently mark the beginning of our discussion. Most of the countries which dominated industrial development in the period 1900–50, while commencing their adoption of mechanical means of production before the 1890s, experienced in that decade a quickening of technological change. It included major transformations, not least the creation of new industries, as well as small-scale innovations in existing ones. It was accompanied by a growth of factory production, the reorganization and speeding-up of work (including the beginnings of scientific management) and the greater discipline of the work-force, and a movement of workers from rural to urban areas. New unions came into existence, as did socialist organizations; old unions had to adapt to new techniques. Some workers physically attacked machinery during industrial disputes: building workers in Dortmund threw sand in a crane; dockers at Le Havre pushed cranes into the sea; grain loaders in Belgium burned a grain elevator. Examples exist of workers trying to prevent machinery being introduced; the unskilled Brittany sardine sorters were unsuccessful in their efforts to stop sorting machinery entering their trade [1]. These were mostly the actions of unskilled, non-unionized workers and perhaps they can be regarded, as are those of the Luddites in Britain in the early nineteenth century, not so much as objections to machines, as 'collective bargaining by riot'—the only way open for workers under stress to express their grievances when other means are not available. Skilled workers, too, could be antagonistic towards new methods. The union of flint glass workers in America was encouraged, by employers in the lamp chimney industry, to oppose the introduction of the lamp chimney machine [2]. During the years of British industrial militancy, just before 1914, groups of dockers, glass bottle makers, and boilermakers objected to the introduction of machinery. But the number of men involved in these incidents was less than a thousand [3].

Sidney and Beatrice Webb, whose *Industrial democracy* was the first systematic study of trade unionism, noted—a little prematurely—that among British unions there was then (in the 1890s) no opposition to machinery as such. Unions, they said, would accept changes provided they were satisfied about the resulting conditions of their members' employment. There was much in this view. The Typographical Association, whose members were affected by the introduction of letterpress machines which replaced hand-composing, decided not to oppose them but insisted they should be operated by skilled men and apprentices in order to 'secure an effective control over the working of these machines' [4]. This was indeed the policy usually adopted by printing unions in many European countries—in Germany, France, and Belgium, for example: 'Everywhere they opted for adaptation, and in so doing they were able to influence, if not control, the course of mechanization' [5]. The reaction of manual glassworkers to the introduction of machinery has been referred to elsewhere (Ch. 22, p. 585).

Skilled unions had usually adopted the policy that certain jobs should be carried out only by their members, traditionally those who had achieved technical competence through an apprenticeship. New machines could upset this arrangement if new types of worker with alternative or lesser skills could work them. The Order of Railroad Telegraphers in the U.S.A., when telephones were introduced manned by men who were sometimes paid a lower rate of wages, responded by extending its membership to the telephone operators and negotiated with some success for equality of pay between the two groups [6]. Other unions were less willing to open their doors to those they considered to be less skilled. The Amalgamated Society of Engineers in Britain did change its rules in the 1890s to permit the recruitment of less skilled machinists and even labourers; the policy was not very successful because skilled men would not accept them into the branches, and the newly eligible recruits were not interested in the union's benefit activities. In 1926 the union permitted the less skilled to join without having to participate in the welfare side; women were admitted in 1943. However, unions were not always free to discriminate in this way. The Australian section of the same union, in order to conform with Federal and State legislation which did not permit restrictions to be placed on the admission of new members, allowed men to join the union purely for industrial purposes without having to take part in its friendly benefits. (This was for the purpose of registration under the Arbitration Acts.)

A somewhat different approach by craft unions was to demand that new types of work, even though it required little expertise, should be done only by their skilled members. In 1897 the Amalgamated Society of Engineers in Hull 'won the right of skilled men to work any new machines which replaced skilled hand labour, and to be paid the full rate'. But they were not always successful: a demand by foundry-workers in the same town that a skilled moulder should operate a machine which replaced three men was denied [7].

Technological change affected union organization in other ways. In many countries the first unions to emerge were occupational and cut across industrial boundaries. Some did not, however: unions of coal-miners or textile workers would necessarily be restricted to their own industries, but even here there were usually separate unions within some industries: cotton-spinners distinctly organized from weavers, for example. Where technical change affected the relationship between these separate unions by blurring existing boundaries new organizational polices had to meet the new situation. The American Federation of Labor (founded 1886), composed mainly of occupational unions, customarily granted charters to its constituent unions giving them exclusive rights of organization. Much American trade union history revolves around inter-union jurisdictional conflicts, often the consequence of unions trying to extend their organizational territory. Disputes would obviously arise when processes and materials changed. The introduction of mechanical type-setting into the printing industry led to struggles between the International Typographical Union and the International Association of Machinists over the right to organize the machinists (engineers) who installed the machines [8]. Similarly in the British shipbuilding industry the translation from wood and sail to steel and steam brought into the industry specialized craftsmen in new techniques, usually members of skilled unions. Each union asserted an interest in a particular occupation, but it was difficult to establish agreed lines of demarcation. The various craftsmen's jobs overlapped—both plumbers and fitters could handle pipework, for example—and where the various occupations were organized in separate unions, each aiming to support the interests of its members, numerous conflicts took place about the right to undertake such work. The shipwrights organized themselves into a national union in order to resist the encroachments of boilermakers and to exercise control over new processes. Demarcation disputes, with unions taking strike action against each other over the right to a particular task, remained a feature of the industry until well after the Second World War. In an industry which, in addition to marked fluctuations of activity, suffered great unemployment for much of the twentieth century the desire to provide employment for its own members was an understandable objective of each union [9].

During the labour ferment in the years before 1914, the response of workers whose expectations and consciousness were sharpened and heightened varied considerably. Some patiently consolidated union organization and concentrated on industrial matters, hoping to influence managerial decisions. Others eschewed this approach, which smacked of acceptance of the prevailing economic order and, as syndicalists for example, demanded far-reaching changes in the ownership and direction of economic life. Intermingled with employment questions were demands for political suffrage as well as for the right to form unions, and strikes took place on these matters as well as on the more usual bread-and-butter issues of demands for higher pay or opposition to proposed reductions. It is, therefore, not easy to discover the precise reasons for overt conflicts between unions and employers: a dispute may have been posed in terms of money but may in reality have been about discipline or machinery. A union might demand higher pay or specify the numbers to be employed when new machinery was introduced in the unstated hope that if granted it would make the machine unprofitable; this might be regarded as a way of objecting to machinery. Even allowing for this, the amount of union opposition to machines in this period seems to have been remarkably small. For this there are many reasons. Despite union expansions in many industries labour organizations were generally weak before 1914. Many industries and occupations were virtually non-union, and even where unions had grown not many included a majority of the labour force. Such extensive occupations as office workers were hardly unionized at all; if they had any views on the introduction of typewriters (and the consequent employment of women) they certainly had no power to affect the matter.

In one or two countries, such as Britain and America, collective bargaining had become normal in a number of industries; in Australia and New Zealand compulsory arbitration gave some force to union bargaining; but elsewhere there was little of it. On the Continent collective bargaining was rare in factory industry, mainly because of employers' opposition to union recognition, and what there was was limited to wages and hours. There was seldom provision for grievance procedures or for dealing with workplace matters [10]. Thus one would not normally expect such agreements as there were to comprehend the introduction of machinery: employers might be prepared to discuss the wages to be paid for it, but not the principle of introducing it. The skilled unions alone had the possibility of power to do anything about it.

It would not be unreasonable to argue that workers adjusted to machinery, either because they had no choice or because in this period the industrial labour force was growing and more jobs were being created. Those who might be made unemployed because of technological change, or whose skills were diluted, had some chance of taking advantage of extended job opportunities. Moreover, the literature records the actions and attitudes of sections of the workforce that were not necessarily representative. By no means all skills were adversely affected by new machinery, and those who worked in new industries—electricity, for example—were by definition accepting innovation. The more important question for some workers was those features which sometimes accompanied the new techniques—the dehumanizing effects of scientific management. The opposition to Taylorism was widespread; American unions were generally antagonistic, and there were strikes in 1912–13 about it at the Renault car factories in France [11].

III. THE FIRST WORLD WAR AND ITS AFTERMATH

The consequences for unions of the First World War cannot be too sharply separated from the general labour upsurge of the first quarter of the century, but they were of sufficient importance to merit separate examination. In general—there were some exceptions—unions and their associated political parties in the belligerent countries supported their countries' war effort; labour leaders, for example, were appointed to positions in government. Union members did not enthusiastically support all the war measures and there was some resistance to the mobilization of labour and to the dilution of skills; that is, the employment of untrained workers to do the work of skilled men. Governments and employers, who not long before were doing their best to deny workers the rights of association, now wooed the labour force. Thus in 1916 the German government constituted workers' committees in industry with certain limited rights, for example to bring workers' grievances and requests to the attention of employers. These committees had a right of appeal to tribunals which included representatives of both employers and employees and could make binding decisions. In Britain, workplace representatives emerged spontaneously, not through legislation, and often in opposition to official union policy, but they were recognized in 1917 by an agreement between unions and employers in the engineering industry. British unions agreed to give up restrictive practices for the period of hostilities but they were able to achieve some major demands: one, a long-term one, was for industry-wide negotiation. The reports of the Whitley Committee (1917–18), officially appointed to investigate how industrial relations might be improved after the war, recommended that collective bargaining and therefore trade unions should be encouraged. Formal Joint Industrial Councils (later called Whitley Councils) should be formed which should have, in addition to normal collective bargaining functions, the objective of joint discussion of industry's general development. Even in countries neutral in the war union status was raised. The war adversely affected some parts of the economy of the Netherlands; to soften the impact of unemployment, the government subsidized union benefit funds. In that country, too, more workers came to have their terms and conditions determined by collective bargaining.

A second strand to wartime union development was greater numerical strength which, combined with such factors as the euphoria which emanated from the Russian Revolution of 1917 and the high cost of living, exploded into a great wave of militancy immediately after the war. Apart from political matters—for example, the German revolution of 1918 and the expansion of socialist parties—this was a peak period for industrial disputes, many, but by no means all, ending in union victories. Perhaps the high point of Italian labour history in the whole of the half-century was the occupation of factories by half a million metalworkers in 1920. The response of the authorities in many countries was to concede some reforms—legal force was given to collective bargaining in France and Germany; in Germany weekly hours were by law limited to 48—as well as efforts to contain the demands of labour. The Emergency Powers Act of 1920 gave the British government extraordinary

powers to adopt in the event of certain strikes. The post-war German government, both in its 1919 constitution and by a statute of 1920, set up Works
Councils which provided workers with rights *vis-à-vis* employers, but also
legislated that the councils' legal duties included assisting employers in
improving efficiency and productivity.

In spite of post-1918 gains, unions for the remainder of the inter-war
period were very much on the defensive. Their strength was curbed, in part
by repressive governments—they were abolished in Italy and Germany under
the Fascist and Nazi regimes—and also by some employers, who insisted on
an 'open shop', that is non-union, labour force, or who dismissed activists,
or while not positively opposing unions did not recognize them. Not many
collective agreements were signed in France. The main reason for this adverse
history is plain. The 1920s and 1930s were years of economic disturbance, of
stagnant international trade, and a general deficiency of demand, together
with technological change and rationalization in industry. Unemployment was
high, union membership fell (except in Russia), strikes were fewer, and unions
generally had to accept wage cuts rather than demanded increases. Unions
could often show their members few tangible improvements apart from their
success in negotiating smaller wage reductions than those demanded by
employers. Both economic conditions and union experience varied from country to country and at different periods. The great lock-out of British engineers
in 1922 was a defeat for the unions, who had to accept a procedural agreement
—the 'York Memorandum'—which explicitly upheld managerial prerogatives. But the skilled members of the Australian section of the (British)
Amalgamated Engineering Union did not suffer the same disabilities and felt
sufficiently secure not to take up less skilled work in new engineering factories
using mass-production methods, despite the efforts of union officials to persuade them to do so (on the familiar ground that such work should be the
property of members of that union). The German unions of the 1920s,
although losing membership and, like other European union movements,
divided along political and religious lines, remained strong enough to control
the Works Councils; they did not in general cooperate with employers in the
pursuit of efficiency [12].

The world-wide slump starting in 1929 affected all countries, although to
varying depths and lengths. The heavy unemployment which accompanied it
produced major, lasting effects on attitudes and policies. American unions in
the 1920s had given up their extreme hostility to scientific management but
this was less evident in the 1930s. Nor did union leaders in that decade repeat

the 1929 statement of William Green, president of the American Federation of Labor, that 'the American labor movement welcomes the installation and extension of the use of machinery' [13]. Unions now became primarily concerned with the problem of unemployment and discussed remedies, both for the economy as a whole and for their particular industries. These included proposals for bringing industries under state control, the establishment of new industries in areas of heavy unemployment, work-sharing, tariffs, and many others. Emigration, the traditional way of siphoning off surplus workers, did not provide the answer; in the 1930s residents of the British Commonwealth returned to Britain. But some alleviation was available to France from which foreign workers, who had immigrated after 1918 to replace the country's wartime casualties, returned home.

Union reactions to technical change depended on a number of variables. One was the kind of union it was; if composed of skilled men it could retain some power even in the face of unemployment. Sumner H. Slichter's detailed study of American union practices [2] gives numerous examples of unions trying, often unsuccessfully, to oppose, or at least control, the introduction of machinery. They were mostly craft unions, and much of the information comes from the 1920s, before the slump. Nevertheless, some managed to be effective even in the 1930s. Painters, for example, were able to prevent the use of spray-guns in many cities right up to the Second World War. A British study of the mid-1930s dealing with restrictive practices recorded that skilled unions were still insisting on various measures of control. It noted that when craft unions tried to stipulate what should happen to their members and to their members' earnings when hand tools were replaced by machines the result was severe tension in industry. The skilled unions, it argued, 'cannot, even if they would (and they are not so charged), prevent the perfecting of mechanical repetition methods and the development of that precision machinery which allows of the fitter being replaced by the assembler'. It was possible, it continued, that demands such as that less skilled work be done only by skilled men at skilled rates of pay might delay mechanization by making the innovation unprofitable, but the unions 'cannot hope to win a lasting victory' [14]. These were of course the conventional responses of craft unions when their traditional all-round skills were affected by mechanization and were not peculiar to the 1930s, by which time factory production was rapidly superseding ancient skills.

It is not at all easy to measure the success of such union policies. Skilled unions might well be able to deploy some influence in particular workshops and enterprises, depending on their local strength and perhaps the absence of competing, unskilled labour. It is most unlikely that coal-miners' unions in Britain, to take a different example, could do very much in the face of severe unemployment and the fact that they lost the great seven months' lock-out of 1926. With a stagnant or falling market for coal, the growing use of machinery must have added to coal-miners' unemployment. They had no power to affect the decisions to introduce machines and the President of the Miners' Federation of Great Britain stated at the annual conference at the height of the depression in 1931 that 'the Federation favours the fullest possible extension of labour-saving machinery'. But since mechanization and rationalization produced unemployment, any proposals should be carried out 'in conjunction with the workers' organization, to secure full protection for those displaced and for the safety of those working' [15]. Since the employers at that period had virtually no contact of any kind with the Federation the request was no more than a morale-booster for the members.

A useful distinction can be drawn between the introduction of machinery into existing plants, which could affect the lives of workers currently employed there, and the creation of new, up-to-date establishments. Where processes were simplified and mass production possible, employers could locate their factories where enough unskilled and semi-skilled workers, who would accept lower wages, were available. The British shoe industry was depressed for much of the inter-war period and the labour force fell. In addition, mechanization enabled factories to be sited in districts away from the industry's traditional locations and cheaper female labour was employed. The National Union of Boot and Shoe Operatives had no power to influence these matters, but proposed that hours of work should be reduced; this was accepted by the employers in the mid-1930s when the demand for footwear rose. In one or two places where the union was strong output was restricted, and in dealing with employers the union concentrated on persistent arguments about the quantity of goods to be produced under piece-work arrangements [16].

Sometimes the impact of technical change was so sudden that even the strongest union could do nothing. In the 1920s large numbers of musicians were employed in cinemas, and the advent of sound films rapidly brought their employment to an end. Sound recording began to loom as a serious menace to movie orchestras in the summer of 1928. By the spring of 1929, about 4000 of the 23000 members of the American Federation of Musicians employed in theatres had lost their jobs. In a few cities strikes were called or threatened to compel the employment of a stated number of musicians but the union realized it had little to gain by this method, because the public wanted the new product. Instead they tried to create among the public an aversion to machine-made music. Their chief method was a national advertising campaign, financed by a 2 per cent tax on the minimum wage [17]. However, the introduction of amplification equipment in the 1930s eased the situation, for it enabled large dance halls to be built, thus providing employment for musicians.

The defensive reactions to change by existing unions are the easiest to document. Less frequently illustrated are the occasions when workers obviously accepted innovation. In many published union histories one finds either no major objections to technical change or a positive welcome for it. The British draughtsmen's union (founded in 1913) was greatly interested in technical matters, to the extent of publishing technical data for its members. It was not at all uncommon for skilled unions to encourage their members to be technically trained, not least to keep up to date with new developments. For example, the electricians' union in the U.S.A. started evening classes to enable its members to become acquainted with radio. The pressmen's union in New York similarly instituted courses for the training of journeymen in the use of high-speed automatic presses. These activities were rather different from the policy of the International Photo-Engravers' Union of North America, which provided technical training for its members so that they could control new types of work. The introduction of photo-offset lithography in the 1920s led the union to encourage the members to learn the new process. In 1937 the union decided to appoint a technical director. 'The services of the technical director are at the disposal, not only of members of the union, but also of employers who are interested in enlarging their activities by the addition of new departments or who are confronted with production problems of a technical nature' [18]. Other workers went along with mechanization in the sense of flocking to man the new mass-production plants, mainly composed of unskilled and semi-skilled labour. The obverse of the depressed industries, in which unions had often existed for some time, were the expanding ones: motor vehicles, aircraft, radios, electrical equipment. The growth of demand stimulated the introduction of mass-production techniques, so that the car industry, for example, changed from being a collection of skilled craftsmen making expensive, hand-built models; cheap vehicles came to be produced by workers who assembled items produced mechanically in large quantities.

It is true that there was sometimes considerable collective action by the workers in new industries including, for example, strikes against the Bedaux system of work measurement and payment by results. But a great deal of union history describes the difficult, preliminary efforts to recruit these workers and to be recognized by employers. In many cases unskilled workers were recruited from agriculture or were immigrants and were thus not particularly promising union material. As against the high degree of unionization on the railways, for example, the new, competing road transport industry was comparatively unorganized. Equally important, where unions were formed, the opposition of many employers was remarkably severe. Some paternal employers did not prevent employees joining unions, but did not negotiate with them and tried to instil notions of loyalty to the company by schemes of profit-sharing or works councils, and the like. Such were Imperial Chemical Industries and Unilever in Britain, or Renault in France. Others were opposed to unions: Morris (motors) in England came to accept them, sometimes after strikes for recognition; the history of the American car industry in the 1930s is one of bloody violence, when unions tried to organize and were opposed by militant employers. It is remarkable that in company histories of the period there is often very little mention of trade unions except perhaps of descriptions of attempts to organize, and some conflicts over pay and conditions. Reference to unions being consulted about the introduction of new policies or methods are absent. At Renault's in the early 1930s some apparently radical proposals were made by a senior executive: that workers should elect representatives to negotiate demands; that paid holidays be introduced; and that long-serving employees be allowed to buy company shares. The last two did not come into operation [19]. In general, apart from some unions which locally and sporadically made some impact on the decisions affecting their members, for many workers, especially in new and expanding industries, the main efforts were devoted to recruitment and, if possible, negotiations over pay and hours. The great demonstrations in France in 1936, at the time of the Popular Front government—whatever their political overtones—were primarily for wage increases and for trade union rights. The chances of unions having much influence on the control of jobs were fairly limited, although whether formal union policy or not it was often possible to limit output, which to some extent reduced the productive efficiency of machinery.

Technological change, including the use of new materials, posed organizational problems for trade unions: who was to recruit and organize these new workers? Craft unions had to decide whether to remain exclusively for skilled men or to open their doors to the unskilled. This was not, as we have seen, a new question in the 1920s and 1930s, but the rapid developments of that period raised it acutely, and it was answered in different ways. By the 1920s it had become almost impossible in Britain for new unions to be formed except for highly specialized groups of workers. To a great extent organizational structures had become fixed and the Trades Union Congress had the task of regulating disputes between unions over recruitment matters. Within any industry, especially the new consumer-durable industries, there were likely to be groups of workers who were members of separate unions, some occupational and others of a more mixed composition. Within the heterogeneous engineering industry there were some forty or so unions, some of them recruiting distinct skills (electricians, pattern-makers, etc.), but others likely to compete with each other. Thus a production worker in a car or radio factory might become a member of the Amalgamated Engineering Union (provided the local branch or district organization was prepared to make the effort to recruit the less skilled); the National Union of Vehicle Builders, a former craft union which did not hesitate to open its ranks; or a general union like the Transport and General Workers' Union. Each jealously guarded what it had and the possibility of forming one union for the car industry hardly arose. Drivers of lorries or buses might be members of the Transport and General Workers' Union, the General and Municipal Workers' Union, the United Road Transport Union, or even the National Union of Railwaymen (in this case because railway companies took over bus companies and on the basis of a common employer the Railwaymen's Union negotiated on their behalf). The unskilled workers in the chemical industry usually joined one of the general unions. A Chemical Workers' Union was formed as a separate body, but invariably incurred the antagonism of the other unions, and eventually, after the Second World War, amalgamated into one of the general unions. The Communist-inspired breakaway unions in the late 1920s achieved even less success and disappeared rapidly.

These questions of union structure were not just matters of a neat organization, or of adaptations to technological change. They were of intense ideological importance. In many European countries British experience of craft, ex-craft, general, and industrial unions was not usually repeated: metalworkers joined a metalworkers' union. But separate unions of metalworkers were formed which were distinctive according to political or religious colouring. Italy (before Fascism) and France also had the special local labour organizations (*bourses du travail, camere del lavoro*) which were centres for unions and for many other labour institutions. They were syndicalist in spirit, organizing local, general strikes of workers of all kinds and unions. The advocacy, early in the twentieth century, of union organization along industrial lines was urgently proposed by those who saw it as a first step towards some form of workers' control of industry: before the workers could take over and run their own industries they needed to discard divisions between them and form a united body. The National Union of Railwaymen in Britain was formed in 1913 with this in mind; but it did not manage to get the locomotive men and the clerical staff to agree. It is possible to see somewhat similar, although less revolutionary, considerations in the development of American unions in the 1930s. The National Labor Relations Act of 1933 provided certain legal rights for unions, at that date mostly the craft unions in the American Federation of Labor, who were regarded as somewhat conservative and exclusive. Some of its more militant leaders organized new sections which offended craft jurisdiction and split off to become the Congress of Industrial Organizations. In a very general sense the two institutions were divided along lines of skill, and the C.I.O. tended to be more radical politically as well as militant industrially.

The C.I.O. unions of the 1930s were primarily concerned with problems of recruitment and organization, of fighting for survival, and of trying to establish collective bargaining procedures. But one American union is worthy of special mention. The old-established United Mine Workers had always accepted the principle of mechanization provided the benefits of higher productivity were shared with the workers. Under John L. Lewis, in the 1940s, the union pursued a militant policy of high wage demands in the expectation that to meet the resultant higher costs the mine-owners would have to become more efficient; this would be done through investment in more advanced machinery. This pursuit of technological change through conflict had the effect of causing large-scale unemployment among miners who worked for unprofitable companies which could not afford to mechanize [20].

IV. THE SECOND WORLD WAR

As well as rapid technical change, union experience during the Second World War followed two opposing directions: in those countries, whether engaged in the war or not, which were not occupied by the Axis forces unions expanded and were further integrated into society. The German occupation of western European countries was followed by the abolition of unions or their collaboration under Nazi leadership. As against the growing membership, the expansion of collective bargaining, the involvement of unions in a whole range of joint committees—on production, workers' health and many others —in countries such as Britain and Sweden, trade unions on the Continent were under the control of a foreign military power. The unions in the unoccupied countries could build, after the war, on their wartime gains and experience: those of Germany, Italy, Japan, and much of western Europe had virtually to start afresh. In part their wartime experiences determined postwar structures and policies. Germany and the Netherlands, both desiring social peace, produced comparatively non-militant organizations. In France the Communist Party emerged, on the basis of its Resistance record, as the major party of the left and came to dominate the main trade union federation, the Confédération Générale du Travail. Moreover in that country, as in Italy, unions still remained less acceptable to employers than those in most other industrial countries. Sweden, for example, had built up a close institutional arrangement between unions and employers following the Saltsjobaden accord of 1938–9. After the war the smooth working of industrial relations in that country was investigated, as a model, by deputations and observers from many countries.

Generally, then, by 1950 unions were much stronger than ever before, although still in the preliminary stages in the countries newly liberated from occupation. In a period of full employment their needs had to be taken into account by both employers and governments; the latter especially, since they were pursuing policies of economic management and social reform. From 1945 to 1951, during the Labour Government's post-war reconstruction period, the British Trades Union Congress actively discussed ways of obtaining higher production, took part in joint visits to America to examine up-to-date methods, and established a Production Department. The wartime joint production committees at plant level were encouraged to continue; in the newly nationalized industries they were made compulsory.

This kind of worker participation in management, imposed by law or introduced voluntarily by employers and unions, became a feature of industrial relations in many European countries, especially after 1950 (for example, the West German co-determination system, with workers' representatives sitting on enterprise supervisory boards). American and Canadian unions, however, continued to rely on collective bargaining for the achievement of their objectives rather than on schemes for sharing with management. Yet their emphasis on job control meant that they, too, had to focus their attention on the workplace, where, for a variety of reasons, industrial relations matters were becoming more important. In Europe, many kinds of local arrangement expanded or were established: local union representatives handled grievances and collective bargaining; and works councils of various kinds also dealt with local problems. The frequency and specificity of technological change encouraged American unions to appoint union officials to handle industrial engineering matters in particular plants [21].

The unions' two roles, as the opposition in collective bargaining and as joint decision-makers in some form of workers' participation scheme, sit uneasily together. Unions, however distinctive the societal framework in which they operate, and whether politically moderate and accommodating or aiming eventually for revolutionary change, cannot but pursue two lines which may conflict with each other. On the one hand they are expressions of conflicts of interest, in which the struggle for workers' rights is endless. On the other they are forces for integration: collective bargaining is essentially the recognition by both sides of the legitimacy of the other. In the early twentieth century French and German syndicalists thus sometimes opposed collective bargaining which they saw as a symbol of class collaboration. In practice, unions, however militant their posture or ideological their stance, have to collaborate in order to achieve their industrial aims. Technological changes, along with other changes, might be received favourably by them or be treated with suspicion. The novelty of the post-1945 years is that whereas for most of the preceding period management and employers could unilaterally introduce technological improvements and then wait for a reaction which might not come, after the Second World War such innovations came increasingly, albeit slowly at first, to be discussed with union representatives, especially at the place of work. For unions the significant question has always been not just that they should produce satisfactory terms for their members. They are more than economic institutions trying to improve wages and the like: they are political institutions whose purpose is to alter the power relationships within employment. Being involved in the making of decisions is the reality of their task and is as important as the content of the arrangement which eventually ensues.

REFERENCES

[1] STEARNS, P. N. *Lives of labour: Work in a maturing industrial society*, pp. 126–7. London (1975).

[2] SLICHTER, S. H. *Union policies and industrial management*, p. 205. The Brookings Institution, Washington, D.C. (1941).

[3] GOODRICH, C. *The frontier of control: A study in British workshop politics*, p. 184. London, G. Bell (1920); reprinted Pluto Press, London (1975).

[4] MUSSON, A. E. *The Typographical Association, origins and history up to 1949*, p. 189. Oxford University Press, London (1954).

[5] STEARNS. *Op. cit.* [1], p. 131.

[6] McISAAC, A. M. *The Order of Railroad Telegraphers: A study in trade unionism and collective bargaining*, p. 238. Princeton University Press (1933).

[7] FYRTH, H. J., and COLLINS, H. *The foundry workers: A trade union history*, p. 112. Amalgamated Union of Foundry Workers, Manchester (1959).

[8] PERLMAN, M. *The machinists: A new study in American trade unionism*, p. 21. Harvard University Press, Cambridge, Mass. (1961).

[9] ROBERTS, G. *Demarcation rules in shipbuilding and shiprepairing*. Department of Applied Economics, University of Cambridge, Occasional Paper No. 14, pp. 10–13 (1967).

[10] STEARNS. *Op. cit.* [1], p. 180.

[11] HOXIE, R. F. *Scientific management and labor*. Appleton, New York and London (1915).

[12] GUILLEBAUD, C. W. *The works council: A German experiment in industrial democracy*. Cambridge University Press (1928).

[13] SLICHTER. *Op. cit.* [2], p. 205.

[14] HILTON, J., *et al. Are trade unions obstructive? An impartial inquiry*, p. 321. Victor Gollancz, London (1935).

[15] PAGE ARNOT, R. *The miners in crisis and war: A history of the Miners' Federation of Great Britain (from 1930 onwards)*, p. 60. Allen and Unwin, London (1961).

[16] FOX, A. *A history of the National Union of Boot and Shoe Operatives, 1874–1957*, pp. 418–27, 431–4. Basil Blackwell, Oxford (1958).

[17] SLICHTER. *Op. cit.* [2], pp. 211–13.

[18] SLICHTER. *Op. cit.* [2], pp. 258–9.

[19] FRIDENSON, P. *Histoire des Usines Renault*. Vol. 1. *Naissance de la Grande Entreprise 1898/1939*, pp. 237–8. Editions du Seuil, Paris (1972).

[20] BARATZ, M. S. *The Union and the coal industry*, pp. 53–4, 71–2. Yale University Press, New Haven (1955).

[21] BARKIN, S. The technological engineering service of an American Trade Union. *International Labour Review*, **61**, 609–36 (1950).

BIBLIOGRAPHY

ANDERMAN, S. D. (ed.) *Trade unions and technological change*. Allen and Unwin, London (1967).

BAKER, E. F. *Printers and technology: A history of the International Pressmen and Assistants' Union*. Columbia University Press, New York (1957).

GALENSON, W. (ed.) *Comparative labor movements*. Prentice-Hall, Inc., New York (1952).

HUNTER, D. *The diseases of occupations* (5th edn.) English Universities Press, London (1975).

KASSALOW, E. M. *Trade unions and industrial relations: An international comparison*. Random House, New York (1969).

KENDALL, W. *The Labour Movement in Europe*. Allen Lane, London (1975).

LORWIN, V. R. *The French labor movement*. Harvard University Press, Cambridge, Mass. (1954).

MAIER, C. S. Between Taylorism and technology: European ideologies and the vision of industrial productivity in the 1920s. *Journal of Contemporary History*, **5**, 27–61 (1970).

MARQUAND, H. M., *et al. Organized labour in four continents*. Longmans, London (1939).

INTERNATIONAL INSTITUTE OF SOCIAL HISTORY. *Mouvements ouvriers et dépressions économiques de 1929 à 1939*. Assen (1966).

NADWORNY, J. *Scientific management and the unions, 1900–1932. A historical analysis.* Harvard University Press, Cambridge, Mass. (1955).

SCOTT, W. H., *et al. Technical change and industrial relations.* Liverpool University Press (1956).

SHERIDAN, T. *Mindful militants: The Amalgamated Engineering Union in Australia 1920–72.* Cambridge University Press (1975).

SHORTER, E., and TILLY, C. *Strikes in France, 1830–1968.* Cambridge University Press (1974).

ZWEIG. F. *Productivity and trade unions.* Basil Blackwell, Oxford (1951).

6

THE ROLE OF GOVERNMENT

ALEXANDER KING

AN aptitude for technological invention has been a characteristic of
Homo sapiens since his earliest arising. From the shaping of the first
crude tools of bone or flint, through the discovery of the wheel, the
lever, the use of fire or the plough, down to the complex, sophisticated
industry of today, technological development has been man's main agent in
the struggle upwards from subsistence. It is not surprising, therefore, that
organized society—first at the tribal level and later at that of the governments
of the sovereign states which constitute the world of today—should have
encouraged the development of technology, both for military purposes of
aggression or defence and through the domestic arts, as a basic attribute and
source of power and, it is to be hoped, of prosperity.

Reciprocally, the possession of novel technologies giving an advantage in
intertribal or international competition has always had a basic impact on
power structures and political developments, although until recently this was
seldom recognized at the time. The invention of the cross-bow, for example,
to replace the then traditional long-bow, had a dominant influence in medieval
politics; and the dangers to humanity of the cross-bow as a weapon of destruction were solemnly enunciated by the Vatican in terms which, without change,
could be read as applying to the nuclear bombs of our own century. Better
means of transportation or communication, improved or quicker methods of
manufacturing pottery or smelting metals had a dominant influence on the
trade and prosperity of countries. The industrial revolution, which triggered
off the chain reaction of contemporary technological development, gave rise
to the great mercantile empires of the last century and technological superiority in defence was the critical factor in the outcome of the Second World
War.

Yet, despite a tacit understanding of the importance of technology and the
need for governments to encourage its development, there was until recently
little understanding of the nature of the process of technological innovation,
of the relation between science, technology, and the economy; and hence no
country had a deliberate and explicit policy for technological development:

indeed, few have such policies today. The government's role in technology has developed piecemeal as an element of industrial, manpower, military, and fiscal policies, of trade regulations and also, very recently, of science policy. Whether intended or not, the policies of governments have powerfully influenced technological progress.

It is only since the 1950s that technological innovation has become the object of detailed analyses, based on the collection of quantitative data, as distinct from unverified speculation. As a consequence, during the first fifty years of this century, very little systematic attention was given to the nature of technological development and the conditions for its success. Before 1960, few systematic compilations were available of the research and development efforts of individual countries. Indeed, governments saw little importance in attempting such compilations and firms were generally unwilling, and often unable, to provide data on their own efforts. Hence it was only in 1965, with the publication of a report by C. Freeman and A. Young to O.E.C.D. [1] that the first tentative international comparisons of research and development resources were made. Since that time there has been rapid progress and regular 'international statistical years for research and development', based on agreed methods and definitions, are now carried out.

In this chapter, therefore, although we shall be concerned with developments during the earlier part of the century, this is necessarily done with hindsight derived from contemporary knowledge of the nature of the process of technological innovation and on the basis of recent international comparisons.

I. THE NATURE OF TECHNOLOGICAL DEVELOPMENT AND ITS ECONOMIC SIGNIFICANCE

Technological development as an intrinsic element of industrial and economic growth is of primary importance to all governments. Not only is it in their immediate interest to encourage it by a whole series of policy measures, but many of them are directly involved in major technological schemes, both military and economic. Such measures to promote technology as evolved in the first half of this century were largely instinctive and related to immediate economic problems; for example, the protection of new manufactures from foreign competition by the erection of tariff barriers, import quotas, and other restrictive measures. At the same time, most governments deliberately created technological facilities to provide a national research and development capacity.

Before considering the various ways in which the governmental role in technology is exerted, it is necessary to discuss the nature of technological innovation and its links with the economy. Technological developments may be originated in a country or a firm (indigenous innovation) or may be imported from other firms or governments by the purchase of patent rights and know-how. Technological innovation may be defined as the first application of science and technology in a novel way with a view to commercial or military success. As has already been noted in relation to economic factors (Ch. 3), the process of technological development has three stages—invention, innovation, and diffusion [2]. Invention is the arising of a new concept as to how science and technology can be applied for a specific purpose. It may either be based on empirical thinking, as have been most of the inventions of the past, or it may result from systematic research, first of a fundamental nature and later applied towards the desired technological objective. Some three-quarters of the inventions which find successful application in production appear to result from market demand; the remainder result from new possibilities arising from scientific discovery and suggest new market opportunities that have to be deliberately stimulated. Inventions of the latter type are often more radical than those which seek to satisfy well-defined market needs, and often give rise to a series of further inventions and innovations.

Innovation is the process which transforms an invention into goods and services; and the agent for so doing in the market-economy countries is generally the industrial firm. In defence technology, however, this stage is the responsibility of government, either within its own establishments or by contract within industrial enterprises. Since there are close similarities between technologies developed for military purposes and those for the civilian market, sophisticated military technology—for example in aircraft, navigational aids, computers, television, or remote-control devices—may have a major influence on civilian developments and international competitivity.

The third stage of technological development is diffusion. This is the means whereby an innovation becomes generalized throughout an industry or, indeed, eventually throughout the world, and thus contributes to widespread economic growth. Many governments have long been concerned to promote the productivity of their industries. For this purpose, the diffusion of existing technology, whether indigenous or imported, is even more effective, and certainly quicker, than original innovation. The mechanism of diffusion involves expansion of the innovating firms; the dissemination of information through a vigorous technical press; by personal contact; and, finally, the purchase of patent licences and 'know-how'.

It is difficult to measure the economic impact of technological developments. Not only has the direct profit which accrues from them to be considered, but also the penalties of not fostering innovation continually and
thus falling behind in international competition. In military technology the
danger of innovating more slowly or less effectively than potential enemy
countries is obvious.

In general, however, since so much technological development takes place
in response to market demand, economists have tended to regard technology
as resulting primarily from the interaction of economic forces. Thus, J. M.
Keynes never discussed technological development specifically and even
today the concept of the 'technological fix' is implicit in the thinking of most
economists; this implies that, when needs arise, technological solutions will
be forthcoming spontaneously.

Of late, however, there has been a good deal of discussion of technology as
an autonomous force. There has been recognition of the importance of scientific discoveries, whose application gives rise to quite new products and
processes which could not have been envisaged by economic thinking and
which have opened up quite new markets. The research scientist—whether
in university, industry, or the defence establishment—tends to think of himself as the originator of new productive possibilities. A scientific discovery
made by the fundamental researcher, concerned only with the extension of
knowledge for its own sake, will be published freely and will become available
internationally. The entrepreneur, possibly in some distant country, and
often much later, will through his scientific assistants see some productive
possibility in the discovery, applied research will be started on it, and, if this
is successful, a pilot plant or engineering prototype will result. Eventually
there will be production for the market, without the original discoverer
necessarily being aware or concerned. Both these lines of development are
operative today, and many developments considered elsewhere in this work
are the consequence of a blend of the two forces.

It was only towards the end of the 1950s that serious attempts were made
to relate technology quantitatively to the economic performance of a country,
first with regard to agriculture and later to industrial development. Thus
E. F. Denison [3], in a study of the growth of the economy of the United
States during the first half of the century, concluded that only 40 per cent of
the increase of G.N.P. could be attributed to additions to the traditional
imputs of capital and labour. The greater part he ascribed to a so-called
residual factor, presumably consisting of a complex of other elements among

which science, technology, education, entrepreneurship, and management skills were seen to be particularly important. Essentially, such studies indicate that qualitative factors are as important as those of a quantitative nature; it is the quality of manpower and of capital utilized which count and which have enabled the G.N.P. to grow so rapidly in many countries. Manpower quality at all levels, from unskilled labour to top management, is raised by education and training; capital utilization becomes more effective through better technology as a consequence of the creation of new materials, processes, and products in opening up new markets and in increasing manpower productivity.

While such approaches were greatly contested, especially in their detailed conclusions, their broad lines were perhaps too rapidly accepted. Both education and science began to be regarded as national investment items and, as such, found it easier to attract resources than hitherto. Both governments and industry assumed, somewhat naïvely, that more research would necessarily mean more economic growth; insufficient attention was given to the intrinsic relevance of the research and to the process whereby its results were coupled to the productive process. Much of the spectacular increase in the resources made available for research and development in all industrialized countries since the last war was due to this recognition of scientific discovery as having been an impulsive factor of growth. It was greatly aided by the first appearance of research and development statistics at about the same time, which by presenting the expenditures of particular countries on research and development as a proportion of G.N.P. provided an international ranking order somewhat like a football league table. This enabled countries low in the table to provide strong arguments for increased resources; the United States, at the top, was the pacemaker.

Until recently, technological development was thought of mainly in terms of specific inventions or of the exploitation of scientific discovery. It is now increasingly accepted that much more is involved. Many discoveries of obvious productive potential have not been exploited in their country of origin; many such are described elsewhere in this work. For example, the discovery of mauveine by W. H. Perkin in the middle of the last century was the quite obvious trigger point for the development of synthetic dyestuffs; a great industry grew up, but not in Britain: in Germany. The discoveries of the basic principles of electricity by Faraday and Maxwell likewise opened up the possibility of a great new industry, but few of the basic electrical machines were invented in their own country.

The basic need, therefore, is for the existence of entrepreneurship, and governments have had to decide the extent to which incentives for this can be developed in balance with other social and economic objectives. Other factors which determine success and failure in technological innovation include the existence of an appropriate fiscal policy, the availability of risk capital, the existence of a basic technological capacity, appropriate education at all levels, management development, and marketing skill. Behind all these lie differences of national psychology and tradition, which favour or discourage innovation. It is necessary for governments to take all these and other factors into consideration if they hope to achieve a high and sustained level of technological development.

There is of course a factor of timeliness. Inventions may appear and be immediately exploitable if the economic, scientific, and social circumstances are propitious; equally, many appear years before such circumstances exist. Many examples are to be found within the period under review. The fork-lift truck was perfected technically many years before labour costs were sufficiently high to justify its adoption. The first gas turbines failed miserably for lack of a good compressor and of high-temperature alloys able to withstand the heat developed. The laser was implicit in Einstein's equations of 1917, but forty years had to elaspe before it could be realized [4].

Another factor is the extremely long lead-time from the initial scientific discovery, through applied research and technological development to production on a significant scale. This is usually not taken into account by governments. A gap of as much as twenty or thirty years, or even more, from discovery to generalized use has been encountered in some major innovations during the present century (see Ch. 3, Table 3.1, p. 52). Of course this lag can be greatly reduced for crash programmes if governments are willing to provide sufficient resources. It took only six years from Otto Hahn's discovery of nuclear fission to the dropping of the first atomic bomb in 1945, and the extraordinarily complex task of placing men on the Moon was achieved very quickly but at very high cost.

However, crash programmes must remain the exception and it is doubtful if the lead-time for technological development in general has decreased significantly, although many assume that this is so. For example, in France some twenty-five years had to elapse between the first experimental reactor 'going critical' and the first regular flow of nuclear electricity into the grid, even though considerable priority was given and little novel technology was involved. It is reasonable to assume that we have not yet seen the concrete results of some basic research conducted before the middle of the present century.

This very long lead-time of technological development was accepted without difficulty in the earlier years of this century, a period of relatively slow change, but it may become exceedingly awkward in the present situation. If the period between two distinct sets of economic and social circumstances is shorter than the average lead-time of technological development, it is probably that the 'technological fix' concept will be unreliable. The danger today is that technological solutions to urgent problems will always come too late to resolve economic difficulties. Thus serious development work on non-traditional energy sources must be given priority now if the results are to be available in the early years of the next century.

Although technological development is now accepted as being a major source of economic growth, there is no evidence that a country's success in originating new technological processes has any direct influence on its productive performance or trade position, at least as between industrialized countries which have a technological capacity above a certain ill-defined threshold. It seems that the diffusion of technology across frontiers is sufficiently quick for such nations to compensate for inadequacies in their own research performance. The outstanding example of a country achieving great economic success on the basis of imported technology is, of course, Japan. But this case illustrates very well the reality of the threshold of technological awareness just mentioned. The Japanese economic miracle—which began only at the end of the period that is our present concern—was very deliberately planned at a time when, after the Second World War, much of her industry had been destroyed and it was inevitable that a new start would have to be made, entailing new objectives and new thinking. The first step was to create a modern education system (see Ch. 7), which was accomplished so thoroughly that Japan is now one of the most literate countries in the world. As part of the educational system, widespread fundamental research in the natural sciences was strongly supported and management structures and attitudes were revolutionized. As a result of all this, and with the help of an excellent technical information system and the employment of information consultants who had worked in the most highly industrialized countries, and particularly in the United States, the world technological possibilities were quickly and thoroughly scanned and the most appropriate innovations purchased and put to productive effect. As is the case with all effective innovation, each development suggested improvements and new products so that after years of reliance on imported technology. Japan then increasingly initiated many new developments. A further element of this success story has been the extraordinarily harmonious cooperation between Japanese industry and the Ministry for International Trade and Development (M.I.T.I.) which, among other things, saw to it that Japanese industry, despite its heavy reliance on the multinational companies from which most innovations were purchased, retained control. Japan is now the only market-economy country which possesses a real, deliberate, and comprehensive industrial, and hence technological, policy.

Can the Japanese example be repeated by other countries? It is possible that a few which possess excellent natural and manpower resources may be able to do so, notably Brazil and perhaps Mexico. This will depend on the awareness of their governments of the threshold needs—namely to build up a solid educational, research, and technological infrastructure which will give them sufficient technological awareness to enable them to purchase and exploit those items of world science and technology that are appropriate to their needs. Iran is attempting a similar path but will meet great difficulties until the level of research and education has been brought above threshold. For the underdeveloped countries, with little in the way of research capacity and managerial skill and with low levels of general education, quick technological success is unlikely.

One danger of relying too exclusively on innovations purchased from foreign firms is that the industry of a country may find itself directed mainly from outside its borders, which in the end can be a threat to national sovereignty and individuality. This danger Japan has very cleverly avoided. Canada, with its 3000-mile frontier with the United States, has encountered a particular difficulty in this way within our present period of study, and, more recently, a large proportion of its industry has become American controlled. In fact, it appears that Canadian subsidiaries of American firms do more research and development in Canada than Canadian-owned firms, but there is always the fear that such research may be more in the interest of the American parent firms, and hence of the American economy, than of that of Canada. This problem is a matter of concern to both government and the public. The same situation prevails in Australia, where there is a very considerable effort in fundamental research, but less development activity than this research effort would normally induce. Here the suspicion is that good research undertaken in Australia will be developed by British and American firms for their own markets, with little advantage to Australia.

One final general point must be stressed. Technological development is not only costly, but is very risky, although the rewards are great. Only a small proportion of the research undertaken, even when it is scientifically successful, results in commercial exploitation. Hence a considerable research effort per firm or per country is necessary if the statistical probability of success is to be great. One large American chemical firm, for example, has calculated that of all the basic research projects started only about 10 per cent showed sufficient economic promise to be pushed forward to the development stage; of these, only a few finally achieved production. A major electronics manufacturer found a similar situation; only about 10 per cent of its developments finally culminated in production, yet 70 per cent of the firm's production was derived from this small proportion of successful development.

II. TECHNOLOGY AND WAR

Over the centuries, wars have often given a new and powerful impulse to technological development, primarily, of course, for direct military objectives, but with wide repercussions for civil industry when peace returned. The United States is an outstanding example of this. From the very beginning it relied greatly on applied research for its continental development. The founding fathers of the American republic were deeply embued with a belief in technological progress and rational thinking, innovators in the applied sciences as well as in political institutions. However with each new war, it became apparent that the American scientific and technological institutions had become too weak, academic, or inbred to meet the new emergency. Consequently during the Civil War, as well as the First and Second World Wars of this century, new and dynamic research structures were created. From the beginning, the growth of the American scientific agencies took place side by side with those of private enterprise, but the stress was generally on applied, rather than fundamental, research. Even after the Second World War, Vannevar Bush, who had led the wartime technological effort, in his report (1945) to President Roosevelt, entitled *Science the endless frontier* [5], pleaded for the building-up of a major capacity in basic research, from which, he argued, applications would come spontaneously. The implicit assumption was that, even at so recent a date, America was relying too much on European basic science for its technological development. The American situation during the early years of the century was quite different from the later experience of Japan. The United States had built up an immense capacity for technological innovation based on fundamental research done elsewhere.

It imported such research as a free good, but its indigenous invention and innovation were without rival.

During the First World War, the initial successes of Germany owed much to the strength of an industry which had already become science-based. Its chemical industry, for example, based on the researches of a whole series of great scientists, had little foreign competition. Britain, on the other hand, despite the excellence of its fundamental research, had no great capacity for the manufacture of dyestuffs, pharmaceuticals, and other essential chemicals, and such manufacture had to be quickly improvised. The government, realizing this weakness, created in 1917 the Department of Scientific and Industrial Research and, in cooperation with the industrial sector, the first of the Industrial Research Associations, which still exist.

At the outbreak of the Second World War, Germany, who gambled on the probability of a short period of hostilities, more or less dismantled its scientific organization which had worked hard and successfully for several years in preparation for war. By contrast, the United Kingdom, where there had been considerable technological development for military purposes, including that of the aeroplane industry and radar, reorganized its technological effort as soon as war broke out and utilized its scientific manpower, including that of the universities, to the full. Likewise in the United States a total mobilization of industry and the universities launched programmes of military research of enormous technological effectiveness. These were the first technological crash programmes in history and they succeeded magnificently, producing jet aircraft, microwave radar, revolutionary aids to navigation, new explosives, bomb-sights, proximity fuses, and, finally, the atomic bomb. One significant feature was the way in which academic scientists, working in teams with engineers and military experts, proved capable of producing fundamental technological innovations which were determinative in winning the war. Regrettably, this technique was insufficiently fostered in post-war industrial development. But the effort was not restricted to military technology in the direct sense: it also gave rise to continuous production methods for the manufacture of penicillin (a British discovery) and DDT (a Swiss discovery); many tropical disease pharmaceuticals; the tropicalization of equipment; and many auxiliary processes as significant in peace as in war. A further new development was the technique known as Operational Research, an application of the scientific method to particular systems such as tactical air operations; keeping areas such as the Bay of Biscay mine-free; or the maintenance of aircraft. This method, which was essentially an application of simple statistical principles to a carefully studied and formulated situation, depended on the establishment of a deep mutual confidence between the decision-makers and their operational research officers. Later, of course, it found application in civil industry and, indirectly, evolved into the systems approach of tackling complex situations, increasingly necessary today.

The post-war application of these military developments for peaceful purposes led to the greatest upsurge of technology, and hence of economic growth, of all time. Some of the developments could be adopted relatively simply for civilian applications—new types of aircraft, radar-navigation devices, television, computers, new drugs and pesticides, and even nuclear power. Many of the consequences were more subtle, however, as in the use of high-temperature alloys to develop types of propulsion impossible with existing materials or in evolving advanced management techniques for the mastery of complex technological systems. Above all, the fact that science, through technology, had shown itself capable of large-scale, and even determinative, influence in war, suggested that an equally impressive result might derive from its massive application in the peacetime economy.

In Europe, the immediate post-war period demanded massive resources for the reconstruction of cities and industries. This was accomplished very rapidly thanks to the Marshall Plan, but there was little risk capital available for major technological developments. The stress was rather on efficiency: increase in manpower productivity which produced a host of minor innovations, increase in the extent of mechanization of industrial processes, and the development of automation. In the United States, on the other hand, there was an almost explosive surge of new technological development, which western Europe and Japan had joined by the end of the 1950s. In the United States also, the cold war—which stimulated competition with the Soviet Union in the development of highly sophisticated military technology—necessitated a maintenance of the industry–defence relationship which had grown up during the war. This was the period of exponential growth of resources devoted to research and development in all the industrialized countries. American government expenditure on R and D was particularly large, and the greatest proportion of it was expended by contract in industrial firms, especially for the gigantic military technology and, later, space programmes. Much government money also flowed into the universities for both fundamental and applied research, to an extent which seemed to many to imperil the independence of these institutions. There was much debate about the importance of the spin-off from military to civilian technology within the highly sophisticated, science-based industries of the United States. This was an effect impossible to measure, but it was certainly great, even where indirect. It brought industrial firms into touch with a whole range of new techniques and scientific principles, and encouraged the emergence of a new type of highly sophisticated systems management. The argument that much greater gains to the economy might have been achieved if the money had been spent directly on projects of economic importance was rather pointless, since without the military justification the resources would simply not have been forthcoming to industry in an economy so identified with private enterprise.

The need to innovate technologically has become a basic factor in the shaping of the modern world economy, and especially of that of the United States. It has become a compulsive force which necessitates for its continuation the stimulation of artificial demand for a constant series of novel products; it has stimulated waste, built-in obsolescence, and overconsumption of materials and energy. A huge self-perpetuating vested interest was created which admittedly resulted in a high, but precarious, economic growth. President Eisenhower, in his farewell speech on quitting the Presidency, warned against the dangers of the military–industrial complex. The size and power of the highly sophisticated industrial enterprises which have come to believe that they must innovate or perish is a disturbing symptom of the instability of the contemporary technological society, whose roots lie back in the period here under review.

The complex government–industry relationships which have arisen in the field of nuclear energy development, military technology, and space illustrate well the importance of the government role in recent and contemporary technological development. The government supplies the research and development resources and is also the main, and often unique, customer for the products. While this has been particularly striking in the case of the United States, it has been true to a lesser extent of the smaller nuclear powers—Britain, France, and Canada. Germany, with modest if any post-war military programmes or nuclear development, took another path, concentrating with considerable success on new civilian developments. In the Soviet Union, military, space, and nuclear developments were given the highest priority; innovation in basic technology has been rather patchy, with much reliance on the purchase of foreign processes and know-how. In a system working within regular five-year economic plans, with individual enterprises given production quotas which have to be met, innovation—which often necessitates interruption of normal production schedules, especially during the teething stage of a new process—proved much more difficult to achieve. Various incentive schemes have been introduced in Russia to overcome this difficulty, but there is little evidence, as yet, about their effectiveness.

III. THE GOVERNMENT FUNCTION IN TECHNOLOGICAL DEVELOPMENT

As indicated above, the securing of an adequate technological development is a secondary objective of governments in the attainment of their primary defence and economic objectives. In achieving this the activities of governments are extremely diverse, and in many instances indirect, as in their formulation of objectives with regard to education and training, industrial structures, commercial policy, and defence. Their influence is exerted also through their regulation of the labour market, and their measures relating to the rewards and constraints on individuals and institutions in various parts of the technological development process. In countries at an early stage of industrialization, priorities may be given to manufacturing processes which achieve import substitution and aid the balance of payments; at another stage, government may provide incentives to promote industries which will gain foreign exchange through exporting their products, and tariff barriers may be erected to protect vulnerable new technological processes. In yet other instances government action is in the public interest, as for instance through clean air acts, regional policies, safety regulations, and so on. The technology system of a country, involving as it does industry as the main agent of innovation and the universities as sources of new knowledge, as well as government activities, is extremely complex and there can be no generally applicable policy to ensure technological innovation. Fluctuation of commodity prices, changes in labour force availability, the changing nature of foreign competition and the terms of trade—these and many other factors demand a dynamic and flexible attitude by governments to ensure that multiple channels be kept open for the transfer and utilization of new or existing technologies.

However, it is a basic function of government policies associated with technology to ensure that the national infrastructures of research and development, of education and training are maintained in a healthy condition and always looking forward. In some countries, the universities are regarded as part of the governmental system and professors are regarded as civil servants. In this chapter we shall, however, regard them as independent institutions and not as part of government. Nevertheless, we must recognize that, however they are supported, universities cannot work in isolation from their national governments.

In the remaining part of this section we shall discuss a few of the fields in which government action on the technological system is exerted.

Information. The effectiveness of a government in maintaining a viable and innovative technology will depend greatly on its understanding of the process of innovation and its knowledge of the strength and weakness of the national effort. If governments regard industry essentially as a milch-cow to produce the maximum taxation yields, their policies are likely to be short-sighted and not conducive to innovation. In fact, the real and long-term technological (and hence economic) success of a nation depends on a symbiotic relationship between government and industry which combines the sense of practicality of the latter with the longer-term understanding of trends and their analyses of the former. This is achieved in too few countries, and once again the exceptional experience of Japan has many lessons for other nations. If governments are to possess an adequate knowledge of the national situation concerning technological development and be able to communicate this to industry, they have a need for a very high standard of statistical services on, for example, the manpower position; national performance in world markets; scientific and technological achievements and their yield; sector-by-sector production performance; patent sales; the technological balance of payments; and so on.

Education and training. The very important subject of the relationship of education to technological development forms the subject of a separate chapter (Ch. 7) and will therefore be only briefly considered here. During the early part of the century government efforts in this field intended to support technological development were mainly concerned with vocational training of engineers, technicians, and the like, as well as the building up of the great professional schools in the universities, educating scientists and engineers. These naturally cultivated research, partly as a means of education and also for the extension of knowledge. It was only towards the end of the Second World War that such research was seen generally as directly relevant to technology, and cultivated as such.

The rapid advance of technology in the U.S.A. and the U.S.S.R., together with the recognition of the need for large numbers of scientists and engineers for the technologically based and technologically competitive society of the post-war decades of the century, stimulated the European countries greatly to increase their facilities for educating scientists and engineers. Furthermore, recognition of the influence of education on economic growth, through technology and in other ways, also provided a major impetus for educational expansion in the industrialized countries.

Special consideration also began to be given to management education; however, the application of many of the proved techniques for general management are difficult to apply to either research or technological innovation, in view of the inherent uncertainties of the research process and its long time-cycle. The question of cultivating an innovative spirit and performance in technological management has yet to be explored in depth. Existing analyses do not indicate any direct correlation between innovative performance and educational background in various countries, but there is some evidence which suggests that scientific training and experience in R and D, with its usually high degree of specialization and concentration on analytical skills, is not particularly conducive to creative innovation in response to practical needs.

Fundamental research. Another function of governments is to ensure that an adequate research capacity exists to enable the world's repository of new knowledge to be appreciated and utilized to provide the basic degree of scientific awareness without which technological development is difficult to achieve. It is generally accepted that the universities provide the best environment for such research but in many countries the funds for such research are only partly provided through grants for education, and have to be supplemented by a number of means such as specific provision for expensive equipment, research project finance, fellowships, etc. Indeed, with increasing sophistication of the instruments of research, costs have risen steeply and, if research funds were provided through normal university financing, this would lead to an imbalance with the resources earmarked for higher education as such. Furthermore, in times of financial stringency, the high fixed overheads of the universities always appear to have the first call on the available funds, and research tends to be sacrificed, even though it is an activity which calls for continuity of effort. To allow for their research funding needs, many governments have created research councils for the distribution of national funds. These provide a forum for discussion between officials and academics concerning national needs for research. The funds are generally distributed on criteria of merit by the widely accepted principle of judgement by peers. In Britain, for example, the Science Research Council was established in 1965 in succession to the Department of Scientific and Industrial Research (1916).

During the earlier years of the century, when research was still a minor activity in the universities, some countries, and especially those with a federal structure, provided most of their research support through such councils. Furthermore they often created large research institutions of their own, with semi-autonomous status. Examples are the National Research Council of Canada (1917) and the Commonwealth Scientific and Industrial Research Organization of Australia. Founded in 1949, the latter grew out of the Advisory Council for Scientific and Industrial Research (1916). Indeed it became so strong and so competent that it tended to overshadow university research in Australia. Experience tends to show that national laboratories, if given a general rather than a mission-oriented role, tend to concentrate on academic research at the expense of applied; the more so, the higher their quality.

The world's fundamental research, freely and internationally published, is in this sense a 'free good'. However, identification of what is relevant for application, its assimilation, modification, and transfer across the science–technology interface has a high cost and can scarcely be effective without a strong indigenous research capacity. Thus national capacity for technological innovation and development must be matched by equivalent capacities in fundamental research, and it is the function of government to ensure that this is the case. At the level of the most advanced science-based industries, where development is near the frontiers of knowledge, applied research and development continually calls for the opening up of new lines of discovery. Hence an increasing amount of so-called fundamental research, financed through contracts from government and industry is in reality 'oriented research', free in the sense that its lines of progress are left to the research leader in the university, who can accept or reject project suggestions according to his intellectual interest, but inherently part of the process of technological innovation.

Governments often attempt by selective funding to compensate for inadequacies in the free-choice research performance of their universities. This is at times necessary in new fields where specialists and funds are scarce. Thus in Britain the Agricultural and Medical Research Councils have created special research units in such fields, which are situated within universities and may be later assimilated into the normal research fabric of these institutions. Other countries have attempted, by special funding to create or expand 'centres of excellence'; in France, for example, by a number of well-financed *actions concertées* and in Germany by *schwerpunkt* research grants.

Background research by government. Many governments undertake sizeable programmes of applied research within their own institutions. Initially, many of these were justified by the background support they provided to industry as a whole and one of their chief tasks was to create and maintain standards of all kinds from those of dimension to radioactivity. Great laboratories for such purposes were created early in the century, such as the National Physical Laboratory (1900) in England, the Kaiser Wilhelm (later Max Planck) Institute (1910) in Germany, and the National Bureau of Standards (1901) in the United States. Such laboratories often generated important research of a broader type; for instance, the important early work of the N.P.L. on the phase rule which had enormous significance for the metallurgy industry. Governments also tend to maintain institutes for research on public utilities, such as the quality of air and water, fuel research, hydrology, pest control, food preservation, prevention of fire, etc. In addition, many of them have established research laboratories for industries such as building construction, timber, and, of course agriculture, where the average production unit is too small to undertake its own research. Another field in which government research has been especially strong since the end of the Second World War is the peaceful use of nuclear energy; in many instances, activities of this type have been conducted in close cooperation with industry.

Many government laboratories have excellent records of innovation, especially during their earlier years, but there is a general feeling that such institutions have an inherent tendency to become somewhat sterile over the years, partly because they operate too far from the user—isolating R and D from changing requirements and opportunities and from the process of commercialization—and partly because, being within the civil service structure, there is insufficient mobility of research workers in relation to the universities and industry. Large national laboratories, if their original missions are nearing completion, or have become of lesser priority, have proved difficult to reconvert to new tasks.

Industrial research. Although the most effective location for industrial research and development is the competitive industrial enterprise, governments have found it necessary to support and stimulate research, both in relation to their own functions such as defence and as part of a general policy of industrial development. Table 6.1 indicates the expenditure of industrial research and development in a number of countries in the 1960s as a percentage of the net industrial output in each case. The table distinguishes the

TABLE 6.1

Expenditures on industrial R and D as a percentage of net industrial output in thirteen countries

	R and D performed in industry as a percentage of net industrial output				R and D financed in industry as a percentage of net industrial output			
Austria	0·4	(1963)	0·8	(1966)	0·4	(1963)	0·8	(1966)
Belgium	1·5	(1963)	n.a.		1·5	(1963)	n.a.	
Canada	1·3	(1963)	1·6	(1967)	1·1	(1963)	1·3	(1967)
France	2·0	(1963)	3·1	(1967)	1·3	(1963)	1·8	(1967)
Germany	1·9	(1964)	2·5	(1967)	1·6	(1964)	2·1	(1967)
Italy	0·9	(1963)	1·0	(1967)	0·9	(1963)	1·0	(1967)
Japan	2·9	(1963)	2·7	(1967–8)	2·9	(1963)	2·7	(1967–8)
Netherlands	2·4	(1964)	3·2	(1967)	2·3	(1964)	3·2	(1967)
Norway	1·0	(1963)	1·4	(1967)	0·8	(1963)	1·1	(1967)
Sweden	2·4	(1964)	2·4	(1967)	1·8	(1964)	1·9	(1967)
Switzerland	n.a.		2·8	(1967)	n.a.		2·9	(1967)
United Kingdom	3·2	(1964–5)	3·3	(1966–7)	2·0	(1964–5)	2·1	(1966–7)
U.S.A.	7·0	(1963)	6·0	(1966)	3·3	(1963)	2·8	(1966)

Source: *International Statistical Year*, O.E.C.D., Paris.

total R and D performed within industry in each case from the R and D financed by industry from its own resources. The difference is to be attributed essentially to research expenditure by governments within industrial firms by contract or other means.

Although Table 6.1 relates to activities in the 1960s, it also applies in general terms to the situation in earlier years. It indicates the very wide variations in the proportional expenditure of the countries concerned. The United States has by far the greatest industrial research intensity and also the highest proportional impact of government-financed industrial R and D.

The means whereby governments stimulate industrial research are very varied. They include contracts for specific research on a profit basis, general subvention, fiscal incentives, loans for research (reimbursable in the case of commercial success), the carrying out of research in government laboratories, and a whole range of institutional devices. Among the latter, the industrial research association system of the United Kingdom is the earliest and possibly the most comprehensive. These associations, organized on a sector by sector basis, were stimulated by governmental initiative through the then Department of Scientific and Industrial Research, which offered to match with government funds the contributions of the individual firms in a sector willing to cooperate in cooperative research. The research undertaken was regarded as basic to the needs of an industry as a whole and was determined in each case by a board of representatives of the member firms. The D.S.I.R., while having a supervisory role with regard to government interests, never tried to influence the research programmes. The quality of research undertaken by the various Associations varied greatly and in a number of cases the programmes were too limited by financial constraints to guarantee break-through results. However, their contribution towards improvement of the general level of technical efficiency of the member firms was often considerable. One inherent difficulty was that such industrial property as resulted belonged to the member firms as a whole and there was therefore little incentive for individual enterprises to exploit it. Perhaps the most important contribution of the research associations has been through their information and industrial liaison services, which enabled relevant advances in technology throughout the world to be channelled to the individual members, increased the general level of technical competence, and encouraged small firms to employ technical staff even if they were too small to undertake research in the real sense.

Similar schemes were evolved in other countries. For example the T.N.O. Organization in the Netherlands (1932), which is concerned with applied research over a broad field including industry and defence, operates a series of important research institutes for various industrial sectors. In many ways these are similar in function to the British industrial research associations, in that they are concerned with raising the general technological levels of the industries concerned, the results of their work being for general exploitation. In addition, however, they undertake sponsored research for individual firms, under conditions of confidentiality. This gives them insights into the real problems of industry and thus helps to give their general work a more realistic flavour.

Sponsored research institutions, often as in the case of the Battelle Memorial Institute (1925) or the Stanford Research Institute (1946), of a not-for-profit nature have developed as an important part of the American research scene. Not only do they serve the needs of individual firms, but a considerable proportion of their income comes from government contracts. The United States Government has, in fact, found it useful to create a number of institutions which are outside bureaucratic restraint and, in particular, can 'hire and fire' so as to attract a mobile and highly-skilled scientific force. Some of these are loosely attached to institutions of higher education and are situated on or near university campuses. Others such as the RAND Corporation (set up in 1946 as part of the Douglas Aircraft Company), are completely independent, although relying entirely on government contracts. There is a wide range of new institutional experience here, of which little use has yet been made by the European countries. Japan, on the other hand, has emulated American experience by creating a series of 'think-tanks' to undertake research for firms or for the government. Some of these have been stimulated and supported by government agencies such as the Economic Planning Agency or M.I.T.I.; others have arisen from the enterprise of large firms.

Large technological programmes. Direct initiative and involvement of governments in large technological programmes has, in recent decades, been a major feature in defence, space, nuclear energy, and related fields. The organization and successful completion of the wartime Manhattan Project in the United States was a classical example, as has been more recently the landing of men on the Moon. Corresponding crash programmes undertaken primarily for economic objectives are much rarer, but it can be argued that military technological development by government has given a major impetus to improvements of technological and managerial levels in a wide number of industrial sectors; for example, in nuclear energy, radar, solid-state electronics, and satellite communications. Large technological projects with economic objectives have, however, been common in the civil aircraft industry, generally through cooperation between government and individual firms and even, in the case of *Concorde*, between governments. It is argued that such cases may become more common in the future as the next generation of equipment in some sophisticated areas such as supersonic aircraft and computers becomes so costly as to be beyond the financial capacity and risk-taking ability of even large corporations.

There is much criticism of such approaches, however, mainly in relation to the wisdom of choice of particular costly projects or the possible misallocation or distortion of resource allocations which results. Such costly projects will be necessary from time to time, but their success is likely to depend on the development of new forms of government–industry cooperation and improved institutional and managerial policies which will keep to a minimum the degree of bureaucracy which seems to be inherent in large-scale enterprises, especially in the public domain.

IV. THE SOCIAL EFFECTS OF TECHNOLOGY

In the early stages of industrialization, technological development is the main instrument for bringing societies up from the subsistence level and increasing the general quality of life. Experience shows, however, that in the past too little attention has been paid to the unwanted side-effects of technological development. The slums and dirt, the bad working conditions which appeared in the nineteenth century as a consequence of the industrial revolution, persisted, despite public protest, into the twentieth. As the extent of industrialization increased, problems of air and water pollution have gradually appeared as factors decreasing the quality of life. Many measures of control were instituted in industrialized countries; for example, the work of the alkali inspectors in Great Britain, since the Alkali Act of 1863, led to considerable improvements which had to be paid for by industry.

As the present century progressed, many new and insidious forms of pollution appeared, of a different nature from the fogs and dirty rivers of the earlier phase of industrialization: radioactive wastes; chemicals such as DDT which are spread universally and are not biologically degraded; pollution of the upper atmosphere, which may be a threat to biological life as a whole through destruction of the ozone layer; or thermal pollution from nuclear and other power stations. Some of these, if unchecked, might produce unwanted and uncontrollable climatic changes.

But the social effects of technology go much further. Mass-production methods, introduced around the turn of the century, are felt by many to reduce satisfaction in work. Urban problems of direct technological origin have changed, not always for the better, the living conditions of vast numbers of people.

These problems have become internationalized; the pollution of rivers, the oceans, and the air—a growing legacy from the past—is not contained within national frontiers, and their influence can threaten the whole of the biosphere. Increasingly there is a demand, particularly in the richer and most highly industrialized countries, that technology in the future must be socially acceptable.

In the simpler cases of air and water pollution national regulatory measures are ever increasing, for example through clean air and water acts, by specifying the allowable impurities in automobile exhaust gases, or the permissible emission of sulphur from factory chimneys. It is important that such measures be harmonized internationally; otherwise countries which are most meticulous in preserving environmental standards will penalize their industries in international competition. It would seem that industrial pollution of the hitherto free goods, air and water, can be controlled, but at a relatively high cost. The management of technology to meet the real needs of society in more subtle ways is a much more difficult business, which is highly political and involves the changing nature of human values.

V. TECHNOLOGY AND THE DEVELOPING COUNTRIES

Science and technology are virtually the monopoly of the rich, industrialized countries of North America, East and West Europe, and Japan; collectively, these are responsible for over 90 per cent of the world's research and development and possibly as much as 95 per cent.

Since the last war, and as decolonization proceeded, attention has rightly been given to improving the lot of the vast majority of the world's population living in conditions of subsistence, disease, and often hunger, in the so-called developing countries. Technology was obviously one of the main agents for this improvement, as it had been in those countries already industrialized and classified as 'developed', that is, with a *per capita* income of more than about $400 p.a. There have therefore been considerable efforts to transfer existing technology to the developing countries. This has not proved easy and there are many signs that, despite great efforts of international aid, the gap between rich and poor countries is actually increasing, partly because of the relatively higher rates of population increase that result from better health and nutrition which are inherent in the development process.

The transfer of technology to the developing countries has mainly been through the agency of the multinational corporations. While much of it, particularly in establishing basic industries such as steel production, oil refineries, etc., is of basic utility, there has been a tendency to export to relatively primitive societies goods and production methods which have been developed to meet the needs of sophisticated societies which already possess sound industrial structures. One result of this has been that much of the imported industry serves a small élite fraction of the population with appetites and aspirations much like those that are widely diffused in industrialized societies. Thus the new technology hardly toches the vast mass of the population, which in agriculture and handicrafts still uses traditional and age-old methods.

In many of the developing countries with large and rapidly increasing populations, unemployment and underemployment are rife and capital is scarce. The new technologies are in the main capital-intensive and not labour-intesive. There are, therefore, three separate needs for technological development in such countries:

First, a proportion of modern capital-intensive industry capable of high investment yield and export potential, and especially for the building up of basic industrial services and skills for the future.

Secondly, the development of appropriate labour-intensive technologies suited to local labour markets and conforming to local social and cultural conditions.

Thirdly, the application of the scientific and technological method to improve traditional and familiar tools, machinery, and practices.

The multinational corporations, in spite of local suspicions as to their motivations and suspicion of interference in domestic policies, are likely to remain the main agents of technological change in the first of these three categories. For the others, new approaches are necessary.

The place of governments in the transfer of technology is important but still ill-defined. On the donor side, it is they who must provide the resources. They therefore have great concern with the direction and efficiency of the aid they give, which is often tied to political or longer-term commercial objectives. On the receiver side, governments are, of course, responsible for economic planning and hence in the selection of the technologies most relevant to their economic and social objectives. Since aid may come from many sources, such planning is particularly necessary, otherwise there can be, and often is, duplication, unbalanced resources, and general inefficiency in the use of the new resources.

VI. CREATION OF A CLIMATE FAVOURABLE TO TECHNOLOGICAL DEVELOPMENT

In summary, then, government, while not responsible for the operational phase of technological innovation has had a vital and diversified, although often indirect, function in its achievement. This has been partly through ensuring suitable structures of education and research and partly through establishing suitable economic, social, and fiscal conditions. It must be stressed again, however, that many of the policy measures involved appear at first sight to be unconnected with technological development as such; they are important, however, in providing incentives for, or barriers against, successful innovation.

With regard to research and development, it is probably true that where these are insufficient and create a bottleneck, governmental action is essential; otherwise, it is not. To create effective and viable R and D policies, governments have to be fully aware of how the research and the innovative systems operate and to adjust quickly to changing circumstances. Thus the balance between national efforts in fundamental and in applied research is important. Before the Second World War Europe excelled in fundamental research, but was somewhat slow in applying it. The United States, on the contrary, relied greatly for its development on fundamental discoveries made elsewhere, but excelled in its effective application; its pilot plant and prototype development was much greater than in the countries originating the basic ideas. At that period, it was calculated that, in Britain, for every fundamental research worker there existed 1·1 applied research worker or development engineer; in the U.S.A., on the other hand, there were 2·5 such people for every basic research worker.

Another aspect of historic importance is the changing attitude towards technological development. In the highly science-based industries, successful innovation depends on the existence of a sophisticated market for new devices and chemicals. Most European countries feel themselves too small to provide such markets and have regarded penetration of the American market as essential for the success of new technological innovations. Among the most technologically successful European countries, Sweden and Switzerland regard their domestic markets as pilot areas for the working out of new developments to be exported later on a world scale; others, less successful, take a more insular and restricted view: home consumption has been regarded as the primary object, followed by export promotion.

Yet another factor in successful national technology is the degree of personal mobility of scientists and engineers. While this is largely a matter for individual firms, governments can influence it greatly, for example in making it easy for research workers to move in and out of government laboratories and in stimulating industry–university and government–industry relations.

As sizeable customers for the products of many industries, governments have had an important influence on the incentives to innovation and high product quality by their procurement policies, that is, not only on technology itself but on market demand. This is the more important, the more scientifically sophisticated the products demanded. It applies particularly to defence, space, and nuclear needs, arising at the end of the period under review, but also to high-speed transportation, computers, public works, education, new energy needs, and so on. Here it has been essential for governments to develop policies which will favour innovation and high quality. Forward requirements should be precisely defined in the light of knowledge of the existence of new technical possibilities, and there should be encouragement of experimental and demonstration projects.

By these means, through intelligent use of standards, regulations, codes of good practices, through the taxation system, and by a wide variety of other measures, governments can do much to influence the technological success of their countries.

REFERENCES

[1] FREEMAN, C., and YOUNG, A. *The Research and development effort of western Europe, North America and the Soviet Union*. O.E.C.D·, Paris (1965).

[2] *The conditions for success in technological innovation*. O.E.C.D., Paris (1971).

[3] DENISON, E. F. *The sources of economic growth in the United States and the alternatives*. Report to the U.S. Committee for Economic Development, New York (1962).

[4] GABOR, D. *Innovations: scientific, technological, and social*. Oxford University Press (1970).

[5] BUSH, V. *Science the endless frontier. A report to the President on a program for post-war scientific research*. United States Government Printing Office, Washington (1945); reissued by the National Science Foundation (1960).

7

EDUCATION IN INDUSTRIALIZED SOCIETIES

DAVID LAYTON

WHEN the twentieth century opened, Britain and Germany were the leading industrial nations of the world. Despite the discomfiture of the South African War, Britain was still supremely confident, supported by the markets of her Empire, and enjoying power and riches beyond those of any other country. By 1950, after two world wars, a new distribution of industrial and military power had been achieved, with the U.S.A. and U.S.S.R. dominant, and newly modernized nations, such as Japan, ascendant.

This transformation, the rate and scale of which were remarkable in world history, was due in substantial measure to the impact on society of scientific knowledge and technological developments. In turn, the advancement of science and the exploitation of technology had their roots in educational processes, although to assert this is not to imply a direct causal relationship between educational investment and industrial progress.

The extent to which an educational system should service the economy is by no means a matter on which all nations have shared a common view, nor indeed on which particular nations have maintained a constant view over time. Other social goals, derived, for example, from populist or élitist views of the nature of society, have influenced the course of educational developments. Furthermore, in addition to its progressive functions, intended to achieve social change, education has had a conservative role also; it has been an agency for the transmission of culture and for the socialization of the rising generation into the prevailing skills and values of an age. Religious and political considerations, no less than economic ones, have determined the course of educational events. Beyond this, however, and irrespective of social functions, there emerged the conception of education as an individual human right; a certain level of provision was deemed necessary so that all, irrespective of caste and creed, should have an equal opportunity to develop their abilities. The educational picture then is one of diverse and, at times, competing demands on the system, with the twin concerns of industrial efficiency and social justice strongly delineated in educational developments in the period under consideration in this chapter.

I. EDUCATION AND INDUSTRY AT THE TURN OF THE CENTURY

By 1900 the relative timings of industrialization and of educational advance had resulted in substantial national differences in the provision of technical education, particularly in its more advanced aspects. In contrast to the continent of Europe and the United States of America, the first industrial revolution was completed early in Britain. It preceded the establishment in the late nineteenth century of a system of mass elementary education and was accomplished without benefit of a sound foundation of secondary schools. When, in the second half of the century, competition from foreign manufacturers gave rise to a pressure for technical education in Britain, this movement was handicapped by the low general educational attainments of many students.

On the continent, and elsewhere, industrialization came late in the nineteenth century after, or alongside, the establishment of systems of general education which were more highly developed than that in Britain. In such situations technical education was able to build on a period of prolonged general education and hence to assume a more advanced character. When the twentieth century opened Britain could show little to compare with the Berlin Royal Technical High School at Charlottenburg (1884); the Zürich Polytechnikum (1855); the Massachusetts Institute of Technology, Boston (1859); and the Faculty of Applied Science in the University of Liège (1893).

Furthermore, national conceptions of technical education differed, particularly in relation to the inclusion of manufacturing practice. Under the Technical Instruction Act of 1889, technical education in England and Wales had been defined as 'instruction in the principles of science and art applicable to industry'; it was not to include 'teaching the practice of any trade or industry or employment'. Antipathy to early vocational specialization and fear about the loss of trade secrets may have been factors which influenced the adoption of this definition; however, a further and powerful determinant was the *laissez-faire* philosophy which prevailed. Grant-aided instruction in the practice of a particular industry could be seen as equivalent to a direct subsidy, an unacceptable departure from the principle of 'a fair field and no favour', supported by most British businessmen of the time. In consequence, technical education in England became dichotomized, responsibility for training in workshop methods residing with the industrial employer, while the complementary theoretical instruction was given in educational institutions.

A different conception of technical education was held elsewhere, for example in France where trade schools, such as the Ecole Diderot in Paris (1873), had been established as *écoles manuelles d'apprentissage*; in these, and in other lower grades of continental technical school, craft training and practice in workshop methods formed an intrinsic part of the instruction. The interpenetration of industry and scientific and technological education was also noted by the Mosley Education Commission in its report on the state of American education as observed during a visit in the autumn of 1903. In the schools, the Commissioners were struck forcibly by the important part which manual training was beginning to play, its inclusion justified as much on vocational grounds as by educational claims about the development of faculties. At the advanced levels, the close and harmonious relations between manufacturers and academics was noted by W. E. Ayrton, Professor of Physics in the City and Guilds of London Central Institute. In connection with the education of electrical engineers, he reported, 'Everywhere I am told: An engineering apprentice in a factory should be a college-trained man; an engineering professor in a college should be actively engaged in the practice of his profession.'

A related point is concerned with the method by which different countries had built up their systems of higher technological education. In the lower and intermediate grades of instruction, such as the various classes organized under the auspices of the Science and Art Department, England compared not unfavourably with her industrial competitors; it was at the advanced level that she lagged most noticeably behind her European and American rivals when the new century opened.

In Germany, for example, active state intervention had led to the establishment of technical high schools for the training of future engineers, managers, and manufacturers. By the end of the century these institutions ranked with universities, possessing self-government and other privileges, including the power to confer degrees. Apart from industrial chemists, the future captains of German industry were educated largely outside the universities in these polytechnic institutions, the *Technische Hochschulen*. At the time when the demand for engineers and technologists was first manifesting itself, the disposition of German universities had been peculiarly inimical to technology; the tradition of *Wissenschaft* and the doctrine of *Lernfreiheit* had institutionalized a resistance to the practical and the empiric. As a result the teaching of applied science became diverted from the existing university system into a separate and independent stream.

In Englnd, the situation was different. The demand for technological education asserted itself later in the nineteenth century, after the cold shock of the International Exhibition in Paris in 1867 (Vol. V, p. 789). It coincided with a rising pressure for new institutions of higher education to cater for the middle classes, institutions which would not only provide a liberal education for those excluded from the older universities on grounds of wealth, religion, or class, but ones which would also offer a modern curriculum appropriate to the times. Thus, the inaugural prospectus of the Yorkshire College of Science (1874), later to become the University of Leeds (1904), emphasized that the prime object was 'to supply an urgent and recognized want, viz., instruction in those sciences which are applicable to the Industrial Arts, especially those which may be classed under Engineering, Manufacturing, Agriculture, Mining and Metallurgy'. The new institution was designed 'for the use of persons who will afterwards be engaged in those trades as foremen, managers or employers'. In this way, technology became assimilated into the stream of civic university education in England. Simultaneously there were stirrings in one, at least, of the ancient universities. At Oxford physical science languished until the departure of R. B. Clifton (1915) and W. Odling (1912) and no engineering professorship was established there until 1907. At Cambridge, however, alongside the development of the Cavendish Laboratory, a chair of Mechanism and Applied Mechanics was created in 1875; and before the end of the century a Mechanical Sciences Tripos had been introduced.

In the United States, variety was the outstanding feature of institutions offering advanced scientific and technical education. Specialized colleges of technology were to be found, such as the Massachusetts Institute at Boston; elsewhere, powerful schools of applied science had grown up within universities, for example the science and technological departments at Columbia University. Variants of both the German and English patterns existed. By the early twentieth century, gross differences of a quantitative kind were also to be observed. As a Board of Education committee reported in 1906, 'Throughout the States we find that the workshops and laboratories, particularly in the Departments of Engineering, are supplied with apparatus and machinery for the use of students on a scale quite unknown in this country; the provision made for research students is specially complete.'

Germanic in being state-supported and controlled, yet incorporating strong technological faculties on the English and American model, the Japanese Imperial universities, four in number by 1911, were a vital source of administrators and technical experts for the national programme of rapid modernization. Supplementing them were private institutions such as the Meiji Technical School (1909) which aimed to produce graduates who combined the skills of the engineer and the soldier. Despite an interval of almost three centuries between the scientific revolution in Europe and the importation of Western science into Japan, the time-lapse between the institutionalization of science in the West and in Japan was only half a century. Stripped of its philosophical and cultural aspects, Western technique was allied to Eastern ethics and an intense patriotism in the service of industrial and military strength.

For their full development, advanced scientific and technical education depended on an adequate foundation at the secondary level. A further preliminary consideration, therefore, is the conception of secondary education which had emerged by the early twentieth century, and in particular, the extent to which vocational and technical considerations were incorporated in this.

In 1900 the situation in England with regard to secondary education was chronically unsatisfactory. Little had been done to remedy what the Royal Commission on Technical Instruction in 1884 had regarded as 'the greatest defect of our educational system', the lack of good modern secondary schools. In 1895 the Bryce Commission had reported on the need to rationalize the administration of the various types of secondary education which existed, including that in pupil-teacher centres and in 'higher tops' pushed up from elementary schools aided by the Education Department, that given in Organized Science Schools earning grants from the Science and Art Department, and that offered in local endowed grammar schools under the Charity Commissioners. (The 'public' independent schools, with their strong university and professional connections, were to remain as a separate, private sector of secondary education throughout the period under study.) Under the provision of the 1899 Education Act, authority had been unified at the centre through the creation of a new Board of Education responsible for the supervision of primary, secondary, and technical education alike. The question of local control was resolved three years later when the 1902 Act municipalized education in England and Wales; the existing School Boards were then replaced by the county and county borough councils which, since the Technical Instruction Act of 1889, had been exploiting their new powers and resources to foster technical studies.

Faced with the problem of unifying a singularly disorganized provision for secondary education, including, on the one hand, grammar schools with a classical curriculum and, on the other, Organized Science Schools with a strong scientific and technical bias, the Bryce Commission inclined to a view which incorporated technical education as one species of the genus 'secondary'. 'No definition of technical instruction is possible that does not bring it under the head of Secondary Education,' the Commissioners reported, 'nor can Secondary Education be so defined as to exclude from it the idea of technical instruction'.

This view was rejected by the Board of Education under the influence of its Permanent Secretary, R. L. Morant. The Regulations for Secondary Schools, issued in 1904, embodied a conception of general education which largely ignored the experience of practical and vocationally orientated studies which had been built up in higher grade schools and elsewhere; instead a predominantly academic curriculum was imposed on the new state system of secondary schools. The subsequent re-emergence of a technical stream is considered later; at this point it is sufficient to note that secondary education in England at the start of the century, selective rather than open, and offering a narrow ladder to a privileged few rather than a broad highway to all, was exclusive of technical education and, indeed, remained so until after the Education Act of 1944. In fairness to Morant and those responsible for the 1904 Regulations, it must be added that their prescription for the curriculum of secondary schools commanded wide support, not least from those who were pressing most strongly the educational claims of science. 'We are promised exactly that for which men of science have frequently and consistently pleaded,' the scientific journal *Nature* told its readers, in a comment on the new regulations.

Given the historical association of technical instruction in Britain with the educational, rather than the industrial, system—the Department of Science and Art was transferred from the Board of Trade to the Committee of Council on Education as far back as 1856—one result of the reorganization of secondary education in the early years of the century was the diversion of resources from technical education to the development of secondary schools. The Board of Education regarded the secondary school as the appropriate source of supply of better-educated and, it was hoped, more socially representative teachers for elementary schools. To meet the demand for these teachers, local authorities were obliged to invest most of their available resources for post-elementary education in the provision of secondary schools. The expansion which followed is indicated by the figures in Table 7.1: at the same time it is interesting to note the increase in the average size of schools during the war years, a reflection of a growing appreciation of the value of education with a greater ability to pay fees, the result of 'War wages'.

TABLE 7.1

Expansion of secondary education in England and Wales, 1904–20

	Grant-aided secondary schools	Average size of secondary school (Number of pupils)	Total number of pupils
1904–5	575	165	94 698
1914–15	1047	190	198 884
1919–20	1141	270	307 862

Perhaps more than any other single factor, their role as supplier of future elementary school teachers influenced the municipal secondary schools to adopt a predominantly academic, unspecialized, and non-vocational curriculum and accounted for the loss of their scientific, technical, and manual bias by Organized Science Schools as these were assimilated into the new secondary arrangements. For the elementary school teacher a sound training in literary subjects and the humanities was regarded as more important than any systematic study of applied science and technology. Conflict over the educational value of scientific and literary studies can be traced back much further than Morant, but the decisions of 1904 most certainly sustained the view that the technician and technologist were of a lower social status than the man of letters. Reviewing developments over the past half-century, the Ministry of Education Report for 1950, in a masterly understatement, recorded, 'The Education Act of 1902 . . . did not lead to any great immediate increase in the material facilities available for technical education.'

Turning from Britain to the Continent, we find technical education more closely associated with the industrial than the educational structure. Secondary schools in both France and Germany had developed independently of considerations of practical utility. Yet the twentieth century opened with many continental manifestations of what the French termed 'the crisis in secondary education'. A Parliamentary Commission on Secondary Education which reported in 1899 expressed the view that what France needed was not more savants, but more practical men and engineers. The traditional curriculum of the *lycées* was too exclusively literary and intellectual, the *classe de philosophie* being its crowning glory. Reform in the following years attempted to provide *un bain de réalisme* by the reorganization of a modern side based chiefly, though not exclusively, on physical science.

By 1900 diversity was already a feature of German secondary education which included, in addition to the classical *Gymnasium*, two established alternatives, the *Realgymnasium*, where Latin was still taught, and the Latin-less *Oberrealschule*, with a curriculum which emphasized modern languages, mathematics, science, and drawing. Neither imparted what could be called, in a narrow sense of the term, technical education, although the *Oberrealschule* was deemed to 'draw the subject matter of instruction very largely from those spheres of knowledge which are nearest to the pupil's present experiences and to his probable career'. Both were intended to counter what the Kaiser described in 1900 as 'an excessive over-production of highly educated people'; in the same year a royal decree on secondary education, illustrative both of Teutonic efficiency and centralized control, established the principle that the three types of secondary schools were 'to be considered as of equal value from the point of view of general intellectual culture' and in relation to progression to higher education.

Looking outwards from his native Prussia some years earlier, Prince Bismark had warned of the need to 'keep an eye on the United States of America, for they may develop into a danger to Europe in economic affairs . . .' By the early years of the twentieth century, American industrial achievements and educational attitudes had become a significant influence on European education, directing it towards both utilitarian and democratic considerations. In the United States, as elsewhere, there was a deep preoccupation with the proper functions of secondary education in a modern industrialized society. A significant step was taken when the 'Committee of Ten', appointed by the National Educational Association, reported in 1893, propounding a view of secondary education as the natural extension of primary education, and discarding the idea of the secondary school as the antechamber to university or higher studies. In contrast to Britain and most European nations, secondary education in the United States was free; the high school was therefore an institution for all adolescents, for the majority of whom it represented an educational terminus, rather than a staging post. Inevitably its curriculum was subjected to vocational pressures, which, in general, were accommodated by the enlargement of existing institutions rather than the creation of new ones. The idea of continental-type trade schools, associated with early career choice and restricted opportunity for advancement across the strata of society, did not consort happily with the prevailing democratic spirit in

American education. As the demand for technical studies grew, the tendency was to assimilate these vocational courses into a 'polytechnic' high school curriculum.

Bismark might well have warned of economic threats from the East, additional to those from the West. Since the Meiji restoration in 1868, Japan had been strengthening her scientific and technological resources, initially by drawing upon foreign advisers and teachers, and then by the development of educational institutions committed to the goal of 'a rich country and a strong army'. To supply the large number of technicians and 'non-commissioned officers' needed for rapid industrial expansion vocational and technical schools were established at the secondary level, alongside academic secondary schools. By 1903, some 200 secondary technical schools existed, compared with 340 academic secondary schools, providing a distinct vocational channel through the educational system.

The final point in this examination of the relation of education and industry at the start of the twentieth century concerns industrial manpower and the extent to which the educational system was supplying this. Lord Ashby has pointed out that 'in the rise of British industry the English Universities played no part whatever'; the Maudsleys, Arkwrights, Cromptons, Darbys, and Bessemers had no systematic education in science and technology. This was in sharp contrast to the development of German industry; by the close of the nineteenth century, in the *Badische Analin-und-Soda Fabrik* alone 100 scientifically trained chemists and 30 engineers were employed. In all, at this stage, Germany possessed some 4000 graduate chemists, and of this stock 1000 were in the chemical industry. Contemporaneously, Britain could muster about 2400 graduate scientists in all specialisms and employed in all fields, including universities, technical colleges, various branches of the scientific civil service, schools, and industry. Of this total only about one-tenth was engaged in industrial pursuits. The majority of science graduates entered the school-teaching profession at one level or another and this situation remained unchanged up to the First World War.

In the United States the hospitable attitude of industry to graduate scientists and technologists has already been commented on. No better testimony is available to the high regard in which American industry held the trained graduate than the size of its donation to education of equipment and money, the latter alone amounting to £23m. between 1890 and 1901. If further evidence is required, we might look to the founding of major industrial research laboratories; Edison's centre at Menlo Park, dating from 1876, was an early forerunner of major establishments such as the General Electric Research Laboratory (1900), Du Pont's Eastern Laboratory (1902), and the Eastman Kodak's Laboratory (1912).

Lacking state financial support, the English civic universities which had appeared in the second half of the nineteenth century, to be chartered in the first decade of the twentieth—Birmingham (1900), Manchester (1903), Liverpool (1903), Leeds (1904), Sheffield (1905), and Bristol (1909)—made at one time or another very clear statements of their intention, in Haldane's phrase, 'to minister to the wants of the manufacturers'. At that time few British firms had research departments, industrial research associations had not been established, and there were no state centres for industrial research other than the National Physical Laboratory, founded in 1900 in imitation of the *Physikalisch-Technische Reichsanstalt* at Charlottenburg. In consequence, the civic universities were able to fill an important role in relation to industrial research and innovation in fields such as metallurgy, fuel technology, textile chemistry, soap-making, brewing, naval architecture, and the various branches of engineering. Indeed, it has been suggested that, proportionately, they were more important in this regard than possibly ever again.

At the same time, there is considerable evidence that, so far as England was concerned, it was not industrial demand that led to the production of professional scientists and technologists. In 1902 Sir William Ramsay told the London County Council Technical Education Board's Subcommittee, inquiring into the applications of science to industry, that the demand for the excellent chemists which the universities were turning out was not keeping up with the supply. His opinion was that 'Manufacturers are not as yet sufficiently alive to the necessity for employing chemists.' The same story was told of engineers. The newly created Imperial College (1907), England's answer to the Technical High School at Charlottenburg, aimed to produce more than 200 engineers a year; analysis of the job opportunities suggested this would lead to gross over-production. University Appointments Boards, a development of the early years of the century, frequently reported on the difficulty which chemistry and engineering graduates were having in finding industrial employment.

II. THE YEARS BEFORE THE FIRST WORLD WAR

The presidential address to the British Association for the Advancement of Science in 1903, delivered by Sir Norman Lockyer, one of the more prescient scientists of his day, was entitled 'The influence of brain-power on history'. Lockyer's general conclusion was much in line with that of the Mosley Commission which reported the following year on American education. The characteristic British qualities of honesty, doggedness, and pluck, though valuable in themselves, were useless for the changed world of the twentieth century unless accompanied by the ability to apply practical, up-to-date scientific knowledge. Trained intelligence held the key to future industrial competitiveness and industrial efficiency.

In this connection, Lockyer drew attention to the lack of efficient universities in Britain and the national neglect of scientific research. Compared with 134 state and privately-endowed universities in the United States, and with 22 state-endowed universities in Germany, Britain, at the time of Lockyer's address, could muster only 13—six in England, one in Wales, four in Scotland, with two additional universities in Ireland. Furthermore financial support from the State was meagre. In 1903, the grant to English universities and colleges amounted to £27000; for the same year, the state grant for the upkeep of a single German university, Berlin, was about £130000, while the income of the universities of the state of New York, excluding benefactions, was £981300 in 1904–5.

As a result of Lockyer's address, a deputation was organized by the British Association to meet the Prime Minister in July 1904; representatives of the universities old and new, of industry, pure and applied sciences, and the humanities were united in urging increased state support for universities. The outcome was a formal recognition of the universities' contribution to the industrial struggle and the allocation of increased state aid. In 1900 the civil estimates had incorporated a grant of £85000 to the universities and colleges

TABLE 7.2

Numbers of students in the U.S.A., Germany, and the U.K.:
figures for 1913–14

	Students in universities and technical institutions	Population (millions)	Students per 10000 population
Scotland	8000	4·8	17
Germany	90000	65	14
United States	100000	100	10
Ireland	3000	4·4	7
Wales	1200	2	6
England	17000	34	5

TABLE 7.3

Full-time students of science and technology in the U.S.A., Germany, and Great Britain: figures for 1913–14

	Full-time students of science and technology	Population (millions)	Science and technology students per 10000 population
U.S.A.	40000	100	4
Germany	17000	65	2·6
Great Britain	6456	40·8	1·6

of Great Britain; by 1912–13 this had risen to £287000, including £150000 for universities and university colleges in England. Even so, the inferior financial support for state-aided universities in England and Wales, compared with the situation in Germany and the U.S.A., was to be a recurring subject of complaint for many years. The 'league table' of number of students per 10000 population served to emphasize further the comparative weakness of English provision before the First World War (Table 7.2). In terms of full-time students of science and technology Britain's showing was even worse (Table 7.3).

The annual number of students graduating with first or second class honours in science and technology (including mathematics) before the war was little more than 500. Of the annual entry of men students to universities and university colleges in England and Wales, nearly half came from the independent 'public' schools, from which institutions 25 to 30 per cent of those leaving over the age of 16 proceeded to universities. In contrast, of those leaving the state-aided secondary schools at 16 years of age or above, little more than 10 per cent went on to universities. It was generally recognized that any increased supply of university science students would have to be sought in this latter category of schools. As a member of the 1904 deputation expressed it, 'hitherto a university training has been the luxury of the comparatively well-to-do. The aim is to open the door to the choicer spirits of the poorer classes.'

The introduction of the 'free place system' in England in 1907, whereby grant-aided secondary schools had to reserve 25 per cent of their annual admissions for non-paying pupils from elementary schools, was one step in this direction. At the other end of the scale, local education authorities were empowered to make awards to promising pupils in their secondary schools to assist their progress to universities. By 1911, 1400 students were maintained in this way at universities and university colleges. In the same year the Board of Education introduced grants for four-year courses, incorporating a degree and professional training for men and women entering the teaching profession. Of the subsequent development of civic universities and university colleges, it has been said that they 'owed the very existence of their arts faculties, and in many cases their pure science faculties to the presence of a large body of intending teachers whose attendance at degree courses was almost guaranteed by the state.'

With the expansion of secondary education in Britain, universities and university colleges were able to shed some of their elementary work and move towards a curriculum which was predominantly concerned with degree-level work. At the same time, institutions which had begun life with a strong scientific and technological bias experienced a substantial growth in their arts faculties, a development attributable in large measure to the demands of secondary schools for teachers. Within all faculties, but especially in the fields of science and technology, the rapid expansion of knowledge and increased specialization led to the creation of new chairs and departments. Typical examples were Biochemistry at Liverpool (1902); Brewing at Birmingham (1899); Coal Gas and Fuel Industries at Leeds (1906); and Naval Architecture at Newcastle (1907).

The 'more even and undisturbed development of new fields and the much smoother working relationship between academic research and its applications' which was achieved in British and American universities during this period has been attributed in large measure to their departmental structure. In contrast, the German system of chairs, and its associated professorial oligarchy, exerted a conservative influence which tended to work against the expansion of studies and the establishment of new subjects. Despite the frequently voiced fears of German industrial superiority and continued acknowledgement of the quality of German scholarship, by the end of the first decade of the new century there was no gainsaying the pre-eminence of developments elsewhere. Among the pure sciences, physics at Cambridge, and among the applied sciences, metallurgy at Sheffield and Birmingham, and colour chemistry and dyeing at Leeds and Manchester, had wrested the academic leadership in their particular fields. In the United States, the Ryerson Physical Laboratory had been established at Chicago with A. A. Michelson, America's first Nobel Prize Winner, at its head. Furthermore, the structural development in American universities of powerful graduate schools in the arts and sciences made possible the effective training of research workers, for whom the degree of Ph.D. had become the symbol of professional competence. Lacking a central authority for education, and any mechanism for establishing a national policy for universities, the United States provided an environment where competition stimulated innovation. Parallel to the emergence of the graduate schools, there developed post-graduate professional schools, such as medicine, engineering, agriculture, and education, undertaking inquiries into 'field-generated' problems and providing a research basis for practice and training. As Joseph Ben-David has shown, by the first decade of the century there had emerged the conception of the professionally qualified research worker.

If America was successful in meeting the high-level manpower requirements of industry by virtue of developments in its universities, the situation was less favourable lower down. The opening decade of the twentieth century was one of remarkable industrial growth in the United States: from both the factory and the farm there came increased demands for skilled workers. The several million immigrants who entered the country in this period were largely drawn from Italy, Russia, and eastern Europe; in contrast to the previous influx from northern Europe, few possessed the skills which the developing economy required.

In 1906 a National Society for the Promotion of Industrial Education was brought into being as a pressure group whose object was the unification of the various national forces of vocational education. Over the next few years, numerous state programmes of vocational education were introduced, accompanied by criticisms of the existing school systems as too exclusively literary and scholastic. Eventually the National Society adopted a policy of Federal aid for vocational education and its campaign bore fruit when, in 1914, President Wilson signed a bill creating a Commission on National Aid to Vocational Education. The Commission's report, the Magna Carta of vocational education in the United States, led to the authorization, under the Smiths–Hughes Act of 1917, of an initial appropriation of $7m. for the promotion of vocational education, below college level, in agriculture, trade and industry, home economics, and commerce. Most of the full-time study of vocational subjects which followed was undertaken in the high schools.

Vocational pressures were exerting themselves in Britain also at this time. It will be recalled that the 1904 secondary school regulations had prescribed a general education up to the age of 16 including 'instruction in the English

Language and Literature, at least one Language other than English, Geography, History, Mathematics, Science and Drawing'. The new regulations of 1907 opened the way to some limited departures from an academic bias, particularly in the last year or so of a pupil's school life. The inclusion of 'free-placers' was a factor which contributed to the provision of courses for pupils who intended to enter industry or commerce at 16. Such developments were restricted, however. Simultaneously, from the tops of elementary schools there were vocational and technical efflorescences. In 1905 the London County Council established a number of Central Schools intended to give their pupils 'a definite bias towards some kind of industrial work'; by 1912 there were 31 such schools in London, and other authorities, such as Manchester, had followed suit. Junior technical schools, constituting general or specialized trade schools, also arose as institutions in which the future artisan and manual worker might usefully continue his pre-apprenticeship education after leaving the elementary school at thirteen or fourteen. Under regulations issued in 1913, the junior technical schools were recognized as a separate category, independent of secondary schools, and to be encouraged by increased grant. At the same time, they were limited in role, to the extent of not providing 'a preparation for the professions, the universities or higher full-time technical work'; their intake at 13 was from the population of elementary schools after the scholarship children had been 'creamed off'.

Both in this period, and subsequently, there were manifestations of what Lord Eustace Percy later termed a 'curious suspicion of technical education', continuing, 'The demand for a working class education which should be liberal and not vocational was pushed to a point where it practically dismissed the whole range of technical studies as a badge of social servitude.' Not the least vocal among these opponents of vocational studies was the Labour Party, which advocated the transformation of Central and Junior Technical. Schools into one part of a system of free and universal secondary education

III. THE FIRST WORLD WAR

Almost overnight, the outbreak of war in 1914 deprived Britain of a variety of essential imports; at the same time, the dependence of industry upon science and technology was demonstrated in painful fashion. Khaki dye for soldiers' uniforms, pharmaceuticals, magnetos, gunsights, and tungsten for high-speed steels were but some of the commodities which quickly became in short supply. Lamentable ignorance of scientific knowledge characterized establishment thinking: a member of the Government, in a public statement, excused a colleague for not having prevented the export of lard to Germany on the fallacious grounds that it had only recently been discovered that glycerine, a basic ingredient for the manufacture of explosives, could be obtained from lard. In the trenches, also, the lack of scientific knowledge had tragic consequences. One outraged university chemist, recruited as a chemical adviser to the army, wrote at the end of the war that 'We have undoubtedly sustained . . . the majority of our casualties (during gas attacks) from ignorance of the elements of natural science on the part of officers and men.' The most basic scientific facts, such as the effect of sunlight and breezes on the evaporation of liquids, appeared to be unknown to officers and men alike.

It was to the industrial connection that attention was most urgently directed in the early years of the war. The weight of scientific opinion from bodies such as the Royal Society, the British Science Guild, the Chemical Society, the Society of Chemical Industry, and the Institute of Chemistry was behind a demand for a thorough reappraisal of the place of science in the national life. Simultaneously, in December 1914, the universities branch of the Board of Education prepared a memorandum which presented the case for an increased supply of scientific research workers. The best estimates revealed no more than 250 teachers and 400 full-time students in the universities of England and Wales who were carrying out research which had a bearing on industry. Perhaps 50 more such researchers were to be found in London polytechnics and in one or two provincial technical colleges. Comparable figures for research workers in German universities and *Technische Hochschulen* were 673 teachers and 3046 students.

Clearly, any increase in the supply of trained research scientists entailed an expansion in the numbers studying science at secondary schools and undergraduate level. Unfortunately, the conditions of war made university developments difficult, as many staff were redeployed on military and industrial projects. The long-term general solution to the problem of increased supply could not be attempted in these circumstances. Instead the Advisory Council to the Committee of the Privy Council 'responsible for the expenditure of new moneys provided by Parliament for scientific and industrial research' (July 1915), and the Department of Scientific and Industrial Research which was subsequently created (December 1916), concentrated their efforts more on the provision of grant to support specific research projects of industrial relevance; encouragement was also given to the formation of industrial research associations. Effective progress in the support of research training by the award of post-graduate scholarships had to wait until the end of the war.

The longer-term educational problem was not forgotten, however. In 1915, members of the Association of Public Schools Science Masters, a body which had been established in 1900 and which had held annual conferences thereafter, set up a small group to look at possible ways of improving the position of science in their schools. After consultation with Sir Ray Lankester, a distinguished scientist of the day and an ex-President of the Association, a memorandum over the signatures of thirty-six leading men of science appeared in *The Times* on 2 February 1916 and in *The Times Educational Supplement* on 7 March. The central concern of this document was unashamedly the education of civic and military leaders. Once the education of those from whose ranks the higher and lower grades of public service were filled was established on the right lines, it was assumed that 'the education of the democracy . . . would follow the change in the education of the wealthier classes'. Despite the fact that the large public schools had spent over a quarter of a million pounds on laboratories and scientific equipment in recent years, old vested interests still retained their dominance. The memorandum referred to the fact that, of the 35 largest and best-known public schools, 34 had classical men as head teachers. Similarly, the entrance examinations for Oxford, Cambridge, and the military academy at Sandhurst, and the examinations for appointment to the Civil Service all were of a character which biased the curriculum of public schools against the natural sciences.

A Committee on the Neglect of Science was formally constituted on 3 May 1916 at a meeting held in the Linnean Society; 13 000 copies of the report of the proceedings were later distributed. The following day a counterblast from the defenders of classical studies appeared in *The Times*. Over the next few months the academic forces of science and of humanistic studies were progressively marshalled. On the initiative of the Royal Society a Conjoint Board of Scientific Societies had been established in 1916, representing some fifty scientific bodies. Always sensitive to the intrusion of others into fields which it regarded as legitimately its own, the Royal Society had earlier been the source of criticism of the proposals to establish an Advisory Council on scientific and industrial research. Under the Conjoint Board it now established a Watching Committee on Education, which by January 1917 had assumed the role of chief spokesman for the interests of science in the educational debate, superseding the Neglect of Science Committee which played little further part. On the arts side, a Council for Humanistic Studies had

been established in the autumn of 1916, consolidating the education committee of the British Academy and the five subject associations, Classical,
English, Historical, Geographical, and Modern Languages.

Between December 1916 and July 1918 a series of joint conferences was
held, in which representatives from both sides achieved a wide measure of
agreement on the form of a general, non-specialized, and non-vocational
curriculum, to include both science and literary subjects, for all secondary
school pupils up to the age of 16. Thereafter, those staying on might engage
in a gradual specialization. Similar conclusions were reached by the Prime
Minister's Committee appointed in August 1916, under the chairmanship of
Sir J. J. Thomson, to report on the position of natural science in the educational system of Great Britain.

The nature of the science component in the general secondary curriculum
was outlined at the same time by the Association of Public School Science
Masters in a pamphlet entitled *Science for all*, prepared as part of the Association's evidence to the Thomson Committee. It must be borne in mind that in
writing of *Science for all*, the science masters meant all in public schools.
Within this context it appeared necessary to draw a firm distinction between
the utilitarian–vocational and the liberalizing justifications for the study of
science. 'If school science can be shown to appeal strongly to the imagination
. . . it must be made an integral part of every school curriculum,' the authors
of *Science for all* argued; 'if, on the other hand, it is merely to be classed as
useful instruction, it had better remain a "special study" for certain Army
candidates, and for future chemists, engineers and doctors'. The inculcation
of scientific method, the appreciation of scientific knowledge and its value,
and the stimulation of imagination and the aesthetic sense were the primary
aims of the public school science teacher; general science, drawing upon
astronomy, geology, and biology, in addition to physics and chemistry, the
staples of the existing science curriculum, was to be science for future citizens,
rather than science for future scientists. In the event, progress of the general
science movement was slow; as the figures in Table 7.4 for entries in various
subjects in the newly established First School Certificate Examination show,
there was no immediate challenge to the dominance of chemistry and physics
(and botany in girls schools).

The establishment of the First Examination or School Certificate in 1917
was part of a plan to bring order into the uncoordinated and tangled field of
external examinations for secondary schools. In order to preserve the general
character of secondary education up to the age of 16 the Certificate was

TABLE 7.4

First School Certificate Examination: Numbers of science candidates, 1919–26

	Botany	Chemistry	General Science	Physics
1919	8017	9110	513	5089
1922	11841	15939	1133	8443
1924	18524	19962	1266	11064
1926	13627	21527	1340	13255

constructed on a group basis, a candidate being required to satisfy the examiners in each of the three main groups: (1) English subjects, (2) foreign languages, and (3) science and mathematics. No provision was made for the formal examination of vocational or practical subjects, and, although the benefits of the simplified and standardized new examination system were considerable, it also had the effect of stereotyping the curriculum on academic lines.

The Second Examination, or Higher Certificate, was intended for pupils who had continued their studies for two years beyond the First Examination. Its establishment coincided with the introduction of new regulations governing state aid for secondary schools, whereby grants were given to encourage the growth of advanced courses in science, mathematics, classics, and modern studies. Before the end of the War, the foundations of high-quality sixth-form work had been securely laid on a basis of limited specialization, involving groups of allied subjects, and appropriate provision of teaching resources (Table 7.5). With this growth of sixth-form work, particularly on the science side, came an increase in the number of those seeking to proceed to universities, as Table 7.6 shows.

TABLE 7.5

Sixth-form courses in England and Wales, 1917–25

	Science and Mathematics	Modern Studies	Classics
1917–18	82	25	20
1918–19	155	78	27
1919–20	189	118	29
1920–21	216	152	35
1921–22	230	180	37
1922–23	228	179	37
1923–24	230	188	37
1924–25	235	188	37

TABLE 7.6

Number of pupils from grant-earning schools in England and Wales proceeding to universities, 1908–25

	Boys	Girls
1908–09	695	361
1920–21	1674	1214
1924–25	1912	1330

Within the universities themselves the war years saw some steps taken to strengthen the output of graduates with research experience. At Oxford, W. H. Perkin, junior, was successful in establishing research as an essential component of the honours degree in chemistry. A more significant innovation, however, was the adoption in 1917 of the degree of Ph.D. as the recognized qualification for postgraduate work. Intended to attract overseas research students and to encourage postgraduate interchanges, the institution of the degree was strongly supported by the Foreign Office. While it failed in the post-war years to achieve the first objective, it nevertheless proved instrumental in the growth of important research schools.

The Neglect of Science Committee had been concerned with the scientific education of an élite. There remained the problem of the education of the remainder, the rank and file as opposed to the officers, for whom the age of compulsory attendance was still 12, although, permissively, authorities could raise this to 14. Partly in response to the view that since war demanded of all equal sacrifices, hence to all should be accorded equal opportunities; partly from a feeling that there was need for some control of the notions of independence which war and high wages had fostered in the juvenile population; but principally to meet the need for increased economic efficiency, the Fisher Education Act of 1918 raised the compulsory attendance age to 14 and instituted part-time day continuation classes for young people between the ages of 14 to 16, extending later to 18. This was a move to replace the conception of the juvenile as wage-earner by that of the juvenile as 'workman and citizen in training'. It was also an attempt to depart from the principle of voluntaryism which had hitherto characterized continuation and evening school work in England; this was in sharp contrast to the situation in Germany where State legislation had earlier established a system of day continuation schools (*Fortbildungsschule*). As the *Journal of Education* had remarked in 1914, 'how cautiously does Germany guard her young in the field of Continuation. In England a burglar may keep a night school, and Fagin probably had a grant in aid.' In the event, the large-scale development of day continuation schools in England fell a victim to industrial opposition and to the campaign for economies in national expenditure in the early 1920s. Local initiatives by certain firms, and by employers in collaboration with local authorities, established a limited number of day continuation schools on a voluntary basis; from these industry recruited some of its foremen and future management. However, growth was slow. Further education in England and Wales, in both its non-vocational and technical aspects, showed little tendency to depart from its tradition of part-time evening instruction.

IV. THE YEARS BETWEEN THE WARS

War between the great industrial powers resulted in human destruction on a scale which previously had been unthinkable. Within a ten-month period in 1915, in their efforts to breach the German lines, the French and British suffered combined losses of a million and a half men. In certain fields of industrial production, one out of every five French mechanics was killed or disabled. Simultaneously, demographic projections indicated that the size of the working population in France would decline in the post-war years. There was evidence of a qualitative deficiency also in the French labour force; out of 1614000 youths in the age range 13 to 18 years, less than a tenth had received an adequate technical training for specific trades.

In these circumstances, and in the interests of economic prosperity, the fostering of maximum individual efficiency had a high priority; the *loi Astier*, enacted in July 1919, inaugurated a comprehensive reorganization of technical education at all levels. To pay for the reforms, *la taxe d'apprentissage* was imposed on all employers with wage bills greater than 10000 francs per annum. Only by provision of adequate training and certification arrangements for its apprentices could a firm achieve exemption from the tax. Provincial committees, comprised of experts from education, industry, and commerce, controlled both exemptions and disbursements. By 1932, the total product of the tax had grown to 158m. francs, of which 69m. was used by industries themselves for the support of their own technical training schemes for apprentices.

No comparable national incentives to the development of technical education were offered in Britain. Denied financial support from the national exchequer, the proposals for compulsory day continuation schools failed to prosper. By 1928, only one area, Rugby, a centre of the electrical industry, retained a school in operation. Such pressures as there were for technical education came from below, originating in the aspirations of students rather than in the manpower requirements of industry. Throughout the inter-war years in Britain the supply of technical manpower was in rough balance with the demands of industry and commerce. Support by employers for technical education, in so far as it was given, was more in terms of the general benefits which serious part-time study might have on the moral character of the labour force than of specific vocational skills to be acquired. As for the students, although the prospect of unlimited vertical mobility of labour drew them to the classes, progression up the occupational ladder depended upon the efficiency and articulation of prevailing examination arrangements.

With the phasing out of the Science and Art Department 'lower' and 'advanced' grade examinations in 1911 and 1918 respectively, the responsibility for technical examinations passed to bodies like the City and Guilds of London Institute and to regional examining unions, some well established such as the Union of Lancashire and Cheshire Institutes and the Union of Educational Institutions, others newly created such as the East Midland Educational Union (1911) and the Northern Counties Technical Examinations Council (1921). The area bodies were more particularly concerned with lower-grade examinations and craft courses, their regional character enabling them to adapt provision to local circumstances. An important national post-war development, of considerable significance for advanced technical education, grew out of the collaboration of technical colleges with professional scientific and engineering institutes in the training and certification of students. In the case of the pioneering National Certificate in Mechanical Engineering (1921) syllabuses drafted by individual colleges were subject to approval by a National Joint Committee made up of representatives of the Board of Education, the Institution of Mechanical Engineers, and the colleges. Assessors appointed by the Joint Committee scrutinized the examination scripts. Similar arrangements were negotiated with the Institute of Chemistry (1921), the Institution of Electrical Engineers (1923), and the Institute of Naval Architecture (1926). National Certificate schemes in Building and Textiles were added later, while for full-time students National Diplomas were established. An important feature of the development was the strengthened link with the national professional body, which normally granted exemptions from its own institutional examinations by virtue of passes in certificate schemes.

As the figures in Table 7.7 show, the number of certificates awarded increased steadily in the inter-war years, with the major development being

TABLE 7.7

Development of National Certificate Schemes: Numbers of certificates awarded
1923–50

Subject	Certificate	1923	1931	1939	1944	1950
Mechanical	Ordinary	606	974	1833	2536	5614
Engineering (1921)	Higher	122	327	632	837	2435
Chemistry (1921)	Ordinary	57	108	196	167	615
	Higher	46	47	58	64	222
Electrical	Ordinary	—	592	1133	1035	2915
Engineering (1923)	Higher	—	279	421	388	1394

in engineering subjects. When, in the period immediately after the Second World War, Britain's atomic energy programme was inaugurated, the design and engineering team, under the leadership of Christopher Hinton (later Lord Hinton), included a number of key personnel who had risen by the National Certificate route. Hinton himself, before graduating in mechanical sciences at Cambridge, had undertaken his pre-university studies, after leaving Chippenham Secondary School at 16, in evening classes while an engineering apprentice with the Great Western Railway.

During the inter-war years approximately two-thirds of the provision for higher technological education in England was in technical colleges and related non-university institutions. Included in this fraction were courses for London external degrees, taken on a part-time basis, and yielding about 10 per cent of all degrees in technology each year. Within the universities themselves, technology was a declining faculty compared with arts and pure science. Numerically the enrolments in 1938–9 were similar to those in the immediate post-war years, but as a percentage of the total university intake technology fell in the same period from 20 per cent to little more than 11 per cent (Table 7.8).

TABLE 7.8

Enrolments by faculty in English universities

	Arts	Pure Science	Medicine	Technology	Agriculture	Total
1919–20	6148	3827	6073	4202	236	20486
1920–21	6587	4263	6788	4492	282	22412
1924–25	13407	5736	6729	2970	539	29381
1929–30	15945	5847	6051	3271	537	31651
1934–35	16941	6851	9168	3420	512	36892
1938–39	16186	5955	10160	4217	671	37189

A number of points arise. First, by comparison with countries such as Germany and the U.S.A., higher technological studies in England constituted a much smaller fraction of all university level work; with roughly three times as many students in universities (including the *Technische Hochschulen*), Germany offered six times the English provision of higher technological education. Second, the decline in technology contrasted with the rising status of pure science, especially physics, during the inter-war years. Despite the incorporation of the Institute of Physics in 1920 (and there is truth in the proposition that when a scientific discipline is in demand, particularly in fields outside education, this manifests itself in the development of a qualifying association) the prestige of physics in universities, particularly at centres such as Cambridge, Manchester, and Bristol, was high for reasons quite dissociated from considerations of application and utility. The extent of this British predilection for pure science, especially physics and mathematics, is illustrated in Table 7.9 by a comparison of the percentages of the resulting stock of qualified scientists in each of these disciplines in Great Britain and in the U.S.A. by the mid-1950s.

An important factor which worked against the more extensive incorporation of specific technologies and branches of applied science into the university curriculum was the strongly expressed view that industrial involvement would deflect universities from their true educational purpose. Not only overseas observers, such as the American educationist Abraham Flexner (who regarded courses such as Brewing at Birmingham, Gas Engineering at Leeds, and Glass Technology at Sheffield as 'neither liberal nor university quality'), but leading British scientists, as politically diverse as Ernest Rutherford and J. D. Bernal, were outspoken on the incompatibility of research freedom and industrial dependence, of liberal education and technical training. Bernal, for example, saw the more ready application of chemistry and physics to industrial needs as a consideration which had inhibited the advance of certain lines of biological research. The general point was to be made again in a different context in 1943 when Sir Ralph Fowler and Professor P. M. S. Blackett

TABLE 7.9

Proportions of mathematicians and physicists, mid-1950s

	Great Britain	U.S.A.
Mathematicians (%)	22·4	11·8
Physicists (%)	20·5	12·5

called the attention of the Royal Society to 'a danger that development in fundamental physics might be relatively neglected in comparison with applied physics' as a consequence of the successful application of physics to the purposes of war.

The high status accorded to pure science was a distinguishing feature of the compromise between different conceptions of a university education towards which English institutions were working in the twentieth century. From Oxford and Cambridge had been derived the view that it was less important to be learned than civilized; in Mark Pattison's phrase, the fruit of learning was a man not a book. From Germany had been imported the regard for *Wissenschaft*, the commitment to scholarship and disinterested inquiry. From polytechnic institutions overseas and from the newer civic universities in England had come the conception of a university as a staff college for technological experts, a source of specialized manpower for an industrialized society. By embedding technological and professional studies in a university context, it was believed they would find 'not only abundant springs of intellectual nourishment, but also the influences which will keep them expansive and wholesome'. In short, provided they were securely based on a foundation of pure science, technological studies could be both liberal and scholarly. At the same time the distinction between useful and useless knowledge which the establishment of separate technological institutions tended to foster, would become untenable.

Reality did not always match the rhetoric. Increasingly in the inter-war years there was concern over the educational consequences of specialization in science degree courses. The subject was broached at the second congress of Universities in the British Commonwealth in 1921 when A. N. Whitehead spoke on 'Science in General Education'. It invoked scholastic responses, such as the creation in 1923 of a new Department of History and Philosophy of Science at University College, London, and contributed to extra-curricular improvements, such as the increased provision of residential halls for students in the newer universities. In 1934 it drew from H. T. Tizard, in his presidential address to the educational Section of the British Association for the Advancement of Science, a proposal for a new type of degree based on a study of science and technology in their social contexts and designed for those who would become administrators and public leaders. Four years later, it moved the Consultative Committee on Secondary Education, in a report on Grammar and Technical High Schools, to comment on the inappropriately narrow range of single-subject honours degree courses in science, particularly for intending teachers. As science became fully professionalized and education in science more specialized, so was strengthened the view of the scientist as a man 'voluntarily withdrawn from human contact, disassociating himself from personal and societal problems . . . a man who was objective to an objectionable degree'. Aldous Huxley's *Brave New World*, a satire on a scientific Utopia, was published in 1932 and by the end of the 1930s many of the ingredients of the post-war debate over the two cultures had manifested themselves.

In one particular instance, that of Weimar Germany, the tendency of universities to 'train the specialist intellect and not the whole man' was a contributory factor to their future subservience to explicit political ends. From the early twentieth century there had been in Germany a vein of educational criticism which vilified analytical thought as the blight of the creative mind and took its stand on the side of instinct, feeling, and intuition. In this tradition, Oswald Spengler's *The decline of the West*, the first volume of which appeared in 1918, had found the German universities too intellectual and too remote from society to produce the élite which a resurgence of German national strength required. Contemporaneously, the philosophy of Sören Kierkegaard, and, later, its modifications by Martin Heidegger and Karl Jaspers, reinforced the trend toward irrationalism. For Kierkegaard the pursuit of empirical knowledge alienated the individual from his self, leading him into an anonymous world of abstractions. Sad professorial figures were depicted as cultivators of their academic plots, oblivious to the urgent problems of human existence. Salvation entailed a surrender to some form of absolutism, in Kierkegaard's case the tenets of Christianity, in the case of Heidegger and others the political dogma of National Socialism.

The German situation was aggravated by a number of other factors, including the restricted social class base of students enrolling in universities and *Technische Hochschulen*. The figures in Table 7.10 are for 1928. Strongholds

TABLE 7.10

Social class of students enrolling in German universities and Technische Hochschulen, *1928 (expressed as percentages)*

| | Upper class | | Middle class | | Lower class | |
	Men	Women	Men	Women	Men	Women
Universities	32·43	44·97	61·48	52·08	4·03	1·17
Technische Hochschulen	37·82	51·23	56·50	44·15	2·68	1·96

of the middle classes, and perpetuating the traditions of the well-to-do, the German institutions of higher education proved resistant to innovations which might have reconciled social change with the traditional values of a liberal education.

After the rise to power of Adolf Hitler and his accession to the Chancellorship in January 1933, the policy of coordinating German universities to the programme and ideology of National Socialism resulted in the dismissal of many non-Aryan scientists and the resignation of others. Earlier disputes within the scientific community, especially over the merits of Einstein's theory of relativity, engendered claims for a 'German physics' as opposed to a 'Jewish physics'. By virtue of their subject matter, biology and the social sciences were even more vulnerable than the physical sciences to nationalistic and racial treatment. Their lack of any humanistic dimension saved technological studies from direct ideological exploitation through manipulation of content; instead the *Technische Hochschulen* became the typical instrument of technical efficiency during the Nazi regime.

Further to the east educational developments of the greatest significance were taking place within another totalitarian society. From the earliest days of the post-Revolution period, Soviet education was cast in an unambiguously instrumental role. Its political function was the Communist regeneration of society; its economic function, increasingly important as the process of national industrialization got under way, was the training of that vast army of technicians, engineers, and other specialists which the economic basis of Socialism entailed. In two decades, state control and central planning achieved an industrial revolution comparable to that experienced in England over a century and a half.

An early and temporary response of the Soviet Government to the reorganization of higher education was the abolition of all academic ranks and degrees, such distinctions being regarded as survivals of capitalistic inequality. At the same time the doors of the universities were opened to all. However, the importance of a sound foundation of secondary education rapidly became clear. The *rabfaks*, or workers' faculties, were established for adult proletarians and later served as an important recruiting ground for the technical high schools and technicums. Additionally, priority was given to younger applicants with working-class backgrounds as a further means of creating a new intelligentsia of truly proletarian origins. The extent to which this goal was achieved is indicated by the figures in Table 7.11, which relate to the social origins of students in institutions of higher education.

TABLE 7.11

Social origins of students in higher education in the U.S.S.R., 1924–31

	Workers	Peasants	Officials	Others
1924–5 (%)	17·8	23·1	39·8	19·3
1930–1 (%)	46·6	20·1	33·3	0

The Five-year Plans for rapid industrialization, inaugurated in 1928, led to considerable experimentation in the Soviet system of higher education. New institutions were created, particularly for technical instruction. Between 1928 and 1933 the number of schools of higher education increased from 152 to 714, much of the growth being achieved by the division and reorganization of existing schools. The Leningrad Polytechnic, for example, was reorganized into a series of specialized monotechnics and each of the five departments of the Moscow Higher Technical School became the nucleus of a new institution. In the same period, the number of students increased from 176 000 to 458 000. The emphasis on applied studies led to a situation in which, for a limited period, the U.S.S.R. had no universities, all higher education being conducted in Technical and Special Institutes into which the previously existing universities had been dissolved. Gradually the recognition grew that, for maximum efficiency, specialists needed a broad basis of general education and, in the early 1930s, the universities were restored along with the academic degrees of candidate, master, and doctor. The numerical balance between universities and the institutes continued to be heavily in favour of the latter, however. Another feature of note about Soviet higher education was the proportion of women students; between 1933 and 1940 this grew from 36 per cent to 58 per cent overall, and from 22 per cent to over 40 per cent in the engineering and industrial fields.

Although the programme of the Communist Party stated that vocational education should begin only after the completion of a general education, the requirement was relaxed during the period of rapid industrialization in order to increase the supply of skilled workers. Intermediate technical vocational training was provided in specialized secondary schools, the technicums, and these continued to be an important source of sub-professional specialists, particularly those in short supply, such as engineering technicians.

Within the general secondary schools and in elementary schools, polytechnism was a primary goal in the immediate post-Revolution period. Derived from Marx's notion of individual all-round development, as opposed to the specialized development entailed by a capitalistic division of labour, and involving the unification of mental and manual activity, its curriculum manifestation was a stress on applied industrial and agricultural techniques and a high priority accorded to practical scientific and mathematical studies. At a later date, and especially after 1931, when the schools were criticized for failing to prepare 'fully literate persons' equipped to progress to technicums and higher education, a more traditional curriculum prevailed, although the study of basic science continued to be supplemented by the 'practical application of science in the formation of work habits'.

By the time of her entry into the Second World War, the U.S.S.R. had achieved the status of a top-ranking industrial nation. Subsequent Soviet achievements in relation to the supply of scientific and technological manpower for industrial, military, and national power goals were to become the cause of serious concern in the Western World, notably the U.S.A., and to contribute substantially to reform in other educational systems.

Japan also, within a very different political context and employing a very different educational system, achieved industrial maturity over the same period. The University Code of 1918 inaugurated a period of expansion in state-supported higher education. Apart from the creation of a sixth Imperial university, twelve monotechnic government universities were brought into being between 1918 and 1938 to aid the supply of technologists, administrators, doctors, and educationists. Private institutions of higher education also flourished. At a lower level, access to secondary education continued to be restricted and the organization of schools was hierarchical; despite the national emphasis on industrialization, the technical schools did not enjoy high status and were able to supply only a small fraction of the technicians needed for industry.

The emphasis in Soviet schools on the applications of science contrasted strongly with the situation in British secondary schools in the inter-war period. Indeed, it has been suggested that the academic and bookish curriculum of the English grammar school at this time diverted able children from industrial pursuits. Although the evidence does not lend a great deal of support to this proposition, the origins of the widespread preference for black-coated occupations being located more in society than in the schools, there was a growing dissatisfaction with the educational provision at secondary level.

One complaint was that secondary education was still conceived as a special kind of education restricted to a limited section of the community. A powerful statement of an alternative view was delivered in 1922, in R. H. Tawney's exposition of Labour Party policy, *Secondary education for all*. Subsequently, an outline for progress in this direction was provided by the Hadow Committee in its report of 1926, *The education of the adolescent*. The principal recommendations were for a minimum school-leaving age of 15 and the recognition of a distinct secondary stage of education for all children, beginning at 11 plus. Pupils were to go either to selective grammar schools (with a predominantly academic curriculum) or to Modern Secondary Schools (the renamed selective and non-selective Central Schools, in which the curriculum would have a 'realistic' or practical bias in the final two years). The Committee, whose membership included Tawney, had received its remit from the first Labour government to hold office; by the time its report was published, the Conservatives were again in power. Few of the Hadow recommendations were implemented at the time; the school-leaving age, for example, was not raised to 15 until twenty-one years later, the provision of the 1936 Act for a partial raising to 15 in 1939 having been suspended on the outbreak of war.

In part the obstacles to implementation were financial, the decline in world trade and the depression having forced severe economies on the education service. Additionally, however, there was a measure of basic opposition to the extension of the term 'secondary' to the whole range of post-primary schools. 'To attempt to bring all post-primary education to one dead high school level,' Lord Eustace Percy had argued in 1925, 'as it is, I am afraid, in the United States, will do far more to prevent any real higher education in this country than anything else.' Against a background of economic uncertainty, there was a growing concern in the late 1920s and early 1930s with the role of the schools as agents for the efficient selection and cultivation of intellectual talent.

Between 1921 and 1938 the proportion of children aged 12–14 receiving education in selective grammar schools increased from 12·9 per cent to 20·6 per cent, although it must be added that these figures for England and Wales conceal very substantial regional differences. Simultaneously, there were developments in senior schools and junior technical schools. The latter, originally intended as post-elementary vocational schools, recruiting at 13, and preparing pupils for artisan employment, had liberalized their courses to a degree which led the Consultative Committee on Secondary Education in 1938 to recognize them as a distinct and attractive form of secondary education. The bilateralism of Hadow became the tripartitism of Spens, with the intention that each of the three partners in secondary education—grammar modern, and technical schools—should be accorded 'all the parity which amenities and conditions can bestow'.

V. THE SECOND WORLD WAR AND AFTER

It is worthy of note that the three most important pieces of educational legislation affecting England and Wales, the Acts of 1902, 1918, and 1944, all closely followed, or coincided with, periods of war. In the last case it is possible, as the Ministry of Education's Report for 1950 suggests, that the falling off in the volume of day-to-day administration and the exile of the Board of Education officers to Bournemouth in the early years of the Second World War left more time for a return to fundamentals. Public opinion, too, after Dunkirk and the Battle of Britain, had become more favourably disposed to the implementation of the Hadow ideals. By the spring of 1941 outline plans for the post-war reconstruction of the educational system had been drawn up. More than two years of extensive consultation followed, leading to the presentation of a Bill to Parliament in December 1943. By the Act which came into force on 3 August 1944, a Minister of Education was appointed with increased powers of control and direction over the local education authorities. A unified system of primary, secondary, and further education, in three progressive stages, was established and it became the duty of the local authorities to provide facilities accordingly. The provision also included the increase of the school-leaving age to 15, with effect from 1947, and the abolition of fees for all maintained primary and secondary schools.

Although the Act itself made no specific reference to the types of secondary school, the publication in 1943 of the Norwood Report on *Curriculum and examinations in secondary schools* provided an explicit endorsement of the tripartite division and exerted an influence on development plans submitted by local authorities for approval by the Ministry after 1944. Even so, there were those who doubted the ability of the tripartite system to achieve a genuine parity of esteem; influenced by reports of secondary schools in the U.S.A. and Canada, and reflecting developments in the educational thinking within the Labour Party, the Education Committee of the London County Council opted for a fully comprehensive reorganization of its secondary schools even before the 1944 Act had become law. The case for a tripartite division was further weakened by the absence of reliable techniques for the assessment of specific technical aptitudes at the age of 11 plus. Although the Ministry of Education continued to hold out hopes that the alternative secondary technical route would counteract the tendency of bright children to go into professional and clerical occupations, and would provide productive and manufacturing industry with a fairer share of the nation's talent, the growth of secondary technical schools was slow; by 1950 there were less than 75000 boys and girls in such schools, compared with almost 512000 in grammar schools. The general social evaluation of technical schools was discouraging and where they were established with adequate resources and a fair share of able pupils, the curriculum all too often tended to approximate to that of the grammar school and lacked a distinctive emphasis and treatment.

If the sense of common suffering and national unity that grew from the experiences of war helped to move education closer to the goal of equal opportunity for all, the achievements of science and technology within a controlled war-time economy and in the interests of national survival also had significant educational consequences. The Second World War has been called 'the physicists' war'; certainly there was an early recognition that this was, in Stafford Cripps's words, 'a truly scientific war', and that 'the battle would not be won merely by the physical ascendancy of our race, but by the ingenuity of those who had been trained in our schools, technical colleges and universities'. In particular, the existence in 1939 of the Central Scientific Register helped to ensure that scientists contributed to the war effort by their knowledge rather than with their lives.

One immediate effect of this use of scientific manpower was a marked reduction in the research output of British universities, most of the younger scientists having been directed elsewhere to work of direct relevance to the war. In contrast, much of the comparable research in America was undertaken in universities and not in Service establishments. As hostilities drew to a close, it was clear that American industry could no longer depend on its traditional European sources of fundamental research. *Science the endless frontier*, the celebrated report of Vannevar Bush, the first of the U.S. Presidential advisers on science, in 1945 urged a major expansion of fundamental research and the sponsorship of undergraduate and postgraduate schools of science and engineering in all the major American universities. To this end, leading European scientists were recruited to the lusher academic environment of the U.S.A.; with them, the centre of gravity of scientific research moved across the Atlantic. After five years of controversy about its precise role, the National Science Foundation was established in 1950 to support 'basic research' and science education.

To take an account of education in industrialized societies much beyond the end of the Second World War is to enter a new period of expansion and increasing complexity. A limited indication of the changes which lay ahead is provided by the figures relating to full-time students in higher education in Great Britain given in Table 7.12.

TABLE 7.12

Full-time students in higher education in Britain, 1938–9 and 1954–5

	Further education	Teacher training	Universities
1938–9	6 000	13 000	50 000
1954–5	12 000	28 000	82 000

War had demonstrated the country's shortage of scientists, technologists, and technicians, and industrialists were beginning to appreciate the dividends which might come from investment in technical education. In practical terms, there was a greater willingness to grant day-release for attendance at technical colleges and the tradition of part-time evening instruction at long last seemed to be giving way to a more rational system.

As the industrial demand for trained manpower increased, there were consequential effects on the educational system. Whereas in pre-war years a very high proportion of those graduating in science had entered education, after 1945 the pull of other occupations led to a shortage of suitably qualified science teachers. Both the post-war reports on developments in technological and scientific education, the Percy Report on *Higher technological education* (1945) and that of the Barlow Committee on *Scientific manpower* (1946), underestimated the demands for trained personnel. The latter report recommended a doubling of the output of science graduates, a goal which the universities achieved in five years instead of the ten thought necessary by the Committee, and to which the educational reforms of the 1944 Act contributed significantly by the provision of a broader social base for secondary education in grammar schools.

In this expansive phase, scientific and technological education moved into the second half of the twentieth century. 'The future of the world is to the highly educated races, who alone can handle the scientific apparatus necessary for pre-eminence in peace and survival in war.' In making this statement in a broadcast in 1943, Winston Churchill gave expression to a widely held view which was to be frequently reiterated. At the same time he placed a unique and heavy burden of responsibility on education.

BIBLIOGRAPHY

ARGLES, M. *South Kensington to Robbins. An account of English technical and scientific education since 1851.* Longmans, Green and Co. Ltd, London (1964).

BANKS, O. *Parity and prestige in English secondary education.* Routledge and Kegan Paul, London (1955).

BEN-DAVID, J. *The scientist's role in society.* Prentice-Hall, New Jersey (1971).

CARDWELL, D. S. L. *The organisation of science in England.* Heinemann, London (1972).

DE WITT, N. *Education and professional employment in the U.S.S.R.* National Science Foundation, Washington (1961).

FOREIGN OFFICE. *University reform in Germany. Report by a German Commission.* H.M.S.O., London (1949).

HALSEY, A. H. (ed.) *Trends in British society since 1900.* Macmillan, London (1972).

HANS, N. *Comparative education.* Routledge and Kegan Paul, London (1967).

HARTSHORNE, E. Y., JR. *The German universities and National Socialism.* Allen and Unwin, London (1937).

KOROL, ALEXANDER G. *Soviet education for science and technology.* Chapman and Hall, London (1957).

LILGE, F. *The abuse of learning. The failure of the German universities.* Macmillan, New York (1948).

MINISTRY OF EDUCATION. *Education 1900–1950. The Report of the Ministry of Education . . . for the Year 1950.* Cmd. 8244. H.M.S.O., London (1951).

NAKAYAMA, SHIGERU, SWAIN, D. L., and YAGI ERI (ed.) *Science and society in modern Japan.* M.I.T. Press, Cambridge, Mass. (1974).

A discussion on the effects of the two world wars on the organization and development of science in the United Kingdom. *Proceedings of the Royal Society of London,* A, **342**, 439–591 (1975).

SANDERSON, M. *The universities and British industry: 1850–1970.* Routledge and Kegan Paul, London (1972).

8

THE FOSSIL FUELS

ARTHUR J. TAYLOR

I. THE FOSSIL FUELS AND ENERGY PROVISION

THE twentieth-century history of fossil fuels closely parallels that of energy supply. Over the centuries men have looked increasingly to inanimate rather than animate sources for the provision of their major energy needs; in the nineteenth century, in western Europe and the United States, energy derived from the fossil fuels gradually displaced that obtained from wood, wind, and water-power. Quantification of this process of evolution is necessarily difficult, resting as it must on evidence that is often fragile and assumptions that are largely conjectural. It has been estimated that by 1900 some three-fifths of the world's energy requirements were met by the three fossil fuels—coal, oil, and natural gas. Half a century later the proportion exceeded four-fifths. In the intervening fifty years among the non-fossil fuels there had been a relative diminution in the use of wood and an increase in that of water-power, primarily in the form of hydroelectric power [1].

Among the fossil fuels themselves, coal in 1900 held a position of almost unchallenged pre-eminence. Notwithstanding the increasing use of oil in the United States and in Russia, all but 5 per cent of the total energy supplied by the fossil fuels came from coal. Over the next half-century the emphasis in fuel utilization gradually moved away from coal towards oil and natural gas, but as late as 1948 coal still provided more than 60 per cent of the world's energy requirements. By this time oil supplied some 28 per cent and natural gas $9\frac{1}{2}$ per cent of the world's energy.

The closing of the gap between the two major fuels had been slow in the years before 1914, at which time coal still outweighed oil as an energy-provider by more than sixteen to one. The arrival of the motor-car and the widening use of the internal combustion engine brought a growing demand for oil, and the gap between the two fuels narrowed with increasing rapidity from decade to decade. At the middle of the century, however, coal still supplied twice as much energy as oil (Table 8.1).

TABLE 8.1

World utilization of fossil fuels (%)

	Coal	Oil	Natural gas
1900	95·1	3·6	1·2
1913	92·8	5·6	1·6
1929	79·6	15·9	4·5
1938	72·7	21·8	5·5
1948	62·2	28·4	9·4

Source: Reference [2].

This changing balance of utilization was not experienced to an equal degree in all parts of the world. Outside the major industrial areas there was a continuing reliance on animal-power and wood as major suppliers of energy. In western Europe, where known indigenous sources of oil were few, coal throughout the half-century retained a stronger hold on the energy market than it did in the world as a whole. In Britain in 1950 coal still supplied 90 per cent of the country's fossil-fuel energy. It was in the United States that the movement away from coal to oil and natural gas was most evident. This was primarily due to the ready availability of oil and natural gas as well as coal, but also to the rapid development and employment in the United States of the internal combustion engine. In 1900 over 70 per cent of American energy consumption was derived from coal as compared to 5 per cent equally shared between oil and natural gas. Fifty years later oil had moved to a position of virtual parity with coal, and natural gas was also advancing strongly (Table 8.2).

The contrast between western Europe and the United States in the changing pattern of energy consumption is therefore evident, but their distinctive experience in energy supply is still more striking. Even if the area of comparison is widened to include the whole of the Americas on the one hand and the U.S.S.R. in Europe on the other, the sharpness of contrast persists. Whereas in the Americas by the middle of the twentieth century energy

TABLE 8.2

Sources of supply of U.S. energy (%)

	Coal	Oil	Natural gas	Fuel wood	Others
1900	71·4	2·4	2·6	21·0	2·6
1950	36·8	36·2	17·0	3·3	6·7*

* Principally hydro-power.
Source: Reference [3].

obtained from oil and natural gas exceeded that derived from coal in the proportion of three to two, in Europe coal retained an ascendency over the other two fuels of fourteen to one.

By the end of the nineteenth century coal had fully established itself as the all-purpose fuel of the developing industrial communities of Europe and America. As a source of heat both domestically and industrially, of light when transmuted through coke into gas, and of motive power for the driving of stationary engines, locomotives, and ships, it dominated the life of industrial society. It was the vital agent for the smelting of iron ore, for the conversion of pig-iron into steel, and for the making of chemicals; and increasingly in the twentieth century it found a major use in the generation of electricity for the provision of heat, light, and power. In rural areas away from the coalfields, where much the larger part of the world's population was still to be found, older sources of energy supply were still predominant, but the energy requirements of rural societies in *per capita* terms were of a wholly different and lower order from those of industrial communities. Some indication of the special energy demands of an industrial society is provided by the evidence that of 167m. tons of coal consumed in Britain in 1903 only 32m. tons were used for domestic purposes and 135m. tons were put to various industrial uses [4].

Oil as yet had little disturbed the monopoly of world energy provision enjoyed by coal. Only in the United States and Russia was there significant production or consumption of oil. In the United States kerosene (paraffin) had come to offer a strong and largely successful challenge to coal-gas as a source of light and of domestic heat. At the end of the nineteenth century over 60 per cent of American refined oil was used for these purposes, half of it in the United States itself and the remainder in other parts of the world. But whereas in the industrial towns of western Europe coal-gas quickly established itself as the major provider of illumination and the use of kerosene was largely confined to the rural areas, in the United States oil had effectively challenged gas as an illuminant in both town and country.

In 1900, coal had little to fear industrially from the competition of oil. The major industrial use of oil at this time was as a lubricant. This was a market which it had wrested from coal in the closing decades of the nineteenth century, but the prize was not a substantial one. Potentially the most significant use for oil was as a motor fuel, but as yet the demand for petrol was small almost to the point of non-existence. In the United States, already the front-runner in motor-car production, only 4200 vehicles had been built by

1900, and of these only one-quarter were driven by internal combustion engines [5]. More substantial competition to coal was offered by the use of oil, either in its unrefined or refined form, as an industrial fuel. Such use, however, was small before 1900 even in the United States and, except perhaps in Russia, almost unknown elsewhere. Even in the United States in 1900, no more than 5 per cent of fossil-fuel-based energy supply came from oil, compared with 3 per cent from natural gas and 92 per cent from coal [6].

Over the first two decades of the twentieth century the production of oil grew rapidly not only in absolute terms but also in relation to the general output of fossil fuels. Yet, so great was the overall demand for fuel, and so limited the share of oil and natural gas in its supply, that as late as 1920 almost 90 per cent of the world's fossil-fuel-based energy still came from coal. Even in the United States, where motor transport was making most rapid headway, coal accounted for over 80 per cent and oil for less than 14 per cent of the energy provided by the fossil fuels.

The inter-war years saw the first serious invasion by oil of fields hitherto regarded as the exclusive preserve of coal. This was most evident in the transportation field. Notwithstanding the invention of the diesel engine, the railways of the world continued to rely principally on coal as their main steam-raising agent; but the competition which road transport now increasingly offered to the railways implied a parallel challenge of oil to coal. At sea the displacement of coal by oil was more direct and had quicker effect. In 1914 less than 4 per cent of the world's shipping tonnage was driven by oil. By 1918 the proportion had increased to 18 per cent and in the course of the next twenty years over half (51·4 per cent by 1937) of the world's tonnage came to use oil as its fuel [7]. Oil was also increasingly used, more particularly in the United States, for industrial steam-raising and for the generation of electric power—fields in which before the First World War the fuel had established no more than a modest foothold. However, the extent of this incursion into coal's traditional territory should not be exaggerated. As late as 1935, it has been calculated, even in the United States in those fields in which there was direct competition between coal and oil, coal remained in the ascendent in the proportion of at least four to one; and, though more oil was used in American automobiles and trucks than in the whole of her manufacturing industry, the total energy contribution of oil to the American economy was still only half that of coal. In the world as a whole the weighting was even more advantageous to the longer-used fuel. By 1939 world consumption of the fossil fuels had increased by more than 130 per cent since the beginning of the century, and the share of coal in this greatly increased total was still more than 72 per cent [8].

The years of the Second World War, and of reconstruction thereafter, imposed new demands upon the major sources of energy: they also served to magnify the tendencies which had become evident in the two preceding decades. In the ten years after 1938 the demand for fossil fuels rose by a quarter, an increase in which all shared, but from which oil and natural gas benefited more considerably than coal. Many factors contributed to this development. Not least among these were, on the one hand, the rapid growth of transport by land, sea, and air, and, on the other, improvements in technology facilitating the use of diesel oil and the transmission of natural gas. In the United States the output of diesel oil rose more than fourfold in the decade between 1940 and 1950—by 1953 it had displaced coal as the major railroad fuel—while in the same decade the consumption of petrol and diesel oil doubled. Elsewhere change came more slowly, perhaps not so much because of a natural conservatism as of the fact that the economic advantages of a shift from coal to oil were less marked in countries without indigenous supplies of oil. Nevertheless, in the field of transport coal was plainly in retreat; by 1975 its position was to have become as weak as it had been strong at the beginning of the century. Whereas, however, in transport the use of coal had by 1950 declined not only relatively to oil but also absolutely, in other fields— notably in the metallurgical industries and in the generation of electricity— its losses in this period tended at most to be relative. Oil, for example, came to be increasingly used, first in the United States and then elsewhere, in the generation of electricity, but the demand for electricity was such that in the single decade 1945–55 consumption of coal by American power stations almost doubled.

In 1950, therefore, coal, oil, and natural gas were still to be regarded rather as complementary sources of energy than as competing fuels. However, to the extent that they were directly competitive, the varying degrees of ascendency of the three were in part a function of their relative convenience for different purposes and in part an expression of variations in cost. Comparisons between the price of the three fossil fuels at the pithead or the well in the United States show that while the price of coal increased more than fourfold between 1900–9 and 1945–54, the price of crude oil increased little more than two-and-a-half times and that of natural gas by less than 30 per cent over the same period [9]. These figures point to an increasingly advantageous position for oil and natural gas in relation to coal, but in assessing their significance it has to be borne in mind that the cost at the point of initial production is only a part of the charge to the consumer. To it must be added the costs of transportation—particularly heavy in the case of coal but not insignificant in that of oil—and the cost of processing, whether in the refining of crude oil or in the manufacture of coke from coal. Account must also be taken of the effects of governmental discrimination by taxation or other means in favour of one fuel against another. These, as much as changes in the available sources of supply and in the technology of production, are forces which shaped the fluctuating fortunes of the three fossil fuels in the course of the first half of the present century.

II. COAL

Between 1900 and 1950, world production of coal and lignite[1] increased from *c.* 750m. to *c.* 1800m. tonnes per annum, a yearly growth rate of rather less than $1\frac{3}{4}$ per cent. Progress, however, was uneven. Until 1913, by which year output was approaching 1350m. tonnes, the annual growth rate was over 4 per cent, as it had been for at least the preceding 30 years. Over the next 25 years to 1938 production rose by some 250m. tonnes, a growth rate of less than $\frac{3}{4}$ per cent per annum. Between 1938 and 1950 output increased by a further 200m. tonnes, a yearly rate of growth somewhat in excess of 1 per cent. It is clear that the uninterrupted forward thrust which had characterized the industry's progress before the First World War had been lost. In the 21 years between 1929 and 1950 production increased in 13 years and fell in eight: in 1928 output approached 1700m. tonnes, higher than at any time before the Second World War; three years later, at 1300m. tonnes, production was less than it had been in 1913 [10].

The reasons for these fluctuations are various, and, since their incidence varied from continent to continent and from country to country, their full appreciation would require a study in depth of the economy of each of the major coal-producing countries. However, a broad distinction may be made between the short-run factors which influenced the temporary fortunes of the industry, and the longer-term forces which more permanently influenced its growth. Under the first head are to be placed in particular the consequences of war and of the ebb and flow of the business cycle. War brought mixed

---

[1] Coal varies greatly in its energy-producing qualities. International comparisons, therefore, present considerable difficulty. Lignite, or brown coal, is an inferior energy-producer, providing at best only half the energy furnished by even the poorer quality bituminous coals. It is mined extensively in Germany, Czechoslovakia, Austria, and Hungary and to a lesser extent in Russia, the United States, and Canada. The output figures cited are gross and make no allowance for these qualitative differences.

fortunes to coal producers, as it did to industrialists in general. British mines, for example, were producing 20 per cent less coal in 1918 than they had done five years earlier and there was a similar loss of output during the course of the Second World War. German output also fell marginally during the First World War, and production of hard coal declined more sharply—by a quarter between 1938 and 1944—in the Second World War, though output of the inferior lignite increased by one-fifth. In the United States, on the other hand, output increased during both World Wars, in the Second by over 40 per cent between 1939 and 1945. However, although war gave an initial stimulus to coal production, it tended to become a disruptive force the longer conflict persisted.

The influence of general economic movements is easier to chart. Before 1914, though short-term fluctuations affected coal as much as any other industry, the strong upward forces working in the expanding world economy were the major determinants of the fortunes of coal producers. The 1920s, however, lay under the conflicting influences of post-war economic dislocation and of the boom which reached its culminating point in 1929; and there were consequently wide differences of experience from year to year and from country to country. World output reached peak levels in 1929, yet Britain, for so long the world's leading coal producer, was in that year producing 30m. tonnes less than she had done in 1913. The 1930s brought depression to the industry on an almost world-wide scale. Output in 1932 was 40 per cent less than it had been only three years earlier, and even in 1939 world production was still some 6 per cent below the level of 1929. But, whereas the United States and Britain in 1939 still fell far short of their 1929 performance, Germany had fully recovered her earlier position, and the U.S.S.R. had not even experienced the traumatic decline of the depression years. The pattern of fluctuation continued after 1945, with world output falling sharply at the end of the war and pursuing a strong but not wholly uninterrupted recovery to 1950.

Among the longer-term determinants of the fortunes of coal, the emergence of competing fuels obviously has an important place, but no less significant are the differing rates of development of various industrial economies. Almost 90 per cent of the coal produced in the world in 1900 was raised in western Europe and North America and over 80 per cent in three countries—the United States, Great Britain, and Germany. Even in 1950 three out of every five tons of world coal were raised in one or other of these countries. However, by the mid-century Russia, which had produced no more than 15m. tonnes in 1900 and only 40m. tonnes in 1929, stood second only to the United States among the world's coal producers. Russia indeed was set on a course of expansion which in a further ten years would make her the world's leading coal producer; while China, still with 40m. tonnes a relatively minor producer in 1950, in the course of a further decade was to expand her output tenfold and complete a rapid shift from west to east in the balance of world coal-power.

The three major producers of the first half of the twentieth century had been in the play of varied influences. In Britain the general rate of economic growth had slowed down and with it the rate of increase in demand for energy supply. Moreover, although direct competition between coal and oil was limited throughout most of this period, Britain had experienced a sharp decline in the export demand for her coal. This was to a significant degree attributable to the supersedence of coal by oil in the fuelling of ships. British output increased by one-third between 1900 and 1913, but the performance of that year—292m. tonnes—was never again equalled. In 1929 production still exceeded 260m. tonnes but by 1945 it had fallen below 190m. tonnes and in 1950, with an output of 220m. tonnes, Britain contributed in proportionate terms only half as much to total world production as she had done half a century earlier.

The United States doubled its output between 1900 and 1913—from 243 to 517m. tonnes—by which time it was producing almost two-fifths of the world's coal. This was a period of substantial growth in the American industrial economy which brought with it a parallel expansion in demand for energy that coal as yet was alone in a position to supply. After 1918 the position changed radically. Coal not only encountered the massive fluctuations of the inter-war economy, but also met a mounting challenge from competing fuels. As a result American output in 1938 was, at 358m. tonnes, one-third below the level of 1929, and now accounted for no more than one-quarter of world production. Yet such was the flexibility and resilience of the American industry that, under the stimulus of war and post-war demand, output was raised again in 1947 to its highest point of 624m. tonnes—over 40 per cent of world supply. Thereafter, however, the competition of oil and natural gas reasserted itself and production fell to 500m. tonnes by 1950 and to 370m. tonnes a decade later.

Germany, like the United States, experienced strong if not uninterrupted industrial growth throughout the half century and, like Britain, remained until 1950 essentially a coal-based industrial state; but the German economy even more than that of the United States or of Britain, was in these years affected by the pressures of war and war politics. A period of rapid growth before the First World War was followed by two decades of sharp ebb and flow in the inter-war years. By 1938 German production of hard coal at 186m. tonnes was 30m. tonnes greater than it had been a quarter of a century earlier, and lignite production at 195m. tonnes was over 100m. tonnes larger than in 1913. During the Second World War production of hard coal fell, while that of lignite increased until, in the final year of war, both declined sharply, hard coal to 40m. tonnes and lignite to 120m. tonnes. Thereafter a divided Germany revived its coal industry in terms of hard coal to some 110m. tonnes by 1950 (mined largely in the west) and in lignite to 215m. tonnes (two-thirds coming from the east).

Coal production in Britain and the United States had passed its peak by 1950. In Germany, though post-war recovery was not complete, output of hard coal—though not of lignite—never again reached pre-1939 levels. By contrast, later arrivals on the industrial scene like Russia, Poland, and China found themselves in 1950 on a rising production curve. Russia, like Germany, showed marked fluctuations in output between 1900 and 1950 and for similar reasons. Output of coal and lignite reached 29m. tonnes in 1913 but fell below 8m. tonnes in the chaotic years which followed the Revolution of 1917. Production, unimpeded by the world depression of the 1930s, rose to 166m. tonnes in 1940, fell back to 76m. tonnes two years later under the first shock of Nazi invasion, and then recovered to reach its, as yet, highest point of 261m. tonnes in 1950. Poland, already a substantial producer in 1913, with an output of 40m. tonnes, profited from the acquisition of German territory in 1945 and by 1950 was mining 80m. tonnes of coal. China, raising 15m. tonnes in 1913, doubled that quantity in the next quarter-century but was still a relatively minor producer with 40m. tonnes in 1950: her great period of development lay, as we have noted, in the immediate future. Other middle-ranking producers in 1950, all of whom were to increase their output substantially in the next decade included Czechoslovakia (46m. tonnes in 1950, of which three-fifths was in the form of lignite), Japan (38m. tonnes) and India (32m. tonnes). Among older industrial countries close to or past their productive peak were France (50m. tonnes) and Belgium (27m. tonnes).

These diverse experiences, representing varying degrees of expansion and contraction and in no way self-balancing, patently influenced the overall movement of coal production both in the short and the long term. They suggest a tendency, made more evident if events are followed through to 1975, for industrializing countries to base their early development on coal-based energy and thereafter, partly because of increasing costs of mining and partly because of greater sophistication in general technology, to place increasing emphasis on oil. This was, of course, most evident in the case of countries like the United States with indigenous oil resources, but it was also true of others, notably Germany and Britain, who had to rely on imported oil. By contrast, where, as in the U.S.S.R., both coal and oil were immediately available but industrialization came relatively late, emphasis until 1950 was strongly on coal.

One factor, however, which by 1950 was already tending to reduce world demand for coal, was the relatively high cost of its transport. Though some international movement of coal—for example, from the United States to Canada and from Russia into Eastern Europe—was increasing, on trans-oceanic routes the collier was rapidly giving way to the oil-tanker. British exports of coal fell from 73m. tonnes in 1913 to 13·5m. tonnes in 1950. Conversely, in 1950, Norway imported only one-half of the coal that she had before the First World War, and over the same period Argentina reduced its imports by two-thirds. Countries which relied on imported fuel were increasingly giving preference to more cheaply transported oil.

More general, and of no less importance in their effects on the demand for coal, were the considerable economies in fuel utilization made possible by changes in technology at all levels of industrial production. Calculations of savings in the use of coal in the United States during the twenty years between 1909 and 1929 show a decline in unit fuel consumption of 33 per cent across the full range of the nation's industrial economy, with degrees of economy ranging from 21 per cent in the manufacture of cement to 66 per cent in the generation of electricity. In Britain economies of a similar order were achieved, most notably in electricity generation and in the making of steel [11]. While from one point of view such economies lessened demand, they also enabled coal to maintain its competitive position in relation to other fuels. This was particularly true in the case of electricity generation, where the prize was substantial and the competition, from both oil and water-power, most evident. Coal also profited from developments in the process of coke-manufacture. Coke and gas had throughout the nineteenth century been important derivatives of coal, but the manufacture of both also released important raw materials for the chemical industry, principally tar, light oils, and ammoniacal liquor.

The history of the manufacture of coal-gas has been considered elsewhere in this work (Vol. IV, Ch. 9) and the industry remained a major user of coal throughout the period we are now considering. Thereafter it began to decline—though not uniformly throughout the world—as resources of natural gas (see below) were increasingly exploited. In Britain, for example, some 25m. tons of coal was carbonized for the manufacture of town gas in 1947. This corresponded to nearly 500000m. cubic feet of gas, distributed through 68000 miles of mains. Two-thirds of this total was supplied to domestic consumers, almost all for heating, since electricity in the course of half a century had very largely displaced gas for lighting.

The development of the process of coal hydrogenation, stimulated by the needs of Germany's war economy for substitute industrial materials, increased still further the range of chemical products yielded by coal. The primary object in this instance was the production of motor and aviation fuels. Though the cost of production made such fuels uncompetitive under normal market conditions, the hydrogenation process also provided organic chemicals for the manufacture of plastics and synthetic fibres; and thus, while technological advance reduced demand for coal in some areas, it offered prospects of expansion in others.

Of the changes in fuel technology influencing coal consumption, much the most important, however, was the growing use of energy in the new form of electric power and light. At the beginning of the twentieth century the generation of electricity was still in its infancy. In the United States, from the outset the principal user of electric power, the capacity of the existing generating plants in 1900 did not exceed a million kilowatts; by 1950 it had reached 82m. kW. This was rather more than one-third of the total world capacity of 231m. kW, providing 900000m. kWh of electric light and power [12]. Not all of this was to be accounted an addition to the world's energy supply since in part it represented energy hitherto delivered in other forms. How large this proportion was is not readily determinable, nor is simple calculation possible of the gain or loss in efficiency involved in the use of electricity rather than one of the primary fuels. It has been estimated that, between 1900 and 1950, the thermal efficiency of electrical supply increased sevenfold; yet up to 80 per cent of the energy supplied by coal in both Britain and the United States was still being lost in 1950 in the process of electric power generation. Against this, however, are to be set the fact that the generating process put low-grade coal to useful service, and the advantages of cost as well as convenience that belong to a fuel which can be easily transmitted and whose supply can be readily turned on and off.

Until 1950 coal occupied a strong position as the principal supplier of fuel to the world's generating stations. Its chief competitor was not oil but water-power, and each tended to have its own clearly defined area of monopoly. Perhaps one-third—the available statistics do not allow a more precise computation—of the world's electricity came from water-power. In mountainous countries with high annual rainfall like Norway, Switzerland, Italy, and Japan, hydroelectric supply had established a position of almost total ascendency; in countries not so endowed—the Netherlands, Denmark, and Australia among others—its contribution was minimal. In other major industrial countries, notably France, Germany, and the United States, geography had given local advantages to both coal and water, but, overall, coal had the stronger hand. In the United States oil and natural gas also claimed a share in the provision of primary fuel to the nation's power stations, but in 1950 American coal-fired power stations still had a capacity double that of those fired by oil and gas. Outside the United States, oil and gas had made little headway: coal at the mid-century had gained much and lost little by the coming of electricity.

The degree of competition between coal and oil was, however, sufficient by 1950 to cause pessimists in the older industry to foresee coal's rapid decline. Support for this view was furnished by the recent experience of the United States where a decline in output of 30 per cent between 1947 and 1949 had brought with it widespread pit closures and the dismissal of thousands of men. A more substantial indication of the industry's problems was provided by evidence of its constantly rising costs. The price of producing coal at the pithead in the United States had risen eight times as fast as that of producing oil at the well in the first half of the twentieth century. Increases in the cost of working coal were at least as great in western Europe. The economy of the two fuels is more complex than these simple statistics might suggest but they indicate a situation in which the 'terms of trade' between coal and oil were steadily changing to coal's disadvantage.

The increase in the cost of mining coal was largely an expression of the growth in wage-costs in a labour-intensive industry. Throughout the first half of the twentieth century wages on average were responsible for over half the cost of Europe's coal at the pithead. In Britain the ratio of wages to total cost never fell below 60 per cent and on occasions reached 75 per cent. Even as late as 1950, when labour-saving machinery had been introduced on a large scale, almost two-thirds of the price paid for getting each ton of British coal went in wages paid to the miner. Over western Europe conditions showed some variation, but in most coalfields wages tended to account for at least half the cost of production. American experience differed little from that of western Europe. Everywhere the miner's desire to maintain and, where possible, enhance his status in a world of improving living standards by increasing his real income and reducing his hours of work inevitably added to the costs of coal production. To counter the effect of higher wages coal-owners sought to increase labour productivity by capital investment and technological innovation (Ch. 31). In this they achieved varying degrees of success. In the United States, where conditions of working were favourable to high levels of productivity, output per man-shift in bituminous coal-mines almost doubled between 1913 and 1950 to over 6 tonnes, and doubled again to 12 tonnes in the succeeding decade. No European country approached these levels of achievement either in *per capita* output or in rate of increase. In Britain between 1914 and 1950 productivity rose by 18 per cent; in the German Ruhr, with post-war recovery still incomplete in 1950, output per man had grown by 14 per cent over the same period; and in France and Belgium increases of 10 and 29 per cent had been made. In no case had output per man-shift come to exceed 1·25 tonnes by 1950. Whereas in the United States investment could be applied directly to increase labour productivity both at the coal-face and in the conveyance of coal from face to pit-head, in the longer-worked coalfields of western Europe the major problem was the maintenance of existing levels of productivity in face of the difficulties presented by increasing depths of working and the diminishing returns which had become a dominating condition of the industry's existence. In Britain, notwithstanding the advances since 1914, labour productivity measured in terms of output per man-year was still less in 1950 than it had been 70 years earlier; and this was in spite of the fact that by 1950, 79 per cent of British coal was mechanically cut and 85 per cent mechanically conveyed [13]. In western Europe the ability of coal to maintain its place as an energy-provider was increasingly dependent on the injection of public money into an industry coming under a growing measure of national ownership or control. In the United States, where the peculiar economic, social, and political pressures which sustained the industry in Europe were largely non-existent, the industry, though growing in efficiency, narrowed its base and reduced its size in face of the sustained competition of oil, gas, and water-power. By contrast, in newer industrial countries, where coal was in general more easily mined and where labour was cheaper than in the west, it retained its competitive position and shared to the full in the experience of general economic growth.

III. OIL

Oil and coal serve similar ends and compete in the same general market, yet at many points the economies of the two fuels diverge fundamentally. Whereas coal is a labour-intensive indstry, oil is capital-intensive; coal is largely consumed in the country of its origin; oil from the early years of its exploitation served international markets; and while coal has been almost wholly nationally financed, oil has been increasingly developed by extra-national and international corporations.

Coal was the basis of nineteenth-century industrialization, and, because of the high cost of its transport, the possession of coal reserves was the essential prerequisite of an industrial economy. Not only, therefore, was industrialization confined to countries which had accessible coal-seams, such as Britain, Germany, France, Belgium, and the United States, but also within these countries industry largely established itself in the immediate vicinity of the coalfields. The economics of oil production, however, dictated a different pattern of development. Because the production of oil in substantial quantity demands considerable capital investment, only those countries which had grown wealthy in terms of the earlier industrial revolution were favourably placed to finance its exploitation; and, since it is more readily and cheaply transportable than coal, oil was brought to areas of existing industrialization rather than made the basis of new industrial economies in the areas in which it was drilled. Though the increasing use of oil as an industrial fuel and the parallel development of electrical power generation made for a dispersion of industrial activity, world industry at the middle of the twentieth century was still largely concentrated in countries rich in coal, even where industry had become as much, if not more, dependent on oil and electricity than on coal itself for its existence and growth.

At the beginning of the twentieth century the world's oil industry was still in its infancy. Output, estimated in 1900 at rather less than 150m. barrels, supplied energy equivalent to 3·6 per cent of that supplied by coal [14]. By 1913 production had increased to 385m. barrels, equivalent to 5·6 per cent of the energy provided by the year's coal supply. Of this total two-thirds were produced in the United States and one-sixth in Russia, a reversal of the order of precedence of 1900 when, briefly, Russia had led the world in oil production. The only other countries producing oil in appreciable quantities were Mexico with 6 per cent, Rumania 4 per cent, and Indonesia 3 per cent of the world total. Fifteen years later in 1928, with world output increased to 1325m.

barrels—in energy terms equivalent to 18 per cent of the year's coal supply—the United States had fully maintained her ascendency as an oil producer. By this time, however, two new major producers had emerged, one—Venezuela—in the Americas, and the other—Iran—even more significantly, in the Middle East. Oil was now being produced on a commercial scale in fourteen countries though with a continuing emphasis on the Americas, where 83 per cent of the world's oil was drilled. By the middle of the century the kaleidoscope had shifted yet again, and in a more radical fashion, with output increased threefold and the centre of oil production beginning to move away from the United States. Output at 3783m. barrels was now equal in energy terms to half that supplied by coal and was produced in more than a score of countries of which the principal were the United States (52·1 per cent of world supply), Venezuela (14·4 per cent), the U.S.S.R. (7·2 per cent), Saudi Arabia (5·3 per cent), and Kuwait (3·3 per cent). The changing balance in regional terms is indicated in Table 8.3.

Whereas large-scale production and consumption of coal went hand in hand, only two of the world's major oil producers were also major consumers. Moreover, five of the six major coal producers were among the six major oil consumers. Among the consumers, as among the producers, the United States enjoyed a clear pre-eminence. At the end of the nineteenth century the United States had been a considerable exporter of oil products. In 1899 two-fifths of her output had been exported, principally in the form of illuminating oil. Throughout the succeeding half-century the United States remained an oil-exporting nation. By 1950, however, her exports were wholly in refined form, and, to make this export possible as well as provide for her own domestic needs, she had become on balance an importer of crude oil. Nevertheless the degree of dependence on external supplies which this

TABLE 8.3

Output of crude oil by region, 1913–50 (%)

	1913	1928	1938	1950
North America	64·4	68·0	61·8	52·9
South America (including Mexico)	7·4	15·5	15·2	18·7
East Europe	20·0	8·8	12·8	8·0
Middle East	0·5	3·4	6·1	16·9
East Asia	4·9	3·5	3·7	2·2
Other	2·3	0·8	0·6	1·3

Source: Reference [15].

implied was small and contrasted strongly with the position of the industrial nations of western Europe which as yet had no alternative but to indulge in large-scale importation to meet their basic needs.

While, therefore, the history of oil production in the first half of the twentieth century cannot be written wholly in American terms, it is evident that the United States must be the major point of reference. At the end of the nineteenth century the principal product of American refineries was kerosene, the greater proportion of which was exported. By 1914 the situation had changed to the extent that kerosene constituted only one-quarter of American production but still formed three-fifths of her oil exports; in the United States itself the balance in utilization had shifted away from kerosene to fuel oil, which now accounted for half of the total supply (Table 8.4). Twenty-five years later the pattern had changed further, and motor fuel had become the major refined product. Half of the American domestic consumption took the form of motor spirit and, of the remainder, all but one-fifth was fuel oil. Exports now comprised no more than one-tenth of total production and were also largely in the form of motor spirit and fuel oil. In the rest of the world, where by 1939 one-third of the world's oil was consumed, the tendency in industrialized states was for a pattern of consumption to emerge similar to that in the United States itself. The coal-rich states of western Europe emphasized the use of oil as a motor fuel, and less-advanced and coal-deficient Italy and Japan showed a weighting on the side of fuel oil. The U.S.S.R. had a wholly different use-pattern. A quarter of its oil was used in the form of kerosene, almost two-fifths as fuel oil, and less than one-fifth as motor fuel. Equally significantly half of its oil provided energy for the country's agriculture. Though Russia stood second only to the United States as an oil user, with a consumption greater than that of Britain, Germany, and Italy combined, the pattern of her oil utilization marked her out as an industrially less

TABLE 8.4

Major refinery products in the United States (%)

	1899	1914	1939	1946
Illuminating oil (kerosene)	61·2	25·4	5·7	5·7
Lubricating oil	9·9	6·8	2·9	2·7
Gasoline (petrol)	13·8	19·0	49·5	46·3
Fuel oil	15·0	48·7	41·9	45·2

Source: Reference [16].

developed nation, to be compared in this respect with India and Egypt rather than with the more highly developed nations of the west. The Second World War and its aftermath brought some disturbance to this pattern, in particular expediting technological developments in the distribution of fuel oil which made for more effective use of the residual element of the refining process; but the general pattern of utilization with its even balance between motor fuel and fuel oil persisted into the second half of the century.

In addition to these major uses, oil made an increasingly significant contribution to twentieth-century industrial life through the lesser products of the refining process. These included asphalt, vehicle oil, and wax, which also found their way at one remove into products as varied as candles, linoleum, petroleum jelly, and a variety of plastic products. In terms of the quantities of oil consumed these were still relatively minor by-products of the primary fuel, but qualitatively their influence on twentieth-century life was of prime importance. At mid-century the tide turned sharply. Petroleum, and not coal tar, became the principal raw material of the rapidly growing organic chemical industry.

In the early years of the industry's development it was usual for oil to be refined close to the point where it was drilled. However, as the demand for oil products grew both in quantity and in the increasing sophistication of the consumers' requirements, refining tended to move away from the point of production nearer to that of consumption. This development, already in train before the First World War, was accelerated by the changes in demand and in processing techniques which came with the rapid advance of motor transport after 1918. In 1919 refining capacity in the United States had been divided in almost equal parts between locations in the interior of the country close to the oil fields, and locations on the coast well-placed to serve national and international markets: ten years later coastal production outweighed that of the interior by more than two to one. Over the remainder of the half-century this movement continued on a world-wide basis. Refineries were established in the major oil-consuming countries and crude oil rather than refined oil products became the staple of world trade. By 1950 refining on a large and growing scale was being undertaken in all the major countries of western Europe including Britain, France, Germany, Italy, and the Netherlands.

This development had ramifications extending over the whole field of energy utilization. So long as countries like Britain were content to leave refining to producer-countries, they could confine their imports to those oil products whose use was complementary to that of native fuels. Thus Britain, Germany, and France all with substantial indigenous coal resources, looked to oil primarily as a motor fuel, and this predominant interest was reflected in American exports to Europe, which in 1929 consisted to the extent of over 60 per cent of motor fuel and only 13 per cent of fuel oil [17]. Once, however, refining had become firmly established in western Europe, the incentive to seek more varied uses for oil products increased and with it the competition offered directly by oil to coal.

Oil, unlike coal, is greedy for capital rather than men. In 1909 every man employed in the business of primary oil production in the United States represented a capital commitment of $18700; in refining the parallel invest-ment figure was $13000. Ten years later investment per man employed had increased in primary oil production to $26000 and in refining to $19900. These figures point the way to the evolution of the large-scale enterprises which in the course of time have come to be characteristic of the oil industry throughout the world. Nevertheless, as late as 1919, there were almost 10000 distinct enterprises producing crude oil in the United States of which over 8000 each employed no more than five workers. Such figures are mis-leading, however, to the extent that they may suggest an industry primarily in the hands of small firms. The American oil industry almost from its earliest years was dominated by large-scale enterprises, though, like coal-mining, it also offered scope for the smaller concern to play a useful if limited part. In 1919, 32 firms were responsible for almost 60 per cent of crude oil production in the United States, and two-thirds of these, with complemen-tary interests in refining, supplied 45 per cent of the nation's oil. Refining, by its technological needs, has always demanded a greater degree of capital concentration than has primary oil production, and although the number of refineries grew steadily—from 67 in 1899 to 147 in 1909, and then to 320 in 1919—the average investment in each refining enterprise by 1919 was over $3·5m., with far larger sums committed in the more substantial concerns [18].

The dominating role of John D. Rockefeller and the Standard Oil Com-pany in the American industry at the turn of the century had been based not so much on simple weight of investment as on control of the rail-borne carriage of oil from the oilfields to the areas of consumption. Though the Company for more than a quarter of a century survived a succession of attacks at state and federal level, its position of virtual monopoly in American oil production was first eroded by the spread of oil operations from its strong-hold in the east to districts further west in Kansas, Oklahoma, and California where new companies could take root, and then destroyed by the Supreme Court's decree of dissolution in 1911. By 1920, therefore, the threatened monopoly of a single great enterprise was clearly passed, but the economics of the industry, reinforced in the 1920s by new capital-demanding innovations, both in crude oil production with deeper drilling and in refining with the extending use of cracking processes, again strengthened the hands of the industry's major producers. By 1955 over half of the capital resources of the industry were concentrated in the hands of six large companies with total assets of $18 000m. who, like the 23 producers of 1919, provided 45 per cent of the country's refined oil products. Three of these companies had issued from the break-up of the old Standard Oil Company in 1911 and first among them stood the Standard Oil Company of New Jersey—the leading producer of both crude and refined oils—with assets of over $7000m. [19].

The dominance of large companies became even more evident when oil-producing activities were extended from a national to an international theatre of operations. The investment involved in the work of exploration and development in itself restricted the taking-up of concessions to all but the more substantial enterpreneurs. To this initial commitment was added not only the cost of operations in the field and in refining, but also the expenditure involved in the provision of pipelines and tankers for the transportation of oil from the point of production to that of consumption. A breakdown of the foreign investments of American oil companies in 1939—totalling $2·5m. in all—shows that 35 per cent of the capital invested was used in exploration and production, 15 per cent in refining, 6 per cent in transport, and 31 per cent in marketing (with a residual unappropriated investment of 13 per cent) [20]. Until 1939 oil production outside the United States and the U.S.S.R. was largely in the hands of concessionaries from one of four countries: the United States, Britain, the Netherlands, or France, with France very much a junior partner. The Americas were preponderantly the preserve of the United States, while the Far East and Middle East were dominated, though not to the point of American exclusion, by British and Dutch interests. In the rapidly developing Middle East, however, American involvement was growing fast, and by 1950 was approaching a position of preponderance. By this time the international oil companies were at the zenith of their oligo-polistic power. The complexity of their relationships with each other was matched by the strength of the vertical integration which each had achieved in its own market sector. But the days of such dominance were numbered. Already, in Bolivia and Mexico, American and British companies had seen their oil assets expropriated. Before the end of 1951 similar action by the Iranian government was to end the Anglo–Iranian concession and place oil operations in Iran under governmental ownership and control. It was the first of a series of such expropriations in the Middle East which in the next quarter-century would radically alter the structure of the international oil industry.

IV. NATURAL GAS

Though natural gas is frequently found in close proximity to oil, existing as in the North Sea in the form of a cap of gas on top of the oil, its exploitation until the middle of the twentieth century was largely confined to the United States. Here it was obtained not only in the oilfields but also, and indeed more commonly, from fields yielding only gas. In 1950 over 85 per cent of the world's natural gas was produced and consumed in the United States. Only two other countries, Venezuela and Russia, made significant use of natural gas, the first producing some 8 per cent and the other 4 per cent of the world's total supply. Smaller quantities were forthcoming in a dozen other countries, but in no case did natural gas make a substantial contribution to national energy requirements. As a result, in 1950 the fuel provided no more than one-tenth of the energy supplied by the fossil fuels to the world's consumers [21].

In the United States, however, gas, though ranking third behind coal and oil, was an important energy-provider. Its uses were more limited than those of the other fuels—it had, for example, little application in transportation—but it offered compensating advantages. Unlike oil, it needed little processing, it was cheaper to transport than coal, the heat it supplied could be easily regulated, and it created no storage problems for the consumer. Yet gas was slow in realizing its market potentialities even in the United States. At the beginning of the century it provided 3·4 per cent of the energy supplied by the fossil fuels, and even by 1920 this proportion had increased only marginally to 4·2 per cent. Thereafter, with only a check in the depression years of the 1930s, production rose steadily from decade to decade. In 1940 gas contributed 10·6 per cent of the energy supply of the United States; ten years later the proportion had reached 17 per cent and was still growing [22].

The key to growth lay, in part, in the discovery of new sources of supply, but even more in innovations which made possible the construction of a lengthy and complex system of pipelines linking the areas of gas production to the populous industrial areas of the American industrial north. By the mid-1930s gas could be readily piped over distances of 1000 miles, and during the Second World War two major pipelines were constructed linking the highly productive gasfield of Texas to the north-east. At the end of 1950 almost 300000 miles of pipes were carrying the United States' gas supply, a length of pipeline exceeding that used to transport the country's oil. Once ready access had been provided to the market, the advantages which gas offered in terms of cost and convenience asserted themselves and the fuel gained ground rapidly on both its competitors.

The proportion of gas moving to domestic and to industrial markets fluctuated over time, but the emphasis throughout was on industrial use. In 1930 the proportion of gas serving industry reached 80 per cent; twenty years later it was still almost 75 per cent. Though comparison is difficult this probably represented a greater proportionate contribution of energy to industry than was the case with either coal or oil but, by comparison with the other fuels, its industrial uses were somewhat restricted. In fact, the principal outlet for natural gas in the early years of the industry's development was in the provision of energy for drilling and pumping gas and oil wells. This proportion, however, declined as gas came to be used on an increasing scale for a widening variety of industrial and domestic purposes, and for the generation of electricity.

Like coal and oil, natural gas yields chemical by-products for industrial use. In 1945 some 10 per cent of gas was used for the manufacture of carbon black, a product of particular importance in the making of motor-car tyres; but by 1950 this market had been substantially lost to oil, which also has carbon black as a by-product of the refining process.

V. CONCLUSION

Between 1900 and 1950 output of the world's energy-providing fossil fuels increased threefold. With world population rising by 50 per cent in the half-century this represented a doubling of consumption in *per capita* terms. This statistic, however, substantially understates the growth in energy provision. Developments in fuel technology made for the increasingly efficient use of each of the fossil fuels. The precise extent of these cannot be calculated for there were losses as well as gains in the displacement of coal by electricity and in the changing balance among the three primary fuels, but overall the gains far outweighed the losses. The use of oil and the development of electric power enabled fossil-fuel energy to reach communities which hitherto had had to rely on wood, wind, and water. In particular agricultural societies experienced benefits which coal had been unable to provide. By 1950 the horse was giving place to the tractor not only in North America and western Europe but also in eastern Europe and, more slowly, in South America.

Yet though these developments made for a wider availability of mineral-based energy, the dominant role of the major industrial powers as energy consumers was in no way diminished by the extending employment of the fossil fuels. The established industrial nations consumed the newer fuels as voraciously as they had consumed the old. Consumption *per capita* of mineral fuels and water-power was still ten times as great in the United States as in Italy or Japan, and ten times as great in Western Germany as in Greece or Turkey. Greece and Turkey, in their turn, used twice as much energy per head as India and four times as much as Indonesia [23].

Such figures underline the wide differences existing in the availability of mineral energy, and in 1950 the gap between rich and poor was continuing to widen absolutely and perhaps relatively. It was the insatiable appetite of the major industrial nations for energy which prompted increasingly serious discussion of the need for fuel conservation and stimulated the search for new sources of energy.

REFERENCES

[1] WOYTINSKY, W. S., and WOYTINSKY, E. S. *World population and production. Trends and outlook*, pp. 930–1. New York (1953). [The estimates are those of potential energy and fuels without allowance for differences in degrees of utilization.]

[2] WOYTINSKY, W. S., and WOYTINSKY, E. S. *op. cit.* [1], p. 930.

[3] SCHURR, S. H., and NETSCHELT, B. C. *Energy in the American economy, 1850–1975*, p. 36. Johns Hopkins University Press, Baltimore (1960).

[4] *Colliery year book and coal trades directory*, p. 422. London (1962).

[5] SCHURR, S. H., and NETSCHELT, B. C. *op. cit.* [3], p. 116.

[6] U.S. DEPARTMENT OF THE INTERIOR, BUREAU OF MINES. *Mineral Yearbook* 1948, pp. 284–6.

[7] INTERNATIONAL LABOUR OFFICE. *Technical tripartite meeting on the Coal-mining industry* Part I, p. 86. Geneva (1938).

[8] WOYTINSKY, W. S., and WOYTINSKY, E. S. *op. cit.* [1], p. 931.

[9] SCHURR, S. H., and NETSCHELT, B. C. *op. cit.* [3], pp. 545–7.

[10] Except where otherwise stated all statistics in this section are from *Colliery year book and coal trades directory, op. cit.* [4].

[11] INTERNATIONAL LABOUR OFFICE. *op. cit.* [7], pp. 96–9.

[12] WOYTINSKY, W. S., and WOYTINSKY, E. S. *op. cit.* [1], pp. 966–7.

[13] *Colliery year book, op. cit.* [4], pp. 392, 402, 404; International Labour Office *op. cit.* [7], p. 218.

[14] WOYTINSKY, W. S., and WOYTINSKY, E. S., *op. cit.* [1], p. 930.

[15] WOYTINSKY, W. S., and WOYTINSKY, E. S., *op. cit.* [1], p. 899.

[16] WILLIAMSON, H. F., ANDREANO, R. L., DAUM, A. R., and KLOSE, G. C. *The American petroleum industry 1899–1959*, pp. 168, 805, 810, Northwestern University Press, Evanston (1963).

[17] WILLIAMSON, H. F., *et al. op. cit.* [16], p. 512.

[18] WILLIAMSON, H. F., *et al. op. cit.* [16], pp. 61–3, 111.

[19] DE CHAZEAU, M. G., and KAHN, A. E. *Integration and competition in the American petroleum industry*, p. 30. Yale University Press, New Haven, Conn. (1959).

[20] DE CHAZEAU, and KAHN, *op. cit.* [19], p. 736.

[21] WOYTINSKY, W. S., and WOYTINSKY, E. S. *op. cit.* [1], pp. 917–23.

[22] SCHURR, S. H., and NETSCHELT, B. C. *op. cit.* [3], p. 36.

[23] WOYTINSKY, W. S., and WOYTINSKY, E. S. *op. cit.* [1], p. 941.

BIBLIOGRAPHY

COURT, W. H. B. Problems of the British coal industry between the wars, *Economic History Review*, **15**, 1–24 (1945).

CHANTLER, P. *The British gas industry: an economic study.* Manchester University Press (1938).

DE CHAZEAU, M. G., and KAHN, A. E. *Integration and competition in the petroleum industry.* Yale University Press, New Haven, Conn. (1959).

DEWHURST, J. F., and associates. *America's needs and resources.* New York (1955).

——, COPPOCK, J. O., YATES, P. L., and associates. *Europe's needs and resources.* New York (1961)

EAVENSON, H. N. *The first century and a quarter of the American coal industry.* Pittsburg (1942).

INTERNATIONAL LABOUR OFFICE. *Technical tripartite meeting on the Coal industry.* Geneva (1938).

POGUE, J. S. *Economics of the petroleum industry.* New York (1939).

Political and Economic Planning. *The British fuel and power industries.* London (1947).

POUNDS, N. J. G., and PARKER, W. N. *Coal and steel in western Europe.* Faber and Faber, London (1957).

SCHURR, S. H., NETSCHELT, B. C., and associates. *Energy in the American economy 1850–1975.* Johns Hopkins University Press, Baltimore (1960).

WILLIAMSON, H. F., ANDREANO, R. L., DAUM, A. R., and KLOSE, G. C. *The American petroleum industry. The age of energy 1899–1959.* Northwestern University Press, Evanston (1963).

WOYTINSKY, W. S., and WOYTINSKY, E. S. *World population and resources. Trends and outlook.* New York (1953).

9

NATURAL SOURCES OF POWER

LORD WILSON OF HIGH WRAY

PART I: WATER POWER

I. CIVIL ENGINEERING

BY far the greatest cost of any scheme for the development of water power usually lies in the civil engineering work and not the machinery. This is, perhaps, an over-generalization, since it ignores the cost of transmitting the power; this is usually not realized by the general public and is often deliberately ignored by politicians. Although, during the early years of the twentieth century many water turbines drove machinery direct and not through the medium of electricity generation, it will be assumed for present purposes that all water power is, in fact, hydro-electric power, and that the electricity is used to supply a load which may fluctuate from full to nil in a matter of seconds.

Until 1900 most hydro-electric schemes were, to all intents and purposes, run-of-the-river; that is to say, their maximum and minimum output depended upon the flow of water available to drive them. This even applied to the Niagara Falls power station described by J. Allen earlier in this work (Vol. V, Ch. 22). However, here the amount of water taken by the turbines at full load was a small fraction of the total flow of the Niagara River, and the Great Lakes supplied all the necessary storage. The same applies to hydro-electric installations on many of the great rivers of the world, but as these feed into large networks, their output can be controlled according to the rate of flow of water available.

It is impossible to give a global estimate of the increase in the use of water power between 1900 and 1950; this is closely associated with the development of long-distance electrical transmission and the manufacture of turbines developing some thousands of horsepower for heads of from 10 to 1000 m. It was only after the end of the First World War that rapid progress was made. Few statistics are available before 1925, although there is no doubt that tens of thousands of water turbines developing from 5 to 500 horsepower were working throughout the world.

The first reasonably reliable estimates are contained in *Power Resources of the world* (1929), giving figures for 1927. From 1933 information was made available in the *World Power Conference statistical year books*, but even here there are important omissions where some countries have failed to provide information. The situation is also complicated by changes in geographical boundaries which have taken place, mainly since the Second World War.

In Table 9.1 it has been possible only to select certain countries using a substantial amount of hydro-electric power which have been consistent in providing information. The table must be taken to indicate only a general

TABLE 9.1

Hydro-electric power development in selected countries, 1927–50

	Power-station capacity (MW)			Increase on 1927 (per cent)
	1927	1933	1950	
Algeria	Nil	—	111	—
Austria	243	642	1250	410
Belgium	Nil	—	24	—
Brazil	373	—	1536	310
Canada	4590	5191	9212	86
Chile	85	—	387	355
Finland	164	288	667	300
France	1490	2700	4739	220
Germany (including Saar)	945	1269		—
Germany (West)			3398	—
Greece	6	6	34	466
Iceland	Nil	—	30	—
Japan	1305	3684	7549	475
New Zealand	45	298	633	1300
Norway	1420	1740	3008	117
Portugal	7	—	138	1900
Spain	746	1250	—	—
Sweden	1000	1156	2566	256
Switzerland	1380	1882	3057	320
United Kingdom and Irish Republic	186	—	793	325
United States of America	8744	11208	19733	125

Notes: A. K. and M. R. Biswas, Department of Environment and U.N. Agencies consultants, Ottawa, Canada, give the following information in *Water power and dam construction*, May 1976 for the proportion of world power generated by water: 1925, 40 per cent; 1970, 25 per cent; 1985 (estimated), 14·4 per cent; 2000 (estimated), 11·4 per cent. Only units rated at 1 MW and over are included.

trend. Many important users of hydro-electric power—such as the U.S.S.R. and Warsaw Pact countries, Italy, India, Australia, and some South American republics—are omitted because reliable statistics are not readily available.

The majority of large hydro-electric schemes are supplied by rivers which may vary from a small trickle for many months in the year to great floods in the rainy season, and to make efficient use of the water, storage reservoirs are necessary. They may store water only in the wet season for use in the dry season, or they may have to store enough to cover deficiencies over several dry years. One thing is certain; every such scheme requires a reservoir, and every reservoir requires a dam.

Most of the reservoirs built before 1900 were for water supply purposes, but it was basically the demand for hydro-electric power which led to the building of the vastly bigger dams during the first half of the twentieth century. The importance of dams in relation to water supply is considered elsewhere (Ch. 55).

II. LARGE DAMS

Large dams may be conveniently listed as earth dams; rockfill dams; gravity dams; massive buttress dams; multiple-arch dams; single-arch dams.

Earth dams. These follow the traditional pattern of most of the water supply dams and 'Mill Lodges' of the nineteenth century. A central core, usually of clay but sometimes of concrete, is incorporated in the centre of the dam and makes it watertight. A 'cut-off' trench is excavated down to impervious rock, and a 'cut-off' wall of concrete is built from the bottom of this trench to just above ground level. The core of puddled clay is keyed to and built up from this wall to the top of the embankment.

The purpose of the embankment is to support the core. Earth is spread in layers (or deposited hydraulically) upstream and downstream and was originally compacted by (steam) road rollers. Later, highly sophisticated and very heavy compacting machines were designed. As the science of soil mechanics was developed from the late 1930s, there was a substantial swing back to this type of dam when the necessary materials were readily available. The normal height of such dams is from 20 to 100 m.

Rockfill dams. These became popular in the U.S.A. for sites from which large quantities of broken rock could be obtained. In general they are similar to earth dams, but the mass of the dam is made up of broken rock and it is

FIG. 9.1. Owen Falls hydro-electric scheme; an example of a massive buttress dam.

made watertight by means of a reinforced concrete facing on the upstream side. One trouble with these dams is that it is impossible to compact the fill as well as can be done with the earth dam. As the water level builds up the pressure on the upstream face tends to shift individual rocks, causing the facing to deflect until leakage occurs. Repair work after this initial settlement can be expensive in terms of money and lost water.

One of the highest rockfill dams is that at Salt Springs, California, which is 100 m high; it was completed in 1930.

Gravity dams. These were first built only for water supply purposes, with hand-dressed stone on the upstream and downstream faces. They are triangular in section and rely upon their weight and correct design not to overturn or slide downstream as the water pressure builds up. Some are designed with a separate spillway, usually through a tunnel, so that they are never overtopped, and others have a lowered portion over most of the crest so that flood water overspills into a stilling basin at the toe of the dam.

Probably the most famous example is the Boulder Dam, 320 m from high water level to foundation, completed in 1930.

Massive buttress dams. These comprise a heavily reinforced concrete upstream face, inclined at a small angle to the vertical, and supported downstream by a series of buttresses. They are more economical in the use of concrete than gravity dams, but require complicated shuttering involving the use of costly skilled labour. The Loch Sloy dam, built for the North of Scotland Hydro Electric Board, and designed by James Williamson and Partners, was the first to be built in the United Kingdom. This dam, 49 m high with a span of 354 m was completed in 1950.

Multiple-arch dams. These are similar to buttress dams, but the upstream face comprises a number of inclined arches having a greater span and consequently a smaller number of buttresses.

Single-arch dams. These are used in mountainous regions where a high dam is to be built across a narrow rocky gorge with sound rock on either side and in the foundation. Viewed in plan, the dam is arched facing upstream; as the pressure of water tries to flatten the arch it is transmitted to the side of the valley which contain it. These dams are often remarkably thin, and call for the highest possible quality of design, materials, and workmanship. The only ones built in the United Kingdom were for the Galloway hydro-electric scheme during the 1930s. In western Europe conditions are much more suitable. Of the dams completed by 1950 the highest was Santa Giustina in Italy, with a height of 152·5 m and a crest length of only 90 m. In the construction stage at that time was the Vajont Dam, also in Italy, with a height of 207 m and a crest length of 130 m. These were ideal sites for dams of this type.

III. TUNNELLING AND UNDERGROUND POWER STATIONS

By 1900 many long and difficult railway tunnels had been completed (Vol. V, Ch. 21), but, apart from the Niagara Falls power plant, no exceptional tunnelling had been undertaken in connection with hydro-electric schemes.

Engineers soon realized that when, as often happens, a river turns through 180° rounding a shoulder or joining a tributary, much the shortest way of getting the water from a high level to a lower level is by driving a tunnel through the hill, and carrying the water through a steeply inclined 'pressure penstock' (pipeline) from the downstream tunnel portal to the power house on the river's edge. Hence a conventional hydro-electric plant layout became:

river–reservoir–low-pressure tunnel–high-pressure penstock–turbines in powerhouse–tail-race–river.

Tunnelling techniques (Ch. 36) improved slowly, an important factor being the air-operated drills for making the holes into which the blasting charges were placed. The long tunnel through hard rock was soon a normal feature which could be accurately costed. The next stage was to put the whole scheme underground. Here Sweden pioneered the way. A private company built a 12-MW underground power station at Mockfjärd, in northern Sweden, which was completed in 1910; it was soon followed by the Porjus station on the Lule river in Lapland, north of the Arctic Circle. This was designed to electrify the railway which connected the Malberget and Kiruna ore fields, the cost of running which was very high because all the coal had to be imported. The station was designed to house five 10-MW sets working under a head of 58 m and was built underground (under appalling conditions so far as the workmen were concerned, owing to the low temperatures and long winter). It had all the features of a large, modern underground station. The machine hall was 90 m long, 12 m wide, and 20 m high. The pressure penstock was carried vertically downwards by a shaft lined with steel tubes, and the turbines discharged into a tail-race tunnel 1150 m long. Work was started in 1910 and the first generator put on load in October 1914.

During the Second World War the Swedes became expert at excavating in hard rock when they built huge underground air-raid shelters and stores in or near their main cities. By the late 1940s engineers throughout the world were planning and building underground power stations instead of surface stations for economic and not necessarily for strategic reasons.

IV. PRESSURE PENSTOCKS (PIPELINES)

The use of the term pressure penstock instead of pipeline for the closed conduit leading water to the turbine(s) has become usual during recent years. The pressure, or high-pressure, penstock is that portion which is open to atmospheric pressure at the top, and to the full pressure driving the turbine at the bottom. As has been explained in the previous section, it is usual to convey water from the reservoir or intake to a point above the turbine house through a tunnel, open channel, or low-pressure pipe down the hill side. Every effort is made to keep the pressure penstock as short as possible, in order to reduce the difficulties of governing the turbine when there are heavy and sudden load changes. At the end of the tunnel or low-pressure penstock a surge chamber is built; its top comes above the water level at the point of supply and it is open to the atmosphere. The surge chamber can be a simple pipe or a complex system of pipes and orifices of various diameters. If a heavy load is suddenly thrown off the turbine the gates will close and the flow of water down the pressure penstock will be reduced, resulting in a pressure wave passing up the pipe which can reach dangerous dimensions. When this pressure wave reaches the surge chamber the water will overflow and it becomes, in effect, a safety valve.

Pressure penstocks are usually made of steel and are carried above ground on concrete piers. In large schemes the thickness increases at the lower end where the pressure is highest. Any bends must be firmly anchored and provision must be made for expansion.

For power stations in mountainous regions, where the head may exceed 1000 m, the tubes forming the pressure penstock are strengthened by reinforcing rings or prestressed binding wire. This avoids using very thick steel plate which could present welding problems.

One of the objections to the building of hydro-electric power stations in areas of natural beauty is the sight of a number of large pipes, usually painted with black bituminous paint to prevent corrosion, supported on concrete blocks defiling the beauty of an unspoiled mountainside. For example, in the Loch Sloy scheme in central Scotland, the reservoir and dam are situated in a remote and little-visited valley, but the four steel pipes which feed the turbines are regarded as an eyesore by the thousands of visitors who drive along the shores of Loch Lomond every year.

This matter of amenity—which over the years has become increasingly a matter of public concern—and the simplification of the design and manufacture of large pressure penstocks, are additional arguments for the construction of underground power stations. The pressure penstock then becomes a vertical shaft, and its length is approximately the gross head on the turbine. It may be lined with a steel tube embedded in concrete, but the tube need only be thin, for the surrounding concrete and rock will carry the stress.

V. VALVES

Turbine-isolating and emergency valves can be regarded either as a part of the civil or mechanical engineering side of a hydro-electric scheme. Emergency self-closing valves are installed at the top of the pressure penstock, their basic purpose being to shut off the water if the penstock should fail. They are usually butterfly valves held in the open position and can be tripped to close if the flow through the valve becomes excessive owing to a failure downstream. Closing is effected either by dead-weight or hydraulically. The upstream pressure on an emergency valve is usually low, not more than a few metres of water, and these valves represent a small element in the cost of a complete scheme. However, with underground power stations they can be omitted altogether, and this represents a saving.

Turbine-isolating valves are essential, and their design has become highly sophisticated. It must be possible to shut off water from the turbine when maintenance is required, and, in general, when the turbine is not generating power. The frictional loss through the valve must be kept to a minimum, since this represents lost power; and almost inevitably friction is the result of eddies which set up a scouring effect which will damage the seating faces.

Rotary, as opposed to sluice, valves are nearly always used if the diameter is more than 1000 mm and the head exceeds 100 m. They are hydraulically or electrically operated and have flexible seatings which close on the moving element of the valve when it is in the closed position. Butterfly valves are also used when a slightly greater friction loss can be accepted and capital cost must be reduced to a minimum. The English Electric 'Straight Flow' rotary ball valve has been very successful, but such valves form a significant proportion of the total capital cost of the hydro-electric machinery. It must however be realized that in power stations which are used to carry peak loads, and

particularly in pumped-storage stations, large and heavy valves may be operated several times in a day, and must be of the highest quality.

VI. WATER TURBINES

It would be quite untrue to say that there has not been a great development
in the field of water turbines since the beginning of the century. Nevertheless,
with the exceptions which will be mentioned later, the two basic classes of
water turbine—namely, reaction and impulse turbines—were firmly established by 1900, had achieved high efficiencies, and were well designed. Many
small turbines installed between 1870 and 1900 were still working in 1950.
The changes which have taken place have been mainly in the size of units
and the steady increase in the specific speed of reaction turbines.[1]

This has not produced a great increase in efficiency, as has taken place
with the development of large steam turbines (Ch. 41), but has resulted in
vastly bigger hydro-electric stations being built. In 1900, 5 MW was regarded
as a very large turbine unit, but by 1950 the size of individual turbines had
risen to 120 MW, and much more powerful machines were on the drawing
board.

The move towards higher speeds has followed the normal technological
trend. As a general principle, the faster a prime mover runs, the smaller and
cheaper it becomes. This applies particularly if it drives an electric generator.
There are, of course, limiting factors. The 50-cycle steam turbo-generator
cannot run at more than 3000 rev/min nor the 60-cycle machine at more than
3600 rev/min. The speed of a hydro-electric generating set is dictated by
design factors and the fact that if the full load is suddenly removed, and the
governor should fail to act correctly, it will 'run away' and in a few seconds
may be running at from 60 to 120 per cent above its normal operating speed.
This imposes very high centrifugal stresses upon the windings of the electric
generator driven by the turbine, and greatly increases the cost of the machine.

In the following sections it is convenient to deal with turbines according
to their types, but there are many features that are common to all.

Francis turbines (Vol. V, p. 529). These take their name from J. B. Francis,
who carried out extensive practical work on inward-flow reaction turbines at

---

[1] Specific speed is a term used by turbine designers to designate the speed of rotation of a family of turbines of similar design but widely differing size. A turbine developing 1 MW under a head of 30 m and
running at a speed of 450 rev/min will have a specific speed of 244. A similar but larger machine developing
2 MW would have the same specific speed but would run at 340 rev/min under this head. If a turbine
could be designed for a specific speed of 310 it would run at 428 rev/min.

Lowell, Massachusetts, in the 1850s, and are in general use for powers of from 25 to 120 MW and heads of 20 to 400 m. For units over 10 MW the hydro-electric set is nearly always of the vertical shaft arrangement with the generator above the turbine; a substantial Michell (or Kingsbury) tilting-pad bearing mounted above the generator carries the weight and hydraulic thrust of the turbine runner (wheel) and shaft, and of the generator rotor. The flow of water to the runner is controlled by a series of streamlined guide vanes operated by the turbine governor. Speeds may vary from 900 to 75 rev/min according to the size of unit and operating head.

All reaction turbines, and these include Francis and fixed- and movable-blade propeller turbines, may suffer from cavitation if the specific speed or the suction head, that is to say the 'negative' head between the turbine runner and the water level in the tail-race, is too high. Cavitation, like specific speed, is difficult to describe without going into technicalities. Water, under a high pressure and velocity, gives up its energy and loses pressure as it passes through the turbine guide vanes and runner into the suction or draft tube leading to the tail race. If, owing to unsatisfactory design, the pressure of the water falls below the vapour pressure, a vapour-filled bubble or 'cavity' will form and will later collapse when it comes into contact with the metal surface of the turbine runner or water passages, delivering a powerful blow over an area which may not exceed that of a pin-point. As millions of these blows are struck in the same area, the metal will break down and be eaten away, forming first 'worm holes' and later reducing the metal to a spongy form leading ultimately to breakdown. Turbines working at high specific speed and low head are most likely to be affected by cavitation; the designer is under constant pressure to increase the speed of the turbine to reduce the price of the generator, and the civil engineers press him to raise the power house for the sake of safety in the event of exceptional floods. In fact, protection from cavitation can often be obtained only by placing the turbine some metres below the tail-race level.

Large Francis turbines can have full-load efficiencies above 90 per cent, and good efficiencies to below half load. They are extremely reliable if, as is usual, they are supplied with relatively clean water from a reservoir.

Propeller, Kaplan, and bulb turbines. These all have runners which somewhat resemble the propeller of a ship. They have a high specific speed, and are designed for low-head operation (about 10 to 40 m) and outputs up to 50 MW or sometimes more.

The fixed-blade propeller turbine appeared to give the answer to the problem of developing large power stations with low heads, usually run-of-the-river plants where a series of rapids would be bypassed. The very high specific speed, over 50 per cent above that of a Francis turbine of high specific speed, brings about a general reduction in size, but one disadvantage of the propeller turbine is its high runaway speed. A further disadvantage is the low efficiency, from 75 per cent full load downwards, and the danger of cavitation.

The problem of low part-load efficiency was sometimes overcome by installing up to ten propeller turbines in a single station and operating as many as possible at full load according to the flow of water available. However, there was a strong move to reduce the number of units and increase their size, and the Kaplan turbine gave the answer.

The Kaplan turbine was developed by Victor Kaplan, a German engineer who carried out his design and experimental work at Brunn in Czechoslovakia. He realized that if the propeller blade angle could be reduced as the quantity of water dropped, much better part-load efficiency would be obtained. He worked from 1910 to 1924 on the problem of fixing the blades in the hub so that their angle could be adjusted without stopping the turbine; by operating the runner blades and guide vanes in unison astonishingly high efficiencies could be obtained even down to very low proportions of full load. The fact that each runner blade can be individually profiled on a special machine gives great accuracy in design and by giving a very smooth surface reduces the danger of cavitation.

Kaplan turbines are expensive to build, and the economic advantage of their operation must be carefully weighed against the high capital cost. They are being used in powers of over 50 MW and heads up to about 30 m.

Bulb turbines are a direct development of the Kaplan turbine, but are axial-flow machines comprising a propeller-type runner, usually with movable blades, placed in a pipe with the water directed on to it by guide vanes. The electric generator is directly coupled, and may be of the waterproof type surrounded by the water discharged from the turbine. The initial purpose of the design was to provide a complete hydro-electric set which could be built into a dam or barrage, possibly one already in existence, with the minimum of difficult concrete foundation work and without a formal power house as such.

The design was developed in France and Switzerland during the Second World War, and formed part of a plan to improve the navigability of the large rivers of southern France by building barrages with locks and simple, compact, hydro-electric units. It was decided that reversible bulb-type turbine generator sets would be suitable for the Rance Tidal Project (see below under Tidal Power, p. 216) and the huge St Michel scheme which was subsequently shelved indefinitely.

The Deriaz turbine, called after Paul Deriaz, who was the chief designer for the English Electric Co., is a cross between a Kaplan and a Francis turbine. It has movable runner blades set at an angle to the hub, and is designed for heads up to 100 m and outputs up to 75 MW. Like the Kaplan turbine it has a high efficiency throughout its range of operation but is costly to build.

Impulse turbines have changed relatively less than reaction turbines. The principle of the bifurcated bucket of ellipsoidal form had been worked out by Lester Pelton (Vol. V, p. 530) and Abner Doble in the United States, and the latter had also developed the spear or needle nozzle which gave a well-formed jet with low flows of water. The multi-jet Pelton wheel (to give a higher speed) was in use in a crude form before 1900, and improvements were mainly in size, methods of construction, and slightly increased efficiency.

Up to about 1920 all Pelton wheels had bolted-on buckets, but as the power of individual units increased trouble was experienced with bucket-lugs or fixing bolts breaking owing to fatigue resulting from the hammer blow of the jet striking the bucket. Gilkes, a firm of water turbine manufacturers at Kendal, England, pointed the way to solving this problem by making an integrally cast Pelton-wheel runner in 1920. The casting technique was difficult, but other manufacturers followed the lead that Gilkes had given, and by 1950 all high-powered Pelton-wheel runners were being cast in one piece.

The Pelton wheel has a low specific speed (about 25 compared to a high-head Francis turbine of 100) and for units of high power vertical-shaft machines were built with up to six jets developing some 50 MW.

A completely new design of impulse turbine was produced by Eric Crewdson, managing director of Gilkes, in 1920. His invention comprised an axial-flow impulse turbine in which the jet struck the runner at one side and was discharged at the other. Fig. 9.3 shows the basic difference in principle between the Pelton and turgo impulse turbine. The result of Crewdson's invention was that a jet of much larger diameter could efficiently be applied to a runner of the same mean diameter as a Pelton wheel, giving a higher speed with negligible loss in efficiency. This single-jet impulse turbine of simple construction has found a wide market for a medium-head machine

FIG. 9.2. A water turbine, Pelton wheel type.

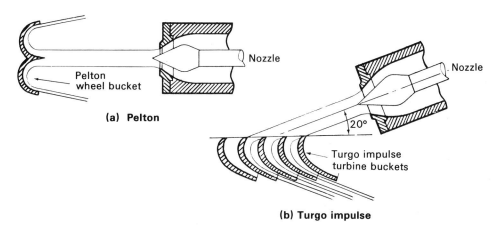

(a) Pelton

(b) Turgo impulse

FIG. 9.3. The difference between the principle of operation of the Pelton wheel and the turgo impulse turbine.

throughout the world. The power range is from 0·1 to 4 MW, with heads of from 75 to 300 m.

VII. WATER-TURBINE GOVERNORS

Water-turbine governors have changed completely during the last fifty years, and are now highly sophisticated pieces of control equipment, the cost of which is a very substantial proportion of the total cost of the hydro-electric equipment.

Except in rare cases when the turbine is supplying power for a fixed load, such as a centrifugal pump or induction generator, the speed must be kept constant over the full range from no load to full load. Water wheels driving textile factories were provided with crude mechanical governors which, using heavy fly-balls and a reversing gear mechanism, slowly closed the inlet sluice or shuttle if the speed rose as looms or spinning machines were stopped, and opened it when the load increased. Similar mechanical governors were used for early water turbines, and were reasonably satisfactory when they were driving factory machinery. However, as more and more turbines drove electric generators the governing problems increased. If a hydro-electric set was supplying electricity to a town, the load could vary quickly, and if there were a fault such as a power line breakdown or a main circuit-breaker tripped, the full load might be thrown off instantaneously. The average water turbine will 'run away' under such conditions if not restrained, and the governor must very rapidly check the rise in speed.

With a steam engine, steam turbine, or internal combustion engine, governing is easy; all that is necessary is to control the fuel (or steam) supply as the load varies, and this can be done by a simple centrifugal governor. With a water turbine the problem is much more difficult. When water is flowing at a speed of 2–3 m/s down the pressure penstock supplying a large turbine the energy stored is comparable to that of a moving ship weighing thousands of tons, and the danger and difficulty of trying to stop it suddenly is comparable. If the flow is suddenly checked by closing the turbine guide vanes or spear nozzle, a pressure wave will build up which may well burst the pipe or blow out a joint with catastrophic results. A great deal of power is also required for a few seconds to operate the turbine guide vanes.

The basic conception of a modern water-turbine governor is a hydraulic servomotor which opens and closes the turbine control mechanism. Early in the century such servomotors were supplied with water under pressure from the penstock, but dirt tended to clog the valves and build up on the main

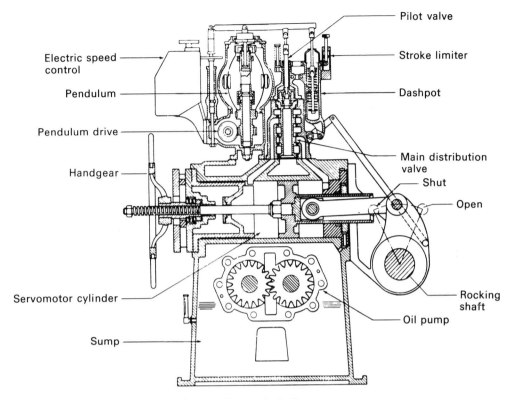

Electric speed control

Pendulum

Pendulum drive

Handgear

Servomotor cylinder

Sump

Pilot valve

Stroke limiter

Dashpot

Main distribution valve

Shut

Open

Rocking shaft

Oil pump

FIG. 9.4. Section of a gateshaft oil-pressure governor.

cylinder and piston. By about 1910 oil servomotors supplied by a pump worked from the turbine shaft were becoming standard practice (Fig. 9.4). Sensitive centrifugal pendulums of the Pickering type, or with knife-edge suspension, replaced the heavy Watt-type fly-balls, for all they had to do was to operate a small relay valve which allowed oil under pressure to flow to one end or the other of the main servomotor.

The pressure rise in the penstock is limited by the closing time of the turbine guide vanes; the longer the penstock the more slowly must the guide vanes close. At the same time the rise in speed has to be controlled by the use of heavy flywheels. The length of the pressure penstock is a dominating factor, and this question has been dealt with above (p. 201). When the pressure penstock is long relative to the head, a governor-operated relief valve is usually installed. This valve is opened as the guide vanes close, thus maintaining approximately constant flow down the penstock. The relief valve then closes slowly, keeping the rise in pressure to an acceptable figure.

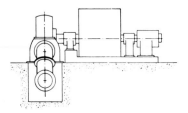

(a) Pelton wheel
Head: 400m
Speed: 750 rev/min

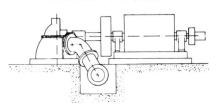

(b) Turgo impulse turbine
Head: 200m
Speed: 1000 rev/min

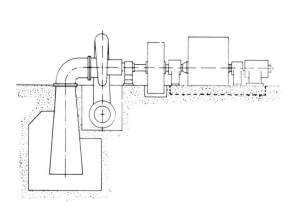

(c) Horizontal Francis turbine
Head: 120m
Speed: 750 rev/min

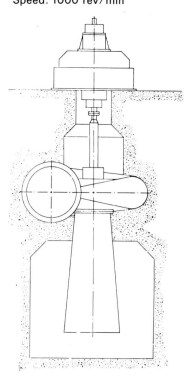

(d) Vertical Francis turbine
Head: 40m
Speed: 428 rev/min

0 1 2 3 4 5 6 7 8 9 10m

FIG. 9.5. Four of the most usual hydro-electric generating set layouts—each develops 2 MW.

With impulse turbines the problem is less difficult. The governor operates a jet deflector, which diverts some of the jet from the turbine runner when load is reduced. This does not affect the flow of water down the penstock, and as the deflector can be swung into the jet in a second or so, there is no need for a heavy flywheel. The spear can be closed slowly by means of a servomotor.

VIII. GENERATORS DRIVEN BY WATER TURBINES

Generators driven by water turbines have some unusual features to be taken into account. The first is that the generator must be designed to withstand safely the runaway speed of the turbine which has been referred to in the previous sections; it has proved impossible to design a really reliable emergency governor for a water turbine which will check its speed if the ordinary governor fails. Such emergency governors are provided for thermal prime movers, so that the generator need be designed to be safe only up to, say, 25 per cent above its normal speed. As the centrifugal stresses rise very fast with speed, generators driven by water turbines are usually much more costly than generators of the same power driven by thermal units. In addition, for reasons of good governing, as already set out, additional flywheel effect may have to be built into the generator rotor. This applies particularly to vertical shaft units, where it is difficult to fit in a separate flywheel.

Typical layouts for hydro-electric generating sets are illustrated in Fig. 9.5. As has been explained earlier, all large hydro-electric units (those over, say, 20 MW), both reaction and impulse, are of the vertical shaft arrangement, usually with one or two generator journal bearings, one turbine journal bearing, and a thrust bearing at the top of the generator which carries the entire weight of the rotating parts together with the hydraulic thrust from the turbine runner. This load may run into hundreds of tonnes and the design of the thrust bearing is of critical importance. All are of the tilting pad (Michell or Kingsbury) type, and when starting up oil under pressure is pumped through the pads to raise the thrust collar clear before the machine starts to rotate. Brakes are usually fitted to bring the machine quickly to rest on shutting down and prevent damage to the bearing pads at slow speeds of rotation, when the oil film may break down.

Increasing use is being made of medium-powered hydro-electric sets (0·5 to 10 MW) incorporating induction generators. These generators are similar to squirrel-cage electric motors driven backwards. They must be excited by an alternating current supply, and must feed into a relatively large network. In other words they cannot run as independent units, and may be described as machines to 'pump' electricity into the mains. The unit is run up to synchronous speed and the breaker is closed, connecting the generator to the network supply. If the turbine is then opened up, the generator will feed into the network, its contribution increasing until the turbine is fully opened. The great advantage of such units is that the turbine governor can be eliminated, and the generator itself is of very simple construction. They are usually designed for unattended operation, and can be made fully automatic. For example, at Loch Cuaich, in the north of Scotland, a 2·5-MW induction generator set is fed from a small loch, the level of which is liable to vary quickly. When the loch is full, the turbine starts up by float control; is automatically synchronized and connected to the mains, and continues to run until the loch is drawn down to minimum level. The unit is then disconnected and stops until the loch fills up.

IX. PUMPED STORAGE SCHEMES

One of the main problems facing the electricity supply industry is that of coping with peak loads. In the early years of this century most of the electricity generated for municipal distribution was used for electric lighting; from about 7 to 9 a.m. in the winter the lights would be switched on, and again there would be a peak load from 4 to 10.30 p.m. The average load for 24 hours might be 250 kW, but the peak load might rise to 1000 kW, and the supply authority must be able to supply the peak and not just the average. At night the load might be as low as 25 kW for 6 or 7 hours. This is one reason why direct current, as opposed to alternating current, was then so often used. The municipal undertaking could install a storage battery to carry the night load and help to cope with the peak. If alternating current was supplied the generating capacity had to be able to cope with the peak load and run continuously.

A number of small rural electricity undertakings in the United Kingdom— Keswick, Windermere, and Lynton and Lynmouth are typical examples— had to generate alternating current because they had long transmission line problems. They were able to carry the base and night load with small hydro-electric units backed up with steam or diesel generators to deal with periods of peak load.

The Lynton and Lynmouth Company in 1890 installed two 110-kW a.c. generators driven by Francis turbines, but as the use for electricity increased these could not take the peak load. Later the company installed a Bailey pump driven from one of their turbines when the load was low and built at the top of the waterfall a reservoir which was filled by the pump. This was 244 m above the power station and when the peak load came on a generator driven by a Pelton wheel, drawing water from this high-level reservoir, was used to help to carry it. This scheme was still working in 1950.

This was an exceptional case; the general theory of pumped storage was developed in Germany and Switzerland during the late 1920s and early 1930s. The pumped storage scheme was, in fact, similar to a storage battery. Thermal power stations run efficiently only at or near their full load, and if, for example, the normal maximum load on a network is 400 MW, with a peak of 500 MW for a few hours, it may well prove economic to provide a 400-MW base-load thermal power station and a 100-MW pumped storage station to carry the short-term peak loads.

The essentials of a pumped storage station are a low-level and high-level reservoir as close together as possible, but with a maximum difference in level, and a water turbine–pump–generator–motor unit. At night, or at any time when the load on the system is low, the reversible motor–generator will drive the pump and transfer water from the low-level to the high-level reservoir. As the peak load period approaches the turbine is started, the motor runs as a generator, and water flowing from the high-level to the low-level reservoir provides the power to carry the load.

The economics of the pumped storage scheme are complex; the site must be such that the maximum head can be obtained, and the capital cost must be kept to a minimum. The first large pumped storage station in the United Kingdom, at Blaenau Ffestiniog in north Wales (1961), is expected to be an economic success as it is close to the atomic power station at Trawsfynydd.

X. POWER RECOVERY TURBINES

Power recovery turbines may be roughly described as those which work with an artificial head not provided by gravity. Examples are:

Chemical process plants. Water under high pressure is pumped into scrubber towers to absorb carbon dioxide in the manufacture of ammonia. The water, containing carbon dioxide in solution, is discharged through a turbine which is coupled to the motor driving the pump which feeds the scrubber tower. In a continuous process up to 50 per cent of the power of the motor can be recovered.

Waterworks. All London's water, provided by the Metropolitan Water Board, is pumped. At many of the pumping stations there are high-pressure and low-pressure mains. In such situations, it is often convenient to by-pass water from the high-pressure to the low-pressure main through a turbine and make use of the power so generated.

Thermal power stations. At some power stations, for example, the 'Back o' the Bank' Power Station at Bolton, Lancashire, condenser cooling water is pumped up to cooling towers and flows back to the cooling pond. It passes through turbines which assist the motors driving the pumps. Power-recovery water turbines are also used when 'dry cooling towers' are built to save loss of water by evaporation in hot climates.

BIBLIOGRAPHY

BINNIE, G. M. Dams. Presidential Address to Junior Institution of Engineers, 1955. *Water & Water Engineering.* February 1956, pp. 56–66.
BROWN, J. GUTHRIE. *Hydro-electric engineering practice*, Vols. 1-4 (1958).
CREAGER, W. P., and JUSTIN, J. D. Dams. In *Hydro-electric Handbook* (1949).
SANDSTROM, G. E. *The history of tunnelling* (1963).
WILSON, PAUL N. *Water turbines.* H.M.S.O., London (1974).

The monthly technical periodical *Water Power* publishes relevant articles.

PART II: OTHER NATURAL SOURCES OF POWER

The principal sources of power have been considered in earlier chapters of the present work, but their general study would be incomplete without some reference to minor sources, notably the wind, tides, the heat of the earth's crust, and the sun. During the first half of this century these were relatively unimportant, but it is possible that in the future—as the world's energy problem changes—they may become more significant.

I. WIND POWER

The use of wind power has never been popular and it declined steadily during the first half of the twentieth century. It can be said that it has not been used if there was a viable alternative, but such is the sentimental attraction of the conventional windmill used for grinding corn or for raising water in fenlands that the general public have possibly tended to think that the value of windmills as power generators is considerable, and that in times of power shortage and increasing fuel costs an important source of natural power is being neglected.

The very fact that thousands of windmills in western Europe have vanished, and those that have been preserved are maintained only at great cost, proves this point. Their history has been told in earlier volumes of this work

FIG. 9.6. A wind pump from the American west.

(Vol. II, Ch. 17; Vol. III, Ch. 4). If wind power is to be used on an appreciable scale in present circumstances there must be very special reasons.

Wind pumps (Fig. 9.6), sometimes referred to as wind motors, are used extensively in South Africa and in other sparsely populated agricultural regions where winds tend to be steady and underground water sources can be tapped. Aesthetically they have little to commend them. A circular wind-wheel with some twenty pressed-steel sails, the whole up to 4·25 m in diameter, is mounted at the top of a lattice-work tower 7–10 m high. The wind-shaft, mounted in ball bearings, drives a reciprocating pump rod through reduction gearing and connecting rods. All these working parts are contained in an oil bath so that the minimum of maintenance is required. The pump shaft is carried down the centre of the tower and works a simple pump cylinder at the bottom of the borehole or may pump from a spring or stream to a high-level tank. The volume of water pumped is small, sufficient only for domestic supply or watering cattle. A tail vane keeps the wheel turned into the wind, and will deflect it if the wind speed rises above about 11 m/s; the starting speed is about 4 m/s. The power of such wheels is about 750 W, but given sufficient water storage several consecutive windless days can be tolerated.

The utilization of wind power for the generation of electricity was greatly stimulated by fuel shortage during the Second World War. Small direct current (d.c.) generators driven by aeroplane-type propellers mounted on well-guyed masts from 10 to 13 m high were installed in remote country districts, usually in coastal areas with a reasonably constant prevailing wind. With wind speeds of from 7 to 15 m/s they could deliver possibly 1000–3000 W, charging storage batteries; 32 volts was usual. This was sufficient to supply electric light and to drive a few small electric appliances.

In Denmark, where the fuel shortage was acute, the engineering firm of F. L. Smidth experimented with wind generators having outputs of from 5 to 100 kW, but they did not continue this work when conditions returned to normal. In the United States an alternating current (a.c.) generator with an output of 1250 kW was erected at Grandpa's Knob, Rutland, Vermont; and in the United Kingdom, a 100-kW unit was installed at Costa Hill in the Orkney Islands. Similar experiments were carried out in the U.S.S.R. and some German designs for even larger machines were produced in the period 1930 to 1950. So far the problem of designing a wind generator which will produce electricity at wind speeds of from 5 to 20 m/s and not be destroyed by a gale or hurricane remains to be solved; the generators at Grandpa's Knob and Costa Hill did not survive for long.

Wind generators with an output of from 1 to 10 MW would be enormous and ugly; by comparison, 100 m-high electricity pylons would appear to be works of art.

11. TIDAL POWER

Man has harnessed the power of the tides since before the Norman Conquest of Britain. A tidal basin, usually part natural and part artificial, is constructed in the estuary of a river where the tidal variation is considerable, having a mean range preferably not less than 4 m. A watermill with an undershot, low-breast, or Poncelet waterwheel is built into the dam which divides the basin from the sea, and tidal sluice gates allow the basin to fill as the tide comes in. When the level of water in the basin is at a maximum the gates are closed and the miller waits until the tide has fallen below the bottom of the floats or buckets of his waterwheel. He then admits water from the basin to the wheel, and works it for as long as he can, probably reducing the feed to his millstones as the water level in the basin sinks.

This is the basis of all simple tidal power schemes, but with the advent of the Kaplan adjustable-blade propeller turbine (p. 204) which can operate over a wide range of heads at high efficiency and can also be designed to raise the level of the water in the basin to an artificially high level by pumping, a highly sophisticated system of operation can be worked out which enables electricity to be generated for a very much longer period than is possible with the simple system.

The generation of tidal power suffers from the following basic disadvantages:

(i) The power generated is dependent upon the tidal regime.

(ii) The power generated must be fed into a network that is sufficiently large to absorb the maximum amount of power at any hour of the day or night.

(iii) The mean tidal variation must be as great as possible and the physical characteristics of the tidal basin must also be suitable.

(iv) The capital cost will inevitably be high, but may be offset if the tidal power scheme can produce other advantages.

With steadily increasing sizes of network the first two disadvantages are becoming less serious, and in countries like France where there is a high tidal variation on the Channel coast and a number of large conventional hydro-electric stations below the Alps, the power from the latter can quickly be reduced as the tidal power comes into the network. By contrast, in the United Kingdom, dependent largely upon thermal power stations which only reach maximum efficiency at a high load factor, the difficulties are likely to be considerably greater.

It is surprising what a tiny fraction of the coastline of the world has a mean tidal variation of more than 6 m. In a comprehensive study, N. Davey (see bibliography) listed only fifteen substantial stretches of coastline where this occurs. The highest mean range is in the Bay of Fundy (Nova Scotia) with a mean range of 14 m; the Sam Sa Inlet in China has a range of 6 m. It is possibly because the tidal range in the English Channel and up the west coast of Britain is reasonably high (8·75 m over the Severn Tunnel and 7 m in Morecambe Bay) that citizens of the United Kingdom tend to think that such tides are normal round most ocean coasts. In reality, this is not so.

FIG. 9.7. La Rance tidal power station, France. The photograph shows the station nearing completion in 1966, the coffer dam of sand-filled cylinders will be removed when the station is complete. The lock for shipping is on the right. A road across the works shortens the distance from Saint Malo to Dinard by some 30 km.

A number of investigations for a tidal power scheme have been carried out for the Bristol Channel, but have been shelved, usually because of high capital cost and the number of interests—shipping, fishing, effluent disposal, etc.—which become involved.

Much more detailed survey work has been undertaken by Electricité de France for a scheme using reversible-pump turbines with a barrage only some 750 m long across the Rance south of Saint Malo (Fig. 9.7); this will house twenty-four 10-MW turbines which can work on the incoming and outgoing tides. The distance from Saint Malo to Dinard will be reduced by some 30 km and shipping facilities up the estuary will be considerably improved. These are ancillary advantages which help the economics of the project.

Probably the best site in the world is that at the Bay of Fundy, close to the Canadian–American frontier. Preliminary survey work has gone on for a number of years, and if the search for natural sources of power is speeded up a vast tidal power scheme may be built there, feeding electricity into both of these power-hungry countries.

III. GEOTHERMAL POWER

The only existing large and successful scheme for the generation of power from underground steam is at Larderello in Tuscany, Italy. Here, since written records first mention them at some time B.C., fumaroles, which are crevices in the ground in volcanic regions, have poured large quantities of steam, heavily charged with chemicals, into the air. The discharge pressure varies from 5 to 27 kg/cm^2, and the temperature from 140° to 260 °C. Boreholes are drilled to tap the steam, a difficult and often dangerous undertaking; they vary in depth from 300 to 600 m. Deeper boreholes will probably be sunk. The approximate composition (weight per cent) of the steam is as follows: H_2O, 95·50; CO_2, 4·20; CH_4 and H_2, 0·01; H_2S, N_2, NH_3, Boric acid (H_3BO_3), He, Ar, Ne, 0·29. Of these the boric acid and ammonia are at present recovered and processed, thus making a contribution to the economy of the project.

Efforts to harness the steam for power were first made in 1904, and a year later a 25-kW steam-driven generator was installed. It was only with the coming of the steam turbine that any significant power development took place. In 1913, a 250-kW generator driven by a steam turbine was installed, followed by three of 1250 kW in 1914. By the outbreak of war in 1939 a total of 16 sets were operating, generating 135 MW.

The Germans, in their retreat from Italy, destroyed most of the installations, but these were soon replaced, and by 1950 20 sets generating 254 MW were making a substantial contribution to electricity supply in Italy.

Fig. 9.8 shows diagrammatically the two systems generally adopted. On the

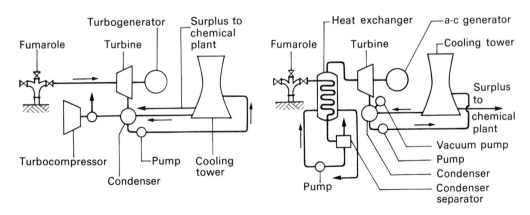

FIG. 9.8. Alternative methods of power development from geothermal sources.

left steam from the fumarole drives the turbine direct, and the gases and liquid to be processed are drawn off from the condenser or cooling tower. On the right a more costly arrangement is depicted: the steam passes through a heat exchanger and only pure steam feeds the turbine.

Small geothermal generating sets have been installed in Ischia, but as the temperature difference between the hot spring water (55 °C) and the cooling sea water (25 °C) is small, ethyl chloride is used in a heat exchanger to drive a 300-kW unit.

The next significant developments are expected to take place in New Zealand, where conditions in the area of hot springs near Taupo, are similar to those at Larderello, though it is expected that the boreholes may have to be much deeper. It must be appreciated that both in Italy and New Zealand the only natural power resources are hydro-electric, and there is a limit to their exploitation. In New Zealand, in particular, most of the hydro-electric power is in the South Island, but the main industrial development is in the North Island. These are important factors to be taken into consideration, for the capital cost of geothermal power development, like that of hydro-electric power, can be very high.

Bearing in mind the fact that power and heat are interchangeable, it would be unrealistic not to include in this section the direct use of underground sources of heat for domestic and industrial use.

The heating system at Reykjavik, the capital of Iceland, is unique, and, thanks to the natural conditions, the capital cost has not been excessive. The initial experimental work was carried out in 1930, when water at a temperature of 95 °C, was piped from hot springs at Pvottalaugar, 3 km east of the city, to a service reservoir and used for limited domestic heating.

The present scheme was started during the Second World War, and is still being developed. Conventional boreholes were sunk to hot springs at Reykjalaug, 17 km east of Reykjavik, and the water pumped to an insulated reservoir above the city. The pipes are insulated with a mixture of chopped turf and pumice, both of which are available on the island. They are enclosed in a trunking of concrete slabs, and are mostly laid above ground. The insulation is so good that there is a drop of only 5 °C between the pump outlets and the inlet to the consumer's heating system. Allowing for about a further 20 °C drop in temperature in the average house, the cooler water is passed on for use in swimming baths, laundries, etc.

Most of the hot springs in the island come to the surface at a temperature below 100 °C, and wherever there is a plume of steam, even in remote country districts, it is almost certain that there will be large greenhouses and a simple, covered swimming pool. For a country very deficient in natural resources the hot springs make a valuable contribution to the national economy.

IV. SOLAR POWER

It could be argued that all our power derives from the sun, but we are here concerned with the utilization of the sun's rays for the generation of power or controlled heating in a form in which it can conveniently be used. Early in the twentieth century experiments were carried out, mainly in the U.S.A., to concentrate the sun's rays by means of mirrors or reflectors in order to generate steam from water at low pressure and use it to drive a conventional steam engine. These were not successful.

Shuman, in about 1910, used ether instead of water as the motive fluid, with a glass-covered box as the boiler, and ran a small toy steam engine from it. After this Carpenter of Cornell University and, independently, A. S. E. Ackerman of Westminster carried out experiments on generating steam at atmospheric pressure, relying on a high vacuum to obtain a reasonable amount of power. Near Cairo in Egypt a set of reflectors 60 m long was used to concentrate the sun's rays on to a tubular boiler; the reflectors were automatically rotated to face the sun. Unfortunately the very sparse records do not state how much power was generated but the purpose of the plant was for irrigation. Immediately after the First World War what might be described as the petroleum age opened, with easily transportable petrol, paraffin, and diesel oil readily available at a low price. Vast numbers of cheap internal combustion engines, usually in the range of from 5 to 20 kW, were sold for driving irrigation or borehole pumps in tropical or sub-tropical countries throughout the world. There was no incentive to continue experiments with the direct use of solar power, and when work was resumed after the Second World War the approach was entirely different. Interest was naturally greatest in regimes enjoying long hours of strong sunshine, such as Israel, India, and the south of France.

The present line of development is by the use of cadmium sulphide cells energized by sunlight; these will generate low-voltage electricity and can be coupled in series or parallel. Much work lies ahead, but there is little doubt that for specialized purposes where other forms of energy are not readily available the solar cell has an important future. It has, of course, a special role in space research (Ch. 35).

Some mention should be made of the use of solar rays as a source of heating in temperate climates. The great importance of domestic house insulation is slowly being appreciated in the United Kingdom, as it is in those countries where very much lower winter temperatures are normal. The waste of heat in the United Kingdom is enormous, and since the last war architects have not only begun to concentrate on insulation, but on the siting of houses and the construction of windows so that the maximum of solar energy can be converted to internal heating. Even on dull days, with the shade temperature around 15 °C, a quite remarkable amount of heat can be transmitted through a correctly designed south-facing window.

Some outdoor swimming baths are heated—or partially heated—by solar boilers, comprising thin hollow panels, suitably inclined to face the sun, through which the water in the bath is circulated by a small centrifugal pump. Loss of heat during the night can be reduced by covering with a polythene sheet.

BIBLIOGRAPHY

BELLISS and MORCOM LTD. Power generation from a hot spring. *Engineering*, 21 November 1952.

DAVEY, N. *Studies in tidal power*. Constable, London (1923).

DONATO, GUISEPPE. Natural steam power plants of Lardarello. *Mechanical Engineering*, September 1951, p. 710.

GIORDANO, A. Geothermal power. *Electrical Times*, February 1949.

GOLDING, E. W. *The generation of electricity by wind power* (1955).

JOHNSON, V. E. *Modern inventions*, Ch. 11 Sun motors. T. C. and E. C. Jack, Edinburgh (1912).

PENNYCUICK, J. A. C. Power without fuel. *The Electrician*, January 1948.

Power from steam wells. *Power Engineering*, October 1950.

THURSTON, T. G. Steam turbines without boilers. *The National Engineer*, July 1952.

10

ATOMIC ENERGY

LORD HINTON OF BANKSIDE

PART I: THE EARLY HISTORY

I. ATOMIC STRUCTURE

WHEN John Dalton published his atomic theory in 1808 he postulated that every element consisted of its own distinctive variety of atom and every compound consisted of a specific combination of atoms. This theory was accepted throughout the nineteenth century: Ernest Rutherford, who was born in 1871, said 'I was brought up to look at the atom as a nice hard fellow, red or grey in colour according to taste' [1]. Before he died, Rutherford and his school of physicists had evolved the theory that, far from being a very small lump of hard matter, the atom was like a tiny solar system in which negative particles (electrons) moved in orbits around a central nucleus which was made up of other particles (neutral neutrons and positive protons).

The revolution in knowledge started with W. K. Röntgen's discovery of X-rays; this led to J. J. Thomson's concept of the electron and to the discovery of alpha and beta radiation by Rutherford and of gamma radiation by P. Villard. 'Thus the close of the nineteenth century found physics concerned with the spontaneous emissions of radiation from the atoms of certain heavy elements. The nature of the radiations had been established but the cause of the phenomenon was still a mystery' [2].

In the next ten years Rutherford and F. Soddy showed that alpha and beta radiations changed the chemical identity of the atom (in other words they resulted in transmutation of elements), and extension of this work established the fact that two atoms which were chemically identical could have different masses; Soddy called such atoms 'isotopes'.

By now, experimental methods of following the tracks of emitted particles had been discovered and it was found that when emitted particles collided with atoms it sometimes happened that the trajectory of the particle was deflected far more than had been expected. To explain this, Rutherford proposed at atomic model which consisted of a central, positively charged nucleus surrounded by electrons moving in relatively distant orbits and carrying a total negative charge equal to the positive charge in the nucleus. This simple picture of the atom was made complete by J. Chadwick's discovery of the neutron, an uncharged particle in the nucleus which had a mass similar to that of the proton. The physicists then had an atomic model in which the nucleus was made up of neutrons and protons. The number of protons determined the number of positive charges in the nucleus and, because a normal atom is electrically uncharged, it determined the number of negatively charged electrons moving in orbits around the nucleus. The chemical properties of the atom depended on the number of electrons and the arrangement of the orbits in which they moved. Chemical properties consequently depended on the number of protons in the nucleus but the number of neutrons in the nucleus could vary so that chemically identical atoms could have different masses and different physical properties. Today, the atom is known to be more complex than this, but it was on the basis of that simple model that nuclear power was developed.

II. ATOMIC FISSION: CHAIN REACTIONS

Irène Curie and Frederic Joliot had shown in 1934 that some light elements could be made radioactive by bombarding them with alpha particles and Enrico Fermi, working in Rome, realized that because the neutron was electrically uncharged it would be an effective missile with which to bombard a nucleus. He found that neutrons were most effective as missiles if they were slowed down by allowing them to collide elastically with atoms of light elements. Materials which slowed neutrons down in this way were called moderators.

In the course of his work Fermi found that, under neutron bombardment, uranium, which is the heaviest element occurring in nature, became radioactive and he explained this by suggesting that even heavier (trans-uranic) elements had been formed. Fermi's work was repeated by others who came to the conclusion that, improbably, one of the radioactive elements formed by neutron bombardment of uranium was radium. In 1938 O. Hahn and F. Strassmann repeated earlier work in which they had isolated the 'radium' so produced, by co-precipitation with barium but they were unable to separate the 'radium' from the barium and were forced to the conclusion that radioactive barium was produced by neutron irradiation of uranium. Hahn, who had collaborated with Lise Meitner before she was forced to leave Germany, wrote to tell her of his conclusions; she was not happy about them and discussed them with her nephew Otto Frisch. Early in 1939 Meitner and Frisch concluded that what happened was that uranium atoms had been split into two roughly equal halves; they calculated the amount of energy that would be released by such fission and found that it was very large. In April 1939 experiments by Joliot-Curie's group in Paris showed that, when fission occurred, secondary neutrons were emitted—more than one per fission—so that a chain reaction was possible.

Natural uranium consists of two isotopes, one with an atomic weight of 235 and the other with an atomic weight of 238: there are 140 atoms of ^{238}U for every single atom of ^{235}U. N. Bohr and J. A. Wheeler showed that fission was more likely to occur in the ^{235}U atoms than in atoms of ^{238}U and that slow neutrons were more likely to cause fission than fast neutrons. Their paper was published two days before the Second World War broke out in 1939; by reading it every physicist in the world could know the basic theory of atomic energy.

It was a fateful coincidence that these fundamental discoveries were made in the year preceding the outbreak of war. It was thought that the immense release of energy in a fission chain reaction might be used to generate power; could it also be used as an explosive? Research began in the United States, Britain, France, the U.S.S.R., Germany and possibly in other countries.

Measured values of neutron cross-sections showed that a chain reaction could not be established in a lump of natural uranium however large but there were two other possibilities. The first was that the chance of making the available neutrons cause further fissions in the rare ^{235}U atoms could be increased by slowing them down. This could be done by arranging rods of metallic natural uranium in a moderator and by making the assembly so large that the loss of neutrons from the surface would still leave enough within the assembly to continue a chain reaction. This would not give a reaction fast enough for a bomb, but it held out the prospect of nuclear power. It was suspected that in such a system neutron absorption might produce a transuranic element which, like ^{235}U, would be fissionable and the existence of this element, plutonium, was demonstrated experimentally in the United States early in 1941. It was a possible material for an atomic bomb.

The second possibility of achieving a chain reaction was to separate the fissionable ^{235}U from the non-fissile ^{238}U. If natural uranium was enriched in ^{235}U a chain reaction might be possible. In a lump of pure uranium-235 every neutron collision would cause fission. Neutrons of any energy would do this; there was no need to slow down the neutrons and the chain reaction would be fast enough to make a bomb possible. This second possibility seemed at first to be out of the question because of the difficulty of separating ^{235}U from ^{238}U; no chemical process could be used because the chemical characteristics of the two atoms are the same and physical processes for separation seemed industrially impossible.

III. THE ATOMIC BOMB

In all the countries pursuing atomic research the balance of effort was different. About the Russian project we know so little that it is not considered here. German scientists wanted to produce a bomb but, although various methods of separating ^{235}U were studied, German scientists were throughout the war mainly preoccupied with building a slow neutron reactor, using heavy water from Norway as the moderator. A British commando raid was mounted to cut off this source of supply. In spite of this the Germans made a good start in building an experimental reactor but after some surprising scientific errors the project foundered and got nowhere near to the production of plutonium or ^{235}U or an atomic bomb.

The French scientists directed their work to slow neutron reactions in natural uranium with the long-term objective of generating nuclear power; even before the war they had taken out patents for certain processes. When France fell in the summer of 1940 two of her atomic scientists, H. von Halban and L. Kowarski, came to England with their stock of heavy water. Their arrival led to the inclusion in the British research programme, which had been vigorously launched, of slow neutron chain reactions in natural uranium.

Soon after the war began, British scientists, although admitting the remote possibility of an atomic bomb, had come to the conclusion that, as a war project, atomic energy was probably not worth pursuing. However, early in 1940 two refugee scientists, R. E. Peierls and O. R. Frisch argued in a short and brilliantly reasoned paper that a mass of only 5 kg of pure ^{235}U could be made to produce the explosive power of several thousand tons of dynamite. They suggested an industrial process for separating the ^{235}U, they showed how the bomb could be detonated, and they gave warnings of the radiation effects.

This paper was transmitted through personal channels to the British government and in April 1940 a small committee was set up to study the paper, to consider the prospects and advise on further action. Known under the code name of MAUD, the committee had most of the distinguished British nuclear physicists as members but, because they were foreign, Peierls and Frisch were not at first members but were only consulted. The original focus of the committee was on ^{235}U: on methods of separation and the physics of making a bomb. The slow neutron work of the French scientists who were now in England was believed to be useful primarily for power although the outside chance of making the new element plutonium in such a reactor and of using the plutonium for a bomb was recognized. In July 1941, only 14 months after it had been formed, the MAUD Committee reported that a scheme for a uranium bomb was practicable and was likely to lead to decisive results in the war.

The United States had not pioneered the great developments in atomic physics but they had been working since 1939 on slow neutron reactions in natural uranium, on the plutonium which might be created, and on the separation of ^{235}U. However, America was neutral until the attack on Pearl Harbour in December 1941 and the object of their work was not clearly defined. They wavered between an interest in nuclear power—primarily for submarine propulsion—and atomic bombs but the stronger emphasis was on power. Their interest in ^{235}U was primarily in the enrichment of natural uranium for slow neutron reactors, and they did little work on fast neutron reactions. The pace of their research was desultory and its organization unsatisfactory. The British kept the Americans fully informed of the MAUD Committee's work and sent them a copy of its report in draft. It was the MAUD Report's clear demonstration of the practicability of a bomb that pushed the American project off the ground even before urgency had been bred by Pearl Harbour. The American atomic effort then multiplied wonderfully and within six months had far outstripped the British project (which was known by the code name of Tube Alloys). In June 1942, industrial development began and the U.S. Army, with General Groves in charge, assumed responsibility for 'the Manhattan Project'. There was collaboration between the British and American projects until the end of 1942, but for nine months in 1943 the British were excluded. Co-operation was restored by the negotiations between Roosevelt and Churchill which resulted in the Quebec Agreement of August 1943 and the British were readmitted as junior partners. Britain's project virtually closed down for the duration of the war and most of her nuclear scientists left for America. The Anglo–French slow neutron team had gone to Canada at the end of 1942 to set up laboratories there and were later joined by other British scientists and engineers. Apart from anything that may have been happening in Russia, the industrial development of atomic energy, both for weapons and power, was a North American—primarily a United States—monopoly until after the war.

REFERENCES

[1] ANDRADE, E. N. DA C. Rutherford Memorial Lecture (1957).
[2] GOWING, MARGARET. *Britain and atomic energy 1939–1945*. Macmillan, London (1964).

BIBLIOGRAPHY

GLASSTONE, SAMUEL. *Sourcebook on atomic energy*. Macmillan, London (1952).
GOWING, MARGARET. *Britain and atomic energy 1939–1945*. Macmillan, London (1964).
HEWLETT, RICHARD G., and ANDERSON, OSCAR E. *History of the U.S. A.E.C.*, Vol. I. *The New World, 1939–1946*. Pennsylvania State University Press, University Park, Pa. (1962).
IRVINE, DAVID. *The virus house*. William Kimber, London (1967).
KRAMISH, A. *Atomic energy in the Soviet Union*. Stanford University Press (1959).
MODELSKI, GEORGE A. *Atomic energy in the Soviet bloc*. Melbourne University Press on behalf of the Australian National University (1959).

PART II: URANIUM ENRICHMENT

I. POSSIBLE METHODS OF SEPARATION

When, after Pearl Harbour, the United States directed the massive power of their research and industrial organizations to the production of atomic bombs, it was still thought that a ^{235}U bomb offered the best prospects of success. There are several ways in which the fissile ^{235}U can be separated from the non-fissile ^{238}U; it was thought that of all these alternatives, four gave promise and could be used on an industrial scale. These four were:

(i) *Centrifugal separation:* In a fluid salt of uranium, molecules which contain the ^{235}U isotope are lighter than molecules that contain the ^{238}U isotope. If such a salt is centrifuged, some separation of the ligher molecules from the heavier molecules is achieved in the same way that cream is removed from milk in a separator.

(ii) *Thermal separation:* If a fluid salt of uranium is introduced into the narrow annulus between two pipes and the outer pipe is cooled while the inner pipe is heated, the molecules containing the lighter isotope (^{235}U) tend to move towards the wall of the hot pipe while the molecules containing the heavier isotope (^{238}U) move towards the cold pipe. With vertical pipes, the lighter molecules will rise towards the top and the heavier ones fall towards the bottom. The longer the pipe, the greater will be the separation.

(iii) *Electromagnetic separation:* By exposing the molecules of a gas to an electric discharge they can be ionized (that is, given an electric charge). Such ionized molecules can be accelerated through another electric field and their direction of movement can be

FIG. 10.1. The principle of the electromagnetic separation cell.

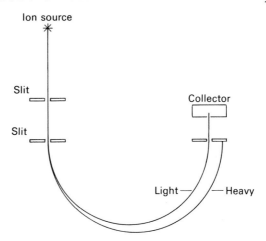

controlled by passing them through slots. If this 'jet' of ionized molecules passes across a magnetic field its trajectory is bent into a circular path (Fig. 10.1). The initial velocity given to all the molecules is the same and the same force is exerted on all the molecules by the magnetic field. But if some of the molecules are heavier than others they will have greater momentum and will be less easily deflected. Their path will therefore have a larger radius than that of lighter molecules so that the heavy and the light molecules will hit different targets; in this way molecules containing the light isotope ^{235}U can be separated from those containing the heavy isotope ^{238}U. The principle had been used for many years on a laboratory scale in the mass spectrometer.

(iv) *Gaseous diffusion:* This was the process that had been recommended by British scientists in the MAUD Report. The kinetic energy of each molecule in a gas is the same and depends on the temperature of the gas; it follows, therefore, that in a mixture of gases light molecules must have a higher velocity than heavy molecules, otherwise their kinetic energies would not be the same. If a gas composed of molecules of different weight is enclosed in a porous container the gases will diffuse through the walls and more light molecules than heavy molecules will pass through the porous barrier or membrane because, since they are travelling faster than the heavier molecules, they are more likely (in random movement) to 'hit' a hole. If the gas which has diffused through the membrane is not immediately swept away there will be a similarly different rate of back diffusion but if the gas that has diffused through the membrane is immediately swept away it will have an increased content of light molecules. It follows that if a gaseous compound of uranium is treated in this way the gas which has diffused through the membrane will be slightly enriched in the lighter molecules which contain ^{235}U.

The uranium compound used in all these processes (other than electromagnetic separation) is uranium hexafluoride, UF_6. There are two reasons for this choice; firstly, the hexafluoride is liquid or gaseous at convenient temperatures and pressures; secondly, fluorine does not have a number of isotopes whose differing weights would mask the difference between the weight of ^{235}U and ^{238}U. The difference between the weight of UF_6 molecules containing the two different isotopes is less than 1 per cent so that separation is difficult.

Working with overriding priority and virtually without regard to cost, the American Manhattan Project used all but the first of these four alternative processes. In fact, the atomic bomb which was dropped on Hiroshima was made from ^{235}U which had been partly enriched in thermal and gaseous diffusion plants and then fed to an electromagnetic plant in which the final enrichment was done. In scale of effort, scientific skill, courage and determination it is doubtful whether, when the time factor is taken into account, anything can be found in the history of technology which matches the work which was done by the Manhattan Project and their consultants and contractors to produce ^{235}U.

Separation of isotopes by centrifuging had been proposed by F. W. Aston and F. A. Lindemann in 1919, and in 1939 work on the enrichment of uranium by centrifugal separation was sponsored by the U.S. Navy at the University of Virginia. When the Manhattan Project was formed hope for this process was high and even in 1942, when the Oak Ridge site was chosen, there was some doubt as to whether centrifuges or a gaseous diffusion plant should be built. A contract for development was placed with Westinghouse but, with the materials of construction and the techniques that were then available, the attempt to develop centrifugal separators on an industrial scale was unsuccessful. Motors, shafts and bearings all failed and, at the speeds which could then be achieved, it was calculated that 25 000 units would be required. In the autumn of 1942 the development contract was cancelled because it was not thought that the project could make any contribution during the war. It was not until the late 1960s, when new techniques and new materials became available, that the process began to look promising; it is likely to be extensively used in the future.

Work on the thermal diffusion process had been started in the National Bureau of Standards as early as 1940. It was financed by the Navy Department who were interested primarily in obtaining an enriched fuel which could be used in power reactors for submarine propulsion. The experimental plant was soon moved to the Naval Research Laboratory where high-pressure steam was available. By January 1942 it was thought that the method would compete with gaseous diffusion or with centrifuging but experience showed that if it were used to produce highly enriched ^{235}U the plant would have a long hold-up time and would be extravagant in its use of uranium. It was, however, pointed out that partial enrichment could be done in a thermal diffusion plant and that the partially enriched material could be used as feedstock for the electromagnetic separation plant so as to increase the output of that plant. With this in mind, and in view of the trouble that was being experienced with all the other separation plants, it was decided to build a large thermal diffusion plant at Oak Ridge. The plant was completed in 90 days; it did everything expected of it but it was justified only under wartime conditions when money was no object and it was scrapped immediately the war ended.

Of all the four alternative processes for the production of ^{235}U the only one which was well established and widely used on a laboratory scale was electromagnetic separation. Mass spectrometers had been used by scientists for many years and the electromagnetic separation units were mass spectrometers scaled up to industrial size. The extrapolation in size was, however, so great that the plant was complex, expensive, and troublesome. The 12-ft diameter magnets were so large that the Manhattan Project could not obtain enough copper, and the windings were made of silver from the U.S. Treasury vaults. The ancillary plant was complex and there was much more scattering of molecules than had been expected so that, initially, a large part of the product was spattered on the walls of the chambers and was lost. That problem was cured partly by re-design and partly by careful control of the chemical recovery processes. Gradually the troubles were overcome but time and output had been lost and it was only by feeding the plant with partially enriched uranium from the thermal and gaseous diffusion plants that enough material for the first bomb was produced.

When the war ended the Manhattan Project closed all the less efficient separation plants. By September 1945 the second electromagnetic separation plant, which was of improved design, was performing well but by then the final stages of the diffusion plant were coming into operation. The electromagnetic separation plant depended for its feedstock on the low separation stages of the gaseous diffusion plant which are the most expensive in capital and operating cost and it was obviously not worth while to keep both plants when, by operating the high-enrichment stages of the diffusion plant, the work could be done in one plant. On 23 December 1945 it was decided to close down all but one track of the electromagnetic separation plant; 'A stirring chapter in the history of the wartime project was drawing to a close' [1].

II. GASEOUS DIFFUSION

The one uranium enrichment process that survived the war was gaseous diffusion and it is still the only process that remains in industrial use although recent developments make centrifuging a strong competitor. It is interesting that although it has so far outlived its wartime competitors it was the most difficult of all the plants to develop; the process is still classified and description of design details is forbidden.

In their original report which triggered investigation by the MAUD Committee, Peierls and Frisch had suggested separation of ^{235}U by thermal diffusion but the Committee, impressed by the arguments of the Oxford physicists, recommended gaseous diffusion. By April 1941, the Oxford team —with whom Peierls and his Birmingham associates were now working— had a half-scale model of a single stage in use and a larger model was being built. At the end of May, an order was placed with Metropolitan-Vickers for the design of two ten-stage pilot plants but the problem of producing the membrane or barrier was far from being solved. In Oxford the team were trying to produce a membrane by rolling down fine metallic gauze.

It has already been explained that the enrichment of uranium in its fissionable ^{235}U isotope depends on the fact that light molecules diffuse through a porous membrane more quickly than heavy molecules. The difference in the rate of diffusion is proportional to the square roots of the molecular weights and because the gas used is the hexafluoride the greatest degree of enrichment theoretically possible in a single stage is only $\sqrt{(352/359)}$ or 1·0043 times.

It is obvious that thousands of stages are necessary in order to give the enrichment factor of 1260 which was needed to produce uranium suitable for the early atomic bombs. These thousands of stages have to be arranged in a 'cascade' which can best be visualized by analogy with the fractionating columns used in distillation. In such a fractionating column separating, let us say, alcohol from water, steam (which is the driving force) is introduced at the bottom of the column while the alcohol–water mixture is fed on to one of the plates part-way up the column. The hot gases which are passing up the column cause evaporation of the alcohol–water mixture but, because alcohol is more volatile than water, the vapour passing up from one plate to the next higher plate is enriched in alcohol while the liquor overflowing from the plate to the next lower plate is depleted in alcohol. As the liquid passes down the column this process of depletion is continued so that the liquor extracted at the bottom of the column is virtually pure water while the vapour passing up the column is progressively enriched, so that at the top of the column, virtually pure alcohol vapour is drawn off. The diameter of the column can be made smaller towards the top where the volume of vapour decreases.

Each stage of a gaseous diffusion plant acts as if it were a plate in a fractional distillation column. The gas which passes through the membrane, and which is slightly enriched in ^{235}U, passes upwards to the next stage while the gas which does not pass through the membrane and which is slightly depleted in ^{235}U, passes downwards to the next stage below. There is obviously a pressure drop through the membrane, and the compressors which are used in each stage to provide this pressure difference can be thought of as the analogue of the steam which provides the driving force in a fractionating column.

In order to make sure that there is no condensation of uranium hexafluoride (HEX) on the membrane, the plant is worked in a high ambient temperature and under vacuum. If there is any inward leakage of atmospheric air into the plant the moisture contained in this air would hydrolyse the HEX and form solid compounds of uranium which would plug the pores in the membrane; the pipes and stages have, therefore, to achieve a standard of vacuum tightness previously unknown in such a large plant. It was thought (wrongly) that, if there were even a short failure of the power supply to the compressors, HEX would condense on the membranes and that it might take weeks to restart the plant; an absolutely reliable power supply was therefore considered to be essential. The materials used in the construction of the plant had to be such that they would not react with HEX; even slight surface reaction (as distinct from corrosion) was unacceptable because this would result in loss of product. The characteristics of HEX as a gas were not known and were so far from normal that compressor design was difficult. Elegant mathematical methods had to be evolved to optimize the 'cascade' of stages.

Responsibility for the development, design, and construction of the wartime diffusion plant was given by the Manhattan Project to the Kellogg Corporation, who formed a subsidiary known as Kellex. This subsidiary was headed by P. C. Keith, and it relied on Columbia University for much of its membrane research work. The plant was to be operated by the Union Carbon and Carbide Corporation on a site at Oak Ridge where, at a reasonable distance away, the electromagnetic separation, the thermal diffusion, and other plants were being built. Construction of the plant was authorized in December 1942 and construction was started in May 1943. As originally authorized (it was later extended) the buildings covered an area of 2 million square feet. The design of the compressors and in particular of their glands was extremely difficult but these and other problems were insignificant compared with the problem of producing a satisfactory membrane. In fact, construction of the building and mass manufacture of the stages went ahead before it was known that a satisfactory membrane could be manufactured. It was a risk that could be taken only under wartime conditions.

The membrane (or barrier as it was called in America) needed to have billions of holes each reasonably uniform in size and each with a diameter of about one ten-thousandth of a millimetre; the distribution of the holes had to be uniform. The membrane had to be chemically resistant to HEX and (in the event of accident) resistant to its breakdown products. It had to be strong enough to resist both the pressure difference across it and the considerable vibration that might be set up by the gas flow. It had to be suitable for manufacture by mass production methods and it had to retain its quality after subsequent handling and fabrication.

Many different men and many different organizations worked to evolve a process for the mass production of membranes. The original method of manufacture was suggested by E. O. Norris and E. Adler and a mass-production plant using their method was built; the membrane produced in it was brittle and lacking in uniformity. An alternative process, starting from nickel powder, was suggested by the Bell Telephone Company but membranes produced by them gave poor separation. In August 1943, three months after construction work had started at Oak Ridge, Keith had to say that he had not yet produced a satisfactory membrane. At the end of the year Groves invited a British team to New York to see if they could help but Keith felt that they were too much concerned with other problems in the plant—problems to which Keith felt that he already had acceptable solutions —and that they were not useful in solving what was his major difficulty in producing a satisfactory membrane. The British team stayed for a short time, leaving behind only two men who acted as liaison officers for a few months.

Meanwhile, the Bakelite Company, which was a subsidiary of the Union Carbon and Carbide Corporation, suggested a development of the Bell process and by April 1944, nearly half the membranes produced in the laboratory were up to specification. By June, the stages (though still without membranes) were flowing off the production lines at Chrysler's works and the first six stages had already been erected at Oak Ridge; by December a satisfactory barrier was being made on the production line and by January some stages were charged with HEX. The first output was obtained on 12 March 1945.

The plant did not produce the ^{235}U which destroyed Hiroshima; it was one of a chain of plants—the thermal diffusion plant, the gaseous diffusion plant and the electromagnetic separation plant—which, working in sequence, produced that material. The plant was extended after the war to produce highly enriched ^{235}U. It was the plant which had been favoured by the British at the beginning of the war and it was the only one of the four alternative processes tried by the Americans which survived it. The diffusion plant did more than play a major part in the manufacture of the Hiroshima bomb; it provided the enriched uranium which made nuclear submarines possible and which encouraged the Americans to develop light-water reactors. Given the limitations of time there are, in all the history of science, technology and engineering, few projects which can match it.

Although shortage of resources and vulnerability to attack had made it impossible for the British to deploy a large effort, work on the diffusion plant had continued in England during the war. As we have already seen, a contract was placed with Metropolitan Vickers for the construction of two ten-stage pilot plants. Scientific advice for the design came mainly from the Oxford group of scientists and from Peierls and his associates at Birmingham but it soon became clear that they needed the collaboration of an organization which was experienced in the design and construction of process plants and it was arranged that I.C.I. should help. Two other large British electrical firms, British Thompson Houston and the General Electric Company, collaborated with Metropolitan Vickers and construction of the pilot plant was started in a wartime factory in North Wales. Delivery of the plant was late; there was trouble with the first stages and the two ten-stage lines were never operated.

In spite of these shortcomings, I.C.I. did work which was to prove of great value to the British project in the post-war years. They evolved the elegant mathematics by which the performance of the plant was optimized, they designed an extremely neat arrangement for the membranes and (although the process was not perfected until after the war) they devised a continuous method for manufacturing membranes of high quality.

In 1946, when the British atomic energy project was set up within the Ministry of Supply, no decision to make a weapon had been taken but there was little doubt that defence was the first objective and the comparative merits of the ^{235}U route and of the plutonium route had already been considered. The plutonium route was selected but it soon became obvious that a diffusion plant would be needed if only to re-enrich uranium which had been depleted by irradiation in the piles. Before the end of 1946 it was decided to start design and development work for a low separation plant which would produce uranium with twice the normal content of ^{235}U. Instructions to proceed with the construction of this plant and to prepare designs for a high separation plant were given in 1949. The low separation plant was completed in 1953. Soon afterwards it was considerably enlarged and high separation stages were added.

The British plant could not compete with the American plants in production cost mainly because of the higher price of electricity in the United Kingdom but otherwise the British plant was successful. The French commissioned a plant of comparable size in about 1970 and there must have been earlier plants in Russia and China. It now seems likely that few more gaseous diffusion plants will be built and that they will be outdated by centrifugal separation plants.

REFERENCE

[1] HEWLETT, RICHARD G., and ANDERSON, OSCAR E. *History of the U.S. A.E.C.*, Vol. 1. *The New World, 1939–1946.* Pennsylvania State University Press, University Park, Pa. (1962).

BIBLIOGRAPHY

GOWING, MARGARET. *Britain and atomic energy 1939–1945.* Macmillan, London (1964).
——. *Independence and deterrence*, Vol. II. *Britain and atomic energy 1945–1952.* Macmillan, London (1974).
HEWLETT, RICHARD G., and ANDERSON, OSCAR E., *History of the U.S. A.E.C.*, Vol. 1. *The New World, 1939–1946.* Pennsylvania State University Press, University Park, Pa. (1962).
SMYTH, K. D. *Atomic energy for military purposes.* Princeton University Press (1945).

PART III: THE DEVELOPMENT OF NUCLEAR REACTORS

Late in the afternoon of 2 December 1942 Arthur Compton, President of the Massachusetts Institute of Technology (M.I.T.), telephoned from Chicago and said to James Conant, the President of Harvard, 'Jim, you'll be interested to know that the Italian navigator has just landed in the New World'. There was no need for him to be more explicit; both Compton and Conant were leaders in the scientific team which was engaged in America's wartime effort to produce atomic bombs. During that December afternoon Compton had been one of a small group of men who had stood on the gallery of the racquets court at the Chicago University sports stadium watching the instruments while E. Fermi, the Italian expatriate physicist, had loaded slugs of uranium into the channels which ran through a pile of carefully machined blocks of

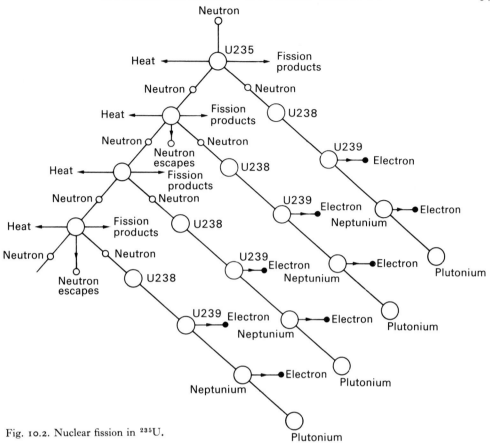

Fig. 10.2. Nuclear fission in ^{235}U.

the purest available graphite. After each batch of slugs had been added to the assembly, the 'control rods', which were made of neutron-absorbing material, were withdrawn and the instruments were watched to see whether the neutron intensity in the pile increased or died away. At 3.20 p.m. the instruments showed that the activity was divergent; the world's first nuclear reactor was working.

The date will go down in history, but it is possible to overrate its scientific importance. We have already seen that although ^{235}U is fissionable it is impossible to establish a chain reaction in a lump of natural uranium because most of the neutrons are absorbed by non-fissionable atoms of ^{238}U. We have seen that it was known that the probability of this happening could be decreased by slowing down the neutrons to 'thermal' velocities in a moderator. There was, therefore, a possibility of establishing the chain reaction shown in Fig. 10.2. The neutrons absorbed by atoms of ^{238}U caused changes which

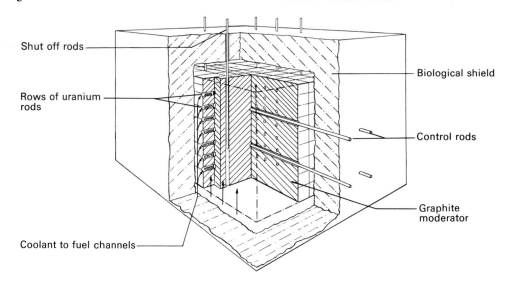

Shut off rods

Rows of uranium rods

Biological shield

Control rods

Graphite moderator

Coolant to fuel channels

FIG. 10.3. Principle of core arrangement in a graphite-moderated reactor. The channels in which the uranium rods lie penetrate all the way through the graphite structure.

resulted in the formation of a new element which did not exist in nature and had been named plutonium; this new element was fissionable and might be used in an atomic bomb. What was not known was whether sufficient neutrons were emitted in each fission to maintain a chain reaction. It can be seen from Fig. 10.2 that the theoretical minimum is two per fission but more are needed in practice because some neutrons are uselessly absorbed in impurities and in the materials of construction and some are absorbed in the moderator. The ideal moderator was known to be 'heavy water' (composed of deuterium, a rare isotope of hydrogen, in combination with oxygen), but this was not available in sufficient quantities. The next best moderator was graphite, but even the purest graphite absorbed neutrons. In addition, the nuclear reaction could be built up and controlled only if there were excess neutrons which could be used to increase the activity in the pile and which could be absorbed in 'control rods' to hold the reaction at its proper level. An elementary pile would therefore be arranged as shown in Fig. 10.3.

I. THE FIRST NUCLEAR PILES

In spite of all of the unknowns and uncertainties, there had been so much hope that a nuclear pile was a practical possibility that on 1 December 1942 General Groves, who was in charge of the Manhattan Project, had sent a letter of intent to Du Pont de Nemours authorizing them to design and construct a nuclear reactor. What Fermi achieved on 2 December was to prove that this hope was not ill-founded; that the possibility of building a nuclear reactor with available materials was real; and that the reactor could be controlled.

Fermi's reactor had done all that had been expected of it but it could do little more; it had no biological shield to protect the operators from neutron irradiation and it had no cooling system. It was obvious that a larger pile was needed to provide not only more information for the pile designers, but also the irradiated slugs which the chemists needed for development of a separation process. On 4 January 1943, Groves gave Du Pont and Company a contract for the construction of an air-cooled pile with a heat output of 1800 kW. The pile was to be built at Oak Ridge and Du Pont undertook the work while emphasizing that they were acting as handmaiden to the scientists in doing the design and construction and that they were not prepared to operate the pile; operation, they said, must be done by the Chicago scientists. Construction was started in April 1943 and the pile became divergent in the following November.

Meanwhile, using the nuclear constants which had been determined in Fermi's pile, all sorts of alternative schemes for the full-scale production piles had been considered. Cooling could be by hydrogen, helium or air; by water; by organic fluids; or by liquid metals. Ultimately water was chosen as the coolant but C. H. Greenewalt of Du Pont's was worried about safety; neutrons would be absorbed by the cooling water and if there were a failure of the water supply to any of the channels, the water in that channel would be expelled as steam and the neutrons that had previously been absorbed in the water would be available to cause additional fissions and so make the pile supercritical. The temperature might then rise and cause the uranium to burn, releasing the fission products and plutonium which had formed in it. These fission products and plutonium might be so disseminated as to cause danger to life and health perhaps as far as thirty miles away. It had already been decided that the piles and the chemical separation plants should be built on the remote site at Hanford on the Columbia river and the location of the piles and of the new town site recognized the hazard that existed. The use of water as a coolant also led to problems with the slugs; in the large experimental pile at Clinton near Oak Ridge the uranium had simply been enclosed in aluminium cans but, with the higher rating that was possible in a liquid-cooled pile, a heat-conducting bond between the uranium and the can was necessary.

It had been decided to build three piles, each with a thermal output of 250 MW. The basic designs were received from Du Pont's headquarters at Wilmington (Delaware) in October 1943, but already there were 25 000 men doing preparatory work on the site at Hanford. Some of the early arrivals had been obliged to live under canvas but by the summer of 1943 there were 130 barracks for men and 45 for women, as well as standings for 1200 trailers. Even so there was a shortage of labour, particularly of skilled men because the standards demanded were unusually high. Each pile was built from blocks of the purest graphite obtainable; these blocks were machined on site in order to minimize the danger of surface contamination. Through this assembly there were horizontal channels and through each of these ran an aluminium tube into which the slugs were inserted. The cooling water passed through the annulus between the slugs and the tube and ran away to storage ponds where it was tested for radioactivity before being returned to the Columbia river. Control rods and emergency shut-off rods were admitted to the graphite core through other channels. The whole of the core assembly was enclosed in a thick concrete biological shield, the holes through which had to match exactly with the channels in the graphite.

In spite of all difficulties, construction of the pile went ahead but the problem of making bonded slugs was intractable; ultimately Du Pont undertook responsibility for manufacture but it was not until the end of 1943 that a process was evolved in which the uranium rod was bonded to the aluminium can with a silicon–aluminium alloy, and even then there was great difficulty in mass-producing a satisfactory product. As late as January 1944 there were doubts about the possibility of perfecting the process and it was not until August that it was certain that bonded slugs could be used. In September, Fermi inserted the first slug into the first pile and it was run up to power on 27 September. To the surprise and consternation of the physicists it soon shut itself down; they had not discovered in their work on experimental and prototype piles that one of the short-lived fission products which is formed is xenon; this avidly absorbs neutrons, therefore extra activity has to be provided to give what is now known as xenon-override. Fortunately, Du Pont had provided extra channels in the pile and by using these it was possible to bring the piles up to the designed output; the first irradiated slugs were discharged on Christmas Day 1944.

II. CANADA: CHALK RIVER

It was the Canadians who followed immediately on the heels of the Americans in that they started design and construction of a large experimental reactor before the end of the war. The French had been leaders in experimental work on slow-neutron reactors and when France fell in 1940 two of their leading physicists, H. von Halban and L. Kowarski, who had been working on slow-neutron reactions, escaped to England, bringing with them the heavy water which they had been using as a moderator. For some time the two Frenchmen had worked at Cambridge but by 1942 it was clear that the Americans were moving ahead very fast in their slow-neutron work. The British wanted the Anglo–French team to move to the United States and to work as part of the Manhattan Project but the Americans refused to have them. 'The team would have been very much out of the picture if it had stayed in England and therefore Canada, so near the United States laboratories, had seemed a good home. The National Research Council of Canada had welcomed the team and had financed first of all a laboratory at Montreal and then the construction of nuclear piles at Chalk River. The team had finally become an integrated Anglo–Canadian one together with five French citizens. . . . In the last eighteen months of the war the project flourished under the leadership of Professor Cockcroft.' [1]

The engineering design of the reactor that was built at Chalk River was done by I.C.I.'s Canadian subsidiary, to which the mother firm had lent two extremely able engineers, R. E. Newell and D. W. Ginns. The pile which they brilliantly designed was commissioned in 1947 and remained the best general-purpose research reactor in the world for at least a decade. It was a calandria[1]-type reactor: using heavy water as the moderator and ordinary water as the coolant. The fuel elements were of natural uranium metal canned in aluminium and arranged in a vertical lattice. The calandria which contained the heavy water moderator was made of aluminium and the coolant was circulated through the annular space between the calandria tubes and the fuel elements. A second and larger research reactor using heavy water both as moderator and coolant was built later and these were the parents of the industrial CANDU reactors which have been extremely successful in Canada and have been sold to developing countries.

---

[1] A calandria is a closed vessel with internal tubes sealed into its walls so as to form channels to keep coolant (passing through the tubes) separate from moderator (in the vessel).

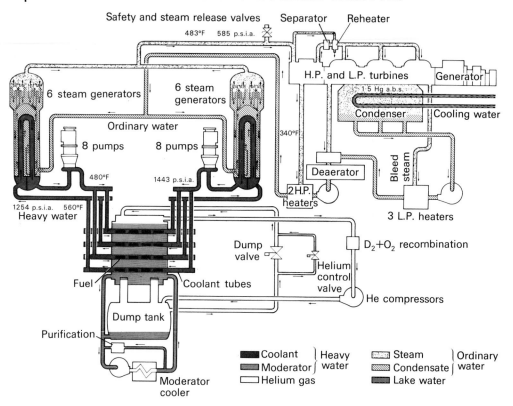

FIG. 10.4. Schematic plan of CANDU 500-MW(electrical) power plant.

The advantages of the CANDU reactor (Fig. 10.4) are that because heavy water absorbs so few neutrons the fuel elements can be made of unenriched uranium and high 'burn-up' of fuel can be achieved: in simple terms, more heat can be got out of the uranium before the fuel elements have to be replaced. The main disadvantage of the CANDU reactor is that, largely because heavy water is expensive, the capital cost is high. Like the American light-water reactor types, the CANDU reactor can generate steam only at modest temperatures and pressures and its long-term future may be doubtful.

III. THE U.S.S.R.

Reactor development in Russia was unexpectedly and commendably fast but, although technical information has been published in scientific papers, the inside story has not been told in as much detail as those of the American and British projects.

The Russian physicists knew as much about the theory of nuclear fission as was known in other countries. In the autumn of 1941, Peter Kapitza said, in a speech which was reported in Pravda, that an atomic bomb 'could easily destroy a major capital city'. Despite this knowledge, the Russians apparently made no attempt to develop bombs during the war: perhaps they felt that the problem could not be solved in time for atom bombs to be useful; perhaps their intelligence organization was able to assure them that the progress of research in Germany was so halting that they need not fear atom bombs. They must, however, have known something of progress in the U.S.A because Klaus Fuchs (the atom spy) started to send information to them in 1942.

There is no evidence that Russia started to build even a research reactor until after the American bombs had been dropped on Japan, but by then they must have done much background research. They claim that they built the first reactor in Europe. Gleep, the low-energy reactor at Harwell, became divergent in August 1947. If, having started to build a similar reactor at the end of 1945, the Russians completed it within eighteen months, they must already have done research on such problems as the purification and metallurgy of uranium. The first Russian atomic bomb test, detected in August 1949, suggests that they must have had a production pile working by 1948—a remarkable achievement in speed of design and construction.

In 1949 the Russians started the construction of a 5-MW (electrical) demonstration power-producing reactor which was in operation four and a half years later. This reactor was fully described in a paper given at the first Geneva Conference on atomic energy in 1955 (Fig. 10.5). In principle it was not dissimilar from the Hanford reactors but it used enriched uranium as fuel and the cooling water which passed through stainless steel tubes in the pile was pressurized to 100 atmospheres. The water in this primary cooling circuit was circulated through heat exchangers in which steam was generated and conveyed to the turbines. The graphite moderator, enclosed in an atmosphere of helium to avoid oxidation, was kept at a high enough temperature to avoid build up of Wigner energy. The Russians may have received some information from spies, but the design of this reactor shows that in some respects the scientific advice available to their engineers was better than in other countries.

In September 1955, Tass announced that a 100-MW (electrical) station had started to operate and a film of this station was shown at Geneva. It did not give the impression that there had been any great advance in design but it was said to be the first of six identical units to be built. The Russians also

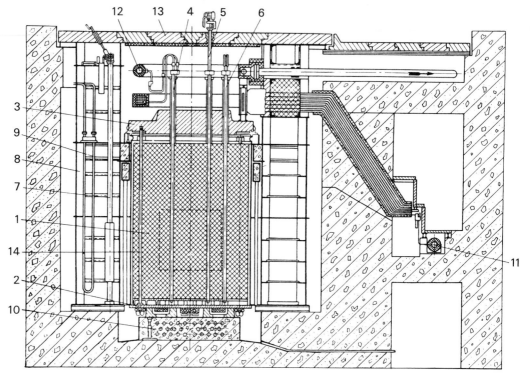

1. Graphite brickwork	8. Side shielding (water)
2. Lower plate	9. Refrigerator.
3. Upper plate	10. Refrigerator
4. Fuel channel	11. Distribution (inlet) header
5. Safety rod channel	12. Outlet header
6. Automatic control rod	13. Top shielding (cast iron)
7. Ionization chamber.	14. Cooled reflector stand

FIG. 10.5. The first Russian power-producing nuclear reactor.

developed boiling water reactors (B.W.R.s) and pressurized water reactors (P.W.R.s), and had used a P.W.R. to power an icebreaking ship, the *Lenin* (1958), which Admiral Rickover (not an easy man to impress) described as 'a fine and creditable job'. The first Russian fast reactor started up in 1959 and their work in this field is abreast of that in France and Britain.

IV. BRITAIN: THE ATOMIC ENERGY ORGANIZATION

The British Industrial Group of the Atomic Energy Organization, which started work in February 1946, was told that its first responsibility was to produce plutonium for use in bombs. Compared with the speed achieved by

the American organization during the war its initial performance appears
pathetically slow but, while the Manhattan project had overriding priority
and could call on any or all of the great American industrial companies for
help, the British team at Risley started with only five engineers; it was in the
Civil Service and could offer only Civil Service rates of pay; industry was
reluctant to help and, until 1950, the British group had no effective priority.
Its first task in reactor construction was to design and build a gas-cooled
graphite-moderated experimental reactor for Harwell. The parametric design
for this had been done in Montreal during the war but initially Risley had
great difficulty in getting from the physicists the information which they
needed for engineering design. This experimental pile at the Harwell Atomic
Energy Research Establishment was similar in design to the Clinton pile
but, whereas the Clinton pile was built in nine months, the construction of
BEPO took about two years.

It had been the Government's initial intention that the British production
pile should, like the Hanford piles, be graphite-moderated and water-cooled;
consulting engineers had already been commissioned to look for sites at
which the necessary quantity of water (which had to be exceptionally pure)
was available. Not only was 30m. gallons a day of pure water needed but, for
reasons of safety, the site had to be remote. The Americans had specified that
their piles should be 50 miles from a town of 50000 inhabitants; 25 miles
from a town of 10000 inhabitants; and 5 miles from a town with 1000
inhabitants. Such a site was not easy to find in Britain and there was bitter
argument before it was agreed that relaxation of the American siting criteria
could not be permitted. Only one site, between Malaig and Arisaig, met the
requirements and that site had no facilities for construction work; access was
bad; all labour would have to be imported and quick completion would be
impossible.

While the search for a site had been going on, the technical staff at Risley
who knew that the Americans had considered such piles during the war, had
been examining the possibility of building a gas-cooled production pile.
Siting considerations made gas cooling attractive because gas-cooled reactors
did not give rise to the district hazard which had to be faced with reactors of
the Hanford type. The Risley engineers realized that heat transfer from the
fuel element to the gas could be greatly improved by using cans with fins on
them and that, because of the high conductivity of aluminium, these fins
could be thin and would not seriously increase neutron absorption. To
improve heat transfer they proposed to enclose the reactor core in a steel

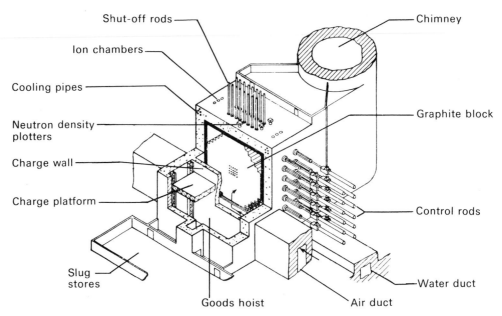

FIG. 10.6. Isometric diagram of a Windscale pile.

shell and to use carbon dioxide under pressure as a coolant. Harwell engineers extrapolated the Risley designs down to atmospheric pressure and showed that it would be possible to produce the specified quantity of plutonium in two graphite-moderated piles through which air would be blown and discharged from a chimney high enough to disperse the active argon produced by neutron irradiation.

After much argument this design was adopted; it made the choice of site far easier and a disused wartime TNT factory at Sellafield in Cumberland was used and was called 'Windscale'. The general arrangement of the Windscale piles is shown in Fig. 10.6; these piles could be thought of as production units which had been scaled up from BEPO but, in fact, many problems had to be overcome in the design and construction of those 'monuments to our initial ignorance'. It was learned from experience at Clinton that it would be necessary to filter the discharged air; construction was then so far advanced that the only possible place to put the filters was on the top of the chimney. A worse difficulty arose from lack of information about the behaviour of graphite under neutron irradiation. It was known from American experience that neutron irradiation modified the crystal structure of the graphite and caused dimensional changes and internal stresses. As information about the dimensional changes filtered through to Risley the design of the graphite moderator had to be changed three times and each change involved the preparation of several hundred new drawings. No warning was given by the physicists that, if the temperature of the graphite increased, the internal stresses would be relaxed with a further rise in temperature which could be dangerous. Because of this limited information the designers predicted only a short life and, in 1957, there was a serious fire in one reactor as a result of which both were shut down. The nuclear data on which the design of the pile had to be based was so far wrong that, even after modifications to the fuel elements, the first pile produced only 70 per cent of its designed output. However, these two reactors did produce the plutonium for the first British atomic bomb trials.

V. FAST REACTORS

From the earliest days, physicists had envisaged fast-neutron reactors as the ideal way of producing power from nuclear fission. We have seen that when using natural or slightly enriched uranium it is necessary to slow the emitted neutrons down to thermal velocities in order to ensure that they cause further fissions in atoms of ^{235}U and are not absorbed in the non-fissionable atoms of ^{238}U. If a pure fissile material (plutonium, ^{235}U, or ^{233}U) is used as the fuel there is no need to slow the neutrons down in a moderator and virtually all the emitted neutrons are available to cause further fissions or to breed fissile atoms from a suitable 'fertile' material such as ^{238}U. In this way it is possible to produce more fissile atoms than are destroyed. Reactors in which this is done are therefore called fast breeder reactors.

When the Windscale reactors had been built, the pile design office at Risley was left without work. Studies of a thermal power-producing reactor had already been done but the known reserves of uranium were limited and it was felt that it was useless to develop a programme of power-producing thermal reactors unless fast reactors, which would generate about 60 times as much power from a given quantity of uranium, could be developed. It was therefore agreed that, after the defence programme, the fast reactor should have highest priority. Work by a joint Harwell–Risley team started in 1951. From every point of view the problem was a difficult one. The heat release in the core of a fast reactor is intense: in a large experimental reactor between 100 and 200 MW of heat would have to be conducted away from a core about 2 ft in diameter and 2 ft long and it was felt that this could be done only by using a liquid metal coolant. Sodium could be used but it solidifies at ambient temperatures; a sodium–potassium alloy which remains liquid at room temperature was therefore chosen. In the early 1950s the techniques of making ceramic uranium oxide fuel elements had not been developed; metallic fuels with a much lower melting-point had therefore to be used. It was feared that this metallic fuel might melt if the flow of coolant were interrupted and that having melted it would fall to the bottom of the containment vessel and form a supercritical mass which would cause a minor explosion.

Neutron irradiation would make the liquid-metal coolant radioactive and because the heat had ultimately to be transferred to water with which, in the case of leakage, there might be a violent chemical reaction, it was decided that a secondary, inactive liquid metal circuit was necessary and that there should be 'double-wall' separation between this liquid metal and the final cooling water. At that time, no centrifugal pumps had been developed in Britain for use with hot liquid metals and it was necessary to use a number of electromagnetic pumps. Because of all the uncertainties that existed it was decided to build the research reactor on an isolated site at Dounreay on the north coast of Scotland and to enclose the reactor within a steel sphere which would contain the fission products in case of an accident.

Construction of the Dounreay experimental reactor (Fig. 10.7) was started in 1955. After the troubles with the Windscale reactors in late 1957, staff were diverted from the project and the reactor did not reach its designed output until 1959. It had been built to determine whether fast reactors were controllable; to give experience with the use of liquid metals and to give facilities for the development of fuel elements. It did all these things and made possible the design of a 250-MW (electrical) prototype reactor which was commissioned in 1976.

VI. COMMERCIAL POWER-PRODUCING REACTORS

While design of the fast reactor had been going on, a group at Harwell had been working on designs of power-producing thermal reactors. If every type of moderator is considered in combination with every possible fuel in every possible form, more than a hundred different types of thermal reactor can be conceived. Harwell considered a large number of these; they were not enthusiastic about reactors which used light water as a moderator and as a coolant, and ultimately they concentrated their work on a reactor cooled with liquid sodium which they thought might be used for submarine propulsion; a heavy water reactor similar to NRX; and a graphite-moderated reactor enclosed in a pressure shell and cooled by carbon dioxide under pressure. Risley had

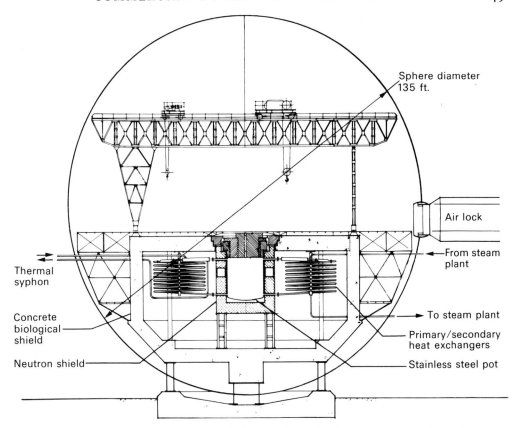

FIG. 10.7. Schematic arrangement of the 60-MW (thermal) Dounreay experimental reactor.

already prepared schemes for this last type of reactor but had been hampered by the fact that, when using aluminium cans, a chemical reaction between aluminium and uranium limited the surface temperature of the can to 350 °C. By 1951 the Harwell metallurgists had developed a can made of magnesium alloy which could be used at temperatures of 400 °C, and this made it possible for two Harwell engineers, with the help of the Ministry of Works and two industrial firms, to prepare a realistic scheme for a power-producing reactor which they called PIPPA. There was considerable opposition to the construction of a PIPPA reactor but, in 1952, the defence requirement for plutonium was increased and, after long discussion, it was agreed that it could best be met by building reactors of the PIPPA type. The design had to be re-optimized because these reactors had to produce plutonium as a primary product and power as a by-product. The project (for which Risley was made responsible) was sanctioned in March 1953 and work on Calder Hall, a site adjacent to the

Windscale piles and chemical plants, was started immediately. The reactor was divergent in May 1956, generated power for use by the factory in August, and, on 17 October, the Queen connected it to the Central Electricity Board's transmission network (Fig. 10.8).

By late 1953 Risley was satisfied that PIPPA-type reactors would be successful and that the problems of the fast reactor were soluble; they therefore prepared a programme for the long-term development of nuclear power in which, after experience of PIPPA-type reactors had been gained, more advanced reactors would be built and the programme would proceed by cautious steps to the fast reactor. This plan was considered by a committee under Treasury chairmanship which decided that it was too modest; they recommended a larger programme which would give 1800 MW of nuclear power by 1965. The nuclear stations were to be owned by the electricity boards and built by the industrial firms which were already being trained by Harwell and Risley. This enlarged nuclear programme was probably already too ambitious but a shortage of coal was expected and the first Suez crisis cut off the main supplies of oil; in the near panic that followed Suez, the programme was enlarged to give 5000 MW by 1965. It was an over-optimistic decision which was damaging to the development of British technology.

To meet this programme the electricity boards built seven nuclear stations. When calculated on the basis of the cost of construction and the fuel prices which were current at that time, all those stations produced power more expensively than the fossil-fuel-fired stations that were built and operated concurrently. In an attempt to bring down the cost of nuclear power, larger reactors were built until there was anxiety about the integrity of the steel pressure vessels. This anxiety was removed by the development of the integral type of reactor in which the pressure vessel and biological shield were combined as a post-stressed reinforced concrete structure. All these stations used metallic uranium as fuel and because this was enclosed in magnesium alloy cans they were called Magnox reactors.

Pressure of circumstances had forced Britain along the path which led to the use of gas-cooled, graphite-moderated reactors but, in the spring of 1957, C. Hinton read a paper in Sweden in which he pointed out that the cost of fossil-fuel-fired power stations had come down as it had been possible to achieve higher top temperatures in the heat cycle. He suggested that the same thing would happen in the development of nuclear power and that it would be easier to achieve high temperatures in gas-cooled than in water-cooled reactors. This set Britain firmly on the road of developing gas-cooled reactors

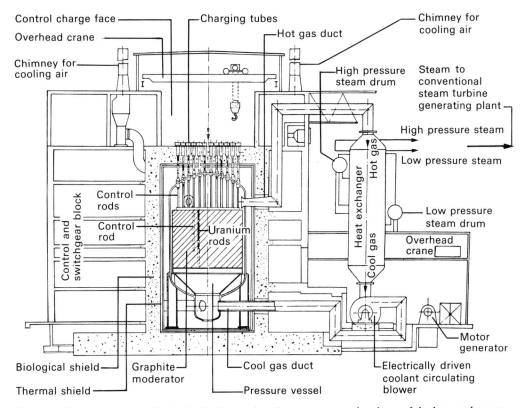

FIG. 10.8. Cross-section of a Calder Hall pile showing the pressure vessel and one of the heat exchangers.

of more advanced design. These reactors used a slightly enriched uranium-oxide fuel and this fuel was contained in stainless steel cans. The Atomic Energy Authority built a successful prototype of the Advanced Gas-Cooled Reactor (A.G.R.) at Windscale, but when the electricity boards placed their first order for a commercial station they pushed up the pressure and temperature of the coolant far beyond the figures that had been tried in the prototype and they accepted a tender which proved to be unsatisfactory. The trouble which resulted has been damaging to the British reactor industry but it still remains to be seen whether gas-cooled or water-cooled reactors will be best in the long run.

The French, like the British, built gas-cooled, graphite-moderated reactors but their early designs gave trouble and were expensive. The French had always been open-minded about reactor types and even while concentrating on the construction of magnox reactors they had built a light-water reactor based on American designs in order to gain experience of the alternative system. When Magnox reactor design reached the limit of its potential and the British changed to A.G.R.s, the French decided to drop gas cooling and to build light-water reactors of the type developed in America.

American enthusiasm for reactor development had faded at the end of the war. The piles at Hanford had produced the plutonium used in the atomic bomb which had been dropped on Nagasaki but the dimensional changes in the graphite which had been predicted by E. P. Wigner made the piles unsafe and they were closed down. (These piles were later rehabilitated and continued to work until the mid 1960s.) When Lord Portal (who was Controller of the U.K. Atomic Energy Organization) visited America in the Spring of 1946, General Groves told him that the best advice he could give about piles was not to build them. There was no pressing need for an industrial nuclear programme in the United States because it had large reserves of oil, natural gas and coal; only the Navy was interested in nuclear power and not everyone in the Navy was interested. The credit for creating the American nuclear reactors goes primarily to one man, Captain (later Admiral) H. G. Rickover who, in the early stages of development, was supported by Admiral E. W. Mills.

The Navy had shown interest in nuclear propulsion before the United States entered the war but Roosevelt had excluded them from participation in the Manhattan Project and there was some hostility between the Navy and General Groves. In spite of this, Groves signed a contract authorizing a paper study of a liquid-metal-cooled reactor for a destroyer and Captain Rickover was seconded to Oak Ridge to participate in this. Rickover was a career engineer officer in the U.S. navy; during the war he had shown great practical ability, tireless energy, and close attention to detail in correcting faults in the design and construction of naval vessels; he was not popular but he had great tenacity. The position in America was one in which it was difficult to generate enthusiasm; the Manhattan Project had been disbanded in 1946 and responsibility for atomic energy had been handed over to the Atomic Energy Commission; most of its members were not highly technical and they did not wish to see control of important developments, which had been wrested from the Army and given to their civilian organization, pass back to another branch of the armed forces. Industry could see little economic motivation for participation in atomic energy work and lacked interest.

During 1948 studies of a high-flux water reactor and of a sodium-cooled fast breeder reactor were being done by W. H. Zinn at the Argonne laboratories near Chicago. Other studies of a sodium-cooled reactor, working at

FIG. 10.9. Layout of power
plant for submarine propulsion
and schematic diagram of a
pressurized water reactor
(P.W.R.).

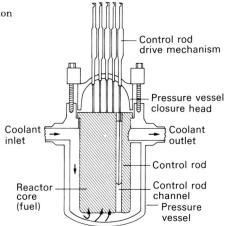

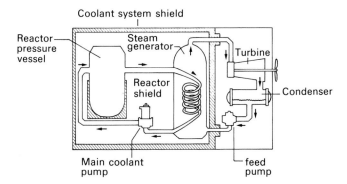

intermediate neutron energies, were being done by General Electric who had
built an atomic energy research establishment at Knolls, near Schenectady.
Rickover was not satisfied that any of these projects would meet his require-
ments for nuclear propulsion and work was started to develop a Pressurized
Water Reactor at the research station at Bettis which had been built for the
American A.E.C. by Westinghouse.

The principal features of this reactor are shown in Fig. 10.9. The core,
made from rods of an enriched uranium alloy enclosed in zirconium cans,
was enclosed in a steel pressure vessel through which light water was circulated to act both as moderator and coolant. This coolant gave up its heat to
boil water in a secondary circuit and the steam from this drove the turbine.
For submarine propulsion, there were objections to the sodium-cooled reactor
and it was the P.W.R. that was chosen for general use in the U.S. submarine

fleet. By 1954 the U.S. Navy had committed itself to nuclear submarines and the first of them—appropriately named *Nautilus*—was launched in January 1955. The fact that it was a notable success was due almost entirely to the skill, zeal and enthusiasm of Admiral Rickover.

Rickover wanted to build a larger reactor to propel an aircraft carrier but the idea was turned down, as also was a proposal by a group of American firms who wished to collaborate and build a land-based power station. Congress Members preferred that the first land-based nuclear station should be a Government project. Argument went backwards and forwards but, in July 1953 the A.E.C. gave responsibility for the design and construction of their first land-based nuclear station to Rickover. Even so there was hesitation and it was not until 22 October 1953 that the intention to build the plant at Shippingport in Pennsylvania was announced.

The American plant was not unnaturally a scaled-up version of the P.W.R. submarine reactor. The reactor had a 'seeded' core in which the central fuel elements were made of an alloy of highly enriched uranium with molybdenum and niobium; this central section of the core was surrounded by uranium oxide fuel elements canned in zirconium. The Shippingport reactor was first run in December 1957; Calder Hall had run eighteen months earlier and, for those who were engaged in the construction of the British plant, it is interesting to contrast the orderly progress of design and construction at Calder Hall with the similar work at Shippingport which, as described by R. G. Hewlett and O. E. Duncan, gives the impression of an unorganized stampede.[1]

Westinghouse continued to design reactors which, in principle, were like Shippingport but a seeded core was never used again; as in all the later industrial reactors, the fuel was made of slightly enriched uranium oxide. While Westinghouse developed the P.W.R., General Electric abandoned development of sodium-cooled reactors and developed the Boiling Water Reactor (B.W.R.). This reactor, as shown in Fig. 10.10 dispenses with the secondary cooling circuit; water is evaporated in the core and the steam which is generated there is passed through separators, where most of the water is removed and the reasonably dry steam goes to the turbines. Although the system would appear to be inherently cheaper than the P.W.R., it has some

---

[1] In their book *Nuclear Navy*, Hewlett and Duncan say: 'Shippingport had much greater impact on nuclear technology than Calder Hall because, being non-military, every aspect of design and operation could be de-classified.' This is a surprising statement because full details of Calder Hall were given at the first Geneva Conference in 1955. British industrial firms had been trained in the design and construction of Magnox reactors and they had submitted tenders for the construction of the first British industrial nuclear stations nearly a year before Shippingport first generated power.

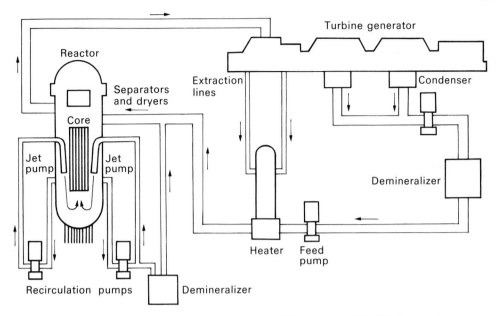

FIG. 10.10. The direct cycle boiling water reactor (B.W.R.) system. The Windscale primary.

disadvantages—one of which is that the steam going to the turbine is slightly radioactive.

The research work on fast reactors which had been done at the Argonne laboratories and at Knolls was followed up by a consortium of American industrial firms who built a prototype liquid-metal-cooled fast reactor. The site chosen was surprisingly close to Detroit; there was delay in obtaining an operating licence and, when this had been obtained, an accident closed down and seriously damaged the reactor which was ultimately abandoned. Steady development and vigorous selling enabled Westinghouse with its P.W.R., and General Electric with its B.W.R., virtually to corner the world's reactor market and light-water reactors are now the preferred type in most countries. Improved techniques of pressure-vessel manufacture, and uprating of fuel elements make light-water reactors with a capacity of 1050 MW (electrical) common and even larger ones with an output of 1300 MW are being commissioned. How long these reactors, which generate steam at a comparatively low temperature, will retain their grip on the market is a question on which it is possible to speculate. Indeed, today the whole future of nuclear power must be in doubt. The world, which will soon be short of fossil fuels, has a real need of nuclear power and the history of nuclear power shows it to be one of the safest, if not the safest, of all the process industries.

Yet, regardless of the frightening risks of a global energy shortage, professional and amateur environmentalists are so stirring up an emotional fear of all things nuclear that the whole future of this great source of power in nature has been thrown into question.

REFERENCE

[1] GOWING, MARGARET. *Independence and deterrence*, Vol. II. *Britain and atomic energy 1945–1952*. Macmillan, London (1974).

BIBLIOGRAPHY

GOWING, MARGARET. *Independence and deterrence*, Vols I and II. Macmillan, London (1974).
HEWLETT, RICHARD G., and ANDERSON, OSCAR E. *History of the U.S. A.E.C.*, Vol. I. *The New World, 1939–1946*. Pennsylvania State University Press, University Park, Pa. (1962).
——, and DUNCAN, FRANCIS. *History of the U.S. A.E.C.*, Vol. III. *Atomic shield, 1944–1952*. Pennsylvania State University Press, University Park, Pa. (1969).
——, and ——. *Nuclear navy*. University of Chicago Press (1974).
HINTON, CHRISTOPHER. Axel Ax-son Johnson Lecture. Royal Swedish Academy of Science (1957).
KRAMISH, A. *Atomic energy in the Soviet Union*. Stanford University Press (1959).
There are, in addition, many articles in the proceedings of the Geneva Conferences and of professional institutions which describe reactors and reactor development.

PART IV: CHEMICAL PROCESSES FOR ATOMIC ENERGY

Laymen imagine that the most intractable problem in the development of atomic energy was reactor design. This is far from the truth. Inaccurate or inadequate scientific information limited the performance and the useful life of the early reactors but, such as they were (and they did all that was asked of them), these reactors presented designers with far fewer and far simpler problems than those of the diffusion and chemical plants.

I. FUEL ELEMENT MANUFACTURE

Uranium is the only fissionable material occurring in nature and it is the unique raw material of atomic energy plants. It occurs in its primary form as pitchblende (an oxide of uranium), which for many years had been mined in Zaire, Czechoslovakia, Canada, the United States, and elsewhere for use as a source of radium. Where the primary ores have been leached out of lodes or disseminated deposits, much of the uranium has reached the sea in solution but some has migrated through relatively porous rock and been deposited as a secondary mineral, mainly as uranium phosphate. The valency of the uranium is changed in this process and the metal is fixed in much the same way as it is in ion-exchange plants. These secondary deposits, many of them in rock which has been subject to metamorphic change, may contain only small quantities of uranium. In such circumstances ion-exchange or other benefaction plants are built near the deposits to minimize transport costs.

There were large stocks of Congo pitchblende in Belgium at the beginning of the Second World War and these were jointly bought by the British and the Americans and shipped to the U.S.A. out of German reach. This jointly owned stock of pitchblende, together with Canadian ore, was the raw material used in the early American and British plants and its division between the two countries was an important factor in post-war negotiation.

In the radium extraction plants uranium had been virtually a waste material, of which a small amount was used as a yellow pigment for decorating pottery. The metal had first been isolated by E. M. Péligot in 1841, but there was no industrial process for production of uranium having the purity which was needed for the atomic energy programme. New chemical processes are usually evolved in test-tubes and beakers on the laboratory bench, that is, they are developed as batch processes. When the engineer has to scale these processes up to industrial size, it is easiest for him to design batch-operated plants; the test-tubes and beakers of the laboratory become tanks, and liquids are transferred not by pouring but through pipes. It is only later, when experience makes this possible, that more elegant and economical continuous processes are developed. This is what happened in the manufacture of pure uranium metal.

In America, various firms were employed to manufacture the billets of pure uranium which were shipped to Hanford where the canned fuel elements were made and irradiated in the reactors. In Britain, on the other hand, all the processes for the manufacture of canned fuel elements were carried out at Springfields. There was no interchange of industrial information between the two countries, but the processes used were broadly similar. One of the decay products of pitchblende is radon, which is a radioactive gas. The risk arising from this, as well as the risk of dust from the pitchblende itself, required that the drums in which it arrived at the factory were opened remotely, in ventilated plant where the ore was crushed and sampled. The wet crushed ore was ground to a slurry and dissolved in a mixture of sulphuric and nitric acids. The radium was co-precipitated with barium and, together with other impurities, it was removed from the liquor by filter pressing. The uranium in the clear liquor from the filter press was precipitated as uranium peroxide; this was dissolved in nitric acid and the solution evaporated so that uranyl nitrate was formed. This uranyl nitrate still contained trace impurities and these were removed in an ether purification process which took advantage of the fact that uranium is more soluble in ether than the impurities. The purified uranium was washed out of solution in the ether with de-mineralized water and precipitated as ammonium diuranate.

After precipitation the diuranate was put into graphite trays and placed in an electric furnace where it was made into uranium tetrafluoride by calcining, reducing with hydrogen, and converting with hydrofluoric gas. The uranium tetrafluoride was mixed with chips of pure calcium metal and placed in a reaction vessel which was lined with purified calcium fluoride; the mixture was ignited in a closed and ventilated chamber where it reacted with pyrotechnic violence.[1] The heat of reaction melted the uranium, which flowed down to form a billet at the bottom of the reaction vessel.

At Hanford the uranium billets were extruded into rods. These rods were cut into suitable lengths and inserted into aluminium cans to which they were bonded with an aluminium–silicon alloy. The development of a satisfactory bonding process gave the Manhattan Project a considerable amount of difficulty.

In Britain, the rods were made, not by extrusion, but by vacuum casting. There were two reasons for this; one was that the British expected difficulty in avoiding health hazards in an extrusion plant but, more importantly, casting gave a random grain structure which caused less distortion when the rod was irradiated in the pile. Because the British fuel elements were to be used in an air-cooled pile where the metal temperature was higher and the heat rating of the fuel elements was lower, it was decided not to use bonding but to obtain good contact between the rod and the aluminium can by external pressurization during manufacture and to improve heat transfer by introducing helium into the can. As in America, a great deal of difficulty was experienced in developing a satisfactory canning process. The one used at Springfields, with all its testing procedures, involved fifty different operations.

Growing demand for fuel in the industrial reactors made it necessary to build bigger plants. Freed from the pressure of time and with more operating

---

[1] When sufficient experience had been obtained both the American and the British plants used magnesium as a reducing agent. Calcium was used in the early days because the $Ca–UF_4$ reaction is more exothermic than the similar $Mg–UF_4$ reaction, but magnesium was cheaper than calcium and gave a purer product.

experience, design practice followed the usual pattern of the process industries and batch processes were replaced by continuous processes. Other solvents were used to replace the dangerously inflammable ether; counter-current purification was used instead of batch solvent extraction and the conversion of ammonium diuranate to uranium tetrafluoride was carried out in continuously operating fluidized beds.

The old batch plants were scrapped and are remembered today only as the clumsy first efforts of those early years. But they served their purpose; at the time when they were built they presented formidable problems of design, of safely handling radioactive materials in tonnage quantities, and of achieving a far higher level of purity than was normal in bulk chemical plants. It was on these first plants that the chemists and engineers of the British organization cut their teeth: 'Let not ambition mock their useful toil.'

The re-designed British plant is still used in a developed form to manufacture fuel elements for Magnox reactors, but, because metallic uranium has a comparatively low melting-point and because a phase change that takes place at an even lower temperature can cause nuclear instability in a reactor, metallic uranium is no longer used in modern reactors where high operating temperatures and high ratings demand high fuel-element temperatures. The light- and heavy-water reactors and the A.G.R. all use ceramic fuel made of compacted uranium oxide and the cans are made from more sophisticated materials than the aluminium that was used in the dawn of atomic history.

These and other plants, such as those for converting uranium tetrafluoride into the hexafluoride which was the feedstock for the diffusion plants, and the plants for converting the enriched hexafluoride back to the oxide, seemed difficult in those pioneering days, but these difficulties paled into insignificance when compared with the difficulties of designing the chemical plants for extracting the plutonium from the irradiated fuel elements. It is still argued within the atomic energy industry whether these chemical plants or the diffusion plants were the more difficult job. The fact that, more than thirty years after the first atomic bomb was dropped, there are only two successful industrial chemical separation plants in the western world (and neither of these is in America) suggests that the chemical separation plants top the list for difficulty.

II. CHEMICAL SEPARATION

What the Manhattan Project needed during the war was a process, however inelegant, to extract plutonium from the irradiated fuel elements. The plutonium was present in the raw material in little more than trace quantities: one kilogramme of plutonium was contained in about three tonnes of irradiated fuel elements and these fuel elements contained an almost equal quantity of highly radioactive fission products. In America, G. T. Seaborg started laboratory research in 1942 and explored a number of different chemical reactions that might provide the basis for an industrial process; he and his assistants tried an oxidation–reduction process using lanthanum fluoride as a carrier but they were worried that the fluorine might cause corrosion of vessels in an industrial plant; they tried precipitation by the addition of hydrogen peroxide but found that the volume of the precipitate was too large; they studied solvent separation but came to the conclusion that the process could not be developed sufficiently quickly. At the same time, chemists in other American universities tried separation by ion-exchange and by dry reactions using fluorine at a high temperature. Meanwhile, B. L. Goldschmidt, the French ex-patriate chemist, and others studied the composition of the mixed fission products to see if any of them had chemical properties so similar to those of plutonium that they would follow it through a solvent extraction.

All this was laboratory research carried out by academic scientists. In August 1942, it was agreed that Du Pont de Nemours, who had great experience of developing new industrial plants from laboratory research should participate in Seaborg's work. C. M. Cooper, with a small group of Du Pont men, worked with Seaborg's team. A large part of Du Pont's normal business was in the manufacture of conventional explosives and they had no wish to become interested in atomic bombs in a way which would strengthen the accusation that they were 'merchants of death'; they did not enter into formal agreement until December 1942 and then with the condition that they should be paid only a nominal fee and should be released at the end of the war.

Before the end of 1942, Cooper came to the conclusion that the lanthanum fluoride route gave most promise and he started to design a pilot plant at Chicago. Following the classic principle that this is the easiest way of proceeding from bench research to the industrial scale, he planned to build a batch plant. Meanwhile Seaborg had been studying an alternative route in which the plutonium was co-precipitated with bismuth phosphate and Cooper's small pilot plant was modified to test this process. Du Pont were anxious to start work on a larger pilot plant at Oak Ridge and in June 1943 it was agreed that, although there was little to choose between the processes, there must be anxiety about plant corrosion if the lanthanum fluoride route was used and, for this reason, the bismuth phosphate process was chosen.

FIG. 10.11. Schematic plan of chemical separation plant.

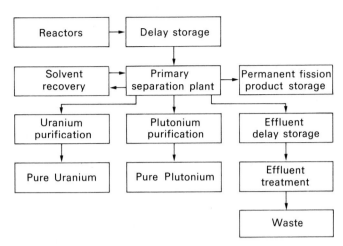

It had already been decided that, because of district hazards arising both from the piles and from the chemical separation plant the plutonium factory should be built in a remote site and Hanford on the Columbia River had been chosen. There was a serious shortage of labour in this remote and inhospitable area and priority had to be given to construction of the piles; besides this, there was a shortage of design information on chemical separation because irradiated fuel elements were not available from the pilot plant at Oak Ridge for research on the chemical extraction processes. Work on the full-scale plants did not start until the end of 1943.

The principles upon which chemical separation plants were based is shown in Fig. 10.11. Each separation plant at Hanford was designed to include a separation building where the bismuth phosphate precipitation process would be carried out; a concentration building where the plutonium would be separated from the phosphate carrier and from other gross impurities; a ventilation building; and a storage area for solid and liquid radioactive wastes. The separation buildings were huge: approximately 250 m by 20 m in plan and 25 m high. Inside each of them was a row of forty concrete cells, each about 5 m square and 6 m deep. The cell walls, which provided biological shielding, were 2 m thick and a concrete lid weighing 35 tons completed the shielding. The lid could be lifted by a crane to give access for plant repairs or replacements which were carried out by means of rudimentary remote-handling equipment.

The concentration buildings were of lighter construction because most of the fission products had been taken out in the precipitation plant. In these concentration buildings relatively large amounts of bismuth phosphate had to be removed and Seaborg had originally proposed that this should be done by dissolving in hydrochloric acid and precipitating the plutonium with a rare-earth fluoride. This method could not be used on the full-scale plant because of corrosion by hydrochloric acid; concentration had to be done by the original lanthanum fluoride process. Combination of the two competing processes had the advantage of giving a purer product. Further purification was necessary but since radioactivity at this stage was small, heavily shielded plant was not needed. The plutonium was separated from the lanthanum fluoride, dissolved in nitric acid, and precipitated with sodium peroxide.

The plants which were completed before the end of 1944 only separated the plutonium from the irradiated fuel elements; they did not separate the fission products from the slightly depleted uranium so that it could be re-enriched and used again. It is probably true to say that never before or since have such heavy, expensive, and cumbersome plants been built to produce so small a quantity of product; yet the miracle is that they were built in time to produce the second atomic bomb which was dropped in 1945.

Because it did not separate the fission products from the slightly depleted uranium the Hanford batch process could be regarded only as a stop-gap; a better process had to be found. One of the cardinal principles of chemical plant design is that, whenever possible, reactions (and particularly those involving noxious materials) are best carried out in the liquid phase; it followed from this principle that a solvent separation process was desirable. Seaborg had worked on this in 1943; Goldschmidt and R. Spence had continued the work at Montreal and Chalk River. Goldschmidt returned to France at the end of the war and Spence continued the work with a small team of assistants. He had only a few irradiated fuel elements (supplied by the Americans) with which to work and these contained less than 20 mg of plutonium (as much as would cover a pinhead). Moreover, these fuel elements had 'aged' after irradiation so that there had been fission-product decay. An immense amount of work had to be done to find a suitable solvent and a suitable process. The choice of solvent depended not merely on extraction characteristics but also on the relative specific gravities of the solvent and aqueous phase; the variation of this with temperature; the viscosity of the solvent; its liability to form emulsions; the flammability of the solvent; and the possibility of its forming corrosive by-products. The process had to be one which used plant which could be contained within a biological shield and operated safely by remote control without access for modification or repair. Spence naturally assumed that counter-current extraction (a process which, in principle, is similar to fractional distillation) would be used. He had to determine how many 'theoretical plates' were necessary at each stage of the process and he could do this only by simulation with a series of test-tube extractions. After each series of experiments, all materials had to be recovered for use in the next series. Spence chose dibutyl-carbitol (butex) as the best solvent; the irradiated rods were dissolved in nitric acid, the uranium being in its hexavalent and the plutonium in its quadrivalent form. Both the plutonium and the uranium were extracted into the solvent in the first column, leaving the fission products in solution in nitric acid, which was stored as 'highly active effluent'. The solvent stream from the first column was neutralized and the plutonium was reduced to the solvent-insoluble trivalent state leaving the valency of the uranium unchanged. Thus, by a series of valency changes and counter-current extractions, the fission products, the uranium, and the plutonium were separated into three separate aqueous streams while the solvent was released for re-cycling.

The British 'master programme' showed that, to meet the date for the bomb trial at Monte Bello, the British design team at Risley would have to start work on the primary separation plant by the end of 1947. J. D. Cockcroft had arranged that J. P. Baxter and his I.C.I. team at Widnes would collaborate with Spence to interpret the bench-scale work for use by the designers at Risley and that they would also carry out chemical engineering development. In September 1947 Baxter, C. Hinton, and C. J. Turner (who was in charge of the chemical design office) visited Chalk River; G. R. Howells, the Risley liaison officer, was already there. This team went carefully through Spence's flow-sheet and came to the conclusion that it could form the basis of a successful plant.

The full-scale chemical separation plant at Windscale in Cumberland (Fig. 10.12) cost about £10m. in 1950. The decision to base this continuously operating industrial plant on research work which had been done with 20 mg of plutonium and 'aged' fission products was in itself a hazardous one but, even if more complete and more soundly based research information had been available, the problems which faced the engineers would have been frightening. They felt fairly certain that access to those parts of the plant (that is, the dissolver and the first extraction stage) where bulk fission products were present would be impossible; access for the operation of less active parts of the plant would be impossible and it could not be assumed that access even for repairs and maintenance would be safe. All operation had to be remote; there could be no valves or moving parts within the shield walls and special arrangements had to be made for instrumentation and sampling.

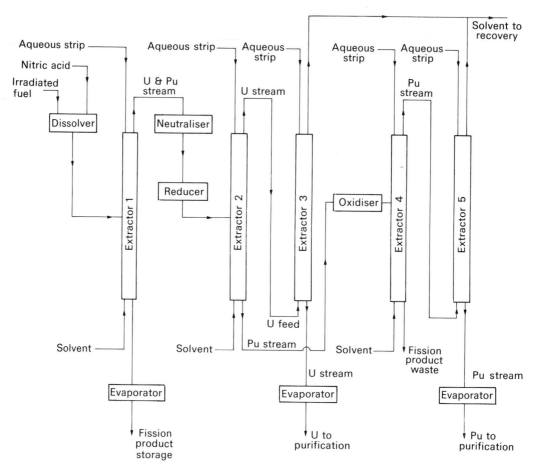

FIG. 10.12. The separation of plutonium from uranium by solvent extraction at Windscale.

In addition, vessels had to be of such a shape and size, and operation so precise, that it was impossible to accumulate a supercritical mass of fissionable material in them.

Counter-current solvent extraction can be carried out in several different types of plant; in this case vertical columns packed with rings were the most simple but were inconveniently high; pulsed columns were shorter but had moving parts within the active cell; mixer settlers had the same disadvantage; air-lift separators might cause a carry-over of active spray. In the light of all of these considerations packed columns were chosen as being best. Even for packed columns, no information existed for the theoretical stage height and extraction data were scanty.

On their return to Risley, Turner and Hinton prepared a master programme for design and construction. This programme demanded that, to meet the requirements for the Monte Bello test, the flow-sheets for the chemical plant should be ready by the end of 1947 and major alterations excluded by April 1948; the plant should be ready for inactive trials by July 1950. Spence's work at Chalk River had been brilliant, but shortage of active material, shortage of laboratory facilities, and shortage of time had made it impossible for him to do more than provide the skeleton of a process. When Baxter's team examined the information with a view to preparing a flow-sheet they found that there were sixty separate topics on which further research work had to be done. Progress was delayed by the fact that freshly irradiated fuel elements were not available from the Chalk River reactor until the middle of 1948, and even then Spence could work only with trace quantities because the 'hot' laboratories of the Atomic Energy Research Establishment at Harwell were not completed until the middle of 1949. However, it was possible for Spence's team to set up a small column at Chalk River to test plutonium separation and for the I.C.I. team to build columns working with uranium as an analogue to determine the extent to which column diameter affected the theoretical plate height. Risley decided that design must go ahead largely on the basis of unchecked data but Hinton was so worried about the position that in 1949 he insisted that Calder Hall farm should be bought as a site for an American-type bismuth phosphate plant if all else failed. There was, in the event, no need to build such a plant and Calder Hall farm was used as the site for the world's first industrial-scale nuclear power station.

In spite of all the difficulties, operation of the primary separation plant with inactive solutions was started in mid-1951. One major difficulty was experienced during commissioning. It had always been feared that traces of butex would be carried over with the nitric acid from the first column and would react vigorously, causing activity to be dispersed into the atmosphere. A mild reaction of this sort did, in fact, occur during the inactive trial runs but the risk was removed by evaporating the Column 1 effluent. Experience so gained made it possible to do further evaporation of the fission product solution, greatly reducing the volume which had to be stored.

The plant was a brilliant success. Its output in the months before the Monte Bello trial was 12 per cent more than the design figure; the extraction efficiency both for plutonium and uranium was over 99 per cent; and the plant worked without major difficulty until increased demand made it necessary to build a bigger plant twelve years later. The Americans built a solvent

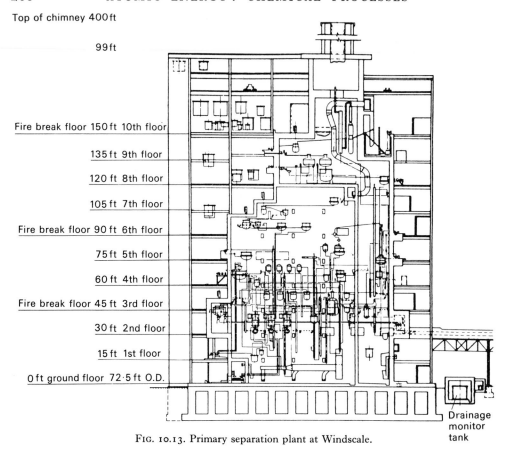

Top of chimney 400 ft

99 ft

Fire break floor 150 ft 10th floor

135 ft 9th floor

120 ft 8th floor

105 ft 7th floor

Fire break floor 90 ft 6th floor

75 ft 5th floor

60 ft 4th floor

Fire break floor 45 ft 3rd floor

30 ft 2nd floor

15 ft 1st floor

0 ft ground floor 72·5 ft O.D.

Drainage monitor tank

FIG. 10.13. Primary separation plant at Windscale.

extraction plant which was commissioned about a year after the Windscale plant but, in their process, it was necessary to 'salt out' the fission products with aluminium nitrate, and this made concentration of the fission product solution impossible.

The product streams from the primary separation plant required further purification but there had been no time to develop an industrial process for plutonium purification when design of the plant had to be started. The first plutonium purification plant was therefore based on laboratory practice; the plutonium liquors were first purified with a further solvent process using tributyl phosphate to reduce activity levels, and finally extracted and backwashed using TTA (thenyltrifluoroacetone) in benzene as the solvent. This was a hazardous stopgap process, carried out in glass vessels which were later replaced by pulsed columns, which are still in use after a quarter of a century.

The uranium purification plant was deferred until it was possible to obtain reliable design data. When built it used a butex cycle in a mixer-settler plant. It operated efficiently and reliably. Mixer-settlers were also used when a larger primary separation plant had to be built in the 1960s.

BIBLIOGRAPHY

BENEDICT, M., and PIGFORD, T. H. *Nuclear chemical engineering*. McGraw-Hill, New York (1957).

DAVEY, H. G. Primary separation plant at the Windscale works. *Nuclear Power*, *1* (1956).

GOWING, MARGARET. *Independence and deterrence*, Vols I and II. Macmillan, London (1974).

HEWLETT, RICHARD G., and ANDERSON, OSCAR E. *History of the U.S. A.E.C.*, Vol. I. *The New World*, *1939–1946*. Pennsylvania State University Press, University Park, Pa. (1962).

HINTON, CHRISTOPHER. The chemical separation process at Windscale Works. Castner Lecture, Society of Chemical Industry, 23 February 1956, *Chemistry and Industry*, July (1956).

HOWELLS, G. R., HUGHES, T. G., MACKEY, D. R., and SADDINGTON, K. The chemical processing of irradiated nuclear fuels from thermal reactors. Second International Conference on the Peaceful Uses of Atomic Energy, **17** (1958).

NICHOLLS, C. M. Development of the butex process for the industrial separation of plutonium from nuclear reactor fuels. *Transactions of the Institution of Chemical Engineers*, **36**, No. 3 (1958).

SPENCE, R. Chemical process development for the Windscale plutonium plant. *Journal of the Royal Institute of Chemistry*, May (1957).

II

DEVELOPMENT OF THE NUCLEAR WEAPON

E. F. NEWLEY

I. THE WARTIME PERIOD

To put the development of the nuclear weapon into proper perspective it is necessary to begin this section by recapitulating the state of affairs in 1939, at the start of the Second World War. The fundamental discoveries in nuclear physics in 1938 and 1939 were hailed by the popular press and some technical journals as harbingers of an era of cheap power on the one hand and a holocuast on the other. In September 1935, C. P. Snow, writing in the journal *Discovery*, said:

Some physicists think that within a few months, science will have produced for military use an explosive a million times more violent than dynamite. It is no secret . . . It may not come off. The most competent opinion is divided upon whether the idea is practicable. If it is, science for the first time will at one bound have altered the scope of warfare. The power of most scientific weapons has been considerably exaggerated; but it would be difficult to exaggerate this.

In private, scientists in various countries were alerting their governments to the possibility that one day nuclear weapons might be developed, although no one could be sure how or whether this would be achieved.

In the United States, scientists persuaded Albert Einstein to sign a letter to the President drawing his attention to the danger, and in due course an Advisory Committee on Uranium was set up. In general, however, American scientists were more impressed by the prospects of harnessing nuclear energy for power production (Ch. 10) and in most cases tended to be sceptical of, or to ignore, the prospects of an early weapon application and all it might mean [4].

In Germany, the War Office reacted quickly in setting up an Office of Nuclear Research in the Army Ordnance Department and, step by step, acquiring control of all nuclear research in the Reich. Laboratory centres for nuclear research were set up by the Army at Gattow and in the Kaiser Wilhelm Institute of Physics at Dahlen. Eminent scientists, including O. Hahn, F. Strassmann, and W. Heisenberg were involved in the work, which later came under the control of the Reich Research Council. The work was badly fragmented, however, and little if any work appears to have been done on the development of a weapon, although the possibilities of such a development were fully appreciated [6]. The German interest in nuclear energy development was well known to the British and the Americans. It was apparent that heavy water was being obtained from Norway and that stocks of uranium had been commandeered when Belgium was overrun. Throughout the war in Europe, Britain and America were haunted by the fear that, if a nuclear bomb was possible, Germany might be successful in developing one. The overall result was to act as a spur to British, and later United States, interest in their own development programmes.

In Britain, the Minister for Co-ordination of Defence, acting on a suggestion put forward by G. P. Thomson, took action to investigate the purchase of all uranium ore remaining in Belgium as a precaution. In fact, the deal was not completed and, as mentioned above, the material was commandeered by Germany, as Thomson had feared. A more profitable result of this initiative was that a programme of research and investigation was started under the control of the Air Ministry.

Many British scientists were sceptical of any early development of a nuclear bomb; J. Chadwick, asked for his views, pointed to the lack of data on which to make a sound judgement. He felt that fission due to thermal neutrons and fission due to fast neutrons could both be developed but that large quantities of uranium, between 1 and 40 tons, might be required [2]. However, in the spring of 1940, O. R. Frisch and R. F. Peierls, two refugees from Europe working in Britain, submitted an analysis of the problem remarkable for its clarity and perception of the essential issues. From the consideration of the very short time interval in which large amounts of energy would be released in a practical bomb, they used simple theoretical argument to discount the use of any chain reaction in which slow neutrons were an essential link, for example, in any 'moderated' system. They also discounted the use of natural uranium in an unmoderated system because of neutron absorption in the ^{238}U isotope. The absorption problem could be overcome by separating the uranium isotope by methods they described and using pure ^{235}U. Although the behaviour of ^{235}U with fast neutrons had not been determined, they concluded that almost every collision would produce fission and that neutrons of a range of energies would be effective. Once started, a fast-neutron chain reaction would develop with great rapidity so that a considerable part of total available energy would be liberated before the reaction stopped owing to the expansion of the material. They suggested that the energy liberated by a bomb using 5 kg of pure ^{235}U would be equivalent to several thousand tons of dynamite. (For the full text see [2].)

Stimulated by this approach, the Air Ministry set up a committee, known by the code name MAUD, under G. P. Thomson to 'investigate the possibilities of uranium contributing to the war effort'. Entering the project with more scepticism than belief, the members of the committee made a comprehensive survey of the essential features of a weapon and the proposed methods for separating ^{235}U. When the committee reported in July 1941 they were convinced that a very powerful weapon was possible and could be developed in three to four years, depending on the supply of separated ^{235}U.

The supporting evidence in their report includes calculations based on papers by Peierls of the key parameters in a nuclear weapon, together with results of such measurements of important data as it had been possible to carry out in order to check the theoretical deductions. The Committee suggested that a practical bomb could be made using 10 kg of pure ^{235}U. This would be fashioned into two projectiles mounted one at each end of a gun barrel. Suitable arrangements would be incorporated to ensure that the projectiles remained separated at a safe distance until the device was required to explode. Charges placed behind the projectiles would then be fired, propelling the fissile projectiles together to form a supercritical mass. They estimated that a chain reaction in the fully assembled material would cause about 2 per cent of the ^{235}U to be fissioned before expansion of the material quenched the reaction. In terms of blast damage, they suggested the resulting explosion would be equivalent to about 1800 tons of TNT and that there would also be large amounts of energy emitted as thermal radiation.

The need for a high velocity of approach was to reduce the chance of a pre-detonation, that is, a chain reaction occurring in the material when it had reached a critical state but before it was fully assembled. They foresaw that if the rate of assembly was low, adventitious neutrons of natural origin—cosmic rays, spontaneous fissions, etc.—would have a high probability of starting an early chain reaction in the incompletely assembled fissile material, causing the bomb to give a low, or even trivial, yield.

The Committee were confident that, given the practical means to separate the uranium isotopes on an adequate scale, a weapon built on the lines they proposed would be successful.

To Lord Cherwell, Churchill's principal scientific advisor, the implications of the report of the MAUD Committee were that the U.K. must go forward to prepare for a weapon development programme. In his minute to Churchill, he wrote: 'It would be unforgivable if we let the Germans develop a process ahead of us by means of which they could defeat us in war or reverse the verdict after they had been defeated' [2]. The Chiefs of Staff concurred, and in due course a project organization was set up in the Department of Scientific and Industrial Research under the cover name of the Tube Alloys Directorate.

During 1940 and 1941 the British Government kept the U.S. Government informed on the work of the MAUD Committee, including sending copies of the draft and final reports. In the U.S.A., however, similar assessments made by the National Academy of Sciences tended to focus on the potential civil applications of nuclear energy and were less concerned about the prospects of an early development of a nuclear weapon.

Some U.S. scientists were, however, convinced by the cogent arguments of the MAUD Committee, and President Roosevelt was impressed by the positive attitude taken by the British Government. In October 1941 he authorized a full-scale investigation of the feasibility of nuclear weapon development, anticipating by several weeks a further report from the National Academy supporting, but in more conservative terms, the conclusions of the MAUD Committee.

The U.S. nuclear programme was initially co-ordinated by Vannevar Bush of the Office of Scientific Research and Development. Powerful teams of scientists and engineers were rapidly assembled to tackle the problems of producing fissile material suitable for weapons. Responsibility for the weapon itself was given to A. H. Compton of the University of Chicago. He was responsible for all fundamental physics for the project including the demonstration of a chain reaction in uranium and, assuming this to be achieved, for the scientific evaluation of the production of 'element 94' (that is, plutonium), from nuclear piles. Minute quantities of this new element had been produced by bombarding uranium with deuterons in a cyclotron and tests had shown it to be more readily fissioned than ^{235}U. In 1941 it offered an alternative route to a nuclear weapon in case the separation of ^{235}U proved too difficult.

In the summer of 1942, a major organizational change occurred when the Manhattan Project (p. 233) was set up to take executive responsibility for the whole nuclear programme. At that time the fast neutron data work essential to understand the functioning of the weapon was being conducted at various sites over the United States. J. R. Oppenheimer, who had been appointed by Compton to co-ordinate this work, was alarmed by its slow progress. He foresaw that if a weapon was to be developed by the time the fissile material was scheduled for production, a greater concentration of effort and skills would be needed. As the design of the weapon would be the most secret part of the whole project he argued that a separate laboratory should be set up in an area that would permit of a secure and closely controlled perimeter.

As a result of Oppenheimer's proposals, a site was selected on an isolated mesa in the Pajarito Plateau about 40 miles by road from Santa Fe in New Mexico. This became the Los Alamos Laboratory, charged with the specific responsibility for developing a weapon to use either ^{235}U or plutonium, scheduled to be produced some two years ahead. As Director of the Laboratory, Oppenheimer arrived at Los Alamos in March 1943, with a nucleus of scientific staff drawn from various universities across the U.S.A. Later in the year he was joined by a team of scientists sent over from Britain under the terms of the Quebec Agreement between the U.S.A., Canada, and Britain. By this time the idea of an independent British project had been shelved and Britain became a junior partner with America.

From the outset it was apparent that the weapon programme would be dominated by two factors: the supply of fissile material and the character of the weapon itself. Regarding the first, supplies of both 'pure' ^{235}U and plutonium would be very restricted initially, but it was essential to complete the development of the weapon by the time explosive amounts of fissile material became available, for any subsequent delay would lead to the war being unnecessarily prolonged at terrible cost. On the second point, the nature of a nuclear weapon is such that a chain reaction could not be sustained without at least one critical mass of material, that is, there would be few, if any, opportunities to conduct an overall experimental test of the bomb and if any such test were planned it would have to have a large *a priori* probability of success. Great reliance would have to be placed upon theory to understand in detail the functioning of the bomb and this, in turn, would depend on reliable data on the physical characteristics of fissile and other materials involved.

For critical mass calculations the theory of neutron diffusion in fissile material and any surrounding tamper had to be refined and account taken of the energy of distribution of fission neutrons, as well as the dependence of nuclear cross-sections upon these energies. For calculation of the efficiency of the nuclear explosion, a meticulous study of the hydrodynamics of the explosion was an essential factor.

The experimental nuclear programme planned for the first year included the determination of the following data:

(i) Neutron number: the average number of neutrons per fission for ^{235}U and ^{239}Pu.

(ii) Fission spectrum: the energy range of neutrons from fission of highly enriched ^{235}U.

(iii) Fission cross-sections: for ^{235}U and ^{239}Pu, for a range of neutron energies.

(iv) Delayed neutron emission: previous experiments had been limited to measuring delay after 10 microseconds, far too long to be of interest in a bomb. Measurements down to a small fraction of a microsecond were planned.

(v) Capture and scattering cross-sections in ^{235}U, ^{239}Pu, and a number of potential tamper materials.

Programmes on the Chemistry and Metallurgy of ^{235}U and ^{239}Pu were established, but these were hampered by shortages of material, in particular plutonium. One problem was that the purity of the fissile materials, and in particular the plutonium, would need to be very high. Alpha-particles emitted by the plutonium would react with light-element impurities to produce neutrons; these, in turn, would fission the plutonium and could lead to predetonation. The metallurgy programme included the development of methods of metal reduction of uranium and plutonium, and of casting and forming components in these materials and in various possible tamper materials and in various possible tamper materials.

The problem of assembling a critical mass, or more, of fissile material was examined in the MAUD report, and the main line of development at Los Alamos was based on a similar solution, namely the use of a 'gun' technique. Different arrangements of single and double-ended guns were considered, but eventually a single gun and target arrangement was chosen. To fit into an aircraft-carried weapon, the gun needed to be fairly short and as light as possible, but the projectile velocity would need to be high, particularly in a bomb using plutonium, to minimize the problem of predetonation.

For the first year the specification of the gun for the plutonium bomb was based on achieving the highest velocity (about 3000 ft/s) attainable within known gun technology, after taking account of the weight and size limitations. This was considered just adequate to achieve an acceptable predetonation probability, providing the light-element impurities in the plutonium were kept low. It was not until the summer of 1944 that the first samples of pile-produced plutonium became available and measurements showed a higher than expected rate of spontaneous fission. Furthermore, the rate of fissioning increased if the sample were returned to the pile and re-irradiated. This indicated that the probable source was the isotope ^{240}Pu formed by neutron capture in ^{239}Pu. Since the rate of spontaneous fission appeared to be proportional to the number of neutrons to which the plutonium had been subjected in the pile, it was apparent that the product of the large reactors then being built would have too high a neutron background to be usable in the gun device under development without running an undue risk of pre-detonation. In July 1944 the use of the gun assembly method with plutonium was therefore abandoned.

The end of the plutonium-gun programme in July 1944 brought great pressure to bear on the development of the alternative 'implosion' technique that had begun in a very small way in the spring of 1943 at Los Alamos. When possible bomb designs were being discussed, a physicist, S. H. Neddermayer, had proposed the use of an implosion technique as an alternative to the gun. His ideas were not well formulated, however, and at first his proposals received little support. At a later stage it was realized that with careful design the technique would have two useful advantages: the rate of change from subcritical to critical could be made more rapid than in the gun; and less fissile material would be needed. Both factors were seen to be important and had led to an acceleration in the research programme. Now it appeared to offer the only hope of achieving an acceptable predetonation performance with plutonium.

The basic concept of the implosion system is to surround a spherical, sub-critical arrangement of fissile and tamping material with a thick spherical shell of high-explosive. In the original scheme the explosive was detonated at its outer surface at a number of points simultaneously. The force of the explosion was intended to produce a spherical ingoing shock wave in the tamper and fissile material, driving it inwards towards the centre of the system. A simplified analysis of the system indicated that most of the kinetic energy of particles of the imploded material would be converted, at the centre of the converging mass, into potential energy of compression. Under the extreme pressures created, the density of the material would be increased so that a mass of material subcritical under normal conditions would, for a brief period, be raised to a supercritical state.

In the summer of 1944, the prospects of success with the implosion system did not appear to be high. Unlike the gun, which was based on well-established technology, the implosion concept was entirely new and beyond all past experience. The theoretical physicists were faced with new problems of understanding the behaviour of materials at hitherto unrealized conditions of pressure and temperature. This led them to numerical calculations far beyond the hand-computing methods then available. Attempts to arrive at simplified expressions for the equation of state of materials under these conditions proved unreliable. Eventually the efforts of the powerful team of theoreticians; the use of a battery of early computers; and some brilliant experimental work combined to produce data in which considerable confidence could be placed.

The theoretical treatments of implosions could, however, deal only with spherically symmetric conditions, whereas it proved difficult to realize such conditions in experimental firings. Extensive investigations were carried out to find the cause of, and the remedy for, the irregularities and inconsistencies that were observed. To this end improved methods of casting high-explosive charges were devised to ensure good homogeneity. Machine tools were used to fashion massive blocks of high explosive accurately to shape so that they could be assembled without leaving gaps that could lead to the formation of jets. In general, high accuracy and great consistency in production of all the components of the implosion devices were found to be essential.

In the original design, the main high-explosive charge was detonated at a number of discrete points on its external surface. The spherically diverging waves emanating from each point of detonation were expected to link up and form a single converging front in the explosive charge. In practice, the results obtained were inconsistent and disappointing. A parallel line of development was begun in the summer of 1944, using an explosive lens between each detonator and the high-explosive charge. The curvature of the detonation wave was reversed as it passed through the lens, so that it became spherically converging at the interface between the lens and the high explosive. This development also called for highly accurate fabrication of explosive components.

A further problem was the determination of the actual performance of experimental systems; for example, the detection of irregularities in the implosion, and the measurement of velocities, pressures, etc. A very large effort was devoted to devising methods of measuring these and other parameters, using a wide range of techniques too numerous and too detailed to go into here, but which contributed to the eventual success of the project.

The intensive development of the implosion system continued through the winter of 1944–5 until in April the design was frozen in order to prepare for a nuclear test of the overall performance of the system. A site for the test had been chosen on the Alamogordo Bombing Range, some 150 miles south of Los Alamos, and all preparations were made in great secrecy. Before the test, extensive studies were made of the likely damage effect of a nuclear explosion, including the spread of radioactive fission products, and a charge of 100 tons of chemical high-explosive was detonated at the test site to rehearse the observation and administrative procedures.

The nuclear explosion, given the code name TRINITY, took place at dawn on 16 July 1945. From the dramatic visual effects it was immediately apparent that the test had been successful. Within a few hours it was known that the energy released was equivalent to the explosion of about 20000 tons of TNT. This yield was in agreement with that which theory had suggested, but there were many reasons why the yield could have been less, and perhaps much less.

Meanwhile, the development of the ^{235}U gun-assembly device had advanced without serious hitch. Tests of the gun mechanism had given consistent results, and by April 1945 enough nuclear data had been accumulated to lay down the final specification for the ^{235}U components. By June, sufficient material was available to begin the fabrication of the first weapon. Although no experimental nuclear test of the gun device was possible, there was general confidence that it would function satisfactorily, particularly after the successful TRINITY explosion had confirmed the validity of the underlying nuclear theory.

The first phase of the development ended with the assembly of combat versions of the gun and implosion devices. These weapons were given the code names LITTLE BOY and FAT MAN respectively. The LITTLE BOY weapon was 71 cm in diameter, 305 cm long, and weighed 4100 kg; the FAT MAN weapon was 152·5 cm in diameter, 325 cm long, and weighed 4550 kg. The weapons were designed to be carried in the U.S. B29 type bomber, and were fused to detonate high above the target in order to reduce radioactive contamination.

The first use of a nuclear weapon in war was on 6 August 1945 when the LITTLE BOY weapon was dropped on Hiroshima.[1] Four days later the FAT MAN weapon was dropped on Nagasaki. A detailed account of the damage done by these explosions is given in [1]. At Hiroshima some four square miles of lightly constructed buildings were devastated, and another nine square miles badly damaged by blast and fire. Faced with weapons of such destructive power, the Japanese had no alternative but surrender. Unprepared for attack by nuclear weapons, the Japanese casualties had been high but they would have been greater if the war had continued and the planned invasion of Japan carried out.

[1] For an account of President Truman's consideration of relevant political issues, prior to authorizing use of the nuclear bomb, see [4], Ch. 11.

II. THE POST-WAR PERIOD

After the dramatic demonstration of the power of nuclear weapons at Hiroshima and Nagasaki, a period of intense political activity ensued whilst the U.S. Government, supported by Britain and Canada, endeavoured to negotiate some means of international control over atomic energy. This led to the setting up of a United Nations Commission to study means for ensuring that atomic energy was used for peaceful purposes and not for war. At the first meeting in June 1946 America put forward proposals (the 'Baruch Plan'), offering to give up their nuclear weapons when control was established; the Russians submitted counter-proposals to which America could not agree. It soon became apparent that agreement on a control plan was very unlikely, but negotiations continued until 1948 when the Commission reported a complete impasse.

Meanwhile, as hopes of agreement faded, the programme in the U.S.A. was accelerated and both Russia and Britain established independent weapon development programmes, all leading eventually to the development of 'H'-bombs. In the following section, the programme of each country is considered separately.

The post-war American programme. When the war ended, the American Government decided to replace the war-time Manhattan Project organization with arrangements more appropriate for the control of their nuclear programme in peacetime. In due course, the U.S. Atomic Energy Act of 1946 was passed, authorizing the creation of the U.S. Atomic Energy Commission with very wide responsibilities for both civil and military applications of nuclear energy.

The Commission assumed its responsibilities in January 1947, at a time when hopes of international control were on the wane. Its first priority was to ensure the provision of fissionable materials and nuclear weapons for U.S. national defence. An early action was to give a new direction to the Los Alamos laboratory, where morale had slumped badly after the war. Many scientists had left the Laboratory, including J. R. Oppenheimer, its first Director. However, there was no shortage of ideas for further weapons research and in due course the laboratory was authorized to proceed with a programme, including nuclear tests, for improving the reliability of the existing types of weapon and developing new designs of greater efficiency.

A small team of theoreticians at Los Alamos were also looking into the feasibility of developing an entirely new type of nuclear explosive in which most of the energy came from the thermonuclear fusion of deuterium, the second isotope of hydrogen. The idea had been put forward as early as 1942 by Edward Teller and studies had continued at a low level of effort during the intervening years. Deuterium exists in combination with oxygen as 'heavy' water, a naturally occurring minor constituent of ordinary water, from which it can be extracted by known processes at a fraction of the cost of producing fissile material. Theoretically, the energy released from the fusion of a given mass of deuterium would be some three times that from the fissioning of the same mass of ^{235}U. But the most important characteristic of such weapons appeared to be that the amount of thermonuclear fuel, and hence the yield, would not be limited by the twin problems of availability and criticality that affected fission weapon designs. The scientists could, therefore, think in terms of weapon yields measured in megatons rather than kilotons of TNT equivalent.

Three fusion reactions were important:

$$D_2 + D_2 \rightarrow He_3 + n + 3 \cdot 27 \text{ MeV},$$

or

$$D_2 + D_2 \rightarrow T_3 + H_1 + 4 \cdot 03 \text{ MeV},$$

and

$$D_2 + T_3 \rightarrow He_4 + n + 17 \cdot 6 \text{ MeV},$$

where protons (H_1), deuterons (D_2), and tritons (T_3) represent individual nuclei of the three isotopes hydrogen, deuterium, and tritium. To induce the thermonuclear fusion of deuterium it would be necessary to raise the temperature of the material to a very high value, of the order 10^8 degrees Kelvin, and maintain it at that temperature for a sufficient time to 'burn' an appreciable fraction. To trigger such a reaction in a mass of deuterium would itself require a large amount of energy, of the order of that of a fission bomb.

Teller firmly believed that such weapons could be developed, given enough effort and imaginative leadership. Other scientists were far less confident of success, and considered that the available effort should be concentrated on improvements in fission weapons. Many people were (and still are) appalled by the prospect of megaton weapons and were reluctant to see America take the lead. Others passionately believed that, if such weapons were possible, America could not afford to let another nation develop them first.

The debate over the future of the work on thermonuclear explosives continued for many months until, in 1949, the American public were shocked and alarmed by the news that the U.S.S.R. had succeeded in exploding a nuclear fission device. The realization that the American monopoly of nuclear weapons no longer existed helped to turn the tide of opinion towards the view that America had no alternative but to go on with the development of all types of nuclear weapons, including the thermonuclear or H-bomb. In January 1950 the President directed the Commission to proceed [5, Ch. 13].

During 1950 the team working on the thermonuclear 'super' bomb was greatly augmented, with Teller playing a very active role, urging and inspiring the theoreticians to explore new ideas. At first progress was difficult, with many problems to be solved and uncertainties to be explored in searching for a route to a practical device. Early in 1951, a leading mathematician, S. M. Ulam, put forward a new concept which appeared to be very promising and inspired Teller to suggest further possibilities [5, p. 36].

From this point onward progress was rapid, and in 1952 an experimental device was prepared for an overall test of the Teller–Ulam ideas. This device was assembled on the island of Elugelab in the Eniwetok Atoll and was successfully detonated in shot IVY-MIKE on 1 November 1952. The yield was 10 megatons, that is, several hundred times larger than that of the first American fission device fired at Alamogordo seven years earlier.

The device fired in the IVY-MIKE shot used liquid deuterium as the thermonuclear fuel. This simplified to some extent the analysis of the results, but it was apparent that other forms of fuel, for example, lithium deuteride, would be more suitable for military weapons. In 1954, a series of six thermonuclear devices was tested. In his book *The advisors*, Herbert York [10] states that the first device of the series gave a yield of about 15 megatons, which he believes to be the highest yield of any American test device. At the time, York was the Director of the Lawrence Radiation Laboratory, Livermore, California, set up to share with Los Alamos the increasing load of nuclear weapons research and development.

The further story of the developments made by these two laboratories in nuclear explosive devices for military and civil use, is a matter for the history of the second half of the twentieth century.

The post-war Soviet programme. The Soviet Government have not revealed details of their military nuclear programme,[1] but the general trend of its development can be traced from the occurrence of nuclear tests, occasional political statements, and other evidence. A detailed analysis based on unclassified sources was published by Kramish [7], and more recent comments by York [10] are relevant.

---

[1] In contrast, the American Government published an unclassified but comprehensive account of the Manhattan Project immediately after the Japanese surrender [9].

Before the U.S.S.R. became involved in the Second World War, nuclear research programmes existed at several physics institutes, notably at Leningrad and Moscow, and a cadre of young and able scientists had been attracted to this new and expanding science. The Nazi invasion in 1941 resulted in the abandonment of these programmes, and the scientists were redirected to work on non-nuclear problems regarded as of more immediate interest to the Soviet war effort. In 1943 Soviet policy appears to have changed, and a modest level of nuclear work was resumed and continued to the end of the war period.

During the post-war negotiations, Soviet leaders made speeches at various times declaring the monopolistic possession of nuclear weapons to be a 'menace'. In September 1946 Stalin, after speaking in these terms, also declared that the American monopoly could not last long. A year later, Molotov claimed that 'the secret of the atom bomb . . . has long ago ceased to exist' [7]. The first evidence that a Soviet nuclear explosive device had been successfully exploded came in 1949, when a U.S. Air Force aeroplane in the Pacific picked up radioactive debris identified as from the explosion of a Soviet nuclear fission device.

York [10] deduces from biographical statements on Russian scientists that thermonuclear studies were also in progress from an early stage in their programme and that a determined attack on the problem was mounted shortly after the 1949 Soviet nuclear test. In August 1953, Malenkov indicated to the Supreme Soviet that the Russian scientists had solved the secret of the hydrogen bomb (Russian press reports). Four days later, a nuclear device was exploded that involved both fission and fusion reactions. Both Kramish and York advance arguments to suggest that this device was not itself a usable weapon. York also deduces, however, that it probably used lithium deuteride as a more convenient form of thermonuclear fuel than the liquid deuterium used in the American device at the IVY-MIKE test in 1952.

In 1955, a second thermonuclear device was exploded in the air, after release from an aircraft. In a political speech Khruschev claimed that the yield was very high and that Russian scientists had demonstrated important new achievements in producing a device capable of an explosion of several megatons using only a relatively small amount of fissile material.

In later years, beyond the period covered by this account, the Russians fired a number of high-yield thermonuclear devices, including one which they claimed as equivalent to 58 megatons of TNT, although they have never mentioned this as a stockpiled weapon [10].

The post-war British programme. When the war in Europe ended, the British Government began the foundation of a domestic British nuclear programme by setting up a central laboratory for general nuclear research at Harwell and creating within the Ministry of Supply a Controllerate of Production, Atomic Energy, to replace the wartime Tube Alloys organization (U.K. Atomic Energy Act, 1946). At the same time, Britain sought to preserve full collaboration with the U.S.A., as agreed between Churchill and Roosevelt at the Hyde Park talks in September 1944 [2]. In this they were disappointed, and the wartime partnership ended when the McMahon Act became law in America in 1946.

The decision to proceed with an independent British programme to develop a nuclear weapon was formally taken in January 1947, although the man chosen to develop the bomb, W. G. Penney, was not appointed until some months later [3].

Penney had worked at Los Alamos as a consultant on hydrodynamics and on the blast effects of nuclear explosions. In 1946 he had been appointed Chief Superintendent of Armament Research in the Ministry of Supply. His own ability, experience, and position, with the backing of the existing armament research teams, made him an excellent choice to direct the development of the weapon. In the initial stages he had the support of scientists at Harwell, some of whom had previously worked at Los Alamos. Subsequently, a separate establishment was set up at Aldermaston specifically for the development of the nuclear weapon.

Penney based the design of the first British nuclear device on the implosion type weapon that he and his British colleagues at Los Alamos had helped to develop and test. Modifications were introduced to meet operational requirements of the Royal Air Force and careful reworking of the performance calculations indicated ways of improving the efficiency of the device. A major effort was involved in setting up facilities for fabricating components to the high quality and consistency demanded, together with complex non-nuclear techniques for checking the overall performance of the implosion system. A particular problem was the design, construction, and commissioning of the special facilities needed for fabricating the radioactive components.

All these activities took time, but the overall development of the nuclear device was completed in phase with the first deliveries of plutonium from the production reactors at Windscale (Ch. 10) and was successfully detonated on 3 October 1952, at the HURRICANE test in the Monte Bello Islands off the west coast of Australia. Military versions entered service with the Royal Air

Force some 18 months later and Britain became the world's third nuclear power [3].

By this time the United States had disclosed the success of their first test of a thermonuclear device, and it was clear that nuclear weapon technology had taken a further leap forward. The question to be faced was whether Britain should follow. The formal decision to proceed with a thermonuclear weapon programme was not taken until 1954, that is, after the Soviets had also exploded a nuclear device involving thermonuclear reactions. The decision was part of a policy of maintaining a British nuclear deterrent capability. This decision more or less coincided with the U.K. Atomic Energy Act of 1954, which created the U.K. Atomic Energy Authority as the body responsible for British nuclear projects.

Exceptional measures were taken to expand the weapon development team at Aldermaston, partly by the temporary or permanent transfer of staff from Defence research establishments. One notable recruit was (Sir) William Cook, whose ability and drive as deputy to Penney had a major influence on the timely progress of the development programme. In the comparatively short time of three years after the formal decision to proceed, a satisfactory design had been evolved and thermonuclear devices in the megaton yield range were being successfully detonated after release from V class aircraft flown from Christmas Island in the Pacific in 1957.

The trend of political events and the progress achieved by the British in developing a nuclear arms capability led to the restoration, by stages, of the earlier collaboration between America and Britain in this field. In 1955, an agreement was concluded for Co-operation Regarding Atomic Information for Mutual Defence Purposes [11]. This agreement related to information for developing defence plans and for training, but excluded information relating to the design or fabrication of nuclear weapons. In 1958 a further agreement was concluded between the U.S.A. and the U.K. for Co-operation on Uses of Atomic Energy for Mutual Defence Purposes [12]. This agreement extended the scope of the exchange to include classified information concerning atomic weapons, thus restoring the partnership that had existed during the war.

III. POSTSCRIPT

It is appropriate to round off this account of the growth of nuclear weapon technology with a brief reference to the attempt made in the late 1950s, after protracted discussions on general disarmament, to negotiate a treaty banning nuclear tests for weapon development. Formal negotiations began in 1958 and eventually achieved partial success in 1963 when a Treaty Banning Nuclear Weapon Tests in the Atmosphere, in Outer Space and Underwater was signed by the U.S.A., the U.S.S.R., and Britain, among many other nations [13].

This Treaty has had the desired effect of reducing atmospheric pollution and fallout of radioactive products from nuclear explosions. The nuclear powers have, however, continued with weapon development programmes using underground emplacements which fully contain the products of the explosion.

REFERENCES

[1] GLASSTONE, S. *Effects of nuclear weapons.* U.S. Atomic Energy Commission (1964).

[2] GOWING, M. *Britain and atomic energy 1939–45.* Macmillan, London (1964).

[3] GOWING, M. *Independence and deterrence, Britain and atomic energy 1945–52,* Vols I and II. Macmillan, London (1974).

[4] HEWLETT, R. G., and ANDERSON, O. E. *The new world, 1939–46, History of the United States Atomic Energy Commission.* Vol. I. Pennsylvania State University Press, University Park (1962).

[5] HEWLETT, R. G., and DUNCAN, F. *Atomic shield, 1947–52, History of the United States Atomic Energy Commission.* Vol. II. Pennsylvania State University Press, University Park (1969).

[6] IRVING, D. *The virus house.* Kimber, London (1967).

[7] KRAMISH, A. *Atomic energy in the Soviet Union.* Stanford University Press (1959).

[8] PIERRE, A. J. *Nuclear politics.* Oxford University Press, London (1972).

[9] SMYTH, H. D. *A general account of the development of methods of using atomic energy for military purposes under the auspices of the United States Government, 1940–45.* H.M.S.O. London (1945).

[10] YORK, H. *The advisors.* W. H. Freeman, San Francisco (1976).

[11] Command 9555. *Agreement for Co-operation regarding Atomic Information for Mutual Defence Purposes.* H.M.S.O., London (1955).

[12] Command 537. *Agreement for Co-operation on Uses of Atomic Energy for Mutual Defence Purposes.* H.M.S.O., London (1958).

[13] Command 2118. *Treaty Banning Nuclear Weapon Tests in the Atmosphere, in Outer Space, and Underwater.* H.M.S.O., London (1963).

N.B. Publication of this chapter does not necessarily imply the agreement or approval of the U.K. Atomic Energy Authority, the Ministry of Defence, or Her Majesty's Government with the contents or the views expressed.

12

ELECTRICITY

BRIAN BOWERS

I. THE ELECTRICITY SUPPLY INDUSTRY IN 1900

BY 1900 the electricity supply industry was firmly established in several countries (Vol. V, Chs. 9 and 10), but as a multiplicity of local undertakings rather than national networks. In the United Kingdom there were about ninety electricity supply undertakings, operating almost two hundred generating stations. Two-thirds were owned by the local authorities and the remaining third by private companies. The first Census of Electrical Industries in the U.S.A. took place in 1902 and recorded 3620 electricity supply undertakings in operation. Of those 80 per cent were privately owned and 20 per cent were municipal. In the same year the United Kingdom had 258 generating stations, of which 41 per cent were privately owned and 59 per cent were municipal.

Most supply undertakings began on a small scale and served only a limited area. As demand grew each local system would be expanded without any coherent national policy. There was little or no standardization on such matters as the voltage of supply or, in alternating current systems, the frequency used. The American journal *Electrical World* lamented in 1929 that 'many existing systems appear to conform to no logical plan of evolution'. In the United Kingdom the electricity supply legislation had the effect of encouraging the development of small local concerns by making it difficult to cross local authority boundaries. The Electricity Supply Act of 1882 had provided that local authorities could purchase supply undertakings in their area after 21 years. The amending Act of 1888 extended the period to 42 years and made the financial terms less unfavourable to the owners. Whether those Acts really discouraged investment in the new industry in Britain is debatable, but they certainly had the effect of making it develop on a very localized basis. Initially, at least, technical considerations also favoured small-scale operations, with each undertaking supplying an area no greater than could be served by a single small generating station. The technical limitations were imposed by the losses in transmission and the practical difficulties of operating two or more generators in parallel. These difficulties were soon overcome and larger generators were developed. It became technically possible, and economically desirable, to supply a much greater area from a single power station; progress was then impeded by outdated legislation.

Initially, some undertakings gave d.c. supplies, others a.c., and the consumer usually had no choice. There was, however, one exception; in London, as a result of the report of the committee headed by F. A. Marindin, in 1889, official policy was that each area should be supplied by two undertakings, one giving a d.c. supply and the other an a.c. one. Marindin thought that a measure of competition between the supply undertakings would be a good thing, but the result was not the simple competition he had envisaged between a.c. and d.c. The variety of different systems is illustrated by the fact that in 1917 the Electrical Trades Committee reported that 'in greater London 70 authorities supply electricity to the public and own 70 generating stations with 50 different types of system, 10 different frequencies and 20 different voltages'. The unification of London's electricity supply took years to achieve, and led Lloyd George to remark to the consulting engineer Charles Merz that electricity supply in London was not a matter of engineering, but of politics. The problem was one of resolving the conflicting interests of the London County Council, the Borough Councils, and the supply undertakings, each of which had a vested interest in preserving its own system. A similar problem had to be faced in many other places too, before a standardized electricity supply could be made available nationally.

The electricity supply undertakings of 1900 generated electricity almost exclusively for lighting. The incandescent filament lamp (Vol. V, p. 213) had been developed twenty years earlier by Joseph Swan in England and, independently, by T. A. Edison in the U.S.A. It was in widespread domestic use, at least in those households able to afford it and fortunate enough to be in an area where there was an electricity supply. The electric arc light was really too bright for domestic use, although it was occasionally used, but it continued to serve for lighting large areas, such as railway stations. Both types were used for street lighting.

Although d.c. motors had been available as long as there had been electricity supply, they were not used much except for traction. The induction motor invented by Nikola Tesla in 1888 made possible the industrial use of a.c. motors but, except for very small machines, the induction motor requires a polyphase supply. The first three-phase supply in Britain was provided by the Wood Lane generating station in West London, which opened in 1900. The Neptune Bank Power Station, opened in Newcastle upon Tyne in 1901, also provided a three-phase a.c. supply. The frequency of operation was chosen to be suitable both for lighting and for induction motors. The proprietors anticipated a motor load in the heavy industry of the area and, in fact, the demand grew so rapidly that they began work on extending the station within months of its opening.

From an early date the supply undertakings encouraged domestic consumers to adopt electric heating and cooking. The advantage to the undertaking was that these provided an 'off-peak' load. The lighting load was switched on mainly during the evening. Any load during the hours of daylight meant that the supply undertakings could increase the amount of electricity generated, and hence their revenue, with no additional capital investment and only a marginal increase in running costs.

The fact that a non-lighting load was likely to develop was noted in 1898 in the report of a Parliamentary Joint Select Committee in Britain under the chairmanship of Viscount Cross. They observed that when the Acts of 1882 and 1888 were before Parliament they were chiefly concerned with electric lighting, but the evidence they heard in 1898 showed that 'although electric light is at present the predominant feature of the enterprises now before the public and Parliament, the application of electrical energy in the form of power to an infinite variety of other purposes is likely to be in the near future the predominant feature and function of these undertakings'.

II. ELECTRICITY SUPPLY IN BRITAIN

The general growth of the industry may be exemplified by British practice; developments elsewhere will be considered later.

The long transition from a large number of local undertakings to a single electricity supply organization covering the whole country was already beginning in Britain in 1900. Private Bills were being introduced into Parliament with the object of setting up 'Power Companies' authorized to generate electricity on a large scale for selling to authorized supply undertakings in their area. The Electric Lighting (Amendment) Act of 1909 permitted the Board of Trade to allow the establishment of power companies, and these companies were not subject to any clause enabling the local authorities to purchase them compulsorily after a number of years. Within a few years over twenty such companies had been established, covering between them most of the rural parts of Britain. The towns which had municipally owned supply undertakings usually preferred to maintain their independence rather than take supplies from a power company.

After the First World War the British Government decided that some more centralized organization had to be imposed on the electricity supply industry. During the war the installed generating capacity had increased by 39 per cent (from 1120 MW to 1555 MW) but the number of units sold had increased by 106 per cent (from 1318m. per year to 2716m.). The lesson was clear: wartime controls had increased the efficiency of the supply industry. Furthermore, electricity supply was becoming increasingly significant in British industrial and economic life. Several bodies reported to the Government on the need for post-war reorganization, and the consensus was that 'a new and independent Board of Commissioners free from political control and untrammelled by past traditions' should be established with extensive powers. The Electricity Supply Act, 1919, established Electricity Commissioners with the duty of 'promoting, regulating and supervising the supply of electricity'. The Act was a step in the right direction, but the Commissioners' powers proved inadequate, and further legislation followed.

During the early 1920s there was pressure on the British Government to establish a transmission network linking the whole country, so that electricity could be generated in the most efficient stations and conveyed wherever it was wanted. The cost of generation at that time varied enormously from place to place and between different power stations. It was clearly in the consumers' interest to be linked to the most economical stations. An additional advantage was that interconnection of stations increased the security of supply, since if one generator or one station suffered a breakdown its load could be supplied from elsewhere.

In early 1925 the Government set up a committee under the chairmanship of Lord Weir to 'review the national problems of the supply of electrical energy' and to report on 'the policy which should be adopted to ensure its most efficient and effective development'. The committee sought technical reports from several independent experts and reported in May 1925. They proposed the establishment as quickly as possible of an independent body, to be known as the 'Central Electricity Board' with the duty of constructing a 'gridiron' system of high-voltage transmission lines. The gridiron, which soon became known as the National Grid, was to interconnect certain selected power stations and the existing distribution systems.

The Weir committee recommended that the operation of the selected stations should be left to the existing authorized undertakings but that the new Board should have overall control. The Board was to purchase the electricity generated in the selected stations and resell it to authorized suppliers. The non-selected stations would rapidly be closed down. The Committee thought that 58 stations should be selected—43 existing and 15 new ones—and that 432 existing stations should be closed. The Government accepted the report and its recommendations were incorporated in the Electricity (Supply) Act, 1926.

The Central Electricity Board outlined a preliminary plan for the grid in their first Report, which dealt with the Board's activities up to the end of 1928. It was a scheme for interconnecting areas, not a scheme for point-to-point transmission of power. They adopted 132000 V as the standard for primary transmission lines, with secondary transmission at 66000 and 33000 volts. Before construction could begin they had to design conductors, insulators, pylons, protective systems, and control arrangements.

An eight-year programme for constructing the grid was announced in 1927. By the end of 1935, when the whole of Britain was linked except for north-east England, there were 4600 km of primary transmission lines in operation, and 1900 km of secondary transmission. By 1946 these had been increased to 5900 km and 2400 km, respectively.

III. STANDARDIZATION OF SYSTEMS

The biggest problem facing the Central Electricity Board was the standardization of frequency. In their first report they noted that three-phase 50-Hz a.c. supplies were standard throughout Europe, except in Italy. In Britain 77 per cent of the installed capacity of authorized undertakers in 1926 was 50-Hz plant, mostly three-phase, and the Board adopted that standard. The main exception was the north-east England area around Newcastle upon Tyne, which was supplied at 40 Hz. Glasgow, Birmingham, South Wales, and London had some 25-Hz supplies. It was realized that the cost of changing the frequency of 23 per cent of the national supply system would be considerable; when completed in 1947 it proved to be £17·5m.

A further task which faced the Board was the standardization of supply voltage. A committee under the chairmanship of Sir Harry McGowan investigated this question in 1935–6. They found that of 642 supply undertakings 282 supplied a.c. only; 77 supplied d.c. only; and 283 supplied both a.c. and d.c. There were 43 different supply voltages ranging from 100 to 480 V. The real problem was that, after agreeing on a standard supply voltage, much consumers' apparatus would have to be changed or converted. It was generally agreed, however, that a uniform supply voltage was highly desirable, and until it was achieved domestic appliance manufacturers had to make and stock a range of models designed for different voltages. In the case of radio and television sets with an internal transformer a tapped primary was provided on the transformer so that the set could be adjusted to the local supply voltage. It was only in 1945 that a standard—240 V a.c.—was finally laid down. After the Second World War the British Government decided to take the whole electricity supply industry into public ownership. The Bill was introduced into Parliament in December 1946 and became effective on 1 April 1948. Under the new organization the British Electricity Authority was established to exercise a general co-ordinating function and control the policy and financial structure of the industry as a whole, except for the north of Scotland. The Authority was responsible for the generation of electricity and its bulk transmission to fourteen separate statutory Area Electricity Boards.

In taking control of the power stations and the grid system, the new Authority also became responsible for running the seven grid control districts, each with its own grid control room. In order to give the best possible service it was decided to divide the management and operation of the power stations and the grid into fourteen Generation Divisions, corresponding as closely as possible with the Area Electricity Boards.

The Electricity Act, 1957, further changed the structure of the industry by abolishing the British Electricity Authority and establishing the Central Electricity Generating Board, responsible for the power stations and the grid, and the Electricity Council. The chief functions of the Electricity Council were 'to advise the Minister on all matters affecting the industry and to promote and assist the maintenance and development by the Generating Board and the Area Boards of an efficient, co-ordinated and economical system of electricity supply'. The Council was also given special responsibilities finance, research, and industrial relations.

Table 12.1 shows the growth in generating capacity and in electricity sold in Britain during these fifty years of growth and reorganization. For comparison, the figures for 1970 are also shown, and show that the rapid growth seen in the first half of the twentieth century is continuing in the second.

IV. ELECTRICITY SUPPLY IN OTHER COUNTRIES

In most countries the electricity supply system has remained in private ownership, though with a considerable degree of government control. In the United States, for example, there were still over 3800 plants supplying electricity for public use in 1947. On the other hand, the federally created Tennessee Valley Authority attained a capacity of over 2400000 kW in 1948.

TABLE 12.1
Electricity supply in Great Britain

Year*	Generating capacity (MW)	Units sold (millions)
1900	—	120
1901	—	200
1905	—	450
1910	960	1 000
1915	1 300	1 700
1920	2 400	3 600
1925	4 300	5 600
1930	6 700	8 900
1935	7 900	14 300
1940	9 900	23 800
1945	12 000	30 600
1950	14 600	46 600
1970	49 000	174 000

* Some figures are for calendar years, others are for the twelve months beginning with 1 April of the year specified.

In Germany, there were 2000 public generating stations in 1941. The trend, as in Britain, has been to link up smaller undertakings into larger networks to give as many customers as possible the benefit of the most economical generating stations and also to increase the security of supply. Large numbers of plants were not incompatible with the bulk of the supply being in the hands of the relatively few efficient producers.

The consumption of electrical energy per head of population varied greatly between countries. Table 12.2 shows the generating capacity installed in public supply power stations, the total units sold, and the units sold per head of population in several countries. The use of electrical energy per head is even higher in the U.S.A. and Canada than appears from the table, because

TABLE 12.2
Electricity supply in various countries

	Generating capacity (MW)		Units generated (millions)		Units per head of population	
	1939	1949	1939	1949	1939	1949
Great Britain	8 000	14 000	23 000	51 000	470	850
France	6 500	7 000	13 000	20 000	300	500
Italy	5 700	5 700	16 000	16 000	350	350
Sweden	1 800	2 700	7 500	13 000	1 200	1 900
Switzerland	—	2 000	3 000	6 300	800	1 300
U.S.A.	39 000	61 000	106 000	249 000	800	1 600
Canada	5 500	15 500	23 000	39 000	2070	2870

those countries have more generating plant installed in private industry than does Britain and the rest of Europe. At the middle of the century hydro-electric power was important in Canada (96 per cent of the installed capacity in 1949), Switzerland (94 per cent), Italy (90 per cent), and Sweden (80 per cent), but its relative contribution to world power generation was declining (p. 195).

V. THE USES OF ELECTRICITY

There are no accurate statistics showing the consumption of electricity by different classes of consumer before 1920. In 1900 most of the electricity generated in the supply industry was used for street lighting and domestic purposes. Industrial consumption was very low and the small number of electric railways and tramways usually generated their own electricity. Tables 12.3 and 12.4 show the consumption in Britain by various classes of user from

TABLE 12.3
Sales of electricity to different classes of consumer in Britain

| | Annual sale (millions of units) | | | | | | |
	Industrial	Commercial	Farming	Domestic	Traction	Street lighting	Total
1920	2500	390	—	290	410	48	3600
1925	3700	690	3	630	510	89	5600
1930	5200	1300	11	1500	800	160	8900
1935	7600	2200	34	3200	1030	260	14300
1940	13500	2900	83	6100	1130	17	23800
1945	17100	3400	150	8500	1220	160	30600
1950–1	23000	6400	480	15000	1440	420	46600
1970–1	73000	26000	2900	66000	2200	1500	174000

TABLE 12.4
Proportion of electricity sales (as percentage of total) to different classes of consumer in Britain

	Industrial	Commercial	Farming	Domestic	Traction	Street lighting
1920	69	11	—	8	11	1·3
1925	66	12	—	11	10	1·6
1930	58	14	0·1	17	9	1·8
1935	53	16	0·2	22	7	1·9
1940	57	12	0·3	26	5	0·1
1945	56	11	0·5	28	4	0·5
1950–1	49	14	1·0	32	3	0·9
1970–1	41·6	16·6	1·7	38·0	1·3	0·8

1920 to 1950, and reveal that, with one exception, no class of consumer had reduced its use of electricity from one year to the next. The exception, not unexpectedly, was street lighting during wartime.

The demand for electricity is not uniform throughout a 24-hour period, and in all countries the supply authorities have always sought to make the demand as even as possible. The reason is a matter of simple economics: the capital expenditure is proportional to the peak demand, the revenue is proportional to the average demand. The ratio of the average to the peak is known as the utilization factor.

In the early years, when the main demand was for lighting, the supply undertakings offered a cheaper rate for electricity for cooking to create a demand during daylight hours and thus improve the utilization factor. As soon as an industrial load developed the peak period came during working hours, and the suppliers sought to encourage a night load. In Britain, from about 1960, a substantial domestic night load was created by charging a reduced rate—about one-half of the day rate—for 'night storage heaters', which stored heat during the night and gave it out during the next day. In 1963 the night storage heater load amounted to 1100 MW; by 1970 it had grown to 9400 MW.

Electricity was little used in farming before 1930, but as a matter of public policy rural electrification schemes were adopted even though they could not be economic. The figures in Tables 12.3 and 12.4 show the growth since then. The applications of electricity on the farm include driving small machinery, pumping, and lighting. Electricity makes machine milking possible. Experiments have been made with electric tractors, and ploughing by using fixed electric motors to pull the plough by a rope, analogous with the old-style steam ploughs, but the petrol or diesel tractor is now universal.

VI. DOMESTIC USE OF ELECTRICITY

Although lighting was the main domestic use of electricity in 1900, a wide range of domestic electrical appliances had been on the market for several years. A catalogue published by Crompton & Co. in 1894 includes a variety of ovens, hot-plates, hot cupboards, irons, kettles, heaters, radiators, and coffee urns. A school of electric cookery had been equipped by the same firm in London.

To encourage the use of electricity for cooking and heating, the undertakings hired out appliances and offered a cheaper tariff for electricity used other than for lighting. For example, the City of London Electric Lighting

Company offered electric ovens (retail price approximately £7 to £14) for hire at 7 to 12 shillings per quarter, and they charged only 4 (old) pence per unit for electricity for cooking, which was half the normal price. The practice of hiring out appliances continued until the nationalization of the supply industry in 1947. Since that date customers have been encouraged to buy their appliances and the supply authorities have aided this by providing hire-purchase or credit facilities.

The growth of domestic electricity has not been uniform throughout the period now under review nor throughout all classes of society. In Britain, the period up to 1920 saw only 10 per cent of the community receiving electricity supplies and using the available appliances. During the 1920s and 1930s more than half the households in the country were connected up, but the average spending on domestic appliances fell by half as these new consumers used electricity for little except lighting. In 1920 the Institution of Electrical Engineers discussed a paper by Leonard Milne on 'The electrical equipment of artisan dwellings'. His concern was to get houses then being built wired for electricity as cheaply as possible. He considered that a typical 'artisan's dwelling' needed nine lighting points only, and that such an installation then usually cost between £11 and £20. Milne thought that by simplifying the wiring—by expedients such as combining the switch and the lampholder in a single fitting—he could reduce the cost to £7. Milne's paper and the lengthy discussions which followed illustrate the determination of the supply engineers of the period, and not only in Britain, to win as many customers for electricity as possible. Their success may be seen in Table 12.5, which shows the growth in the number of wired households in Britain. Table 12.5 also shows expenditure on domestic appliances, and the fall in average expenditure per household as electricity spread to the less-well-off members of the community. It can also be seen that there was a very large increase in consumer

TABLE 12.5
Numbers of wired households in Britain

Year	Number of households wired for electricity (millions)	Percentage of households wired for electricity	Expenditure on domestic electrical appliances	
			Total (£m.)	Per wired household (£)
1921	1·1	12	2·7	2–45
1931	3·5	32	4·2	1–20
1938	8·7	65	12·6	1–45
1951	12·2	86	78·0	6–39
1961	16·0	96	279·0	17–44

spending on electrical appliances in the 1950s, the era when every British household came to expect amenities which had previously been regarded as luxuries. At the end of 1959 *The Economist* looked back over a decade and observed that ten years earlier the ordinary wage-earner in Britain regarded any increase in wages as 'meaning a little more money for beer, baccy, pools, and the dogs; the idea had not been generally accepted that an increase in the standard of living meant a change in a way of life. Then, in the middle of that roaring decade, something began to happen in the field of consumer goods that can only be called a break-through.' Expenditure on domestic appliances multiplied more than threefold, though the number of wired households rose by only 10 per cent. *The Economist* added 'we barely yet comprehend what had occurred; indeed it is fashionable to cover our surprise by mocking at it. In the last decade the leisure time of nearly two-thirds of the population had been transformed by the purchase of a television set.' Domestic appliances generally form the subject of another chapter in this work (Ch. 47), but we may briefly note here those of the greatest significance in relation to the utilization of electricity.

After lighting, the electrical appliance most widely adopted from an early date was the electric iron. In 1948, 86 per cent of British households wired for electricity had an electric iron, and the proportion had reached virtually 100 per cent by 1963.

The early popularity of the vacuum cleaner stands out in Table 12.6. It was originally devised in 1901 by H. Cecil Booth, a young civil engineer in London. He observed that the usual practice of sweeping and beating carpets and upholstery was highly inefficient, for most of the dust simply rose into the air and then settled again. His first vacuum cleaner was contained in a horse-drawn vehicle, and the staff of his Vacuum Cleaner Co. Ltd. took long hoses from the vehicle into people's houses in order to clean the furniture and carpets. Small, portable vacuum cleaners were soon made, usually

TABLE 12.6

Percentages of wired households in Britain owning various appliances

	1938	1948	1963
Vacuum cleaners	27	40	77
Fires	*	64	72
Washing machines	3	4	50
Water heaters	*	16	44
Cookers	18	19	35
Refrigerators	3	2	33

* no figures available

powered by an electric motor. In 1907, W. Hoover, an American leather manufacturer, developed another design of vacuum cleaner which was produced commercially by the Hoover Suction-Sweeper Company of Ohio. By mid-century, the vacuum cleaner was a standard piece of domestic equipment.

During the 1920s the U.S.A. and, to a lesser extent, Germany took the lead in the then rather limited market for domestic electrical appliances. In 1931, Britain went off the gold standard and sterling dropped in value compared with the dollar. Import duties were introduced in 1932. These changes increased the price of imported appliances and the effect on British industry was dramatic. Before 1932, 80 per cent of the vacuum cleaners sold in Britain were imported; by 1935 the import figure was 3 per cent. During the 1930s several British firms began manufacturing refrigerators, washing machines, fires, and irons. Hotpoint, Hoover, and the Pressed Steel Company, originally British subsidiaries of American firms, began to manufacture in Britain, and D. W. Morphy and C. F. Richards established Morphy-Richards Ltd. to make, initially, radiant fires.

The programme of house building which led to an average of 360000 houses a year being built in Britain between 1934 and 1938 promoted a demand for appliances. Between 1930 and 1935 sales of electric cookers rose from 75000 to 240000 per year. In response the gas industry organized a large—and successful—advertising campaign to retain its share of the market. In this they were assisted by some local councils who sought to strike a fair balance between electricity and gas. In 1939 there were about 1·3m. electric cookers in use compared with nearly 9m. gas. In the same year there were 220000 electric refrigerators in use compared with 90000 gas refrigerators, although more gas than electric refrigerators were then being sold each year.

VII. INDUSTRIAL USES FOR ELECTRICITY

During the half-century with which we are here concerned, the utilization of electricity by industry increased enormously. In this field we must include not only manufacturing industry but the transport industry, for this period saw a considerable adoption of electricity for railway traction throughout the world; in the towns, electric trams became a familiar and popular form of public transport. Indeed, as we have seen in an earlier volume (Vol. V, pp. 346–8), electric railways were established in the early 1880s. These early ventures included the world's first electric underground railways, both in London: the City and South London Railway (1890) and the Waterloo and City (1898).

The development of electric railways is discussed elsewhere in the present work (Ch. 32) in the general context of transportation. Here, where we are concerned primarily with the electrical industry, we may note that most of the first railway companies adopting electric traction built their own power stations; indeed, they usually had no alternative, for the small public supply companies were inadequate. As an example, we may cite the Lancashire and Yorkshire Railway, which introduced an electric service on its 83-mile route in 1904. From its 10-MW power station at Formby, electricity (three-phase, 25 Hz) was distributed at 7500 V to four substations. These supplied the railway third-rail with current at 600 V, d.c., through four 600-kW rotary converters. Exceptionally, the suburban electric railways system around Newcastle upon Tyne at the same period drew its electricity from the local public supply.

The first general use of electricity in industry was for lighting. R. E. B. Crompton became a leading pioneer of electric lighting and electricity supply in Britain. His interest began when the family ironworks in Derbyshire installed a new foundry which needed to be worked day and night to be economic. Crompton visited Paris in 1878, since electric lighting was then more advanced in France than in Britain. On his return he established a business importing electric lighting equipment, but he soon turned to manufacturing his own.

Electric motors were gradually introduced for small machines and auxiliary equipment in factories where electricity was already provided for lighting. They were appreciated for their flexibility: a small motor could be set up in places difficult to connect to the traditional factory lineshaft. Crompton optimistically thought electric power might bring an end to the factory system: 'England in future, instead of being spoilt by densely populated industrial centres, might be covered with cottages . . . the population would be more evenly spread over the kingdom. The factory hands, instead of having to work under the shafting in factories, should be able by the electrical transmission of power to carry on industrial pursuits in their own cottage homes.'

The iron and steel industry became the largest user of electric power. The rolling mills continued to be driven by steam long after electricity had been introduced for small power work, but the modern automated high-precision rolling mill is possible only with the fine control given by electric drives.

Electric-arc steelmaking (Vol. V, p. 66) was pioneered by Siemens in 1879–80 but suddenly became important during the First World War. Vast quantities of steel borings and turnings came from the munitions factories and could potentially be reformed into usable steel, but the Bessemer converter, which produced most of the high-grade steel before the war, could not accept much scrap. The arc furnace produced high-grade steel of accurately known composition and with no contamination. At the end of 1917 there were 733 electric arc furnaces in the world; 131 of these were in Britain—more than in any other country except the U.S.A. They were small by modern standards; typically of 15 tons capacity, compared with 80 tons in the 1960s.

Induction furnaces have been used since about 1900 for small quantities (usually under one ton) of special alloys and where extreme purity is required. In an induction furnace the material being melted acts as the secondary winding of a transformer. Current induced in the material provides the heating and also effects thorough mixing.

A process introduced during the Second World War was the electrolytic manufacture of tinplate. Tinplate was previously made by cutting the steel sheet into small pieces which were mechanically dipped in molten tin. The electrolytic process was developed because it was economical in the use of tin, then in short supply. It required less labour than the dipping process and was faster, and it gave a better and more consistent product.

BIBLIOGRAPHY

ANDREWS, H. H. *Electricity in transport.* English Electric Co. Ltd. London (1951).

BRITISH ELECTRICITY AUTHORITY. *Power and prosperity.* London (1954).

CORLEY, T. A. B. *Domestic electrical appliances.* Cape, London (1960).

DUNSHEATH, PERCY. *A history of electrical engineering.* Faber and Faber, London (1962).

Electrical trades directory. Benn, London (1883 on, annually).

ELECTRICAL WORLD AND ENGINEER, STAFF OF. *The electric power industry, past, present and future.* McGraw Hill, New York (1949).

ELECTRICITY COUNCIL. *Electricity supply in Great Britain—A chronology.* London (1973).

——. *Electricity supply in Great Britain—Organization and development.* London (1973).

Electricity undertakings of the world, 1962–3. (72nd edn.), Benn, London (1962).

GALE, W. K. V. *The British iron and steel industry.* David and Charles, Newton Abbot (1967).

GARCKE, EMILE. *Manual of electrical undertakings.* London (1895 on, annually).

HENNESSEY, R. A. S. *The electric revolution.* Oriel Press, Newcastle upon Tyne (1972).

HUNTER, P. V., and HAZELL, J. TEMPLE. *Development of power cables,* Newnes, London (1956).

MILNE, LEONARD. The electrical equipment of artisan dwellings, and subsequent discussions. *Proceedings of the Institution of Electrical Engineers,* **58**, 464–7, 476–90 (1920).

SELF, SIR HENRY, and WATSON, ELIZABETH M. *Electricity supply in Great Britain, its development and organization.* Allen and Unwin, London (1952).

SWALE, W. E. *Forerunners of the North Western Electricity Board.* North Western Electricity Board, Manchester (1963).

UNITED NATIONS ECONOMIC COMMISSION FOR EUROPE. *Organization of electric power services in Europe.* Geneva (1956).

13

AGRICULTURE

LYNNETTE J. PEEL

PART I: ANIMAL PRODUCTION

I. AIMS OF PRODUCTION

DURING the second half of the nineteenth century much of the increase in the world food supply came from an increase in the area of land cropped and grazed in the New World (Vol. V, Ch. 1). New technologies were developed to adapt European methods to the extensive farming of these regions, where land was plentiful, and labour and capital scarce. These new technologies allowed high productivity per man though yields per unit of land were often low. The areas farmed, nevertheless, were so great that wheat from low-yielding crops in Canada and the United States of America, wool from sparsely grazed sheep in Australia, and meat or livestock from the scattered herds and flocks of Argentina, the United States, New Zealand, and Australia could be shipped to Europe to compete profitably with local produce. By 1900 the broad-acre farming lands of the New World had been occupied; consequently, an increase in agricultural production in the next half-century, from 1900 to 1950, was dependent on an intensification of production on already occupied land. This was a complex process, strongly influenced by industrializing economies. Growing use of oil, new developments in engineering and chemical technology, and changing consumer preferences dictated change in the very aims of rural production in some instances.

Draught. Throughout the nineteenth century the numbers of horses used for transport on the roads and for draught and other tasks on the farms had increased steadily. One estimate of the horse population of Great Britain by 1900 is 3·3m. But from the beginning of the twentieth century, horse numbers began to decline; by 1924 the British total had fallen to 1·9m., of which 753 000 were in use in agriculture [1]. In 1950 the number used in agriculture had fallen to 347 000, and after 1958 horses had become of such insignificance in British agriculture they were no longer recorded regularly in the agricultural census. Working bullocks, already replaced on farms by horses for the lighter work, similarly declined in numbers.

The reason for the eclipse of the horse and bullock was the development of petrol and oil engines (Ch. 40). These internal combustion engines began to rival animal power from the turn of the century onwards and, in countries as disparate as Japan and Australia, steadily replaced animal power in the 1930s and 1940s. Only in the poorer countries, such as those of the tropics, did animal power (predominantly buffalo power in the tropics) remain pre-eminent. This change in the source of power meant, not a change or development in one technology, but a replacement of technology relating to animals by that relating to machines. Much of the knowledge and expertise concerned with breeding, feeding, handling, and caring for horses and bullocks, and that of making the various products and equipment associated with them fell into disuse. This process was hastened as people brought up in the horse age were succeeded by a younger generation more familiar with mechanized warfare and the power of machines.

Fibre. The production of fibres from animals also suffered from industrial developments. Rayon was the first man-made fibre to be sold in quantity and by 1920 it comprised 0·3 per cent of the world production of textile fibres. Its share of the market then grew steadily after the depression of the early 1930s. By 1950 the cellulosic fibres (rayon and acetate), represented 16·8 per cent of world fibre production; the non-cellulosic man-made fibres (such as nylon) 0·8 per cent; cotton 71·0 per cent; wool 11·2 per cent; and silk 0·2 per cent [2].

Silk had represented 0·4 per cent of world fibre production in 1920 and this share had been increased to 0·9 per cent in 1930 when the industry was expanded in Japan. But the onset of economic depression in the major market, the U.S.A., together with competition from rayon for stockings, and the loss of markets with the outbreak of war in 1939, brought a decrease in the importance of silk in total fibre production. After 1945 new developments of man-made fibres provided further serious competition and helped to keep the quantity of silk produced below pre-war levels. Thus in the course of half a century farmers keeping silk-worms expanded production initially but later were to see their mulberry trees uprooted. Wool producers, on the other hand, were able to improve their techniques of production and encourage improvements in wool processing sufficiently to continue to compete with the new chemical technologies of the man-made fibres. From 1920 to 1950 the total world production of wool increased although the proportion it represented of total fibre production fell from 15·2 to 11·2 per cent.

Food. The production of food in the twentieth century was influenced by rising incomes and a desire for more varied diets (Ch. 56). Unlike the situation in fibre production, competition between products remained within the sphere of agricultural production. As incomes rose the trend was for the consumption of animal products to increase and that of cereals and potatoes to decrease. In the United Kingdom meat consumption, for example, had risen from 91 lb per head in 1880 to 131 lb in the period 1909–13. Increasing prosperity then brought further modifications in the nature of the animal products most desired by consumers. Large fatty joints of meat, sought after at the turn of the century, were gradually rejected in favour of smaller leaner joints from younger animals. Where prosperity had increased sufficiently to encourage an increase in meat consumption, it tended also to encourage a fall in the birth-rate of the population, and smaller families wanted smaller joints. Furthermore smaller joints were eaten several times a week instead of larger joints once only, as previously. In an industrialized society with more sedentry employment less energy was needed by working people and there was less appetite for fat, particularly where more lean meat could be afforded.

In the carcass classes at the Smithfield Show in London 21-month-old Southdown sheep had an average carcass weight of 123 lb in 1840–2, 91 lb in 1893–1913, and 72 lb in 1921–32. These changes reflect the changing views about the most desired type of carcass. By 1939 in the major cities of England, the greatest demand in sheep meat was for 24–36-lb lamb with little fat [3]. By the end of the 1930s 93 per cent of the total world exports of mutton and lamb were imported into Great Britain. Thus the exporting countries were under pressure to produce lean meat from younger animals to suit the changing requirements of this market. In times of stress, particularly during the Second World War, these trends were reversed and maximum weight of total product, including fat, again became of paramount importance. Concern for the niceties of quality returned again only in about 1950.

A further impetus to the demand for animal products came with the recognition of the comparative nutritional value of various forms of protein and the discovery of vitamins. Until the beginning of the twentieth century it was believed that all proteins, whether of animal or vegetable origin (except gelatin) were of similar composition and thus of similar value in human

 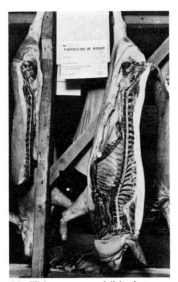 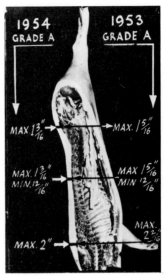

Fig. 13.1 (a). The champion bacon pig carcass at the Smithfield Club Show, London, in 1936 was long and lean.

(b). This carcass, exhibited at the same show, was too fat. During the war quantity was more important than quality and many fat pigs were produced.

(c). After 1951 allowable fat cover for Grade A commercial carcasses was gradually reduced until, by 1954, pre-war standards were reached.

nutrition. Then Emil Fischer and his colleagues in Germany, working in the 1900s, showed that proteins were composed of amino-acids and varied in composition. Experimental work in various countries then gradually revealed that certain amino-acids are essential in human nutrition and not all protein foods provide them. Furthermore it was shown that milk and eggs, closely followed by meat and fish, were the best single sources of these essential amino-acids.

Other quite separate studies, at the turn of the century, also emphasized the nutritional value of animal products. Following the work of Christian Eijkman on beri-beri, C. A. Pekelharing, at Utrecht, showed that milk contained an unrecognized substance, in minute quantities, that was of paramount importance in nutrition. Then shortly afterwards, in 1912, Frederick Gowland Hopkins, at Cambridge, published the results of his experiments which showed the need for certain accessory factors in animal diets, in addition to protein, fat, carbohydrates, and minerals. These factors were later called vitamins.

By the outbreak of the First World War scientists were aware that crude totals of protein and calories for a diet were an inadequate guide to the nutritional value of the diet. But it was not until the post-war period that

FIG. 13.2. Bottled milk supplied to English school children in 1938.

government officials and others responsible for planning diets were fully aware of the importance of the nature of the protein source and the necessity for the inclusion of other substances in small amounts. Studies of rickets and other nutritional disorders in Europe, after the deprivations of war, confirmed the need for vitamins and certain minerals in the diet for good health. Gradually the importance of the 'protective' foods—milk, dairy products, eggs, fruit, and vegetables—was recognized. In 1928 the League of Nations took up the subject of nutrition and various governments began to study the requirements for an adequate diet. The onset of the depression in 1929 further increased the problems of malnutrition among people who were too poor to buy adequate food. The farmers at this time, however, had difficulty in disposing of their produce and, in Great Britain and the U.S.A., surplus milk was distributed to unemployed people and school-children. Already nutritional workers had identified the particular vulnerability of school-children to malnutrition and school milk schemes were promoted to combat this. By 1939 over half the children in grant-aided schools in Britain were receiving school milk (Fig. 13.2). Thus milk, from being a food for infants or a liquid for cooking in the nineteenth century, had become by the end of the 1930s an important component of diet promoted by governments. The other dairy products and eggs also benefited from the growth in knowledge of, and interest in, nutrition in the inter-war years, and consumption of them increased. For example, before 1937 egg consumption in the U.S.A. was about 300 eggs per person per year, by 1945 this had risen to 403.

11. MEANS OF PRODUCTION

The intensification of animal production after 1900 was made possible by the growth in scientific knowledge of the biological aspects of production, and by the adoption of the engineering skills and materials of the industrialized economies. Developments in breeding, feeding, housing, management, and pest and disease control all enabled the farmers to produce animal products more efficiently, and to adapt the nature of the products more readily to the changing requirements of the consumers.

Breeding. At the turn of the century the breed societies dictated the direction of livestock breeding. Founded mainly in the nineteenth century, these societies exerted a declining influence throughout the first half of the twentieth century. Although there was much cross-breeding of animals in commercial production, the pure-bred lines held a dominant position at the apex of the breeding system. And it was the breed societies in each country which decided the physical standards and pedigree requirements that an animal had to meet in order to be registered as a pure-bred member of a particular breed. Those animals conforming to the standards were recorded in the herd books published by the various societies. Animals were judged particularly on their appearance and the major change in the twentieth century has been the change made by farmers from an emphasis on visual features, including details such as coat colour, when selecting their livestock, to emphasis on measurable characteristics related to production. Milk, egg, and wool production were the easiest attributes to measure and use as a guide to animal selection since their measurement did not involve slaughtering the animals.

With the development in 1890 of a reliable test for the butterfat content of milk by S. M. Babcock, it became a simple matter to measure both the volume and the fat content of a cow's production. During the 1890s butter factories were established in dairying districts in a number of countries, often as cooperative ventures by local farmers. When cream was sent to a factory it was tested for fat content and payment was usually made on that basis. This encouraged the farmers to take an interest in the measurable production characteristics of their herds. Government departments of agriculture in Europe, north America, and Australasia also gave considerable attention to the dairying industries at the turn of the century and encouraged the establishment of milk recording schemes. Some schemes were organized by government agencies, others by farmers' associations or by breed societies. In

Holland, for example, the first milk recording society was established in 1899. The aim of these schemes was to sample milk yield and the fat content of the milk throughout a cow's lactation, or for a standard lactation period. This information could then be used to assess management practices and to select animals for breeding.

Poultry farmers soon followed the example of dairy farmers in measuring and comparing the productive characteristics of their livestock. From at least as early as 1901 egg-laying competitions were conducted in which hens were brought together at a central location, housed and fed under the same conditions, and a record kept of the number of eggs produced in a year by each bird.

Egg production and milk production records, however, represented the productivity of only one-half of a breeding partnership. This A. L. Hage-doorn realized when in 1912 he wrote in the first Dutch edition of his book on animal breeding, 'The only way to judge the quality of a bull in a breed, in which fat content of the milk is of great value, is according to the fat content of the milk of all his daughters. And we know that it is important to see to it that in reality all his daughters are examined.' [4] Even before Hage-doorn published his book some attempts had been made to progeny-test bulls. There was a scheme in Denmark, for instance, dating from 1900, in which the milk records of heifers were compared with those of their mothers at the same age. But the greatest benefits were not to come from progeny testing until after 1945 when, with the return of peace, the knowledge of genetics and artificial insemination gained in the 1920s and 1930s could be applied to progeny testing and animal breeding on a large scale.

The most important advance in genetics in the first half of the twentieth century was the rediscovery of the laws of Gregor Mendel in 1900. Work on Mendel's laws showed first that 'identical pedigrees need not mean identical heredity'; secondly, that 'genetically caused and environmentally caused variations were both present and often indistinguishable in the individual but had quite different consequences for its descendants'; and thirdly, that 'it was no longer necessary to suppose that mutations were so abundant or that their nature or causes were so important for the results of practical animal breeding as had previously seemed obvious and unavoidable'. [5]

Much of the experimental work on the application of Mendel's laws was confined at first to plants and small animals. By 1918 the inheritance of coat colour in sheep, cattle, and pigs had been examined but work on the inheritance of characteristics of economic importance had only just begun. One

FIG. 13.3. Dr. John Hammond demonstrates some specimens to farmers on a cattle breeding course at Cambridge University, England, in January 1948. This illustrates the growing application of scientific principles in agricultural practice.

difficulty was a lack of basic information on growth and production in the various breeds of livestock on which the geneticist could work. Without this information for the larger farm animals, the geneticist had to wait some years before he had data on the offspring of several successive generations. Thus progress was slow. In 1939, at the Seventh International Genetical Congress, H. C. McPhee pointed out that 'Much has been written and spoken of the value of genetic research as the basis of livestock improvement, but little has been accomplished. Can the notable advances made by breeders of plants and small animals be repeated with farm livestock?' [6] Two major difficulties were still apparent. One was to distinguish between genotypic and phenotypic improvements—whether, for example, increasing milk yields over several generations were due to better breeding or better feeding, or indeed to both. The other difficulty was the need to be able to measure the productive characteristics of the livestock. This was particularly difficult in meat animals where economy of gain and suitability of the carcasses to the market were important. Research during the 1940s was severely hampered by war, but progress in the statistical treatment of genetic data and in artificial insemination immediately before and during these years made possible the rapid advances in animal breeding that occurred after 1945. These advances were based on an increasingly detailed knowledge of the inheritance of economically important characteristics in livestock.

There had been various instances of successful artificial insemination of animals in the nineteenth century but it was the development of the technique in Russia by I. I. Ivanov and his colleagues which caught the attention of farmers and scientists in various parts of the world. In the 1920s and 1930s many thousands of sheep and cattle were artificially inseminated in the Soviet Union. The first artificial insemination association in Denmark was established in 1936; the first in the U.S.A. in 1938. In 1937–9 interest in the new technique spread rapidly. The war halted progress in some countries, but in the United States the total of 7539 cows bred by artificial insemination in 1939 had increased to 1 184000 by 1947 [7]. With the discovery of a method for freezing semen for long-term preservation, in 1949, and a means of satisfactorily diluting semen, the refinements of technique became available for a substantial post-war expansion in the use of artificial insemination. This in turn provided the geneticist with detailed breeding and production records over numbers of generations, and from single sires over many dams. The farmer had the benefit of mating his herd to the best bulls available (judged by their progeny) at low cost.

Throughout the first half of the twentieth century the simplest breeding practices continued side by side with the most advanced. Change was gradual. In general, selection continued to be within groups of livestock of a recognizable breed. But as the aims of production became more clearly defined, less emphasis was given to purity of breed as recognized by breed societies. This process is most easily recognized in poultry production. At the turn of the century birds from dual-purpose breeds were raised on mixed farms to provide both eggs and meat. As the demand for eggs increased, the farmers selected birds from a breed recognized for its high egg production, as for example the White Leghorn. Further improvement in production then came with the keeping of records on which selection was based. The next development was to cross-breed high-yielding stock, first from strains within a breed, and then regardless of breed. Gradually the purity of breed, in showyard terms, became of less and less significance and the performance record became the dominating factor. In compiling the performance record various measures of egg quality, and of efficiency of feed conversion into eggs, were added to the total of eggs laid. Similarly the efficiency of converting food into poultry meat was studied for the production of meat birds. In the United

States 2 per cent of the 11m. breeding birds in the National Poultry Improvement Plan were crossbreds by 1941–2. By 1952–3 this figure had increased to 16 per cent, and by 1958–9 to 64 per cent [8]. The standards set in the showyards for poultry in 1900 were, by 1950, no longer relevant to commercial poultry production. Instead breeding birds were selected on the basis of statistical computations of breeding and performance data.

In other forms of livestock production the overall trend has been the same, but the change from herd-book standards to total reliance on breeding plans, based solely on measured production characteristics, has not progressed so far. There was of course considerable scope for meeting changing market requirements by changing from one recognized breed of livestock to another, and by selecting for improved productivity within a breed. For instance, the dominant ram breed used in New Zealand for prime lamb production was changed from Lincoln in 1900 to Romney Marsh by the 1950s in response to the demand for smaller carcasses. And careful selection of Danish Landrace pigs, in Denmark, increased the body length from 88·9 cm in 1927 to 93·2 cm in 1953, thereby improving the pigs for bacon production [9].

Feeding. Improvements in feeding have been as important as improvements in breeding for the increase of animal production. Indeed, by 1950, in any particular type of farming, it was still debatable whether increases in production were due more to progress in breeding or to progress in feeding. There was little point in breeding from high-yielding cows if the progeny were then not fed sufficiently well to obtain maximum milk production.

At the turn of the century most poultry scavenged for a large part of their food around the farmyard; pigs consumed much refuse material; and sheep and cattle grazed a diversity of volunteer pastures. This picture was modified in different ways by the farmers growing specific crops or pastures for their livestock such as turnips or lucerne, and supplementing the diets—with grain, for example, or hay—at certain times of the year. In general, only working horses, animals housed during winter in the colder climates, and some breeding and milking stock had their diets closely regulated by the farmer. In intensively farmed countries, such as Britain, livestock on some arable farms were valued mainly as converters of bulky, unmarketable farm produce into farmyard manure which then could be used to maintain soil fertility for crop production.

Progress in feeding was achieved through improving scientific knowledge of the optimum food requirements of livestock (Fig. 13.4), and through

FIG. 13.4. A day's maintenance ration for a herd of pedigree Ayrshire cows in wartime England, January 1943. The carefully calculated ration was 35 lb (15·9 kg) mangolds, 20 lb (9·1 kg) oat and tare silage, and 10 lb (4·5 kg) hay per cow. In addition a concentrates mixture of oats, beans, and National Dairy Nuts was fed at the rate of 3 lb (1·4 kg) per gallon (4·5 l) of milk produced.

FIG. 13.5. Horses provide power to drive a small chaff-cutter used to chaff cow peas and sorghum, grown as fodder crops, to make silage for feeding to cattle. The chaffed material is carried by elevator to the top of the silo. Queensland, Australia, c. 1910.

improving the quality and quantity of the feed available. By bringing these two factors together the farmers, by 1950, were able to provide their animals with more of the right kind of food at the right time than they had been able to do in 1900.

The main chemical method for analysing feeding-stuffs was changed very little during the first half of the twentieth century. The Weende system of proximate analysis, developed in Germany in 1860, was used, with little modification, throughout this period to analyse feeding-stuffs into moisture, ether extract (oil), crude fibre, crude protein, nitrogen-free extract (carbo-hydrate), and ash. Analyses of this nature were simple to perform and provided a rough, rule-of-thumb guide to the nutritional value of a feeding-stuff [10]. They told nothing, though, of the actual utilization by an animal of a particular feeding-stuff. This latter type of information was far more complex and difficult to obtain.

Already, by the turn of the century, scientists were investigating the contents of digestible energy and digestible protein in feeding-stuffs—essentially the amounts of energy and protein retained by an animal. These were deduced by subtracting the amounts of protein and energy left in the faeces from the amounts in the food fed to the animal. A concept adopted by some scientists was that animals required a certain amount of energy to maintain them in a state of equilibrium, and then a further amount above this maintenance requirement according to the work or production expected from the animal. In 1896-1905 O. Kellner, in Germany, suggested a system for assessing the value of a feed in terms of its capacity to produce fat in the animal body. From experiments with animals he calculated the quantity of fat produced by a kilogram of pure starch fed to an animal already receiving its maintenance requirement of energy. He then went on to determine the fat producing powers of various feeding-stuffs and related them to that of starch so as to obtain a value for each material in terms of its starch equivalent. Kellner, and others after him, considerably refined the idea of digestible nutrients as they endeavoured to account for all the possible losses of energy from the original food in its passage through the animal. Kellner, for instance, measured losses in faeces, urine, and gases. Other workers, including H. P. Armsby in America, measured direct heat losses as well, using an animal calorimeter. In 1917 Armsby published his book which put forward a system for evaluating feeding-stuffs in terms of their net energy values. He arrived at these by feeding animals at two levels, both below maintenance, and relating the extra energy taken in from the larger ration to the extra heat given out and the saving of body tissue. The Kellner starch equivalent system became the basis for feeding standards in Germany and Britain, while the Armsby system of net energy values was the basis for much work in America. A third system, published by Nils Hansson in 1908 was adopted in Sweden, Norway, Denmark, and Finland. The Hansson system related the productive value of various feeding-stuffs to that of barley, and the relative standards were obtained from numerous feeding trials. These three systems all required considerable experimental effort to assess the feeding value of any one feeding-stuff; hence various modifications, and alternative systems, have been suggested since the 1900s. For practical purposes, though, food compounders and farmers have relied on results from proximate analysis and simple digestibility trials for assessing a given type of feeding-stuff, incorporating correction factors where appropriate. The tables of total digestible nutrients for feeding-stuffs commonly used in the U.S.A. were of this nature.

During the first two decades of the twentieth century when the work on energy and feeding standards was being done, the variable nature of proteins was being discovered and also the existence of vitamins. The next two decades, the 1920s and 1930s, saw a change from emphasis on the digestibility of energy and protein to emphasis on the amino-acid content of proteins and the supply of vitamins and minerals. During this period much was learnt about nutritional diseases in livestock and how to prevent them. For example, sheep and cattle were shown to suffer from a lack of copper in Holland, Florida, Australia and then in various other copper-deficient areas in the world. Piglets when born and bred indoors in Sweden, away from soil, were found to suffer from iron deficiency and to develop anaemia.

The mineral nutrition of livestock was investigated in three main ways: by analysing the mineral constituents of animal tissues, by feeding diets of pure, known ingredients, and by trying to find a cure for livestock in the field suffering from nutritional diseases. This work, and that with vitamins, was closely linked with advances in analytical chemistry. Vitamins and many elements are required by the animal body only in very small amounts, or traces. For a while research workers did not always recognize the existence of these substances because they passed unsuspected as trace impurities in otherwise 'pure' chemicals. Once their importance for nutrition was recognized, further research was necessary so that these substances could be isolated, and particularly in the case of vitamins, pure samples prepared. The chemical industry was then able to make available, commercially, pure preparations of the vitamins and trace elements for use in compounded feeding-stuffs. This trade expanded rapidly after the Second World War.

Microbial action in the rumen in sheep and cattle produces a variety of proteins and essential amino acids, together with some vitamins, with the result that these animals are not so dependent on their food source for a supply of these essential substances. Pigs and poultry possess no such facility, however, and so investigations were directed particularly at establishing all the essential components of diet for these species. So fashionable did work on minerals and vitamins become that by 1939 Charles Crowther was led to comment that 'The intensive concentration in recent years upon the newer developments, often loosely described as the "new science" of nutrition, has tended to produce, especially in the lay mind, a distorted conception of nutrition as being primarily a matter of supply of vitamins and of minerals. This has led to the indiscriminate addition of vitamin and mineral preparations to all kinds of diets, regardless of whether such additions will remedy an existing deficiency or imperfection of "balance" in the diets . . .' [11]

With a better knowledge of the nutrients necessary for the maintenance of animal health, farmers and stock food merchants could take advantage of a wide range of materials in preparing feeds for housed or penned livestock. Meanwhile the diet of grazing animals, too, was being improved. Until the 1920s the attention of agricultural scientists had been given predominantly to cereal production, fodder crop production, and the question of fertilizers. However, during the 1920s and 1930s the poverty of pastures in many areas prompted investigations into how they might be improved. Scientists in Sweden, Denmark, the United Kingdom, New Zealand, and Australia in particular began to study local ecotypes of pasture grasses and legumes. Men such as Sir George Stapledon, at Aberystwyth in Wales, were quick to demonstrate to farmers the importance of clover in building up soil fertility and in increasing pasture production. High-yielding strains of pasture species were identified, and systems of seed certification were established to provide a guarantee to farmers of the purity of the seeds they were purchasing. The study of pasture nutrition with its recognition of trace-element deficiencies in pastures was a parallel development to the growing understanding of mineral nutrition in animals. The Second World War intervened just when much of this new knowledge was being applied on farms. But after 1945, when fertilizers were again available and seed stocks were multiplied, pasture development became the basis for a rapid increase in stocking rates in temperate parts of the world.

Increased numbers of livestock needed more feed right through the year and not just at the height of the spring growth. Attention was therefore soon given to conserving surplus spring growth for use during less productive seasons. The machinery manufacturers were quick to appreciate the new scope for their skills and by the 1950s a range of new forage harvesters, hay bailers, and other silage and haymaking machines was appearing on the market. Temperate pasture production had entered a new phase. In the tropics, though, research on pastures was only just beginning.

III. MANAGEMENT

Although improved techniques of breeding and feeding provided the basis for the intensification of livestock production in the first half of the twentieth century, much of the increased output would not have been economically feasible without the development of labour-saving techniques and efficient management systems. This included developing mechanical aids so that one man could handle far more animals than previously; regulating the physical environment to provide the optimum conditions for various biological functions; and protecting the livestock from pests and diseases.

The most notable labour-saving device introduced on to dairy farms was the milking machine. Various types of machine had been patented by 1900. They were of the bucket type, that is milk was collected in a bucket beside the cow. During the 1920s the 'releaser' type machine became widely available and slowly replaced the bucket type. With the 'releaser' machines milk was conveyed from the cow in overhead pipes to the dairy. A convenient source of power was essential for the milking machines and the development of small internal combustion engines greatly helped the mechanization of milking, as did later the supply of electricity to farms. Even so, the rate of adoption of milking machines varied from country to country. By 1939 in England only about 15 per cent of cows were milked by machine [12], whereas in 1941 in New Zealand about 86 per cent were milked in that way [13]. The development of machine milking has been associated with a wide range of other technological changes involving milking-shed design, dairy design and equipment, and methods to ensure cleanliness and freedom from bacterial contamination (Fig. 13.6).

In beef cattle and sheep husbandry, farmers improved their stock yards by improving the layouts of large and small pens, races, and drafting gates so that a few men could handle large numbers of stock quickly. This was particularly necessary for drafting, marking, branding, drenching, or inoculating animals. The new power sources were applied to shearing sheep although the basic mechanical system inherited from the nineteenth century was refined

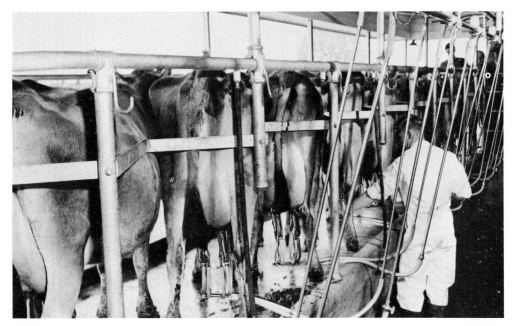

FIG. 13.6. Cows being milked by machine in a herringbone-design milking shed, Victoria, Australia, *c.* 1959. In this design a pit sunk between two rows of cows allows the operator to work without having to bend when attending to each cow.

rather than changed. Shearing-shed design was improved as numerous far-mers modified the customary designs when they built new sheds.

In poultry husbandry after the turn of the century there was considerable experimentation with housing styles. As flocks of birds were increased in size, thought had to be given to protecting the birds either from heat prostra-tion or cold stress according to the prevalent climatic conditions. The basic system adopted during the 1930s was a deep litter system with or without an outside run. Meanwhile much was learnt about the physiology of poultry and the requirements of the birds for floor space, length of feeding trough, watering points, number and style of nesting boxes, light, heat, and humidity. The more intensive production became, with birds housed all the time, the more essential was information of this type if the birds were to be maintained in good health and at maximum productivity. The introduction of battery cages was a further step in this process of intensification of production and reduction of labour requirement per bird. Battery cages for laying hens were first used in the U.S.A. in the 1920s and were produced in Britain in the 1930s. But widespread adoption of the system occurred only after the Second World War. In 1939 in the United Kingdom, for instance, less than 10 per

FIG. 13.7. A small incubator, warmed by a kerosene heater attached to the side, in use in England in 1939. This simple type of incubator was used in many countries.

cent of poultry were kept wholly indoors; by 1968, 90 per cent were so housed [14].

Next to poultry, pigs were the livestock best adapted to intensive, fully housed systems of production. But it was only from the 1950s onwards that scientific investigations revealed the environmental requirements of pigs housed intensively, and new types of housing systems for large numbers of pigs began to appear.

As man gradually increased his control over biological processes, he had also to increase his control over the physical environment surrounding these processes. This is perhaps seen to the greatest extent in the replacement of the broody hen by the incubator. Again, as with milking machines, the provision of a reliable source of power was an integral part of incubator development. In about 1900 small kerosene (paraffin) incubators which held about 50 to 200 eggs were available in the United States. This type of incubator became established in many countries (Fig. 13.7). Gradually incubator size was increased, particularly when an electricity supply became available. Then the growth of broiler-type poultry production stimulated a demand for year-

FIG. 13.8. Dipping sheep in western Queensland, Australia, *c.* 1910. The recently shorn sheep are pushed into the deep end of the dip and left to swim out. Drums of dipping fluid are stacked at the side of the dip. This proprietary fluid probably contained arsenic for the control of lice.

round hatching; and with growing egg consumption as well, special hatcheries were established to supply day-old chickens to producers. In the U.S.A. the average capacity of hatcheries had reached 24000 eggs by 1934, and 80000 eggs by 1953 [15]. For this type of commercial hatchery complete and reliable control of temperature, humidity, carbon dioxide and, oxygen content of the air, egg position, and egg turning are all essential.

Despite the development of very intensive systems of poultry and then pig production, it is important to recognize that up to 1950 by far the most significant proportion of the world's domestic livestock was herded and cared for in an extensive manner. Under these circumstances the most important contributions of science and technology beyond those in the areas of breeding and feeding were in combating pests and diseases. Techniques of pest and disease control through quarantine and slaughter of infected animals were well known before 1900 and continued to be used thereafter. Vaccination, too, was already practised against such diseases as pleuro-pneumonia, anthrax, and tick fever. Lime sulphur, nicotine, and arsenic were among the substances used to control external parasites, and nicotine among those for internal parasites. However, in 1900 the causes of many diseases were not known and they were therefore very difficult to control. Research into the causes of animal diseases thus became a major part of scientific research concerning livestock after about the 1910s. In the inter-war years much more was learnt not only about the nutritional deficiency diseases, already referred to, but also about a wide range of internal and external parasites and bacteriological diseases. Tests were devised to reveal bacterial infections such as brucellosis, and vaccines were prepared against others such as tuberculosis. Soon large industrial companies were preparing these products on a commercial scale for the rural industries. Research on animal diseases was slowed by the onset of the Second World War but this was compensated for by an intensification of research in other areas. Penicillin and DDT were notable products of this war-time research. After the end of the war penicillin and other antibiotics were used on farms to treat bacterial infections such as certain forms of mastitis in cows. DDT and other insecticides were used in dips and sprays against external parasites. From about 1950 the pharmaceutical industry rapidly expanded its veterinary work and, in its own research departments, developed further materials for controlling pests and diseases in livestock. Scientists working in government, university, and industrial research establishments continued to investigate and explain the life-cycles of many disease-causing organisms. With this new knowledge it was then possible to adapt management practices on farms so that the life-cycles of parasitic organisms were broken, risks of infection were reduced, or preventive chemical treatments were adopted. The hypodermic syringe and multiple-dose drenching gun became regular items of farm equipment. Losses from pests and diseases continued to be serious in animal production, but by 1950 the farmer or veterinarian knew much more about the problem with which he was faced than he did in 1900, and was able to act on the basis of his knowledge instead of blindly, in ignorance of cause and effect.

IV. CONCLUSION

During the first half of the twentieth century farmers had to adapt their aims in animal production to a replacement of horses by machines, to competition in fibre production from fibres produced industrially, and to changing consumer requirements for meat, milk, and egg products. This adaptation furthermore had to be achieved within the constraints dictated by alternating prosperity and recession in economic affairs, by two world wars and by an increasing human population. The farmers met these challenges by improving their breeding, feeding, and management practices in ways which allowed them to handle larger numbers of livestock and to control more closely the biological processes of animal production.

REFERENCES

[1] THOMPSON, F. M. L. *Economic History Review*, **29**, 60 (1976).
[2] MOLNAR, I. (ed.) *A manual of Australian agriculture* (2nd edn.). Heinemann, Melbourne (1966).
[3] PÁLSSON, H. *Journal of Agricultural Science*, **29**, 544 (1939).
[4] HAGEDOORN, A. L. *Animal breeding* (6th edn.), p. 197. Crosby Lockwood, London (1962).
[5] LUSH, J. L. in L. C. DUNN (ed.) *Genetics in the twentieth century*, p. 496. Macmillan, New York (1951).
[6] SEVENTH INTERNATIONAL GENETICAL CONGRESS (Section D). *Animal breeding in the light of genetics*, p. 9. Imperial Bureau of Animal Breeding and Genetics, Edinburgh (1939).
[7] RASMUSSEN, W. D. (ed.) *Readings in the history of American agriculture*. University of Illinois Press, Urbana (1960).
[8] CARD, L. E., and NESHEIM, M. C. *Poultry production* (11th edn.). Lea and Febiger, Philadelphia (1972).
[9] AERSØE, H. *Animal Breeding Abstracts*, **22**, 87 (1954).
[10] VAN SOEST, P. J. in D. CUTHBERTSON (ed.) *Nutrition of animals of agricultural importance*. (*International Encyclopaedia of Food and Nutrition*, Vol .17, Pt. 1.) Pergamon Press, London (1969).
[11] CROWTHER, C. in D. HALL (ed.) *Agriculture in the twentieth century*, p. 363. Clarendon Press, Oxford (1939).
[12] HARVEY, N. *A history of farm buildings in England and Wales*. David and Charles, Newton Abbot (1970).
[13] HAMILTON, W. M. The dairy industry in New Zealand. *CSIRNZ Bulletin 89*. Government Printer, Wellington (1944).
[14] ROBINSON, D. H. (ed.) *Fream's elements of agriculture* (15th edn.) John Murray, London (1972).
[15] CARD, L. E., and NESHEIM, M. C. *Op. cit.* [8].

BIBLIOGRAPHY

A century of technical development in Japanese agriculture. Japan F.A.O. Association, Tokyo (1959).
ALEXANDER, G., and WILLIAMS, O. B. (eds.) *The pastoral industries of Australia*. Sydney University Press (1973).
BARNARD, A. (ed.) *The simple fleece*. Melbourne University Press (1962).
CUTHBERTSON, D. P. (ed.) *Progress in nutrition and allied sciences*. Oliver and Boyd, Edinburgh (1963).
DRUMMOND, J. C., and WILBRAHAM, A. *The Englishman's food* (revised edn.). Jonathan Cape, London (1959).
HANSON, S. G. *Argentine meat and the British market*. Stanford University Press (1938).
HUNTER, H. (ed.) *Bailliére's encyclopaedia of scientific agriculture* (2 vols.). Baillière, Tindall and Cox, London (1931).

LUSH, J. L. Genetics and Animal Breeding. In L. C. DUNN (ed.) *Genetics in the twentieth century*. Macmillan, New York (1951).

MARSHALL, F. H. A., and HAMMOND, J. *The science of animal breeding in Britain: A short history*. Longmans, Green and Co., London (1946).

MAYNARD, L. A. Animal species that feed mankind: The role of nutrition. *Science, N.Y.*, **120**, 164 (1954).

MINISTRY OF AGRICULTURE, FISHERIES AND FOOD. *Animal health: A centenary 1865–1965*. H.M.S.O., London (1965).

——. *A century of agricultural statistics, Great Britain 1866–1966*. H.M.S.O., London (1968).

RUSSELL, E. J. *A history of agricultural science in Great Britain*. George Allen and Unwin, London (1966).

TYLER, C. The historical development of feeding standards. In *Scientific principles of feeding farm live stock* (Conference Report). Farmer and Stockbreeder Publications Ltd., London (1959).

——. Albrecht Thaer's hay equivalents: Fact or fiction? *Nutritional Abstracts and Reviews*, **45**, 1 (1975).

UNDERWOOD, E. J. *Trace elements in human and animal nutrition* (2nd edn.). Academic Press, New York (1962).

UNITED STATES DEPARTMENT OF AGRICULTURE. *After a hundred years. The Yearbook of Agriculture 1962*. United States Government Printing Office, Washington (1962).

PART II: FOOD AND INDUSTRIAL CROPS

Crop production in the first half of the twentieth century passed through several phases. In each of these phases one particular approach to production was exploited to the point of disaster—in the form of soil erosion, pollution, or energy shortage—before there was a retreat from that approach and a change of emphasis. Parallel with these more spectacular events, scientifically based improvements in plant breeding and nutrition and in pest and disease control progressed steadily. Developments in engineering and chemical technology led to mechanization in agriculture, and, in some instances, to competition between industrially produced goods and the more traditional agricultural products. The inability of nations to maintain peace and economic stability further complicated the task of the farmer during these years and confounded his long-term planning. Yet the aftermath of war had its compensations. The foundation of the Food and Agriculture Organization (F.A.O.) of the United Nations in 1945 provided, for the first time, a reliable source of statistical information on the production of agricultural commodities of all kinds (Table 13.1).

I. MAN AGAINST NATURE

Soil erosion. The expansion of crop cultivation in the New World in the nineteenth century was largely an expansion into virgin land. This expansion was continued into the twentieth century as more of the land, by then occupied by European settlers, was brought under the plough. Much of this newly cropped land was in areas arid by western European standards, and yields were low. Land, though, was plentiful compared with labour and capital. The pioneering life was a hard one, as not-very-prosperous farmers endeavoured to make a living in an environment that was not yet well understood. They held their land and worked it within land tenure systems dictated by legislators who only partly understood the constraints of the environment on the farmers. Often farms were too small or farmers had too little capital for farm development. Some settlers knew no other way to farm than to exploit the natural soil fertility. Whatever the basis, the intrusion of cropping into new country meant an abrupt break in a long-established ecosystem. Former vegetation was removed, soil fertility reduced, and, in particular, bare loosened soil was exposed to the effects of wind and rain. At first a type of shifting cultivation could be practised, cropped land being abandoned after several crops were taken and new land ploughed up; but as settlement intensity increased this was no longer possible. High prices for produce enticed farmers to crop and re-crop their land. Later, glutted markets and economic depression left them little alternative but to continue the process in an attempt to remain solvent. By the mid-1930s the devastation of the soil was made dramatically apparent as dust storms rolled across parts of the United States and Australia, blackening out the sun and eventually burying fence posts and buildings (Fig. 13.9). In wetter areas, large gullies were cut in denuded hillsides by water no longer held by vegetation; in other areas sheet erosion developed. As people became aware of, and concerned about, erosion it came to be recognized as a problem of varying severity in most countries of the world.

The first reaction to the more dramatic manifestations of erosion was to gather data on the effects of wind and rain on particular soil types and to devise mechanical means of stemming the loss of topsoil and the extension of gullies. The problems were of national importance and government departments of soil conservation were established to deal with them. These departments then developed methods of erosion control and worked with farmers in implementing them. Techniques of terracing, contour ploughing, water

TABLE 13.1
World cereal, potato, and oil-seed production
(ooo acres except for total areas)

	Total area (million acres)	Wheat	Rye	Barley	Oats	Maize	Rice	Potatoes	Ground-nuts	Linseed
World total	3622·7‡	382000	98000	106220	137850	214700	199000	45396	23100	19000
United Kingdom	60·3	2279	80	2216	3771	—	—	1398	—	—
Europe*	1254·6	61800	32000	23860	35200	27200	440	24032	80	950
U.S.S.R.	5273·6	98764†	53125†	22528†	40089†	7041†	—	17569†	—	5805†
Canada	2364·6	23414	487	7350	14393	237	—	508	—	1059
U.S.A.	1937·1	64740	1981	10195	41503	94455	1506	2824	3183	3914
Argentine	690·2	10030	—	1400	2200	9000	122	372	351	3419
North and Central America	} 9887·9	89500	2468	17950	56000	102500	2220	3546	} 4000	5200
South America		14300	—	1510	2563	23180	4700	1200		5000
Asia	6927·7	110000	—	37880	676	8812	187000	1100	15000	3550
Union of S. Africa	302·0	2400	—	—	545†	5909†	—	60	56†	—
Africa	7166·2	12800	—	8925	722	19940	3267	288	4000	120
Australasia	1970·0‡	11284	—	811	2085	305	23†	280†	20	66

* Excluding U.S.S.R.
† Average of 1935–9.
‡ Excluding New Guinea and Oceania.
Source: FAO *Handbook*, 1946.

FIG. 13·9. A dust cloud advances on the town of Fort Macleod in Alberta, Canada, c. 1935.

diversion, and stubble mulching in dry areas were introduced. Gradually knowledge was gained of the intricate relationship between soil, slope, wind, rain, run-off, and plant cover. During the years from 1935 to 1950 emphasis gradually changed from emergency control of soil erosion to conservation of land with maintenance of both soil and fertility. This last approach evolved into an overall concept of land management in which cropping and pasture systems, and cultivation techniques were closely adapted to the soil, land types, and climatic environment of the farm. In short, there developed new ecosystems incorporating the farming activities of man but with due regard for the constraints of nature.

Pesticides. Just as man cultivated land with little heed to the finer points of nature, so too did he eventually apply pesticides with reckless abandon. In the early years of the twentieth century much of the research in agricultural science was concerned with the control of pests and diseases, and more will be said of this subsequently. But until the outbreak of the Second World War the main substances used to control weeds, fungi, and insects were inorganic compounds such as copper salts, arsenical compounds, sulphur, sodium chlorate, and sodium fluoride. Plant extracts were also used against insects, the most common being nicotine, rotenone, and pyrethrum. These substances tended to be expensive and they were used against specific pests in limited areas and usually in small quantities. Nevertheless during the inter-war years use of pesticides did increase and research workers in universities, government institutes, and the larger chemical companies looked for new substances to control pests.

In 1932 dinitro-*ortho*-cresol (DNOC) was patented as a herbicide in France; it was already known to be useful as an insecticide. It was introduced into England and, in 1939, began to be used there widely against weeds in arable crops. After the outbreak of war in 1939 governments became interested in the potential use of pesticides as agents of biological warfare. In the U.S.A. work on plant growth hormones during the 1930s had led to the realization by 1941 that synthetic growth regulators such as the phenoxyacetic acids (of which 2,4-D is one) might be useful in agriculture as herbicides. Scientists then suggested that this work could also be important in terms of the war effort and research continued on these substances as part of the biological warfare programme, so that by the end of the war 2,4-D was recognized as a new and potent herbicide, though one still in need of extensive testing for agricultural purposes. In 1945, 917000 lb of 2,4-D were produced

FIG. 13.10. Spraying apple trees with lime and copper sulphate, and Paris green (copper acetoarsenite) in Nova Scotia, Canada, c. 1908. The spray pump is worked by hand.

FIG. 13.11. A Tiger Moth aircraft flying low to spread superphosphate fertilizer on pasture in eastern Australia, c. 1950.

in the United States; by 1950, this total had risen to 14m. lb in response to its rapid adoption as a herbicide by chemical firms supplying farmers [1].

The chlorinated hydrocarbon insecticides and organo-phosphorus insecticides were developed similarly during the war. Dichloro-diphenyl-trichloro-ethane (DDT) was the earliest and best known of the chlorinated hydrocarbons. First synthesized, by Othmar Zeidler, before the end of the nineteenth century, its insecticidal properties were discovered in 1939 by Paul Müller in the laboratories of J. R. Geigy S. A. in Switzerland, and tested against various insects including pests of vegetable crops. When the new substance reached the U.S.A. and Britain it was quickly adopted by the armed forces for use against such insects as lice, mosquitos, and flies. This was due in part to the wartime difficulties of importing rotenone and pyrethrum. In 1944 a mass dusting of civilians in Naples against lice, after the outbreak of typhus, dramatically demonstrated the powers of this new insecticide. Subsequently it was used with great effect by the forces to control malaria-carrying mosquitos in the tropics. Considerable public interest was created by these events and immediately after the war DDT was used widely against household pests and pests of plants and animals. Once the insecticidal properties of DDT had been recognized other related compounds, as well as many chemicals of different kinds, were screened for the same purpose. By 1945 other chlorinated hydrocarbons, such as the gamma isomer of benzene hexachloride, were being brought into use, and also organo-phosphorus insecticides such as parathion. This latter group was initially developed in Germany during research on nerve gases and the toxicity of the materials was well recognized.

The toxicity of DDT and related substances, though, was not appreciated; indeed, these substances were considered to be relatively harmless to man. In the enthusiasm of the years after 1945 governments and private individuals spread the new insecticides and herbicides with abandon. Aeroplanes had been used to spray DDT to control mosquitos during the war, and in peacetime they were used to spread insecticides and herbicides on agricultural crops and, at times, in urban areas as well. Trucks, tractors, and various mechanical devices were used to apply them on the ground. Man had found a new and very powerful weapon against the pests of his crops. Some pests were controlled most effectively, and yields were increased as a result. But the application was often indiscriminate and this created problems. By 1950 doubts were beginning to be voiced. Residues of insecticides were being found in cows milk and fat tissues. By the mid-1950s it was becoming apparent that pesticides were killing some kinds of wildlife and contaminating waterways. A further serious problem was the appearance of insects showing resistance to the new insecticides. The under-estimation of the danger to humans was also gradually recognized and during the 1960s there was a strong public reaction against the indiscriminate use of pesticides. Pesticide pollution became the new enemy of advanced societies, and ecological relationships the new concern. The farmers' use of pesticides became a matter for national control and popular debate. New systems of pest control had to be devised in which pesticides were used far more judiciously and in combination with other methods of control. In time the more toxic chemicals were restricted in use or withdrawn from the market, while research was continued to develop less noxious substances to replace those previously used. Undoubtedly, immense benefits have been demonstrated: for example, of the 1814m. people living in the originally malarious areas of the world, 1347m. (74 per cent) were by 1970 living in areas where—mainly by use of insecticides—the disease had either been eradicated or reduced to minor proportions. Such success, however, was not to be the excuse for seeking to increase the undoubted toxic hazards.

Fertilizers. Advances in chemical technology also made possible marked changes in fertilizer practice. The industrial fixation of nitrogen, in particular, eventually led to a preoccupation with nitrogenous fertilizers among farmers, scientists, governments, and fertilizer manufacturers to the detriment of the study and exploitation of non-industrially based methods of maintaining and increasing soil fertility.

The elements which need to be returned to the soil in the greatest amounts after the removal of agricultural produce are nitrogen, phosphorus, and potassium. This was well recognized by 1900, and sodium nitrate in Chile, potassium salts in Germany, and phosphatic rock in various countries were by then being mined and exported to farmers. Fears were already being expressed, though, that the supply of nitrate would soon be exhausted. This was seen as particularly critical for the future production of food grains.

Sir William Crookes, in his Presidential Address to the British Association in 1898, emphasized this problem and pointed to a possible solution that might be found in the industrial fixation of atmospheric nitrogen [2]. This did eventually become a commercial possibility with the development of the Haber–Bosch process in Germany in 1913. The raw materials for this process were water and nitrogen and the energy of fossil fuels. When, at the end of the First World War, industrially fixed nitrogen was no longer required for making nitric acid for munitions, attention was given to its possible use in agriculture. Factories for fixing nitrogen, ultimately as nitrates, were built and production was increased during the inter-war years, but it was not until after 1945 that there was a dramatic increase in the use of nitrogenous fertilizers made from industrially fixed nitrogen. F.A.O. statistics indicate that in 1938 2·5m. tons of nitrogen were used in commercial fertilizers throughout the world (excluding the U.S.S.R.); by 1950 this figure had reached 3·9m. tons; and by 1960 10·2m. tons (including the U.S.S.R.). Within the next ten years there was a further threefold increase in the consumption of nitrogen in fertilizers to a level of 31·7m. tons by 1970. Usage of phosphatic and potassic fertilizers also increased, but not at the high rate of that of the nitrogenous fertilizers. In Europe and the U.S.A. more and more mineral fertilizer was applied to crops. In the poorer countries, such as India, people noted the increased yields resulting from these high applications of fertilizer and saw in this approach part of the solution to their food problems. Increased fertilizer production and use were given high priority in national development plans and, by 1965, 33 countries were subsidizing fertilizers at rates ranging up to 50 per cent of their cost. Special strains of rice, wheat, and other crops were bred to utilize the increased levels of fertility being made available to plants. Former tenets of good husbandry, involving rotational systems of cropping and the use of legumes to fix nitrogen, tended to be overlooked as yields were markedly increased even where land was repeatedly cropped.

These events did not pass uncriticized. During the 1940s—as more and more reliance was placed on inorganic, in preference to the organic, fertilizers —a so-called 'humus' controversy developed, in which there was a popular outcry against the 'unnatural' character of the 'artificial' or inorganic fertilizers. This had little or no bearing, though, on the development of fertilizer practice. Reliance on nitrogenous fertilizers increased in many countries until the oil crisis of the 1970s. Then it was suddenly realized how dependent high crop yields had become on fertilizers by then derived from oil and gas. The question then arose whether increased crop production achieved at the expense of non-renewable energy resources could be sustained indefinitely. What had once seemed a very economic way of producing more food for hungry people was suddenly becoming a much more expensive undertaking. This takes us beyond the period of our present concern but we must note that a whole farming technology had by the middle of the century been built up on the basis of cheap industrially produced nitrogenous fertilizers.

The cropping of soil until it was denuded by erosion; the mass application of pesticides until streams and rivers were polluted; and the use of industrially fixed nitrogen until crop yields were dependent on the produce of oil wells; all were an expression of intense endeavour to increase the supplies of food and fibre available to man. And in increasing these supplies man was successful. In the U.S.A., for example, by 1953 37 per cent fewer people were producing 77 per cent more agricultural produce than in 1910 [3].

II. SCIENCE AND CROP PRODUCTION

While the general public was becoming concerned about dust storms, pesticides, and 'artificial' fertilizers in turn, the scientists were making steady progress in the breeding and nutrition of plants, and in understanding the physiology of plant growth and development.

Breeding. The selection of plants from natural or 'wild' populations had for long been a method of obtaining varieties for cultivation, and an allied approach was to select high-yielding plants from within a cultivated crop. During the nineteenth century considerable quantities of plant seeds were exchanged between countries as people sought the best varieties of crop plants for local conditions. The next step was to cross-breed the more promising plants. Marquis wheat, in Canada, was the result of a cross made in 1892 between the main Canadian wheat, Red Fife (Fig. 13.13), and an Indian wheat, Hard Red Calcutta. It was selected by Charles E. Saunders from a collection of cross-bred wheats in 1904 and, by the end of the First World War, was the main variety sown in western Canada. In Australia, in 1889, William Farrer had begun his crossbreeding of wheat. He sought particularly to breed for specific qualities, such as early maturity to suit the Australian climate, and in 1901 his early maturing, short-strawed and high-yielding wheat Federation was released to the farmers.

Farrer and his contemporaries, however, had worked without the benefit of knowledge of Mendel's laws of inheritance. The work of Gregor Mendel had been re-discovered in 1900. Originally published in German in 1866, it had remained neglected and was not understood until the end of the century. But with its rediscovery and reappraisal plant breeders were given evidence of the stability of hereditary units and an explanation of the principles of inheritance of these units (or genes). Thus plant breeders, concerned with improving the economically important characteristics of plants, could now begin to plan their breeding programmes with some knowledge of the range of results they might expect from each crossbred generation. At Cambridge

FIG. 13.12. Cross-breeding cotton at Namulonge, Uganda. (a). The corolla and part of the calyx are cut from the female flower.

(b). The unripe anthers are removed.

(c). The stigma is washed with water.

University R. H. Biffen quickly began a study of wheat and barley varieties and soon was able to show that certain morphological and quality characteristics were inherited in accordance with Mendel's laws. In particular, he was able to show that resistance to yellow rust in wheat was a simple dominant character. He then proceeded to breed a wheat suitable for English conditions which incorporated rust resistance. The result was Little Jos, which was available to farmers by 1910, and by 1919 was one of the main wheats grown in the eastern counties of England.

The science of genetics was developed rapidly and it was soon recognized that the inheritance of many characteristics was a complex matter. This made the plant breeder's task more difficult than it had appeared at first. But with the early demonstration of the value of the science of genetics to the art of breeding crop plants, particularly in so important a matter as rust resistance in wheat, the scientists were soon to receive considerable encouragement from farmers and governments in various countries.

In Russia N. I. Vavilov began, during the First World War, a remarkable series of plant-collecting expeditions to such places as Afghanistan, Abyssinia, China, and Central and South America. The object was to collect domesticated plants of economic importance together with the closely related wild varieties of these species. The collection of wheats, for example, that was built up and maintained near Leningrad amounted to some 30000 varie-

(d). The stigma is covered with a piece of drinking straw.

(e). The male flower is tied up.

(f). The next day ripe pollen is transferred from the anthers of the male flower to the stigma of the female flower.

ties [4]. Vavilov wished to study the world-wide variation in the characteristics of these plants so that plant breeders could utilize this variation in their breeding programmes. This work by Vavilov, and his theories about the centres of origin of cultivated plants and the possible genetic diversity to be found at these centres, inspired plant breeders in many countries, although in 1939 he was discredited in his own country.

In the U.S.A. Mendelian genetics was exploited in a different way, again to the benefit of the farmer. The work there was influenced by the earlier experiments of Charles Darwin on inbreeding and crossbreeding. The procedure developed by American scientists was to inbreed corn to produce strains of decreased vigour but which bred true to type. They then took four of these strains, crossed them in two pairs and then crossed the progeny from the two pairs to give a double cross. Size, vigour, and fertility were shown to be fully restored and certain undesirable features eliminated by this process. Indeed, these new hybrids were out-yielding standard corn varieties by up to 30 per cent in some trials. New hybrid varieties were made available generally to farmers in about 1930, and by 1949, 78 per cent of the corn acreage in the U.S.A. was planted to these varieties. The average corn yield, which had been about 26 bushels per acre in 1900, was increased to 39 bushels by 1950-4, an increase due in part, at least, to the hybrid varieties [5]. The techniques for producing hybrid corn were transferred to Europe in the post-war years, and this method of hybridization has since been used in breeding other crop plants.

The combination of scientific knowledge and practical skill in plant breeding led to improvement in productivity in many species of crop plants besides the more common food grains during the first half of the twentieth century. The several stages that were characteristic of improvement in sugar-cane breeding were probably typical of the stages of progress for many other crops as well, even though timing for these various stages varied greatly. The first stage was the selection of natural or wild varieties; the second the crossing of these varieties; the third breeding for disease resistance; and the fourth breeding for specific soil and climatic conditions [6]. It was this type of progress, too, that led to the so called 'green revolution' of the high-yielding varieties of wheat and rice sown in tropical areas in the 1960s.

Nutrition. High yields from crop plants were possible only when the plants were supplied adequately with nutrients. Thus research on plant nutrition was a parallel development with that on plant breeding, even if the scientists working in these two areas tended, at times, to work in separate institutions. The concern for, and changes in, the supply of nitrogenous fertilizers have already been referred to. These were general matters of fertilizer production and use, and involved farmers, fertilizer companies, and governments at a national level. At a regional level, in order to obtain information on which to base recommendations for fertilizer practice, many field trials had to be carried out. This was a continuation and expansion of work begun in the nineteenth century. In the twentieth century the conduct of these trials was made less laborious and more accurate by the development of statistical designs and techniques of data manipulation. These allowed for the calculation of, first, the minimum size and number of test plots needed for meaningful results, and secondly an estimate of the limits of reliability of these results. The work of people such as W. S. Gosset (known as 'Student') and then of R. A. Fisher on experimental design in agriculture provided statistical techniques which were adopted as a standard tool in the majority of agricultural research programmes. Additionally, it presented to other fields of study, where variability of materials was inherent, a technique for quantifying, and thereby understanding, this variability. Fisher was originally engaged in 1919 to work at the Rothamsted Experimental Station in England on the accumulated data from the field trials dating back to 1843. And it was in fertilizer field trials that his statistical techniques were most quickly and widely applied.

Fig. 13.13. Experimental plots on the Broadbalk wheat field at Rothamsted Experimental Station, England, in 1925. Permanent fertilizer trials with wheat were first laid out in this field in 1834 when the research station was founded by John Bennet Lawes. From 1852 onwards 13 plots on the Broadbalk field were given the same fertilizer treatment each successive year, and it was the accumulated data from these renowned trials that R. A. Fisher was engaged to analyse statistically in 1919.

So much so, that by the 1950s in many parts of the world the Latin square design had tended to become the standard 'recipe' for agricultural advisers when laying out fertilizer trials. It was on the basis of these local trials that farmers were then urged to apply specific amounts of nitrogen, phosphorous, and potassium to their crops.

By the beginning of the twentieth century it was already recognized that the macronutrients nitrogen, phosphorus, sulphur, potassium, calcium, and magnesium, were essential to plant growth together with the micronutrients (trace elements) iron and manganese. Water culture was by then a standard method used for investigating the nutrient requirements of plants under laboratory or greenhouse conditions. Further progress in identifying the essential elements required for plant growth was linked with an increasing ability to purify the culture solutions. The micronutrients are needed by plants in such small traces that impurities in salts and water used to make culture solutions, or contaminants on containers, could pass undetected and yet supply sufficient of a certain element to render the plants healthy. A sufficient supply might also be introduced in the seed. A greater degree of resolution in analysing plant materials was also important, although it was well recognized that the presence of a certain element in plant tissue did not mean that that element was essential for growth. Thus the exploration of the micronutrient needs of plants was associated with a great deal of experimental ingenuity on the part of the plant scientists and the development of more accurate analytical techniques. In 1914 it was shown that zinc was essential for plant growth and work in the period 1910–23 confirmed the same for boron. These findings lent impetus to the search for other elements that might be essential to plants; in 1931 copper was added to the list; in 1939 molybdenum; and in 1954 chlorine. Molybdenum, it was found in one experiment, was needed in no more than one part in one hundred million parts of culture solution to allow for the healthy growth of tomato plants.

As soon as it was recognized that a certain element was essential for the growth of a particular plant, the effect of not supplying it could be investigated for a range of plants. Gradually descriptions were obtained of the symptoms of a particular nutrient deficiency in various plants. Grey speck in oats, for instance, resulted from a lack of manganese and had characteristic visual symptoms. However, the identification of a particular disease in the field as being of nutritional origin was often a difficult matter because secondary fungal attacks and other factors could complicate the evidence. Nevertheless extensive research in a number of countries gradually linked the disease symptoms seen in fruit trees, cereal crops, sugar cane, and other crops with nutrient deficiencies. Once this cause had been established the cure was usually simple and dramatic. In some instances the farmers had already anticipated the scientists: copper sulphate was applied to orange trees suffering from die-back long before the cause of the condition was shown to be a copper deficiency. But in other instances the scientists were able to show the way to making large tracts of land far more productive by, for example, applying copper to reclaimed heath and moorland in Denmark and Holland or to sandy soils in South Australia.

Once the existence of a trace-element deficiency was recognized as being general in a particular soil type or region the fertilizer manufacturers were quick to include this element in their standard fertilizer preparations sold to farmers in that area. The farmers and politicians readily appreciated the value of the trace-element work, and some new development schemes were made possible only as a result of it. However, the curing of gross deficiencies then led scientists to study the possible occurrence of less dramatic chronic deficiencies which caused plant growth to be retarded but not dramatically so. Trace-element excess, or poisoning, was also shown to be a problem, and gradually intricate inter-reactions between elements in plant metabolism were revealed. The complexities that were discovered were such that a chlorosis in flax was shown to result from excess molybdenum if nitrogen was supplied as nitrate or urea, but not as an ammonium salt (excluding nitrate). Thus it was realized that much more remained to be learnt about the process of growth in plants and the way in which nutrients taken up from the soil were utilized.

As attention was transferred from the macro aspects of plant nutrition to studies at the tissue and cellular level, nutritional research merged into the more general studies of plant physiology. Plant growth and function had interested scientists for many years, and the adoption of plant growth regulators as herbicides has already been referred to. Other physiological studies, such as those on the day length and temperature requirements of plants, were shown to have practical implications also, particularly for horticulturists. But in general the practical application of the results of research on plants at the tissue and cellular level belongs more to the period after 1950 than to the years before [7].

Pest and disease control. Insects, fungi, bacteria, and viruses have been destroying man's crops for centuries, and less desirable plants have been competing with them. Between 1900 and the 1940s, when the new organic pesticides were developed, a number of research centres were established for the study of the pests and diseases of crops and much work was done in this field; and this was in addition to the development of disease-resistant strains of crops by the plant breeders. The life cycles of insects and fungi were studied and described, the existence of plant viruses was confirmed, and in time the role of insects as transmitters of viruses established. In short, these years saw above all the identification and detailed description of many of the more important pests and diseases of crops. To identify the enemy was the first step towards its control.

A knowledge of the life cycle of a pest or disease organism immediately provided some hope of reducing the severity of its attack by arranging cropping methods to destroy the pest at a vulnerable point—such as removing vegetation necessary for an over-wintering phase, or by avoiding attack by growing earlier-maturing strains of a crop. Knowledge of the natural enemies of a crop pest in some instances made biological control of the pest possible, especially where a pest had been imported into a country without its natural enemies. An example of this was the control of the cottony cushion scale when, in the 1880s, it threatened to devastate the citrus trees on the west coast of the United States. This scale had been imported accidently into California from Australia in about 1869. In 1887 and 1889 its natural enemy, the vedalia or ladybird, was introduced from that country, and the ladybird then proceeded to reduce the cottony cushion scale population to insignificance. In a similar way prickly pear, which had over-run some 60m. acres in Queensland by 1920, was eliminated by the introduction during the next decade of several insect parasites from its native America, particularly *Cactoblastis cactorum*, which fed on it and destroyed it. But this close reference to the biology of pests and diseases tended to be overlooked with the widespread use in the post-war years of the new pesticides and of the aeroplane and power sprayers adapted to their mass application.

III. MECHANIZATION

The application of new machines and new forms of power to crop production was very much a feature of the first half of the twentieth century (Fig. 13.13 and 13.14). The most important development was that of tractors powered by internal combustion engines. Their development was gradual from the beginning of the century until the First World War, when they became more generally accepted. Shortage of labour during the war years and high prices for farm produce prompted farmers to turn to mechanical aids. This was particularly so in the broad-acre grain growing regions where tractors could be used to haul ploughs and harvesting machinery in large-scale operations. The depression years of the 1930s discouraged investment in expensive capital items on farms, but the outbreak of the Second World War provided a further impetus for the adoption of tractors and other labour saving machines. During the 1940s tractors steadily replaced animals as the main power source on farms in many parts of North America, western Europe, and Australasia. In Japan during the 1930s and then immediately after the last war there was a similar development of small hand-held power-cultivators. These machines, of about 4 to 10 horsepower, were adapted to the small wet-rice fields of Japan and were also well suited to the economic and social situation of the Japanese farmers. The power-cultivators were later to be exported to other rice-growing regions of Asia, just as the large tractors of America were exported to other broad-acre farming areas.

The tractor had two characteristics which influenced the development of machinery to be used with it. It provided greater haulage power and more rapid movement under load than horses, and it also provided turning power which could be transmitted directly to hauled machines to drive the moving parts. Steam engines already provided a self-propelled power source at the

FIG. 13.14. Harvesting Red Fife wheat with reapers and binders in Alberta, Canada, *c.* 1908.

FIG. 13.15. Caterpillar tractors haul combine harvesters through a wheat crop in Idaho, U.S.A. in the 1930s.

FIG. 13.16. Ferguson tractors with tractor-mounted (three-point linkage) cultivator and ridger being demonstrated in England, 1938.

time when tractors were introduced, but they could not rival the new petroleum fuelled tractors in versatility for farm work; consequently, they were superseded except for driving stationary machines such as large chaffcutters and threshers.

At first tractors were simply coupled to existing implements or machines in the place of horses. But with the passing of the transition period, when both horses and tractors were in use, new equipment was designed specifically for use with tractors. This led to the production of larger machines with a broader cut and more rapid rate of transit through the crop or soil. These changes required new designs for various parts of machines, and further studies of machine interaction with crop or soil, though the basic principles of implement and machine operation continued to be those introduced in the latter part of the nineteenth century (Vol. V, Ch. 1). The Australian header-harvester and the American combine-harvester were both machines which brought together operations of reaping and threshing based on principles already recognized and used separately in other machines. Similarly the auto-headers, developed in the 1920s, essentially represented a telescoping together of the tractor and the harvester. These machines came into widespread use only during and after the Second World War.

As tractors and large machines became more widespread, 'mechanization' became a popular concept. Mechanization, or the replacement of draught animals by machines, meant that crops no longer had to be grown to feed the draught animals and instead the land could be used to produce food and fibre for direct human use. More particularly, though, mechanization was seen in terms of economies of scale and the saving of labour. The cure for

FIG. 13.17. An Ann Arbor pick-up bailer at work in England in 1932. At this time the bales were tied by hand

low farm productivity in small farming regions, or where the land was split into small fields, was seen to lie in introducing machinery to aid the farmer. To do this, it was argued, holdings or fields had to be amalgamated to give areas of land large enough for economic operation of the new machines. In England, for example, hedgerows were uprooted in some areas so that a number of tiny fields could be converted to a few large fields. In this way farming technologies that had proved so advantageous on the broad plains of the New World were to have an impact on the very different landscape of the Old World. The same trend was also evident in the more closely settled parts of the New World. In the U.S.A. the average size of commercial farms increased from 220 acres to 336 acres between 1940 and 1954.

This process, though, was not all in the one direction. Small farm and field size influenced the development of the machines as well. The Japanese cultivators are a particularly good example of this. But tractors, too, were made lighter and smaller as design and engineering skills were improved. A power take-off, pneumatic rubber tyres, and the Ferguson three-point linkage with hydraulic lift were incorporated into the design of these small light-weight tractors (Fig. 13.15). It then became possible to introduce a range of tractor-mounted equipment, including items such as ploughs and mowers, suitable for use in small areas. This equipment together with side-delivery rakes, pick-up balers (Fig. 13.16), and forage harvesters then heralded the arrival of the tractor as a versatile addition to a small mixed farm. The power take-off had been developed during the First World War but widespread adoption of these smaller machines occurred only after the Second World War. Thus over the years mechanical equipment and farm scale interacted to bring about modifications to both, and a better adaptation between these and other factors in the farming systems.

IV. MARKET INFLUENCES

The major developments in plant breeding and nutrition, pest and disease control, and mechanization arose in the industrialized countries; they were then transmitted in varying degrees to the less developed countries. This process of transmission reflected a certain enlightened self-interest on the part of the industrialized countries which possessed colonial empires. These countries were coming to depend more and more on the raw materials from their empires for industrial expansion and for diversification of their peoples' diet. It was commercial good sense that some effort was made to improve the productivity of these raw materials in the regions where they were produced, especially when capital was already invested in plantation schemes there.

Early in the twentieth century Britain imported more agricultural products than any other country; in 1929, 23 per cent of world imports went to her shores. Among these were cane sugar, cotton, rubber, tea, and oil seeds from the British colonies in the tropics. In the nineteenth century technical progress in producing these crops had depended on individual enterprise and the work of people employed in botanical gardens, the latter often in informal association with the botanists at the Royal Botanic Gardens at Kew in England (Ch. V, pp. 773–5). From the turn of the century onwards, however, the British Government gradually established a colonial agricultural service to promote and regulate the development of agriculture in the colonies. British-trained personnel were sent to the colonies and eventually research centres were established to investigate the problems of producing tropical crops of interest to Britain. An impetus was given to this work when the Empire Marketing Board of 1926–33 granted funds for commodity research. Similarly, in 1921 the Empire Cotton Growing Corporation was established to study the problems of cotton production. Progress was slow, but gradually the more recent advances in crop breeding and selection techniques were applied to the tropical crops. In time, also, the principles of plant nutrition and pest and disease control were similarly applied. Throughout the first half of the twentieth century this work was almost entirely limited, in the British colonies, to the crops of commercial importance that would be exported to Britain, and similar relationships existed in other colonial empires. Undoubtedly the most important developments during these years were the expansion of the acreage under plantation crops, including the

FIG. 13.18. Trials with a chain-gamma system (*left*) and herringbone system (*right*) of tapping Para rubber trees (*Hevea brasiliensis*) in the Economic Gardens, Singapore, *c.* 1909. The first major improvements on the Brazilian natives' methods of rubber tapping were developed in Singapore.

introduction of these crops into new colonies, and the selection of varieties best suited to specific local environments.

An added impetus to this transfer of crop production techniques also came when new industries were developed which provided new markets for crop products. Rubber for car tyres and vegetable oil for margarine are examples of this. The obverse occurred when substitutes for plant products were produced industrially as in the case of man-made fibres competing with cotton and jute. But that situation arose as a serious challenge to crop production more in the 1960s and 1970s than in the years before 1950.

The production of rubber was an industry that grew in parallel with the motor industry (Vol. V, Ch. 5). Until about 1914, world rubber supplies had come mainly from wild rubber trees in South America and to a lesser extent in Asia (Fig. 13.17) and Africa. However, during the first decade of the twentieth century British and Dutch entrepreneurs had recognized the potential for a rubber industry and had established rubber plantations in Malaya, Ceylon, and the Netherlands Indies. By the outbreak of the First World War these plantations were coming into full production and from then on the rubber industry came to be centred on Malaya and neighbouring countries. European plantations were expanded and native small-holders also established rubber trees. World production of rubber increased from 0·15m. tons in 1909–13 to 1·31m. tons by 1934.

Until 1900 the main raw material for the manufacture of margarine was beef tallow. Then in 1903 a process was patented for hardening oils by catalytic hydrogenation and it became possible to use vegetable oils as a basis for margarine. By the outbreak of the First World War a considerable proportion of vegetable oils and fats was being used in margarine production and thereafter the vegetable products became the predominant ingredients. Coconuts, oil palm fruits, cotton seeds, and peanuts (groundnuts) were the main sources of the edible vegetable oils before the First World War, but by the middle of the century soya beans had become the dominant oil seed crop. This was due to the adoption of soya beans as a crop in the United States in the inter-war years and its development there under a system of mechanized farming during the Second World War. Less dramatic, but also substantial, was the expansion of coconut, oil palm, and peanut production during the same period. It was an expansion based in part on the requirements of oil for margarine, and also of oil for soap and other products, and of oil seed residues for high-protein stock feed. The world production of margarine was increased from 0·4m. tons in 1900, when it was a tallow-based product, to 2·1m. tons in 1950, when some 60 per cent of the raw material was vegetable oils and fats.

V. CONCLUSION

During the first half of the twentieth century progress in plant breeding and nutrition, pest and disease control, and mechanization transformed the technology of crop production. This transformation was not a simple process, for each aspect of technology tended to interact with the other aspects: higher yielding crops needed better nutrient supplies; mechanical harvesters worked best in even, stiff-strawed crops. Furthermore, the biological and ever-changing nature of farming could not be forgotten; if it was, erosion, or pollution, or other gross distortion of the ecosystem was likely to result. Nevertheless, during this half-century man's ability to increase the production of food and fibre was greatly increased in terms of the land and labour used.

REFERENCES

[1] PETERSON, G. E. *Agricultural History*, **41**, 243 (1967).
[2] CROOKES, W. *The wheat problem* (3rd edn.). Longmans, Green and Co., London (1917).
[3] CAVERT, W. L. *Agricultural History*, **30**, 18 (1956).
[4] HARLAND, S. C. *Obituary Notices of Fellows of the Royal Society*, **9**, 259 (1954).
[5] RASMUSSEN, W. D. *Journal of Economic History*, **22**, 578 (1962).

[6] EVENSON, R. E., HOUCK, J. P., JR., and RUTTAN, V. W., in R. VERNON (ed.) *The technology factor in international trade.* National Bureau of Economic Research, New York (1970).

[7] WILLIAMS, R. F. *Journal of the Australian Institute of Agricultural Science*, **41**, 18 (1975).

BIBLIOGRAPHY

A century of technical development in Japanese agriculture. Japan F.A.O. Association, Tokyo (1959).

ÅKERMAN, Å., TEDIN, O., and FRÖIER, K. *Svalöf 1886–1946, History and present problems.* Carl Bloms Boktryckeri A.-B., Lund (1948).

BUNTING, A. H. (ed.) *Change in agriculture.* Gerald Duckworth and Co. Ltd., London (1970).

CALLAGHAN, A. R., and MILLINGTON, A. J. *The wheat indstry in Australia.* Angus and Robertson, Sydney (1956).

CARSON, R. *Silent spring.* Hamish Hamilton, London (1963).

DUNN, L. C. *Genetics in the twentieth century.* Macmillan, New York (1951).

HELD, R. B., and CLAWSON, M. *Soil conservation in perspective.* Johns Hopkins Press, Baltimore (1965).

JACKS, G. V., and WHYTE, R. O. *The rape of the earth. A world survey of soil erosion.* Faber and Faber, London (1939).

LARGE, E. C. *The advance of the fungi.* Jonathan Cape, London (1940).

LEACH, G. *Energy and food production.* International Institute for Environment and Development, London (1975).

MARTIN, H. *The scientific principles of crop protection* (4th edn.). Edward Arnold, London (1959).

MASEFIELD, G. B. *A history of the Colonial Agricultural Service.* Clarendon Press, Oxford (1972).

MEIJ, J. L. (ed.) *Mechanization in agriculture.* North-Holland Publishing Company, Amsterdam (1960).

MELLANBY, K. *Pesticides and pollution.* Collins, London (1967).

RASMUSSEN, W. D. *Readings in the history of American agriculture.* University of Illinois Press, Urbana (1960).

RUSSELL, E. J. *A history of agricultural science in Great Britain.* George Allen and Unwin, London (1966).

SCHLEBECKER, J. T. *Whereby we thrive. A history of American farming, 1607–1972.* Iowa State University Press, Ames (1975).

SIMMONDS, N. W. (ed.) *Evolution of crop plants.* Longman, London (1976).

SOUTHWORTH, H. *Farm mechanization in East Asia.* The Agricultural Development Council, Inc., New York (1972).

STILES, W. *Trace elements in plants.* Cambridge University Press (1961).

TAYLOR, H. C., and TAYLOR, A. D. *World trade in agricultural products.* The Macmillan Company, New York (1943).

UNITED STATES DEPARTMENT OF AGRICULTURE. *After a hundred years. The Yearbook of Agriculture 1962.* Washington (1962).

VAN STUYVENBERG, J. H. (ed.) *Margarine. An economic, social and scientific history, 1869–1969.* Liverpool University Press (1969).

WEEVERS, T. *Fifty years of plant physiology.* Scheltema & Holkema's Boekhandel en Uitgeversmaatschappij N. V., Amsterdam (1949).

WEST, T. F., and CAMPBELL, G. A. *DDT and newer persistent insecticides* (2nd edn.). Chapman and Hall, London (1950).

14

FISHING AND WHALING

G. H. O. BURGESS AND J. J. WATERMAN

LTHOUGH some countries began to collect statistics of their catches
and landings of fish towards the end of the nineteenth century, those
figures that are available are far short of what would be desirable to
give a detailed picture of developments throughout the world during the
period 1900–50. The first sustained attempt to collect fisheries statistics on a
world-wide scale was begun only in 1947 by the Food and Agriculture
Organization of the United Nations [1]. This valuable source, described in
the first volume as a preliminary experimental publication, gives statistics
from 1930 onwards for some countries, but contains large gaps and omissions.
Subsequent volumes in the series, which is an essential source for all concerned with world fishery problems, become increasingly comprehensive and
reliable in more recent years, but unfortunately only after 1950. As an indication of the scale of operations it may be noted that in 1948 the total weight of
wet fish landed in Britain was about one million tons, with a value of £45m.

In the absence of information showing in quantitative terms the changes
in world fishing patterns, the structure and composition of fleets, and alterations in the compositions of catches themselves, it has been necessary to
select developments that have seemed to have been the most important ones;
others, no doubt, would have placed their emphasis rather differently.

The objectives of every fishery have been, until recently, to obtain the
catch speedily and in quantity, and to deliver it in such a condition that the
purchaser will be prepared to buy it at an economic price. Significant developments have been judged to be those that have enabled these objectives to be
more readily achieved. Industrial fishing—that is, the deliberate capture of
fish as raw material for the manufacture of animal feed, a development that
has wrought profound changes in the whole pattern of world fisheries—
occurred mainly in the 1950s and later, and thus does not call for further
mention here.

It has been said that 'fishing is an industry for which it is difficult to draw the line between the medieval and modern periods. Short of the introduction of steam trawlers it is hard to find any change in the methods employed.' There is much truth in this statement. Most fishing methods in use in 1900 were evolved in prehistoric times and the majority were passive, as they often still are. Such methods are highly selective and catch only one or two species of fish. This is an advantage where the catch is to be preserved by traditional methods of smoking, salting, or drying (Vol. IV, Ch. 2), since treatment differs according to species, but it is a positive advantage to be able to offer a wide variety of species where the market is for fresh fish.

The story of fishing technology in the twentieth century is largely one of the extension of methods, usually already known, for the active pursuit of fish, with consequent large catches. At first sight this feature might be attributed to the introduction of steam, but in reality it is more closely related to the existence of suitable markets, as is well illustrated by the growth of trawling. Trawling, which is essentially the dragging of an open-mouthed conical bag of net along the sea floor, was known at least in the Middle Ages but its use increased dramatically in the North Sea during the first half of the nineteenth century. The trawl was used at first by sailing vessels and the growth of trawling resulted from the existence in Britain of large centres of population with the capacity to absorb large quantities of fresh fish [2]. Without adequate markets and suitable supporting facilities on shore, the incentive for change in fishing methods is often small and fisheries remain undeveloped, relying upon well-tried catching techniques and primitive craft and traditional methods of preservation. This explains why many fishery resources, particularly those in tropical and subtropical areas, have only recently begun to be exploited on a considerable scale.

I. PROPULSION

It is interesting to note that steam came relatively late to the fishing industry. At the beginning of this century the U.K., Japan, and the U.S.A. were the leading fishing nations. In the U.K. trawling industry, steam was introduced effectively in the late 1870s. By 1900 steam propulsion was ousting sail in all the major trawling fleets in Europe. The rate of change in the U.K. fleet, at that time the largest and most modern in the world, was remarkable. The Grimsby fleet can be cited as an example: two primitive steam trawlers were fishing out of the port in 1882, but ten years later there were 113; and by 1902 there were 424 of them. Sailing trawlers on the other hand declined

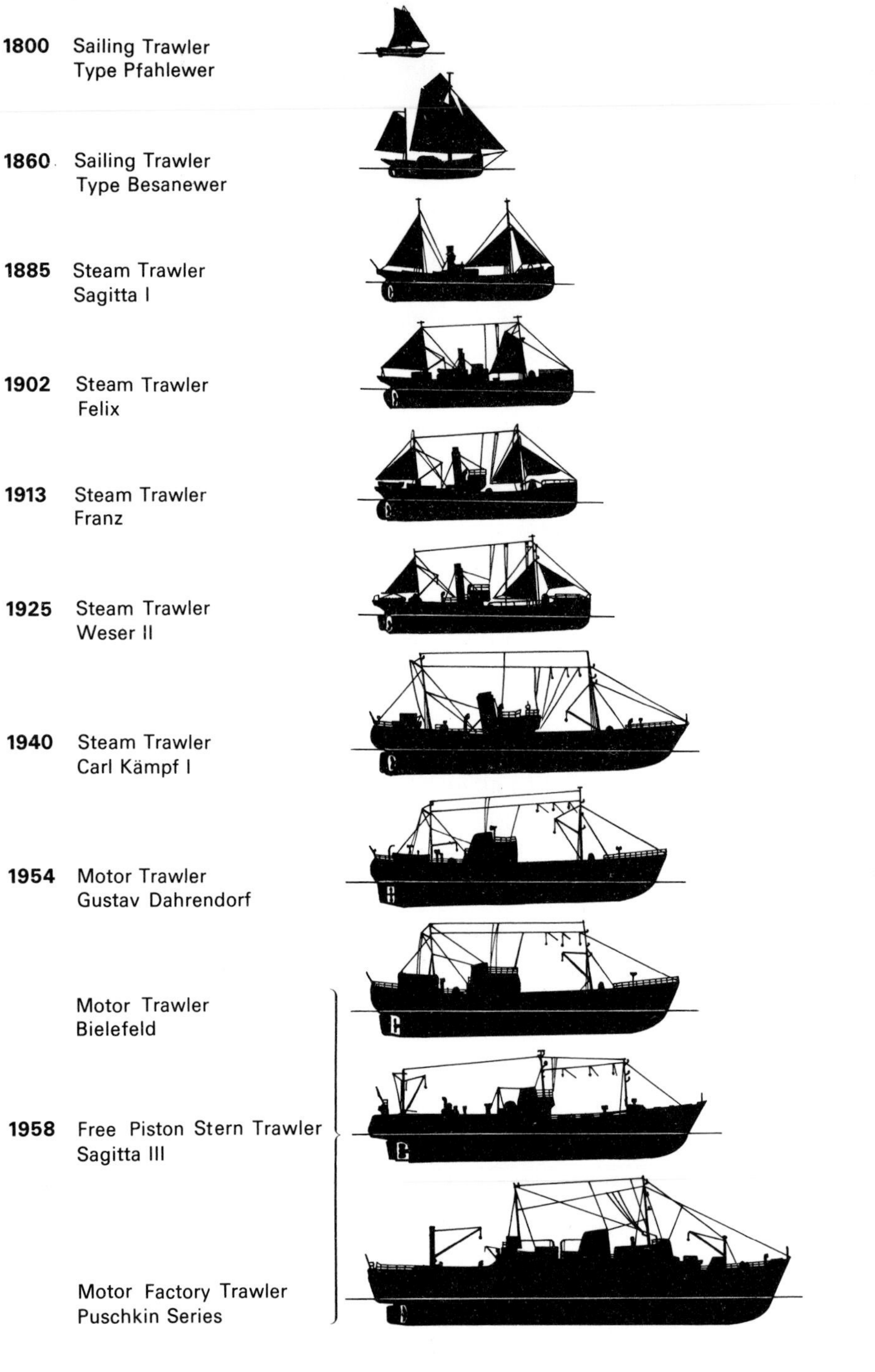

1800	Sailing Trawler Type Pfahlewer
1860	Sailing Trawler Type Besanewer
1885	Steam Trawler Sagitta I
1902	Steam Trawler Felix
1913	Steam Trawler Franz
1925	Steam Trawler Weser II
1940	Steam Trawler Carl Kämpf I
1954	Motor Trawler Gustav Dahrendorf
	Motor Trawler Bielefeld
1958	Free Piston Stern Trawler Sagitta III
	Motor Factory Trawler Puschkin Series

0   50   80 m

FIG. 14.1. Development of the European trawler.

from 820 in 1886 to 686 in 1892 and 29 in 1902. A major reason for this rapid change was the considerably higher productivity of powered trawlers.

At first the introduction of steam propulsion for vessels, using passive methods such as lining or drifting, was not so clearly seen to be advantageous, although capstans, operated by steam from auxiliary boilers, were commonly in use on U.K. sailing vessels in 1900. Where fishing was confined to a relatively narrow coastal region, the change to steam propulsion was slower, but even here the reliability and speed of powered vessels more than offset the greater capital and running costs. Steam propulsion was slowly eclipsed by the diesel engine in larger vessels, and sometimes by the petrol engine in smaller ones, but the change was less rapid and dramatic than the initial switch to steam, which was still in widespread use at the end of the period. The first Japanese diesel trawler was built in 1929 and the U.S. New England trawler fleet was largely diesel by 1930; in the more conservative U.K. fleet, however, no distant-water large trawlers used diesel engines before 1939. Engine power increased throughout the period: in 1900 the largest fishing vessels were no more than about 30 m in length and with a power of a few kilowatts; by 1950 they were over 65 m in length and 1000 kW or more in power (Fig. 14.1).

II. VESSELS AND EQUIPMENT

Space does not allow consideration in detail of the many types of vessel developed but it should be noted that trawling spread all over the world and vessels using seine and purse seine nets (see later) became of dominating importance in certain fisheries. Nevertheless, even in the most advanced fleets, passive methods such as drifting and line fishing remained of importance. A significant development, first apparent in the 1920s in North America, Japan, and Europe, was the use of large vessels as factory ships. These were entirely processing vessels equipped to freeze, salt, or can fish supplied to them by fleets of catchers, and they thus enabled fleets to remain on distant fishing grounds for longer periods than would otherwise have been possible. Factory ships and factory trawlers began to increase in importance after 1945 and are further discussed below.

On the whole, the associated developments in gear were rather of size than basic design. Trawls could be made much larger and the use of powered capstans and winches made it possible to lay more extensive lines or gill nets because they could be more readily brought on board again. The trawl remained the most important method of catching demersal (bottom-dwelling)

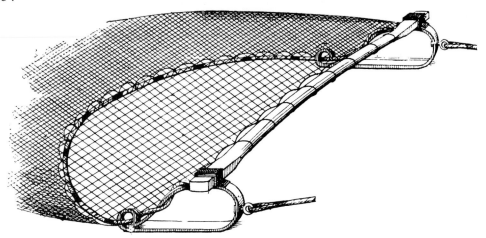

FIG. 14.2. The beam trawl.

fish but underwent some important modifications. The beam trawl (Fig. 14.2), which was suitable for use on sailing trawlers because its aspect on the sea bed was unaffected by the speed of towing, but whose size was in practice limited to a length of beam of about 18 m, was rapidly superseded by the otter trawl on steam trawlers. The otter trawl (Fig. 14.3) owed much of its early development in the late nineteenth century to the British trawling industry, but it was rapidly taken up and improved, particularly by other European countries. The mouth of the trawl was no longer held open by a beam; instead a large otter board was fixed on each side of the mouth of the net, the upper edge of which was kept open by a row of floats (Fig. 14.4). A separate towing warp was attached to one face of each otter board, and under the steady pull of the vessel the boards tended to move apart and hold the mouth open. There is no theoretical limit to the size of an otter trawl and in practice its size is limited by the power available to tow it. The otter trawl was subsequently modified in a number of ways, the first, and perhaps the most significant, modification being that due to the Frenchmen Vigneron and Dahl and introduced in the 1920s. They separated each otter board from the net by a long length of rope or bridle and found that this improved catching power, it was believed, by herding fish from a much wider area towards the net. Further developments have included the use of additional small otter boards to raise the opening of the mouth (Fig. 14.5). Attempts were also made to develop a trawl that would fish in mid-water rather than on the sea floor; a major difficulty to be overcome in such a net is to devise a method of maintaining its depth in the water. The first successful mid-water trawl

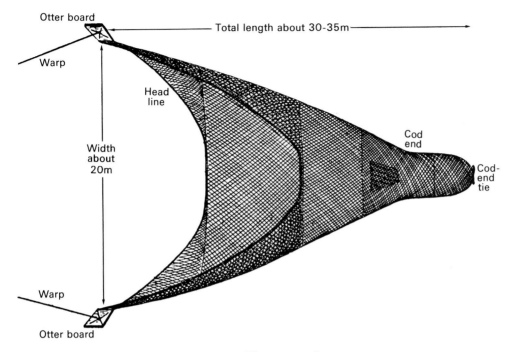

FIG. 14.3. The otter trawl.

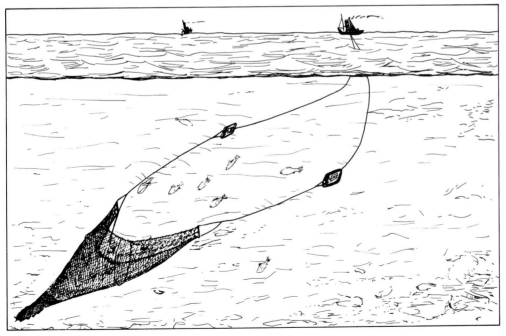

FIG. 14.4. The otter trawl in action.

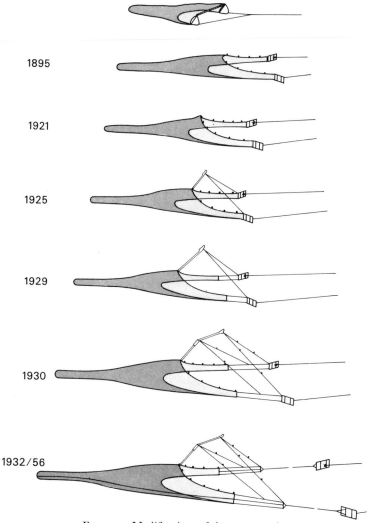

1895

1921

1925

1929

1930

1932/56

FIG. 14.5. Modifications of the otter trawl.

was perfected by Robert Larsen in Denmark in 1948. For the first time it became possible to hunt in the mid-water regions where fish were known to occur in large concentrations, but where hitherto they had been inaccessible [3].

Seining, a method of fishing which has already been mentioned, consists essentially of encircling shoals of fish near the surface with a wall of net. In the purse seine, a line is run through rings fixed at intervals along the bottom of the net and, after a shoal has been surrounded, the line is drawn so that the net is pursed and the catch cannot escape by swimming downwards.

Although the method dates back to the nineteenth century, it was mechanization of boats and winches that vastly increased its catching power and led to its adoption in many fisheries. Indeed, its ability to catch and hold very large quantities of fish made it the method of choice in the post-1950 period in many so-called industrial fisheries, that is, those catching fish specially for the manufacture of meal and oil. It is perhaps in the U.S.A. that the major developments in purse seining have been made. In the tuna, sardine, and mackerel fisheries of California, power brailing and the use of boom and winch allowed greater productivity and reduction in the size of crews. The first attempts at purse seining for tuna were made as early as 1914; by 1918, following the successful outcome of these experiments, the Californian bluefin (*Thunnus thynnus*) canning industry was established. By 1929 purse seiners of up to 100 tonnes capacity were being built for this fishery. In 1950 it was probably the most important method, in terms of size of catch, in use in the world's pelagic fisheries.

Nets were made of natural fibres until synthetic fibres became available in the late 1940s. The first synthetic fibres were developed following W. H. Carothers' discovery of nylon in the early 1930s. In fact, nylon was used experimentally for gill nets in the U.S.A. in 1939, but suitable materials were not commercially available before about 1948; thereafter, they rapidly replaced natural materials. Nevertheless, when they were first introduced there were many who believed that they had a strictly limited role in the construction of nets, for which natural fibres were believed to be better suited, and the potentialities of these strong and hard-wearing materials that did not rot in sea-water were at first appreciated by few people actually engaged in fishing.

Mention of boats and gear would be incomplete without reference to some of the aids to navigation that were developed during the period (Ch. 34). Most of these aids are similar to those used in other types of vessel and do not require detailed discussion here. At the beginning of the century methods of navigation were still much the same as they had been a hundred years earlier; most fishermen found their way about the seas on the basis of experience, supported by what help they could get from the magnetic compass and the lead line with tallow at the bottom of the weight to give an indication of the nature of the sea-bed. The sextant was of little value in the cloudy northern latitudes, even supposing that its use was understood. Communication between vessels, which became a matter of increasing importance as supply was geared to an ever more demanding market, was usually by flags.

The introduction of wireless around 1912 allowed better control of vessels and from about 1920 onwards was a significant factor in improving safety at sea, particularly on the larger boats which fished far afield. Perhaps the most significant advance was the introduction, at the end of the period, of sophisticated systems of radiolocation. These allowed accurate determination of position to an extent previously undreamed of. Skippers could now fish on prolific grounds but avoid known hazards such as wrecks which might destroy or damage gear. Radar, introduced to commercial fishing vessels in the late 1940s, helped to make a notoriously hazardous calling safer.

One aid to navigation which was adapted for the special purposes of fishing was the echo-sounder. This sounding instrument was originally developed for military use and later fitted on merchant vessels. It was observed in 1926, in studies on the Grand Banks, that signals could be received from shoals of fish when ultrasonic transmissions were used. The invention of the recording echo-sounder in about 1935 enabled permanent records to be obtained. Research in this area practically ceased during the Second World War, so far as fishing was concerned, but was taken up vigorously after 1945 in many countries. As a result instruments of greatly increased sensitivity became available and enhanced the capability of fishermen for finding and catching fish. Skilled men began to demonstrate an ability not only for finding fish but even for identifying their species from the nature of the echo [4].

III. PRESERVATION BY CHILLING AND FREEZING

A general trend, already noted, was to make vessels larger and more powerful. The reasons lay not only with the desire to increase the output per man and to fish more safely in distant waters, but also with the need to return rapidly to port before the first-caught fish was spoiled. The maximum economic speed, before fuel consumption begins to rise steeply, is directly related to the length of the vessel. The use of ice for extending the storage life of fish was certainly known in Europe in the eighteenth century, and even earlier elsewhere, but its widespread application to the catch at sea did not occur until the middle of the nineteenth century. Initially natural ice, obtained from frozen lakes in winter, was used but with the development of reliable and more efficient mechanical refrigeration plant, ice factories were erected at major ports [5]. In Hull, for instance, an ice plant capable of producing 50 tons a day was erected in 1891, but 14000 tons of natural ice were still required in 1902.

The preservative effect of ice is entirely due to the lowering of temperature. Fish from temperate waters carry a natural bacterial flora both on the skin and in the gut that grows well at temperatures only a few degrees above o °C. The spoilage of wet fish is in fact largely bacterial and pieces of sterile flesh will remain edible for many weeks when held at room temperature. The reduction of the temperature of fish caught in northern waters will dramatically extend the shelf life. Cod (*Gadus morhua*), for example, will be completely spoiled within 3 or 4 days at 5 °C but will last for over 16 days at o °C. The precise extent of these effects depends upon the species, how and where it was caught, and various biological factors [6].

When used for storing tropical fish, ice has an even more dramatic effect, and a shelf life of 6 or 7 weeks is not unusual. However, the costs of manufacturing and storing ice in tropical climates, and the much larger quantities required to absorb heat from the fish and surroundings, meant that ice contributed little to the growth of tropical fisheries before 1950, and even after that date most schemes favoured freezing rather than icing. Indeed, it has been the development of freezing at sea that has allowed the fuller exploitation of many tropical fisheries since 1950.

Even in temperate and Arctic regions an appreciable amount of heat can enter the hold by conduction through the ship's sides and the engine-room bulkhead, and special care has to be taken in the disposition of the ice to make sure that the catch does not become warmed. In the period before 1910 early attempts were made in Britain and France to fit refrigeration machinery which would help to keep down the temperature of the hold and so save the catch. Furthermore, the practice developed in the late 1930s of insulating the hold with some suitable material, though this has not become universal practice even now. Cork was always used, generally at a thickness of about 150 mm, in vessels fishing in the North Atlantic until the late 1940s when other materials, such as expanded vulcanized rubber and glass fibre, became available.

Even before the end of the last century, however, it was clear to some people that the limitations imposed by the use of ice were preventing the exploitation of more distant fishing grounds. A definitive history of the development of freezing at sea has still to be written, but it is clear that by 1900 workers in many countries were attempting to solve the many technical problems raised by this new technology. Frozen fish, if it is to be of good quality, must be rapidly frozen and stored at −30 °C or even lower. These requirements impose severe problems, especially when they have to be met on a vessel at sea. Early attempts were made long before research in the 1930s, mainly in Germany, Britain, the U.S.A., and Canada, demonstrated the physical and biochemical factors involved in making a good product. Not surprisingly, there were many failures.

Fish can be frozen by one of three methods: by immersion in cold liquid, usually brine; by contact with a refrigerated surface, or in a blast of cold air. All three methods have been used successfully at sea. The earliest technically successful experiments were in the 1920s, and even at this time two quite different approaches can be recognized. The first was to fit a plant into a fishing vessel; freezing is used in place of ice for preserving the catch, which is held in a cold store. The second is to build a large factory freezing vessel which is supplied with fish by catchers; some further processing is usually done before freezing and cold storage. Both methods have their particular attractions and both have proved to be successful lines of evolution.

In 1929 a German trawler, the *Volkswohl*, was fitted with brine freezing equipment with the aid of government funds. Freezer trawlers, all using various methods of brine freezing, were also built or equipped in other countries, including Britain, Italy, and Japan, but it was the French who were particularly active at this time, and they built a series of freezer trawlers until the Second World War prevented further experiment.

A French firm also operated an early factory ship, the *Janot*, in about 1925 and, in 1926, a British company began to use a converted meat carrier as a factory freezer on the Greenland fishing grounds. This was the *Arctic Prince*, later to be joined by a second vessel, the *Arctic Queen*. The fish, mainly halibut, was delivered to the boats by dories which fished by line. A somewhat similar venture was begun by a Newfoundland firm which acquired the *Blue Peter* (4300 tons) in 1927. Most of these experiments with factory ships had been terminated by the mid-1930s for economic rather than technical reasons. The world economic situation, sociological problems with crews, declining yields of fish on the grounds, and lack of a suitable infrastructure on shore capable of accepting and handling frozen fish, all no doubt contributed. Perhaps also the quality of the product fell short of expectations.

One successful use of freezing at sea was on the Pacific tuna clippers, whose development of purse seining has already been noted. Freezing experiments began at sea in the late 1930s, when it was shown that tuna could be successfully frozen in refrigerated brine tanks. Subsequently, after the fish was frozen, the brine was pumped away and low temperatures were maintained by the cooling coils in the walls of the tanks. The canneries were well

FIG. 14.6. The *Fairfree*, an experimental freezer trawler with stern ramp.

able to accept frozen fish, which in a number of respects was better fitted to
their requirements than unfrozen material. The tuna clipper is probably the
first commercially successful example of freezing at sea, and it is noteworthy
because it underlines the importance of a suitable market for the product.

In the late 1940s experiments were being vigorously made in many countries, and one or two of these should be mentioned. Norway, which before
1939 had had virtually no freezing industry, had developed a considerable
interest in the field, partly because of the acquisition of German plant erected
during the period of occupation. The *Thorland*, a large vessel designed to
produce 4500 tons of frozen fish annually, was operating in 1947, but was not
commercially successful. In its first year of operation it landed only 863 tons
of product and it became clear that output was severely limited by problems
of processing and packaging. Also in 1947, however, a British firm at that
time mainly concerned with whaling, Salvesen of Leith, was experimenting
with a converted minesweeper, the *Fairfree* (Fig. 14.6). This vessel was used
for trying out a number of novel ideas and one, the stern ramp, has proved
to be of outstanding significance. In traditional side trawlers the fishing gear

is handled over the side of the vessel, and during towing the warps are brought together into a block or towing eye on the side, but near the stern. This is necessary in order to allow the vessel to manoeuvre. The net is brought aboard over the side and operations such as sorting, gutting, and washing of the catch are done on the open deck. Some manhandling of the net is necessary and the work can be dangerous, especially in rough weather. Furthermore, dealing with the catch on an unsheltered deck can be extremely uncomfortable. The stern ramp, unknown at that time in fishing but universally employed on whaling factory ships, allowed the gear to be hauled and shot over the stern and made it relatively easy to mechanize the whole operation of handling the net. The catch could now be hauled up the ramp and lowered to a sheltered processing deck below the open fishing deck. In the *Fairfree* arrangements were made to fillet the fish, which were frozen in specially designed freezing equipment employing a combination of air-blast and contact freezing [7]. A similar method of freezing had in fact been developed in the freezing plant of a German factory ship of about 10000 tons, the *Hamburg*, which had been sunk in 1941 before it had undertaken a proper fishing operation. The *Fairfree* experiment was one of the first to show that fish could be caught and processed on the same vessel and pointed the direction in which the development of significant parts of the fisheries of the high seas must go if they were to continue to be viable. It had already been demonstrated in the 1930s that operations based on factory mother ships could be successful, although Japanese trials in 1948–9 were terminated because of problems, often encountered elsewhere, of transferring the catch at sea. With freezing at sea the length of a fishing voyage was no longer limited by the storage life of the catch in ice; vessels could sail to any fishing ground in the world, provided it was economic to do so, and remain there until holds were full. It was on the threshold of this development that the fishing industry of the world was poised in 1950; as a result countries such as the U.S.S.R., Poland, and East Germany were able to build large fleets and the whole pattern of world fishing changed dramatically.

IV. CONSERVATION OF FISHERIES

It was noted at the beginning of this chapter that reliable catch statistics are not available for most countries for much of the period. This information is needed in a modern fishery for many purposes, not least to enable assessments to be made of the effects of fishing on stocks. In many European fishing countries fishermen were complaining well before the end of the nineteenth century about overfishing, the damage done by one method of fishing to the prospects of another, or the destruction of spawn by certain types of gear (see [8] for some interesting early examples). Without reliable data, and the scientific knowledge needed for its interpretation, such complaints were unsubstantiated; and where more powerful boats and larger nets were being introduced, catches were actually rising, making it appear to some that overfishing could not be occurring. Nevertheless, by 1900 there had been five Royal Commissions or Committees of Inquiry in the United Kingdom on the state of the industry. The need had been established for some strong conservation measures and, equally important, sound scientific evidence on which to base forecasting and methods of control.

Interest in marine biology increased throughout the nineteenth century and many famous marine stations were established well before 1900. Indeed, the first marine station in the world was opened at Concarneau in France in 1859 [9]. One of the first tasks for biologists was to establish the life histories of some of the common commercial species; here the Norwegian G. O. Sars was a pioneer. Throughout the early years of the twentieth century biologists were identifying those stocks which were being seriously depleted, which fishing methods were most destructive, and which grounds might benefit from catch restrictions. More particularly, they began to find out whether the variations in the abundance of a stock were due to natural causes or to man the predator.

It soon became only too clear that control procedures would have to be extended beyond national territorial waters, since fish knew no man-made boundaries, and methods of conservation, if they were to be effective, would have to be agreed internationally. Bodies like the International Council for the Exploration of the Sea, founded in 1902, and commissions concerned with regulation of particular sea areas or species, attempted to apply the newly found knowledge in a rational manner. Their success has been limited. Controls have been determined for the good of the fisheries as a whole, but have often appeared to individual countries, ports, or fishermen to be of more benefit to others and so have been frequently disregarded. Fishermen are not particularly good at exercising self-restraint when faced with the opportunity of making a quick profit before times are hard again. There were many flagrant breaches of international agreements during the first half of the century. The view expressed here may appear unduly pessimistic, but events after 1950 give it some support; some well-known fisheries collapsed completely and others were seriously over-exploited because of unreasonable and virtually uncontrolled fishing.

Scientists by 1950 were developing more powerful methods of catch prediction based on a deepening understanding of fish population dynamics. From about 1925 onwards governments of the more advanced fishing nations began to set up technological laboratories concerned with research into the handling and processing of fish as food. It was on the results of this research that engineers drew in developing new techniques such as freezing at sea.

V. ORGANIZATION OF THE FISHING INDUSTRY

The interaction between the catching side of the industry and the supporting services on shore had, in the advanced fishing countries, already resulted in considerable concentration of the fleets at fewer ports by 1900. Docks rather than tidal creeks and harbours were needed in order to provide full quayside market facilities; fuel, ice, and engine and hull repairs; and skilled services for making and mending fishing gear and navigational and fishing aids. The crews, back from longer voyages than the inshore fisherman was accustomed to, needed brief but well-earned breaks before sailing again. Investment in sophisticated deep-sea vessels made it more difficult for skipper owners to remain independent; partnerships and, eventually, bigger private and public companies took over much of the business, and these in turn needed the administrative and technical services that were generally practicable only in the larger ports. At the start of the century, at least in Europe and North America, there were still many small fishing communities with fleets of individually owned or shared boats, but by 1950 the smaller bases had either ceased to function altogether, or had settled for smaller-scale operation in inshore or near-water fisheries.

VI. WHALING

The development of whaling (Vol. IV, Ch. 2, pp. 55–63) has mainly occurred independently of fishing and, apart from the stern ramp, there are few examples of one influencing the other. Whales have been caught in the north-east Atlantic from very early times and, even before the introduction of steam and the Svend Foyn harpoon gun in the last century, there were signs of decreasing catches. Processing of the carcasses was carried out on vessels or at shore stations, the latter, for example, occurred in Norway and Iceland, but whaling operations sometimes created opposition, perhaps in part because not all of the carcasses were used, but were allowed to rot. Fishermen in northern Norway objected that whaling was ruining the cod fishing and as a result the Norwegian Government in 1904 prohibited whaling in waters opposite Nordland, Tromsø, and Finmark and also would allow no shore-based operations for processing whales. Norwegian companies were forced to set up plants elsewhere, in Shetland, Faeroe, and Iceland in the north-east Atlantic and in many other countries as well; by 1910 they were to be found in Alaska, Chile, the Galapagos Islands, Western Australia, and many parts of Africa.

It had long been known that whales occurred in large numbers in the Antarctic and C. A. Larsen had attempted to set up a Norwegian company to exploit them after he visited the Antarctic on a German expedition in 1894. He eventually established an Argentinian company on South Georgia where he built his factory. His success encouraged others to follow. In the early days of the century most of the processing was on shore but two factors encouraged the development of factory ships. As whales were killed off in the region around the shore factories, it became necessary to hunt for them further afield. In addition, however, the British Government, concerned for the stocks of whales, introduced in 1908 a regulation making it compulsory to process the whole carcass and not to leave the stripped carcasses on the beach. Furthermore, the U.K. introduced a levy on every barrel of oil landed in the South American sector of the Antarctic.

The early factory ships were mostly converted merchantmen, and the whales were generally tied alongside and stripped in the water by flensers who walked on the carcasses as they cut off the blubber. It was possible to carry out this operation only in calm waters, and hence most vessels anchored in sheltered bays in South Georgia or the South Shetlands. This type of operation had almost ceased by the early 1920s and the Antarctic whale fishery had become largely pelagic, with catchers operating along the edge of the ice in international waters. The first factory ship with a ramp was introduced in 1925, and by the early 1930s factory ships were being purpose-built rather than being made by adaptation of merchantmen. These new factory ships were over 10000 gross registered tons and operations were highly mechanized. The carcasses were rapidly cut up and the oil was extracted in pressure boilers. This oil became an important constituent in margarine manufacture. Meat and bone meal, the other main products of whaling, were widely used for animal feed; for a brief period after 1945 when the world was chronically short of food, frozen whale meat was sold for human consumption in a number of European countries including the U.K. where, however, it was never fully accepted by the public [10].

Antarctic whaling went through what has become an all-too-familiar and depressing pattern of initial development and rapid expansion to be followed, more than a decade after the end of the period discussed here, by collapse. The signs of over-exploitation were well recognized by scientists long before 1950. The lesson was clearly to be learned here, as with other fisheries, that there is a limit to the size of even the renewable resources of the world. Rational exploitation of international resources is impossible without international agreements that are accepted by all and can be policed and enforced.

REFERENCES

[1] *Yearbook of fisheries statistics.* F.A.O. (1947).
[2] *Fish Trades Gazette.* 19 March, 21 (1921).
[3] Von Brandt, A. *Fish catching methods of the world.* Fishing News (Books) Ltd. London (1964).
[4] Cushing, D. H. *The detection of fish.* Pergamon, Oxford (1973).
[5] Cutting, C. L. *Fish saving.* Hill, London (1955).
[6] Burgess, G. H. O., Cutting, C. L., Lovern, J. A., and Waterman, J. J. *Fish handling and processing.* H.M.S.O., Edinburgh (1965).
[7] Lockridge, W. *Transactions of the Institution of Engineers and Shipbuilders in Scotland*, **93**, 504 (1950).
[8] *Report of Commissioners appointed to inquire into the Sea Fisheries of the United Kingdom.* II. Minutes of Evidence and Index. H.M.S.O., London (1865).
[9] Kofoid, C. A. *The biological stations of Europe.* Government Printing Office, Washington (1910).
[10] Vamplew, W. *Salvesen of Leith.* Scottish Academic Press, Edinburgh (1975).

BIBLIOGRAPHY

Fishing

Cutting, C. L. *Fish saving.* Hill, London (1955).
Hjul, P. (ed.) *The stern trawler.* Fishing News (Books) Ltd. London (1972).
Kristjonsson, H. (ed.) *Modern fishing gear of the world.* Fishing News (Books) Ltd. London (1959).
Traung, J.-O. (ed.) *Fishing boats of the world.* Fishing News (Books) Ltd. London (1955).
——. *Fishing boats of the world.* 2. Fishing News (Books) Ltd. London (1960).

Whaling

Mackintosh, N. A. *The stocks of whales.* Fishing News (Books) Ltd. London (1965).
Ommaney, F. D. *Lost leviathan: Whales and whaling.* Hutchinson, London (1971).
Vamplew, W. *Salvesen of Leith.* Scottish Academic Press, Edinburgh (1975).

15

COAL-MINING

SIR ANDREW BRYAN

THE general pattern of growth of the coal industry, worldwide, has been described in some detail in Chapter 8. To exemplify it by British practice, it may be remarked that in 1875, coal-mining, already one of Britain's oldest industries, had become one of its largest, employing some 500000 persons and producing an annual output of around 135m. tonnes. It was a vitally important and expanding industry because coal was the primary source of the energy required to meet and maintain the ever-increasing power requirements of the country's rapidly growing industries. By this time the two basic methods of working coal seams, bord-and-pillar and longwall (Vol. IV, Ch. 3), had been established, though as we shall see the first became increasingly identified with American practice.

I. COAL-CUTTING MACHINES

At the beginning of the century the undercutting of coal seams by power-operated machines, a major advance in the coal-getting process, was still in its infancy, although the first trials of these heavy machines began in the early 1850s and there were already in use three types of heavy mechanical coal-cutters: the disc, chain, and bar machines, respectively (Vol. IV, p. 82). The number installed, however, was so small that their output made little impact on that produced by the miners wielding hand-picks and using explosives and blasting.

The coal-cutting machines were introduced at a time when mining engineers, particularly in Britain and European countries, found it necessary to change the method of working from bord-and-pillar to longwall because the growing demand for coal compelled the exploitation of seams at greater depths. The higher rock pressures on underground workings owing to the increasing weight of overlying strata caused heavy crushing of coal pillars; greatly increased the difficulty of maintaining roadways and efficient airways; and made bord-and-pillar working more and more impracticable. Thus the early heavy coal-cutters in Britain were designed to work on longwall faces to mechanize the process of undercutting coal seams by hand-picks. During the last quarter of the nineteenth century, despite a growing realization of the need for them, the use of heavy power-operated coal-cutters extended very slowly, with the disc-type at first attracting most attention.

Factors that adversely affected the wider adoption of mechanization were the high cost of installation and maintenance of long compressed-air transmission lines, coupled with a high loss of power and general inefficiency in transmission. Fortunately, these factors were largely overcome in the closing years of the nineteenth century by the expanding use of electricity, a more versatile source of power with a much higher efficiency in transmission. The growing adoption of electricity stimulated mining engineers and mining machinery manufacturers to take a greater interest in the design and application of longwall coal-cutting machines. Their use extended, and by the dawn of the twentieth century 180 were operating in British coalmines: two years later there were 345.

In 1913, the annual output of British coalmines reached its peak of nearly 300m. tonnes. The years from 1900 to the beginning of the First World War in 1914, during which there was a heavy demand for coal both at home and abroad, saw significant progress in the design and use of heavy coal-cutters on longwall faces. Machines, now mostly chain-type—which cut a deep slot in the coal face—were designed on a three-unit principle comprising, respectively, a haulage section providing a variable cutting speed and a separate removal speed; a motor section; and a jib or gear head section. Machine maintenance improved and there were fewer breakdowns. In addition, advances in metallurgy (Ch. 18) brought higher-grade and special alloy steels, enabling more robust and compact coal-cutters to be built. These improvements not only increased the safety and reliability of the machines; they also helped to engender in mine management and workmen alike a favourable attitude towards the mechanization of coal-getting operations at the working face. In 1931, the President of the Miners' Federation of Great Britain is on record as saying: 'the Federation favours the fullest possible extension of labour-saving material' (Ch. 5, p. 103).

II. LONGWALL FACE CONVEYORS: CONVENTIONAL MACHINE MINING

With the increasing use of the longwall coal-cutter it soon became plain that to realize more fully its production potential, mechanical help was needed to ensure a more rapid and regular removal of the cut coal. The development of the face conveyor soon followed. The first of these, the Blackett conveyor, comprising an endless steel scraper-chain moving inside steel troughs, was patented by W. C. Blackett and installed in a Durham colliery in 1902. Three years later, Richard Sutcliffe introduced his endless-belt conveyor at a Yorkshire colliery.

The combination of heavy coal-cutter and conveyor on a longwall face became known as 'conventional' machine mining. Usually operating on a daily three-shift cycle, it meant that on the dayshift the machine-cut coal, (prepared for loading by blasting), was shovelled by hand on to the face conveyor. The preparatory work for the next day's coal loading—comprising coal-cutting, ripping and packing, withdrawing back supports, and advancing the conveyor—was spread over the succeeding two shifts of the cycle. While the combination of coal-cutter and face conveyor helped productivity, the introduction of the face conveyor brought notable advances in mining practice. By reducing the number of access roads to and from the longwall face, it cut the cost and risks of roadmaking. By automatically requiring a straight line of face, it eliminated stepped faces with their 'corners' where the roof was unduly stressed and weakened. Finally, it made easier the introduction of systematic and safer systems of roof support, thus significantly reducing the risk from the most prolific source of accidents in mines at that time: falls of roof at the working face.

During the first half of the present century the percentage of coal cut by machine in Britain rose from $1\frac{1}{2}$ to almost 80 per cent of the total output, while that for coal carried underground on conveyors rose from nil to 80 per cent. Despite these significant increases, until the Second World War the technological advances made in coal-getting methods at the working face in British coal mines were, with a few notable exceptions, confined to marginal engineering improvements in the design and construction of coal-cutters and face conveyors. In respect of conveyors, although there was a big extension in their use on the face, and especially in roadways, there was no outstanding advance in design. This, however, did not apply to coal-getting.

In the breaking-down and loading of coal on a longwall face, an outstanding development was the introduction in the early 1930s of the Meco–Moore cutter-loader. In essence this was a coal-cutter which trailed a loading unit behind it. As the machine travelled on the floor between the conveyor and the face, it undercut the coal, which was then prepared for loading by blasting. On reaching the face-end the machine travelled back again, loading the prepared coal on to the face conveyor by means of its loading unit, which had been swung into line with the cutting jib. Although this pioneering machine excited great interest, it was not an unqualified success. The double journey for one 'cut-and-fill' was a drawback, but nevertheless a promising start had been made in the development of that ideal of mining men: a machine that would cut and load coal simultaneously.

III. MINING PRACTICE IN WESTERN EUROPE

In Europe, where the longwall method of working was widely adopted, development (until the Second World War) related more closely to British than to American practice. But because European coal seams, compared with those in Britain, were softer and more steeply inclined, power-operated machines for breaking down coal at the face took the form of hand-held percussive picks powered by compressed air. Face conveyors were developed to operate on the 'jigger' principle, in which the coal is jerked along a steel trough by a reciprocating motion.

As in Britain, with the coming of the Second World War much greater attention was devoted, especially in Germany, to finding a mining system in which the coal was not only broken down by machine but was loaded by the same machine, preferably simultaneously. This led to a major break-through in German longwall mining: the development of the coal plough. This was a steel pick or wedge device tracked to ride on the face-side of an armoured face conveyor positioned close against the coal face. Shuttled repeatedly across the face by rope-haulage, the plough wedges from the face thin slices of coal which are dropped on to the conveyor, thus achieving the simultaneous cutting and loading of coal. As the coal is stripped off, the flexible conveyor is jacked close into the coalface ready for the return of the plough.

IV. MINING PRACTICE IN THE U.S.A.

Coalmining in the U.S.A. during the last quarter of the nineteenth century made notable progress in mechanizing coal-cutting. In American mines, because of vast reserves and favourable geological and mining conditions, coal was mined in relatively thick and flat seams, lying at shallow depths, to which access was gained by roadways driven in the seam from an outcrop on the side of a hill or valley. Under such conditions the bord-and-pillar method of working (known in North America as room-and-pillar) was almost exclusively adopted.

FIG. 15.1. Sorting the coal by handpicking at a coalmine in Virginia, U.S.A., *c.* 1910.

Labour shortage, and the disinclination of Americans to employ manual labour on tasks for which machines could be contrived, led to the early development of machines which could undercut coal. By 1900, 25 per cent of the American bituminous coal output was undercut by machine; by 1947, this had risen to 90 per cent. Since undercut coal was prepared for loading by blasting, machines were also developed for drilling shot-holes. Whilst these machines helped production, their potential could not be realized so long as the hand-loading of coal continued and so, in the 1920s, attention turned to the development of loading machines. In 1923, 1 per cent of American output was loaded mechanically; in 1939, 21 per cent; in 1950, 70 per cent.

The sequence of work in an American mechanized room (or heading) was as follows: first, undercutting, and in some instances also overcutting and shearing, in the coal seam to a depth up to 3·4 m; then drilling shot-holes in the coal face; blasting the coal by explosives or by a safe alternative, such as

FIG. 15.2. Shortwall coal-cutter undercutting a heading in a coalmine in the U.S.A.

Cardox (compressed carbon dioxide); and loading the prepared coal by machine into mine cars, shuttle cars, or conveyors which bridged the gap between the face loader and the outbye transport system. To perform this sequence (except for blasting), a combination of machines (from the formidable array by then available) was selected to match the conditions and methods of the particular mine. These machines included shortwall coal-cutters for undercutting (Fig. 15.2); universal arc-wall-type coal-cutters for undercutting, overcutting, and shearing (Fig. 15.3); shot-hole drilling machines; loading machines of the duck-bill, chain, jigger, or gathering-arm type, which picked up the prepared coal and delivered it to the outbye coal transport system; and a timber-sawing and support-setting machine. Most machines were capable of moving under their own electric power. Trackless machines, caterpillar mounted or rubber-tyred, were introduced wherever floor conditions allowed because they were manoeuvrable and eliminated the unproductive labour and capital cost of laying tracks into every working face (Figs. 15.4 and 15.5).

With the use of so many separate machines the chances of a breakdown which could halt the whole system greatly increased. This led to the development of a new type of combination machine, the 'continuous miner', capable

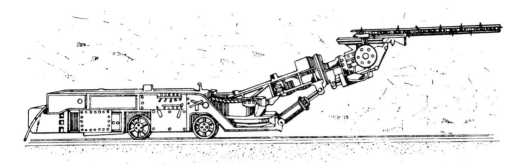

Fig. 15.3. Joy 7-AU coal-cutter: track-mounted.

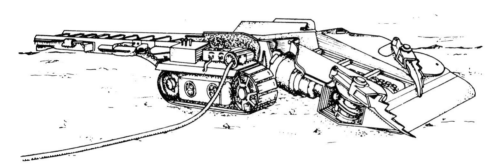

Fig. 15.4. Joy 12-BU loading machine: caterpillar-mounted.

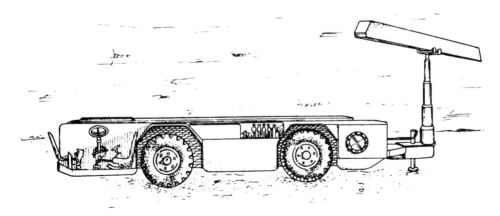

Fig. 15.5. Joy timber-setter.

of breaking the coal from the face and loading it on to the outbye transport system in one continuous operation, thus eliminating the orthodox cycle of cutting, drilling, blasting, and loading. By 1950, 12 types were in existence or under development. Undoubtedly, American practice contributed much to the efficiency of room-and-pillar mining in a number of countries.

V. SIMULTANEOUS CUTTING AND LOADING IN BRITAIN

The outbreak of the Second World War emphasized the need to increase the output of coal at a time of acute manpower shortage; brought an immediate demand for machines that could load coal; and redirected attention to the further development of the Meco–Moore cutter-loader. A trial of this machine carried out at Sneyd Colliery (Staffordshire) just before the war, and especially subsequent trials (in which an AB coalcutter was mounted on a Meco–Moore loader) carried out at Bolsover and Rufford Collieries (Nottinghamshire) in 1942, proved that simultaneous cutting and loading was practicable. A joint reconsideration of the problem by the two manufacturers concerned quickly resulted in the development of the AB–Meco–Moore cutter-loader, the first longwall machine to cut and load coal simultaneously (Fig. 15.6). Developed in 1943, it achieved an immediate and remarkable success.

FIG. 15.6. Meco–Moore cutter-loader at work on a longwall face with individual hand-operated hydraulic props.

Meanwhile, apart from the American continuous-miner, the German combination of coal plough and armoured face conveyor, and the new AB–Meco–Moore cutter-loader, other relatively successful types of cutter-loader were developed. They included, in Canada, the Dosco and, in Britain, the Logan Slab-Cutter, Uskside Miner, Gloster-Getter, and two other machines working on the principle of the wedge: the M. and C. Samson Stripper and the Huwood Slicer-Loader. Further, in addition to the several American loading machines (previously listed), two other loading devices were also developed in Britain: the Shelton and Huwood Loaders, designed to load coal on a longwall face after it had been machine-cut and the coal prepared for loading by blasting.

The end of the Second World War saw significant changes in British longwall coal-getting practice. Although the wide-web machine helped production, it also introduced factors which weakened the roof along the working face, thus increasing the risk of falls of ground. It demanded a wide stretch of roof across the face to accommodate all the equipment: namely roof supports behind the face conveyor, the face conveyor itself, and the coal-cutting machine. Further, the relatively deep undercut (1·6 m or more) inevitably left a wide span of roof between the front row of supports and the back of the undercut. This remained unsupported for a length of time determined by the speed of loading and the time taken to advance the conveyor and roof supports.

These drawbacks were quickly overcome. Trials of the German armoured face conveyor and coal plough showed that while the hard nature of British coal severely restricted the application of the plough, the use of this conveyor offered great scope for increased production with greater safety. Being narrow, flexible, and strongly constructed, it needed little face-room, could be advanced without dismantling, and provided a ready-made track for a narrow-web coal-cutter. The combination of the German conveyor and narrow-web conveyor-mounted cutter-loader offered outstanding advantages: a big reduction in the span of roof from the waste-edge to the back of the undercut and, more particularly, a reduction in the roof span from the back of the face conveyor to the back of the undercut large enough to permit a prop-free-front. This allowed a conveyor-mounted machine to cut the coal-face without interruption from roof supports. Further, because the flexible conveyor could be snaked against the coalface close behind the cutter-loader, roof supports could be advanced quickly to give prompt support to the newly exposed roof.

FIG. 15.7. Shearer-loader mounted on an armoured face-conveyor on a longwall face with powered roof-supports.

The advent of the armoured flexible conveyor gave a strong impetus to the development of narrow-web cutter-loaders. This was quickly met, in the early 1950s, by the development of two machines, the Anderton Shearer and the AB Trepanner. Basically a conventional coal-cutter on which the cutter-jib was replaced by a shearer-drum fitted with picks, the Anderton Shearer (Fig. 15.7) cut a web of coal, about 0·5 m deep, from the face. As the machine travelled along the conveyor, most of the sheared coal fell into the moving conveyor; the face-side spillage was picked up and loaded by a plough attached to the underframe of the machine. The shearer quickly achieved a spectacular success. The other successful narrow-web machine, the AB-Trepanner (Fig. 15.8), was based on the application of the auger principle to a longwall face. Important improvements on the design of these two machines, extending their range in action and permitting bi-directional shearing, quickly followed and added to their versatility and efficiency. A new phase of longwall face mechanization had begun.

VI. HYDRAULIC ROOF SUPPORTS

The new system greatly increased the rate of face advance and called for stronger and more flexible roof supports, incorporating ready and reliable

FIG. 15.8. Trepanner at work on a longwall face with powered roof-supports.

means for rapid setting, pre-loading, speedy withdrawal, and variation of length. This led to an extended use of friction-type yielding props and later, but more importantly, to the introduction by Dowty of the highly efficient hydraulic prop. From this came the development of self-advancing powered roof-support systems which eliminate much of the physical effort required in setting supports. Another important direction, in which hydraulic supports made for safer mining, was that, by providing a strong 'breaking-off' force at the waste edge, they led to a rapid extension of total caving of wastes and the abandonment of strip packing.

VII. UNDERGROUND HAULAGE AND TRANSPORT

From 1875 until the early years of the nineteenth century, little change occurred in established methods of underground haulage. At or near the working face pony-haulage, hand-tramming, and, where practicable, self-acting inclines were the rule. On main roadways wire-rope haulage, employing main-rope, main-and-tail-rope, or endless-rope systems were employed. A major change came in 1902 with the introduction of the conveyor on longwall faces, and its use soon afterwards in gateroads. Improvement in rubber belting in the 1920s brought a marked extension of main-road belt-conveying,

FIG. 15.9. Trunk-belt conveyor transporting coal along a main haulage road supported by steel arches.

leading to trunk-belt installations (Fig. 15.9). In the 1940s, the introduction of the cable-belt conveyor, in which the belt, supported on and carried along by wire ropes, was used only as a carrying medium, marked an important advance. Nevertheless, despite these achievements, 90 per cent of deep-mined output in Britain was transported by wire-rope haulage even in 1947. A major advance, the development of much stronger and fire-resistant belting, followed the disastrous conveyor-belt fire at Cresswell Colliery (Nottinghamshire), in 1950; this led to a dramatic increase in belt-conveying.

Locomotives, which can operate on gradients of less than 1 in 25, were commonly used on the Continent and in the U.S.A., but found little favour in Britain. Diesel types are used on the steeper gradients,and battery-powered locomotives, reliable and easy to maintain, on the flatter parts. The use of the trolley locomotive has been severely limited by considerations of safety and capital cost.

Underground haulage and transport, of course, means more than transporting coal and dirt. It has also to provide for the carriage of miners, the distribution of materials, and the transfer of plant and equipment. For these duties, conveyor transport, efficient in coal clearance from mechanized longwall faces, has not proved satisfactory. To accomplish them, auxiliary rope haulage is often necessary, supplemented where appropriate by overhead monorail systems near the face; by coolie cars and trackless vehicles; and, on occasion, by a form of ski-lift to help men to travel on steep roadways. Studies of transport problems indicated that a colliery layout should, among other things, be designed to ensure quick and continuous man-riding from the surface to the working place, and that the concept that main roads should be reserved for coal transport, and other roads for men and materials, was outmoded.

VIII. SHAFT WINDING

In Britain, until the 1920s, there was little change in the conventional system of shaft winding established in the last quarter of the nineteenth century. This comprised steam-driven winding engines, fitted with parallel drums carrying steel wire ropes attached to cages running in shafts of relatively small diameter. Few balance-ropes were used, payloads were small, and winding speeds rather high. Tubs were usually small and were manually handled during the loading and unloading of cages. As the use of electricity extended, and the capacity of electricity supply systems increased, so also did the number of electrically-driven winding engines. Between 1911 and

1940 the number increased from 17 to 631. The earlier installations were driven by d.c. motors but by the 1920s the geared a.c. induction motor predominated.

Nationalization of the industry in Britain in 1947 brought many major reconstructions to improve shaft-winding capacity; to increase pay-loads, either by replacing cages with skips or by using multi-deck cages carrying large capacity mine cars; and to mechanize the loading and unloading of skips or cages at shaft top and bottom. The end of the war saw a more general acceptance for winders of the friction (or Koepe) drive type, which had already been adopted at many continental collieries for nearly half a century. By 1950 there was a marked increase in the use of friction winders—two-rope, three-rope, and four-rope types—suitable for handling large payloads in deep shafts and for automatic operation. Their introduction required new designs for winders and headgear and called for the development of special decking plant; for example, hinged platforms and cage-steadying gear for shaft top and bottom, and for inset working. As depths, outputs, and loads increased, and the limit of weight which could be safely suspended on the largest practical size of single rope (with drum or friction wheel) had been reached with a net load of 10–12 tonnes at depths of 1000 m or thereabouts, the value and versatility of the multi-rope winder became well established.

IX. PUMPING

In mine drainage, the final decade of the nineteenth century and the first decade of the twentieth saw the replacement of the large steam-driven beam-operated pumps (with their tall engine-houses and their heavy wooden rods suspended in the shaft) by new-type centrifugal pumps. Thereafter developments have largely been a matter of improvements in the pumps themselves rather than in the practice adopted by mining engineers. Apart from ensuring that pumps were compact, durable, reliable, and efficient in design and construction, and endeavouring to obtain as many manpower economies as possible by the use of remote or automatic control, mining engineers concentrated on improving operating conditions and reducing running costs. For inbye or gathering pumps, this included measures to prevent pumps blocking by the provision of suitable sumps and simple water-level controls. For shaft-pumps, it entailed making adequate provision against damage from corrosive water. Notable advances were the development of the borehole pump and the introduction of the submersible centrifugal pump suitable for automatic operation in disused shafts.

X. VENTILATION AND LIGHTING

The aim of mine ventilation is to circulate economically and efficiently sufficient air through the mine to dilute and remove harmful gases; to sweep away air-borne dust from all working places; and to accomplish these with an air-flow of such velocity, temperature, and humidity as will provide safe and comfortable environmental conditions for all persons at work underground. During the nineteenth century mine-ventilation systems tended to develop haphazardly (Vol. IV, Ch. 3, p. 92). Since then, however, because much has been done to obtain a better understanding of basic principles, the proper planning of a mine ventilation system—greatly facilitated by the development of ventilation models in which the airways of a system were simulated by electrical resistances—was now recognized as an essential part of the planning of the mine as a whole. Models (later superseded by digital computer systems) have proved especially helpful in the solution of ventilation network problems, as, for example, in determining the resulting air-flows when the ventilation systems of two or more mines are connected together.

For nearly a century from its invention by Humphry Davy in 1815, the flame safety lamp was the symbol of the coal miner (Vol. IV, Ch. 3, p. 94). Early in the twentieth century it began to be displaced by the electric safety hand lamp which, in turn, was replaced in the 1930s by the far more effective electric cap lamp. The incentive for improvement came from the need for a much higher standard of illumination to improve the health and safety of the miner and, in particular, to minimize the incidence of nystagmus, an eye disease arising from poor illumination which had become a major occupational hazard of British coalmining. With the wide adoption of the cap lamp, nystagmus practically disappeared. It had long been realized, of course, that the lighting afforded by the miner's lamp was unable to provide the degree of illumination necessary for safety and efficiency at particular work locations on longwall faces and on haulage and travelling roadways. Progress in the lighting of roadways from mains electric supply came quickly and there was a wide expansion of its use after the First World War, but not so on coalfaces. In the 1940s, however, several experimental installations showed that coalface mains lighting was practicable. Machines themselves do not need light, of course, but there are periods of installation, inspection, repair, and maintenance when efficient lighting is a necessity.

XI. SAFETY AND HEALTH

Much has been done since 1875 to meet the many, and often unique, hazards of coalmining and to reduce the toll of accidents and incidence of industrial disease. The sources of hazards include falls of ground; explosions of flammable gases and dust; breathing of harmful gases and airborne dust; inrushes of water; underground fires resulting from spontaneous combustion and other causes; use and handling of explosives; mishaps in underground transport and shaft-winding operations; contact with moving machines, chains, belts, ropes, tubs, and trams; electrical faults; and other miscellaneous causes. The measure of success achieved in raising standards of health and safety, has come mainly from statutory regulation and inspection; voluntary codes of practice; higher standards of basic education and training and refresher courses for workmen and management; improvement in the quality and degree of supervision; more scientific and engineering research; application of advanced technology; joint consultation on matters affecting safety and health, and planned involvement of workmen and under-officials in new projects; safety propaganda and campaigns to overcome apathy in matters affecting health and safety; and improvement in social and living conditions to raise morale in mining communities.

Experience and inquiry suggest that a major contribution to healthier and safer coalmining lies in the efficient application of advanced technology. This has as its aims improving the environmental conditions in places where men work or pass to an extent that a casual human error does not inevitably lead to an accident; applying remote control and automation to reduce and, if possible to eliminate, manpower exposure in recognized danger areas; and designing inherently safe machines and safe methods of mining. This implies that in machine designing and mine planning, each element in the system, and its interaction with other elements, is considered comprehensively to determine their combined effect on safety as a whole and that inherent dangers are recognized and either 'designed out' or appropriately guarded against. In short, the technological way to higher standards of health and safety lies in the use of mining machines and methods of work in which safety is either inherent or inbuilt.

BIBLIOGRAPHY

ANDERSON, F. S., and THORPE, R. H. A century of coal-face mechanization. *Mining Engineer*, No. 83, 775–85 (1967).

BRYAN, ANDREW, and HARLEY, F. H. (eds.) *A survey of mining engineering.* Industrial Newspapers Ltd., London (1959). Attention is drawn to the following contributions:

ADCOCK, W. J. *Strata control and support development*, pp. 29–34.

ATKINSON, F. S. *Coal-cutters and power-loaders in British mining*, pp. 21–8.

BROMILOW, J. G. *Review of present mine ventilation practice*, pp. 5–15.

DUDLEY, N. *Winding in prospect and retrospect*, pp. 16–20.

GRIERSON, A. *Conveyor haulage in mines*, pp. 48–54.

ROBERTS, A. *Progress in mine lighting*, pp. 94–7.

SAUL, H. *Developments in pumping practice in British mines*, pp. 73–9.

CROOK, A. E. Presidential Address: The coal mining industry—Change, progress and consequence.' *Mining Engineer*, No. 67, 414–24 (1966).

GALLOWAY, R. L. *Annals of coal mining and the coal trade.* Colliery Guardian Co., London (first series, 1898; second series, 1904).

HOWSE, R. M., and HARLEY, F. H. *History of the Mining Engineering Company Limited, 1909–1959.* MECO, Worcester (1959).

LUPTON, A., PARR, G. D. A., and PERKIN, H. *Electricity applied to mining.* Crosby Lockwood, London (1903).

MINISTRY OF FUEL AND POWER AND BRITISH INTELLIGENCE OBJECTIVES SUBCOMMITTEE. *Technical report on the Ruhr coalfield, by a mission from the Mechanization Advisory Committee of the Ministry of Fuel and Power.* H.M.S.O., London (1947).

Report of a productivity team representing the British coal mining industry which visited the United States of America in 1951. Anglo-American Council on Productivity (UK Section), London (1951).

16

PRODUCTION OF PETROLEUM
OIL AND NATURAL GAS

H. R. TAINSH AND S. E. CHURCHFIELD

THE production of oil, mainly for illuminants and lubricants, had become well established by the end of the nineteenth century (Vol. V, Ch. 5), but it was the first half of the twentieth century that witnessed the growth of the new dynamic industry and the widespread geographical development of a great new source of energy which led to profound changes in many countries. Primarily, oil was the objective in early exploration; until recently, gas discoveries were largely incidental to the search for oil.

World production of oil in 1900 was about 150m. barrels, of which half was contributed by Russia; in 1950 production had risen to 3800m. barrels, of which 55 per cent was North American. The period of growth was largely initiated by a well at Spindletop, near Beaumont, Texas, which started to gush oil at about 80000 barrels per day in January 1901. In early years procedures were simple and wasteful, to a great extent through ignorance of the physical forces involved and because of a lack of proper equipment. In later years, however, there was great progress in the design of equipment, in operational techniques, and in the development of the oil-finding sciences and petroleum engineering.

By 1900 the anticlinal theory of the accumulation of oil in porous sandstones was well established. Geological exploration was being undertaken in many parts of the world, but many wells were still located close to petroleum seepages. With increasing demand, attention was turned to the search for oil in areas where there was no surface evidence, using geophysical techniques which depended upon accepted physical principles.

The first gravity surveys were made, in 1915, in Czechoslovakia with the Eötvös torsion balance. In the 1930s more efficient survey tools, the gravimeters, were developed and these have since been widely employed.

Some seismic exploration was conducted in Europe in about 1919, and the successful use of refraction surveys in the American Gulf Coast in 1923, in Mexico, and later in Persia (Iran), led to considerable employment of this technique. By the early 1930s large areas were being explored by seismic reflection methods, which have subsequently become the most widely used, and have led to many discoveries both on land and beneath the sea. Over the years there were great advances in instrumentation, in surveying procedures, and in techniques for the presentation and analysis of data.

I. DRILLING

Initially, wells for oil were dug by hand in several countries and in Burma, for example, such operations were continued even into the late 1930s. A variety of mechanical drilling techniques were introduced and improved, but all essentially involved one of two basic concepts. The first employed percussion—the cable-tool method—and the other a rotary grinding action.

Cable-tool drilling. The percussion method drills by pounding the rock. A drilling tool, or bit, is attached to a weighted cable suspended from a rocking arm. A reciprocating vertical motion is applied to the cable and, with suitable adjustments of the frequency and amplitude of the movement and of the length of the cable, the rocks are progressively broken up.

In China, brine wells were drilled by cable-tool methods centuries ago, and by 1900 they were being dug to 3000 ft.; the techniques were carried on into the twentieth century with only minor modifications. The derrick was constructed of wooden poles bound with bamboo. The hoist was a horizontal bull wheel turned by water buffaloes, with a drilling line made of bamboo strips bound with hemp. The drilling bits and stems were the only iron materials used. The reciprocating movement was achieved by several workers stepping on to and off a wooden spring board (the walking beam). Many tools were devised for the recovery of broken lines, lost bailers, or drilling bits. Surface water was excluded in the first 125 ft with wooden pipe, but thereafter drilling was in uncased hole.

Many of these brine wells also produced natural gas, which was transmitted over short distances through bamboo pipes for local lighting and heating. This Chinese system developed many cable-tool techniques perhaps centuries before their use in other regions, and it is uncertain how far later developments elsewhere in the world were influenced by them.

The first well drilled purposely for oil in the U.S.A. is traditionally believed to be the Drake well, drilled in Pennsylvania to 69 ft in 1859.

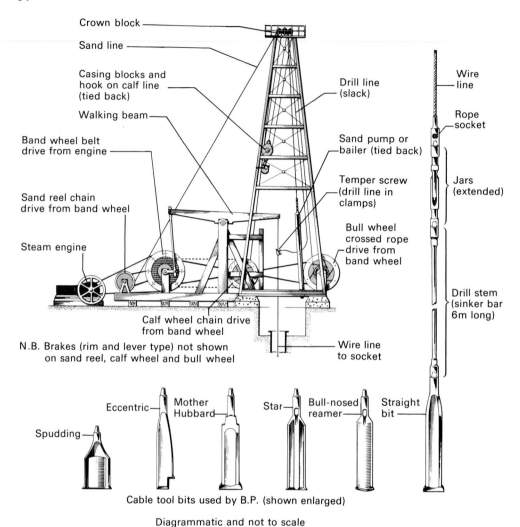

FIG. 16.1. A standard cable-tool drilling rig and bits, early 1900s.

Thereafter hundreds of wells were drilled and the standard cable-tool rig had evolved by the 1880s (Figs. 16.1 and 16.2). The four-legged wooden derrick was 72 ft or more in height. Only small amounts of iron were used in early rigs, although by about 1890 a complete steel rig was designed, and tubular derricks were in use early in France.

Single-cylinder steam engines of 20 horsepower or more generally provided the power to the wooden band-wheel and thence to the other wheels. The drilling cable was wound on to the bull-wheel shaft, the casing line on to the

FIG. 16.2. A cable-tool rig of the type shown in Fig. 16.1, under construction in Persia, 1909. The steel derrick is complete; the steam engine on the right, the bull-wheel, and the walking-beam are being installed.

calf-wheel shaft, and the bailing line on to the iron sand reel. The crank attached to the band-wheel had a wide selection of pin holes so that the correct rocking motion could be imparted to the large horizontal walking-beam. The steel drilling cable was clamped to the base of a screw assembly (the temper screw) suspended from the walking-beam, and as drilling progressed the bit was lowered in the hole by turning the temper screw through a thread box in the assembly. Provision was thus made for variation in frequency, length of travel, and degree of impact of the bit in order to deal with varying drilling conditions.

The casing line was passed over the crown block pulleys to a travelling pulley block designed to accommodate the number of lines required to raise the casing suspended in an elevator collar. In wells where the hole was liable to collapse, it was usually necessary to start drilling at the surface with a large hole and to run pipe of perhaps 16 inches diameter as deep as possible, joint by joint (average length 20 ft), as drilling and under-reaming proceeded. Casing strings of successively smaller diameter were then run to consolidate progress. In early Russian wells the first casing was as much as 42 inches in diameter, made of riveted sheet iron; the sections, only 56 inches long, were also connected by rivets.

For the successful recovery of lost tools, pipe, or cable a variety of 'fishing' tools were developed over the years: hooks, spears, knives, etc.

Figure 16.1 shows some standard drilling bits. Only modest changes were made in the shape of these tools in the period with which we are here concerned, but there were considerable improvements in materials and in the perfection of manufacture. In early years in many countries it was customary for the crude tools to be dressed on the drill site. Later tools, however, were made to standard design, sizes, joints, and quality of steel; bits were then generally dressed in field workshops under regulated conditions.

Normally only a little water was kept in the hole, sufficient to mix the pulverized rock for removal at intervals by means of a bailer. In certain circumstances, however, as for example to offset high pressures or to prevent excessive caving, mud-laden fluid was circulated down the casing during drilling.

The cable-tool technique was suitable for drilling in hard rocks with few water-bearing sands, but the rotary method, introduced in the 1890s, was quickly proved to be far more satisfactory in soft or easily caving rocks, in gas zones, etc., and with improvements it slowly replaced the older system. By 1920, the cable-tool rig had just about reached its maximum use; there had been no major changes in fifty years, although many improvements had been made in capacity and efficiency. By the early 1930s, all-steel rigs had been perfected with rated capacities up to 8000 ft, compared with the 2000-ft capacity of older rigs, but despite this the depth records changed little over the years; the deepest cable-tool well appears to have been one drilled to 7759 ft in Pennsylvania in 1925.

In the 1860s mainly wooden portable machines were extensively employed for shallow wells. Improved designs used an increasing steel content and internal combustion engines as a source of power. In the early 1930s self-propelled or trailer-mounted spudders with heavy A-frame masts, became available with depth capacities ranging from 1000 ft to 6000 ft. These rigs, with modifications, were widely used through the 1940s in shallow oilfields and on workover tasks.

In the earliest drilling operations the percussion tools were supported by drill rods, initially made of wood, but later of iron or steel. American rigs turned, in about 1860, to the use of rope and, in the 1890s, to wire line, but in some regions the use of drill rods continued into the early years of the twentieth century, as in Canada, Mexico, Russia, Poland, and elsewhere in Europe. In some areas tubes were substituted for solid rods, and water was circulated through the rods to remove cuttings.

Rotary drilling. In rotary drilling a hole is bored by the rotation of a column of hollow pipes to which is attached an abrading tool or bit. A stream of liquid, generally colloidal mud, is pumped down the pipe and out through nozzles in the bit. Important functions of the mud are to cool the bit, to lift the rock cuttings to the surface, and to support the walls of the hole.

In the early 1800s French rigs used wrought-iron rods with bits, essentially for dry rotary drilling. In 1844, an English patent by Robert Beart covered the principles of rotary drilling with many of the basic features of the fluid circulation system of later rigs. In the early 1860s the French engineer Rodolphe Leschot developed the rotary diamond core drill, which was the forerunner of present-day machines used in mining and oil exploration.

From the 1880s elementary rotary rigs were being developed, initially for water wells in unconsolidated sands. These led to discoveries of oil in Texas and to the subsequent drilling of many wells by rotary methods prior to 1900. In January 1901 the Spindletop well being drilled at about 1100 ft by a rotary rig blew out as a gusher. Apart possibly from some earlier oil gushers in Baku (and some French water wells) this was the greatest flow achieved to that date. It proved the presence of prolific oilfields in geologically young beds, and it also demonstrated the importance of rotary drilling in unconsolidated rocks. Nevertheless, in the early years of the twentieth century the rotary system was not widely accepted. One reason, was that the bits then available could drill only soft rocks. Much of the equipment was adapted from cable-tool rigs or from stock items of industrial machinery; the special equipment was manufactured locally.

The wooden derrick was up to 84 ft high; steam engines of 20 to 30 horsepower were in common use. Drilling practices in those early days were simple. The column of pipe with a hydraulic swivel at the upper end and a cutting tool at the lower end was supported by the hoisting gear and was rotated rapidly by a rotary table equipped with grip rings. Water, and later a suspension of clay, was pumped down the pipe.

For drill-pipe and well casing both lapwelded merchant pipe and line pipe were used, although special threads and couplings were available for casing. In the early years it was customary to use the casing also as the drill-pipe; a serrated steel shoe was used as a cutting tool and drilling continued until no further progress could be made; drilling was then resumed with the next smaller size of pipe, and so on.

Drilling bits of the fishtail type (Fig. 16.3(a)) for soft rocks, and heavy blunt straight blades run with granular adamantine in the hole for hard rocks, were

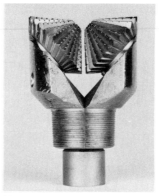

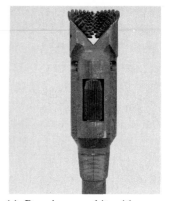

FIG. 16.3. Rotary drilling bits. (a). Early fishtail bit, with only two cutting edges, which quickly wore down flat on hard rock.

(b). First rotary rock bit, with two rotating conical cutters, 1909.

(c). Reaming cone bit, with regular cones and a reamer built into the body of the bit, 1917.

dressed and tempered at a forge near the rig. Wear on these early bits was considerable, and it was necessary to pull drill-pipe frequently to dress or replace a bit.

Even up to the 1920s drilling operations were regarded as tasks for skilled artisans. The driller, with a crew of four or more rig-hands working a 12-hour shift, carried out many duties. There were very few engineers, and little reliable recording of data.

By 1910 rotary equipment was being manufactured by a number of firms, and more or less complete rigs were available. The drill-pipe was becoming a separate tool. The new equipment was developed principally in Texas and Louisiana, where many thousands of wells were drilled, but rotary rigs were sent to Argentina, Peru, Colombia, Venezuela, Trinidad, Mexico, Roumania, Russia, Sumatra, and Borneo, where they were used with varying success.

After the introduction of rotary rigs into California in 1908 many modifications were made to cope more effectively with local conditions, including harder rocks. Equipment was made heavier and there were important developments, as for example in rolling-cutter rock bits, casing cementing, long couplings for casing, and drill-pipe tool joints. Another very important step was the assignment of trained engineers, or engineers-cum-geologists, to assist the practical field men on drilling problems; this was the beginning of petroleum engineering.

With the outbreak of the First World War the demand for petroleum products increased considerably, and the period up to 1930 saw many new

(d). Roller core bit for use in hard rock formations, 1926.

(e). Unit-type bit with anti-friction bearings, eliminating the delay caused in changing cones, 1931.

(f). Tungsten carbide compact bit, with increased penetration rate and bit life in hard, abrasive formations, 1951.

developments in equipment and drilling practices. For rotary operations the single-cylinder steam engine left much to be desired. In 1918 the first two-cylinder engine came into the fields, but the old engines continued in use for a decade; in general, boiler capacities remained inadequate for many years.

Wooden derricks were gradually replaced by steel. There were new draw-works but they still had to be dismantled and reassembled at each well-site. In 1914–15 came the first square kelly (a length of pipe with a square cross-section) which obviated the use of grip rings to hold the pipe in the rotary table; interestingly, this had been described as early as 1844 in the Robert Beart patent.

By the 1920s rotary rigs were rapidly increasing in number; the development of rolling-cutter rock bits and the use of hard alloys on the faces of bits (Fig. 16.3) permitted an encroachment into hard-rock drilling, previously the domain of cable-tools. Many oilmen, however, still questioned the advisability of drilling through an oil-sand with a hole full of mud and it was a long time before rotary tools were fully accepted.

From about 1908 moves had been made in Europe for the standardization of materials, but a great step towards improving the quality and efficiency of oilfield equipment was taken in 1924 when the American Petroleum Institute (A.P.I.) standardization committees commenced their sessions.

From 1926 new rigs were available, capable of increased depths and performance, but there were continuing difficulties because of limited understanding of drilling fluids; crooked holes; drill-pipe failures; the low footage

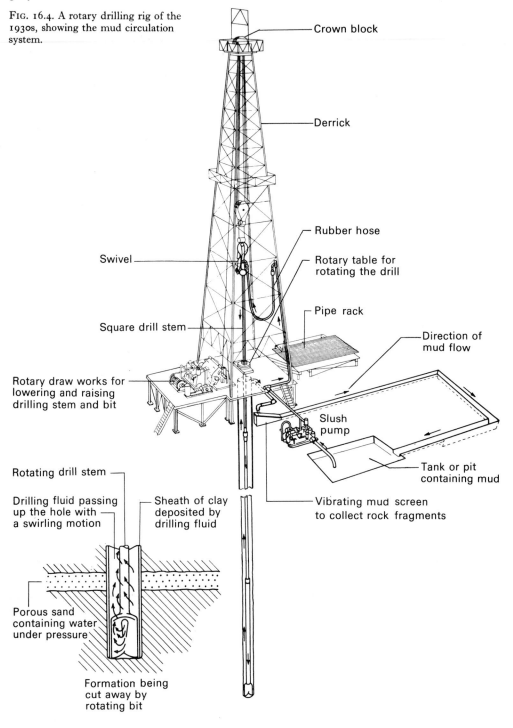

FIG. 16.4. A rotary drilling rig of the 1930s, showing the mud circulation system.

Crown block

Derrick

Rubber hose

Rotary table for rotating the drill

Swivel

Pipe rack

Direction of mud flow

Square drill stem

Rotary draw works for lowering and raising drilling stem and bit

Slush pump

Tank or pit containing mud

Rotating drill stem

Drilling fluid passing up the hole with a swirling motion

Sheath of clay deposited by drilling fluid

Vibrating mud screen to collect rock fragments

Porous sand containing water under pressure

Formation being cut away by rotating bit

achieved before the bit required renewal; slow rigging-up times; and a lack of drilling instrumentation. These were some of the problems that brought the petroleum engineer increasingly into all phases of operation. Some of the large new oilfields discovered in the early 1920s, particularly in California, presented problems which led to further improvements, especially in casing programmes and well-head control equipment. The various 'Beach' fields by the Pacific Ocean fostered the development of controlled, directionally deviated drilling.

In cable-tool wells casing of decreasing diameter was introduced at intervals as drilling progressed, so that on completion the well had as many as four to six strings of casing. This practice tended to be carried over into rotary well procedures in some areas, but it was not until the early 1930s that attention to the physical characteristics of drilling fluids permitted the running and cementing of much longer casing pipe strings to become general practice.

These various problems led, in the early 1930s, to the formation of drilling-practice committees under the auspices of the A.P.I. At the same time detailed studies were widely initiated, based on sound engineering principles, with the publication of a growing volume of technical papers.

The safety of rig floor personnel—to which little thought had hitherto been given—was also taken more into account. Drilling organizations undertook studies of safety practices, and regulations were established under various authorities. The subject was taken up by many study committees and by schools of petroleum engineering developing in universities. This was the period when it was realized that oilwell drilling in all its phases presented complicated and difficult problems which required the best engineering and technical skills possible for their successful solution.

The standard derrick of 1930 was 122 ft high, constructed in both wood and steel, but by 1932, 136-ft steel derricks were being built for deep wells (Fig. 16.4). Derrick substructures were also first developed at this time, to eliminate the digging of cellars for well-head control equipment. Other important improvements were the introduction of boilers operating at higher pressures, with correspondingly larger steam engines and mud pumps. Unitized, shop-assembled draw-works with other rig parts of equal relative capacity, gave for the first time a properly 'balanced' rig. There were many other important technical innovations, among them the introduction of hydromatic brakes on draw-works; a separate rotary table unit drive; and a great improvement in drilling control instruments to show weight on the bit, fluid pressure, rotating speed, and drill-pipe torque.

Considerable attention was paid also to the drill-pipe string. By using heavier pipe immediately above the bit, it became possible to rotate the drill-pipe in tension while keeping the desired weight on the bit. This change in practice reduced pipe failures, led to faster drilling and straighter holes, and greatly extended the life of the pipe.

Although steam predominated in drilling operations in the early decades of the twentieth century, there were many examples of the early use of electric power, or of internal combustion engines run on field gas, gasoline (petrol), or diesel fuel. Operations became smoother from the late 1930s with the introduction of friction clutches and of hydraulic torque converters in the power transmission to multiple-speed hoists. During 1939–40 a complete range of rigs incorporating internal combustion engines was made available, but during the war years manufacture of oil equipment was halted.

After the war very few steam rigs were built and few were left operating. Extensive drilling in arid areas in West Texas, in New Mexico, and in Middle Eastern countries gave a considerable impetus to the demand for power rigs which did not require large quantitites of boiler water and fuel. A considerable number of diesel-electric powered rigs were manufactured, particularly for drilling barges (see later), but the greater number of new rigs were directly driven by diesel engines. These rigs had rated capacities of up to 20000 ft, but were capable of drilling much deeper holes (Figs. 16.5 and 16.6).

About 1920 the first practicable portable rotary rigs were used for studying subsurface geology. Later models have been extensively used for seismic shot-hole drilling, and the larger versions, truck or trailer-mounted with telescopic masts, have been used for production drilling at depths to 5000 ft.

Combination Rigs. In the early days of rotary drilling, with bits capable of cutting only soft rocks, it was common practice in some regions to have a rig in which both rotary and cable-tool equipment were installed. The early fear that rotary drilling fluids would 'mud-off' oil-sands led many operators to cement casing above an oil-sand after rotary drilling, and then to drill through the sand and complete the well with cable-tools. Many combination rigs were used prior to 1920 and some were still in use in the 1930s.

II. OFF-SHORE DRILLING

The first off-shore drilling appears to have been in California at the turn of the century in the Summerland field, where small wells were drilled from wooden piers (Fig. 16.7(a)). This method was used in other Californian fields even up to 1932. Drilling on a piling-supported platform was undertaken in Caddo Lake, Louisiana, in 1911 and in Lake Maracaibo, Venezuela, from 1924. All the early Lake Maracaibo wells were on wooden piles, but these structures suffered rapidly from attacks by the teredo mollusc, and this was countered successfully only by the use of reinforced concrete piles (from 1927) and concrete caissons in deeper water after 1940. In 1934 steam drilling tenders were introduced, and thereafter only a small fixed platform was erected for the drilling rig.

Another area of early off-shore drilling was at Baku, in the Caspian Sea. The first well in open water was drilled in 1925 on an island built on wooden pilings. Subsequently, many wells were completed with a highly organized procedure for piling and for the construction of trestle-type interconnecting platforms of steel or reinforced concrete. There were arrangements for road and railway traffic, for pipelines, etc., and even an entire oilfield and industrial city complex many miles off shore.

During the 1920s, drilling in the swamps on the edge of the American Gulf Coast required considerable piling, but by about 1930 light equipment mounted on barges was used for the drilling of shallow holes in quiet waters. In 1932, the Texas Company designed a submersible drilling barge which, it was found, had been patented four years previously by Louis Giliasso. This first barge, with twin hulls, was designed to drill while resting on the bottom in 15 ft of water. It spudded its first well in late 1933, with power supplied by a boiler barge of similar design. Operations proved very successful, and subsequently many wells were drilled by submersible barges in quiet inland waters.

The first well from an independent platform off the West Coast of America was drilled in 1932. In the following year the first attempt was made to explore in the Gulf of Mexico with the same kind of piled structure, but it was not until 1938 that the first discovery was made, a mile from the shore in 14 ft of water.

In the post-war years there was a great increase in exploration in the Gulf of Mexico. The small platform and drilling tender combination had been used in inland waters, and it is interesting that this concept had been patented by T. F. Rowland in 1869. Kerr-McGee Oil Industries designed the first equipment for use in open waters under severe wind and wave conditions. The platform, on piles, took the drilling rig and small tanks for water, fuel, and mud; all other services, including crew quarters, were placed on the floating drilling tender. In 1947 the first well with the new equipment on a location

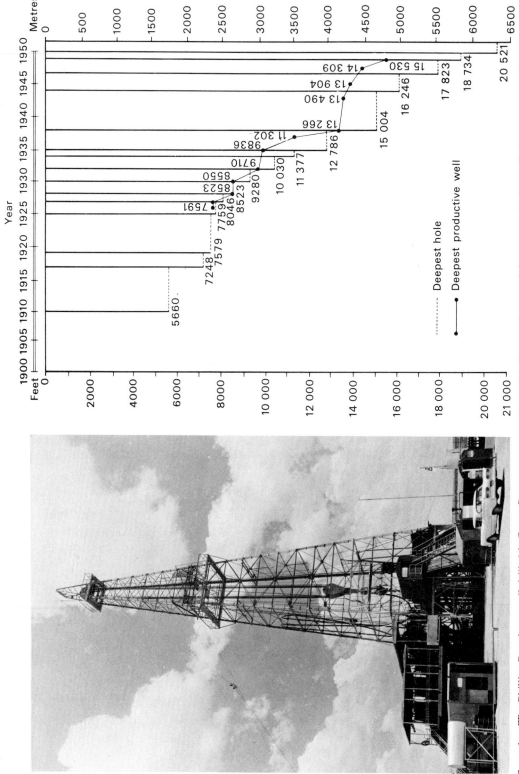

FIG. 16.6. Deepest holes and deepest productive wells in the world's oil industry.

FIG. 16.5. The Phillips Petroleum well drilled in Pecos County, Texas, 1956, to a depth of 25 304 ft, at that time the deepest well in the world. The equipment is similar to that shown in Fig. 16.4.

FIG. 16.7 (a). A view of Summerland Oilfield near Santa Barbara as it appeared in 1903. The field was discovered in 1894 and was the first offshore oilfield to be developed in the U.S.A. Vertical wells were drilled from piers and walkways built out into the sea.

FIG. 16.7 (b). A drilling and production platform in 1948 in the Gulf of Mexico, about seven miles off Louisiana. The platform is self-contained with living quarters, power station, workshops, and all equipment needed to drill and produce wells. At that time the oil was being taken to shore by barges, one of which is seen alongside. Subsequently a pipeline to shore was built.

ten miles from land discovered a new oilfield and demonstrated the efficiency of the outfit. Thereafter many wells were drilled with this type of equipment in the Gulf of Mexico. (Fig. 16.7(b)).

At the end of the period under review, in 1949, there came the first self-contained off-shore drilling barge. The Hayward-Barnsdale barge was in effect a drilling platform supported by columns on top of a barge hull. Under tow the hull floated. On location the hull was completely submerged, resting on the sea bed, but with the platform well above water and only the supporting columns subjected to wave forces.

III. ROTARY DRILLING FLUIDS AND CIRCULATING SYSTEMS

By 1900 mud-laden drilling fluids were known in Texas and Louisiana. It was realized early that apart from cooling and lubricating the drilling bit and lifting cuttings, a clay sheath tended to plaster and support the wall of the hole. It was also recognized that drilling fluids, unfortunately, tended to seal off oil-bearing formations, although they could be used in the control of high-pressure gas, and in 'mudding off' water-bearing sands.

Some technical papers on the subject were published in the period 1914–24, including some on the use of heavy materials, such as barytes, in suspension in drilling fluids as a means of controlling high-pressure gas wells. However, it was not until the late 1920s and early 1930s that oil companies began to examine the physical characteristics of drilling fluids from scientific and technical viewpoints. Since then there has been much research in many laboratories; various test instruments have been devised for the measurement of properties such as density, viscosity, wall-building capability, water loss, and gel strength. The clay minerals have been studied extensively and much work has been done in developing chemical reagents and materials for the control of the physical properties of muds.

In 1935 efforts were first made to prepare oil-based drilling fluids in order to avoid the possible sealing of oil and gas formations, and these muds have been widely used in some areas. Benefit is also derived from their high lubricating qualities, and in drilling through formations such as salt or gypsum which could severely contaminate a water-based mud. Such muds were, however, more costly than the water-based fluids and had other disadvantages.

In certain areas where it was found necessary to maintain a low fluid pressure on rock formations, use was made of compressed air or gas-cut mud as the circulating fluid.

It has always been a necessary practice to recirculate the rotary drilling fluid to remove cuttings (Fig. 16.4). This was done in early years making use of a variety of settling pits and ditches. With the development of power rigs came a standard system of tanks and metal ditches, and the removal of cuttings largely by the use of electrically operated vibrating screens or centrifugal separators.

With increasing depths, and with the demand for higher drilling speeds, better mud-pump performance was required. At the turn of the century, a small general service pump sufficed, delivering 200 gallons per minute at 250 lb/in² pressure; these pumps gave much trouble, particularly when drilling abrasive sandy rocks. In the 1930s pumps were available, operating on superheated steam, to deliver 800 gallons per minute at 340 lb/in² pressure. Together with changes in motive power away from steam, new pumps also incorporated greatly improved heat-treated alloy steels, refinements in moving parts, and greatly improved methods of lubrication.

In some regions abnormally high subsurface pressures are met, and in such situations the present common practice is to use barytes-loaded muds. In the early 1930s however, much work was done, particularly in Persia, on drilling with a closed circulation system under pressure, using a snubbing device for raising and lowering drill-pipe.

IV. DIRECTIONAL DRILLING

In general, the object in drilling is a straight vertical well, but there are circumstances in which a controlled deviated hole is required. Deviated drilling in order to sidetrack a 'fish' blocking the hole had been employed since 1895 using special tools; the whipstock (a form of wedge), knuckle joints in the drill string, and spudding bits. The first instrument for measuring the deviation of a hole from the vertical was the acid bottle containing hydrofluoric acid; the acid etched a horizontal meniscus when the bottle was allowed to rest and so indicated the slope of the hole at the point of measurement. Later, this technique was adapted to give a directional indication also. Many instruments for the measurement of the amount and direction of deviation were designed and tested up to 1930, but only a few have become widely accepted. Use of these instruments indicated that many old wells had unintentionally deviated a long way from the vertical, even as much as 60°, largely because of the earlier drilling practices and of limitations of equipment.

The first controlled directional drilling of oil wells was developed in the Californian Huntington Beach field in 1933. At about this time the technique was also used to tap oil-sands beneath the Irrawaddy river in Upper Burma; some of these wells had an average deviation of about 40° to 3000 ft.

Directional drilling has in recent years become highly developed for reaching inaccessible subsurface locations, as for example in the development of oilfields beneath the sea and town sites; for the correction of the underground position of a well in complicated geological structures; and for re-entering the lower part of a well flowing out of control.

V. OILWELL CEMENTING

In early wells it was common practice to set pipe on hard rock or in tough clay, with cuttings and mud sometimes giving a fluid seal. No cement was used, consequently in many oilfields producing formations were prematurely damaged by inflow of water. In order to seal off and to separate the oil-, gas-, and water-bearing rocks, oilwell cementing became an important factor in sound oilfield practice.

The cementing of the annulus at the base of a string of casing was practised in cable-tool wells for some time prior to 1910. The earliest procedure was to dump cement slurry on to the bottom of the hole, and then to lower the casing into the cement, which was left some days to set before resuming drilling. In the following years, a number of patents were taken out relating to improved methods of pumping cement down the hole and out behind the casing without contamination.

Although much attention was paid in the 1920s to the composition and texture of the cement in order to accelerate the initial setting time and to give high early strength, it was not until the late 1920s and early 1930s that extensive laboratory tests were undertaken on the varied physical problems of cement and cementing in oilwells. It was only in 1947 that the first A.P.I. Code for testing well cements was approved.

In the 1930s there were many advances in well-completion methods, which in turn required sound cementing operations covering hundreds, and even thousands, of feet of annulus between bore hole and casing. The effective and rapid placing of the large quantities of cement involved required heavy-duty high-pressure equipment. The operations were greatly helped by the current improvement in the physical qualities of drilling muds, and the results were more effectively monitored by the introduction of various additional measuring techniques, such as well-temperature surveying in about 1935. From 1940, continuous caliper logging of the borehole diameter also permitted better estimation of cement requirements.

VI. FORMATION TESTING

It had long been recognized that it was important to have some indication of the potential flow of fluid before equipping a well for production. In cable-tool operations, some idea of probable producing rates was generally obtained during drilling, but with the advent of rotary drilling it became essential to have a reliable means of isolating a zone to be tested from the column of mud in the hole. The requirements for such a test were a packer or packers which could isolate a section of the borehole, tubing or drill-pipe on which the tester could be run; and a means of opening and closing valves near the packer so that the produced fluids could be flowed or lifted to the surface for measurement.

The earliest practical device for rotary wells appears to have been patented in 1933; it had a conical packer and a valve operated by rotating the pipe at the surface. From this beginning many improvements were made, and testers were also adapted for packers to operate inside casing, but the early formation testers gave many problems and it was not until the 1950s that they became reliable tools. In 1934, pressure recorders were incorporated in the lower part of the tool.

VII. WELL LOGGING

In many early wells records of the nature of the rocks penetrated, of production data, etc. were poor or non-existent. The cable-tool driller assessed the type of rock penetrated by the rate of progress and performance of the bit, supplemented by rock cuttings brought up by the bailer. Most rotary drillers used the evidence of the performance of the single-cylinder steam engine, the pump, and the rotary table, but some did arrange for the collection of cuttings. In those days the well geologist was rare, but towards the end of the second decade of the century the examination of cuttings became more common; for a time this was the only information available in studies of the subsurface geology.

Cable-tool coring equipment was devised in early years, but it was the mid-1920s before efficient equipment was available. Core barrels had been successfully used by the mining industry on diamond-core drills for many years, but surprisingly they were not generally adapted to oilfield service for rotary drilling until the early 1920s.

In the mid-1930s recording equipment was devised to give a continuous indication of drilling performance, including the rate of penetration, and this was followed by equipment for the detection of oil and gas in the drilling mud and rock cuttings.

The introduction of electrical logging of boreholes has proved to be one of the most important advances in exploration and oilfield development in this century. Experiments in exploring subsurface rocks by means of electrical measurements on the surface were initiated in 1912 by Conrad Schlumberger, but it was not until 1927 that the idea of making measurements in drill-holes was translated into practice. Initial experiments were made in 1927 in Alsace; in 1929–30 the technique was introduced into Venezuela, Baku, and the Dutch East Indies but it was not generally accepted in the U.S.A. until 1932.

Initially the measurement made in a borehole was a point-by-point determination of resistivity of the rock; this was immediately found to be valuable for geological correlations and as an aid in determination of rock types. In 1931 measurements were also made of spontaneous potentials of physical and chemical origin. Both resistivity and spontaneous potentials were then recorded continuously, the measurements being synchronized with the depth of the measuring electrodes.

FIG. 16.8. Signal Hill, California, the richest oil field per acre in the world, during its heyday in the 1920s.

With an intensive research programme improvements followed steadily, and by the 1940s a variety of techniques were available for continuous sub-surface recording of resistivity (including focused and induction logging), radioactivity, etc. There were great advances also in instrumentation and in interpretation in relation to rock characteristics and their fluid contents.

VIII. PRODUCTION

Gas and oil are principally found, in association with water, trapped in the pore spaces of sandstones and limestones. Any free gas occurs in the highest part of the reservoir and the oil concentrates above the water. Pressures of fluids in underground reservoirs vary, but are frequently hydrostatic, and the oil has in solution varying volumes of gas. The flow of oil to the well bore is powered by the expansion of free or dissolved gas, and in some cases by the influx of water into the reservoir. From some wells oil may flow unaided to the surface, depending upon the pressure and energy stored in the reservoir, but in many fields it is necessary to assist the flow by external means.

In the early years little or no serious thought was given to the nature and behaviour of oil reservoirs. With the prevailing 'rule of capture' the policy was to drill a field rapidly with closely spaced wells (Fig. 16.8), and to produce oil, often competitively, at the maximum capacity either by natural flow or pumping, until the pumps could no longer lift oil in economic quantities or until water production was excessively high. Gas was generally regarded as a nuisance, saleable only in small volumes locally. Water was almost every-where regarded as a menace.

Nevertheless, even in those early days some operators undertook various control or stimulation projects, albeit on an empirical basis.

Well completion methods. In early cable-tool wells, if oil tended to flow after the drill had reached the oil-bearing rocks, a casing-head tee was screwed on top of the innermost string of casing, and a side outlet was connected to a flow line extending to a tank. It was possible in some cases to continue drill-ing with the well flowing by fitting a control head with a packing gland, which permitted the movement of the drilling cable. Early wells drilled by the rotary method were induced to flow, if necessary, by lowering the fluid level in the hole by bailing.

Flow tubing was not run into a well until after the initial flowing production had ceased, although in the 1920s equipment was developed for snubbing tubing into wells flowing under pressure. With the higher-pressured reservoirs that were encountered in some areas, as for example in Spindletop, and in many early Mexican wells, oil and gas flowed out of control for days or months, and the oil was collected as far as possible in pits or bunded reservoirs. There were many serious fires.

In some areas the oil-bearing rocks were liable to cave into the hole. It then became routine procedure, until the 1930s, to set casing before drilling through the productive horizons, and to place perforated or screen pipe in the unprotected part of the hole. To gain access to oil reservoirs shut off behind casing, a number of mechanical perforators using knives were patented after 1910, but were used to only a limited extent. Experiments using bullets were carried out in the mid-1920s but it was not until 1932 that Lane Wells satisfactorily perforated casing in a well using an electrically operated gun. This invention, used in combination with electrical logging, revolutionized rotary well completions since it became possible to cement casing opposite a multiple or thick pay zone and to perforate selected ranges either initially or at some later date. In the following years many improvements were made in the explosive charges and in gun and bullet design to give increased penetration and for use in the higher working temperatures of deep wells.

With deeper drilling the simple casing-head control equipment no longer sufficed to withstand the volume and pressure of flowing oil and gas. Early casing heads were designed to suspend tubing, with suitable packing to withstand the appropriate pressure. More complicated assemblies were developed

FIG. 16.9. A high-pressure 'christmas tree' on a deep well in Pakistan, with gas-oil separators in the background, 1950.

in the early 1920s to take a number of strings of casing with threaded connections in addition to the tubing, but later assemblies used slip-type suspensions, which simplified the setting of pipe with the correct tension.

Uppermost in the production well-head assembly is the tubing and annulus flow control equipment known as the 'christmas tree' (Fig. 16.9), which has become more complicated with increasing depths and working pressures.

In some multi-reservoir fields it has been found undesirable, and in some areas of the United States it has been illegal, to produce from more than one reservoir through the same hole. After the mid-1920s a number of wells were completed in such a way that, for example, one horizon could be produced through the tubing and another horizon through the tubing/casing annulus.

Pumping. In the early 1900s the cable-tool outfit was normally left on the well, and when flow ceased tubing with a pump barrel attached was run into the hole. The pump plunger was then lowered on sucker rods, initially made of hickory, and the well was pumped using the walking beam, employing either the original drilling steam engine, or a simple gas engine. If gas was required for fuel, a simple separator was installed in the flow system.

Central pumping units were constructed in the 1870s as a simple means of pumping a number of shallow, low-volume wells from a central point on a lease. Prime movement for individual wells was transmitted by rods from an eccentric powered by a steam or gas engine and various devices were adopted to overcome the problems of terrain. These were widely used until the 1940s; for example, in the Yenangyaung oilfield in Upper Burma.

Wooden sucker rods presented continual problems and by 1900 iron, and later steel, rods with box and pin couplings were replacing hickory. As well-depths increased in the 1920s rod failures became more frequent, but by 1930 fully heat-treated, stress-relieved rods became available.

The walking-beam driven by the band wheel provided the pumping action into the 1930s, using a horse-head (incorporated in Newcomen's water pump of 1705 (Vol. IV, p. 174)) to impart vertical motion to the rods, but some gear reduction units were manufactured even in early years and these became more numerous in the early 1920s. The complex relationship between the movements and the loads in deep-well pumping was not generally recognized until the 1920s, and this led to many innovations being made on the basis of trial and error. Counterbalancing was fairly primitive up to that time, but the introduction of the dynamometer permitted analysis of the relationship between the loading and movement of the beam, the rods, and the pump

FIG. 16.10. Independent beam pumping unit powered by a gas engine; the unit was used for pumping 4000–5000-ft deep wells during the 1940s.

during the pumping cycle and laid the basis for a sounder approach. Many engineering studies on this subject were published in the period from 1928 to 1943.

The development of mobile well-servicing outfits eliminated the need to leave the drilling derrick and equipment on site in many fields and this, together with the development of efficient gear reduction units, facilitated the introduction of the independent beam pumping unit in the 1920s, which became generally accepted in the 1930s. In later years balanced units with pumping strokes up to 12 ft became available. These units brought increased use of the electric motor and the gas engine (Fig. 16.10).

Various designs of pumping units were also introduced, involving pneumatic or hydraulic cylinders with pistons directly or indirectly connected to the sucker rods. These followed a basic concept dating back many years, and used in the mining industry, but serious development in the oil industry awaited the need for high-capacity, long-stroke pumping for deep wells after the 1920s. By 1950 both pneumatic and hydraulic units had been successfully developed with strokes up to 30 ft and capable of operating in wells more than 12000 ft deep.

Gas lift. The injection of compressed air to produce water from wells was practised in the first half of the nineteenth century. The method is reported to have been used to lift oil in Pennsylvania in 1864, and was in use in the Baku fields in 1899. It was introduced into Texas and California at the turn of the century.

There is mention of gas being used instead of air in 1911 in California, but it was not until the 1920s, with the development of the deeper fields, that gas-lift methods became widely employed. The original technique, which is still used, involves continuous injection of gas into the annulus or into a tubing string, the oil being driven up the alternative channel. If the oil level in the well is high, the gas pressure needed to start flow becomes excessive, and as early as 1907 kick-off valves were in use which, when placed at intervals on the tubing, enabled flow to be started with lower surface pressures.

Continuous gas lift tends to inhibit oil flow by imposing a back-pressure on the producing formation. Several techniques have been developed to deal with this problem. In intermittent lift, gas is injected at intervals adjusted to allow the fluid level in the well to rise. The method was used experimentally in 1903 but not mentioned as a regular practice until the late 1920s. In gas displacement, or chamber lift—the principle of which was patented in 1908, but apparently not used until 1927—concentric tubing strings are used with a larger diameter chamber fitted with a standing valve at the bottom of the outer tubing. Oil flows into the chamber and is lifted by intermittent shots of gas from the inner tube. From 1930 onwards the widening interest in the use of gas lift led to considerable theoretical and experimental work on the flow of gas–oil mixtures through vertical tubes and to much practical work on equipment.

A plunger designed to move freely in the tubing and to reduce gas slippage in both flowing and gas-lift wells was marketed in the early 1930s. By 1944 improvements to the plunger resulted in satisfactory results in standard tubing.

Hydraulic pumping. In this method a pump and hydraulic motor, both reciprocating, are installed at the bottom of the well and are powered by oil pumped from the surface down a small tubing string. Numerous patents were issued from 1872 onwards, and various modifications and improvements in equipment have been made since the method was introduced in 1924.

Electric Bottom Hole Pumps. A number of electric pump designs were patented from 1894. The first successful pumps were used in the highly productive fields in Kansas and Oklahoma in 1927. A motor with a long stator and a number of short rotors on a common shaft drove a multi-stage centrifugal pump. Similar pumps with minor modifications continue in use today for pumping volumes of 1000–5000 barrels per day at depths down to 10000 ft.

IX. WELL STIMULATION

In the past various methods have been used to improve the production from individual wells, particularly in formations of low permeability with high resistance to fluid flow.

Well shooting. The use of explosives in fracturing oil-bearing rocks was well known by the end of the last century. In the mid-1920s the 'shooting' of wells with liquid nitroglycerine was a routine completion or production-stimulation practice in some regions of hard sandstone or limestone rocks, and this practice continued into the late 1940s. Various procedures for detonation and tamping were used to improve the effects. Solidified nitro-glycerine was used later in some areas, but the liquid explosive was the most favoured, despite its greater sensitivity to detonation, which led in early years to many fatal accidents.

Acidizing. The experimental use of hydrochloric acid to stimulate production by opening flow channels in oil-bearing limestone formations was recorded as early as 1894, but it was not until the early 1930s, after laboratory work on the development of suitable inhibitors to prevent damage to the well casing and tubing, that acidization became a practical field method. The technique soon became competitive with well-shooting in hard limestones. It entailed injecting hydrochloric acid of about 15 per cent strength treated with inhibitor; volumes ranged from 500 to more than 10000 gallons in a single treatment. During the late 1930s and early 1940s various additives were developed: surface-active agents, de-mulsifying compounds as well as corrosion inhibitors. Acids were also available for the destruction of mud sheaths.

Hydraulic fracturing. As early as 1935 it was realized that it was possible to fracture a rock formation at depth by applying sufficient surface pressure to a fluid column in the well. Research into this concept led to the introduction in 1949 of hydraulic formation fracturing. In this technique the producing rocks are fractured and a gelled oil, carrying sand, is injected into the fractures. On release of pressure the sand grains prop open the fractures, thus improving the permeability, and hence the productivity of the well. Modifications of this technique soon became widely used in production stimulation, leading to considerable increases in production in some fields.

Sand control. Production from unconsolidated formations is frequently adversely affected by the intrusion of sand, which fills the well bore and leads to rapid erosion of equipment. By 1920 various types of slotted and wire-wrapped screen pipe had been developed to hold back the sand. The technique of gravel packing had been in general use for water wells since the turn of the century but it was not until the mid-1930s that the method was generally accepted as a means of sand control in oil wells.

The use of plastics to bind together the sand grains of a poorly consolidated rock, without materially affecting its permeability, was introduced in 1945 and has subsequently been extensively used, particularly on the American Gulf Coast. Two methods were developed: in the first, plastic is squeezed into the formation followed by oil which removes excess plastic and leaves permeable paths; the second method involves squeezing into the formation a plastic which shrinks as it hardens.

X. RESERVOIR ENGINEERING

Reservoir engineering can be defined as the applied science concerned with the transfer of fluids to, from, and within natural underground reservoirs. It may be said that the science was born in the 1920s with the growing awareness of the wastage of oil and gas, and of the need for controls.

Although attention had been drawn as early as 1865 to the role of natural gas dissolved in oil as a major source of energy in production, it was many years before the oil industry generally appreciated its importance. The first co-ordinated efforts to understand the physical processes of oil production started with the formation of a Petroleum Division of the United States Bureau of Mines in 1914; this division undertook many theoretical, laboratory, and field-engineering studies and also provided many well-trained engineers and scientists. Much valuable work was done in the 1920s, leading to a considerable expansion of studies in reservoir behaviour and control in the latter part of the decade by many companies and research establishments.

The properties of reservoir fluids are profoundly influenced by temperature and pressure, which affect the phase relationships of oil and gas, their viscosities, and their compressibilities. Temperature surveys had been made in earlier years, but it was from the 1920s that instruments were developed for the measurement of pressure in well bores, and the first attempts were made to collect samples of the produced fluids under their reservoir conditions. Thereafter better measuring devices and sampling equipment were developed, and much laboratory work was done on the effect of varying temperature and pressure on the physical characteristics and phase behaviour of the complex mixture of hydrocarbons in oil and gas reservoirs.

The physical characteristics of rocks—porosity, permeability, etc.—were studied in the later years of the nineteenth century in relation to the movement of ground-water, but again it was not until the early 1920s that similar studies were made for cored samples of oil-bearing rocks. Thereafter, until the end of the period under review, there was a considerable research effort, and many papers were published relating to the permeability of different rocks, to the content of connate water in oil-sands, to the flow of multi-phase systems, and to the concept of relative permeability. The work was greatly aided in the 1940s by the data derived from improved electrical logs of boreholes.

The derivation and solution of mathematical equations describing the flow of reservoir fluid systems started in the early 1930s and many important papers have since been published on this subject. During the same period much work was done on material-balance equations, which are the mathematical statements of the law of conservation of mass during production; the balance between the original fluids in a reservoir, the produced and remaining volumes, and any incoming fluids.

Detailed solutions of the simplest reservoir problems require an enormous number of computations. The analogy between the flow of electricity and the flow of fluids was early recognized and analogue computers were used from the mid-1930s onwards to resolve specific reservoir flow problems. In more recent years high-speed digital computers have become invaluable for this purpose (Ch. 48).

As a result of the vast amount of theoretical and experimental work undertaken since the 1920s it is now possible, in preparation for the development of an oil or gas reservoir, to plan effectively the collection and analysis of essential data and then to assess the recoverable oil and gas in the reservoir, the required spacing of the wells, the maximum efficient rate of production, and the production techniques and controls to be adopted in order to obtain the highest possible economic recovery of the hydrocarbons originally in place.

XI. SECONDARY RECOVERY

Even under controlled conditions, the depletion of an oil reservoir by primary methods can leave 50 to 90 per cent of the original oil in place in the reservoir, and during the early 1900s the failure to restrict gas production resulted in an even larger proportion of the oil being left in some reservoirs. The residual amount depends on many factors, including the physical characteristics of the rock, the nature of the reservoir fluids, and the degree of natural water drive or gas-cap expansion.

The application of a partial vacuum to the casing heads of depleted wells was introduced in the 1860s and continued into the 1930s. The method resulted in increases in oil and gas production for a time in some fields, but had many undesirable features.

Two techniques are in common use to recover additional oil from depleted reservoirs and to maintain pressure and so improve recovery factors in new reservoirs. The first is injection of gas, usually, but not exclusively, into the gas-cap or the highest part of the reservoir. The second is water injection with the object of sweeping oil to the producing wells. Proper application of recovery techniques may lead to the recovery of all but 30 to 50 per cent of the original oil in a reservoir.

Gas injection. The potential of gas injection was reputedly first realized in the 1880s; a number of injection schemes were subsequently initiated, frequently using air rather than gas. By the 1920s the importance of gas in the production of oil was becoming better appreciated. The years 1925–7 saw the start of the modern approach, in which gas is re-injected early in the life of the reservoir in order to maintain pressure.

Gas recycling is an important form of gas injection in those reservoirs in which the hydrocarbons are in the gaseous phase at the original temperature and high pressure, but became liquid as the pressure is lowered. In order to avoid the loss of valuable liquid hydrocarbons in the reservoir, the gas is produced at high pressure and stripped of liquid fractions at the surface. The resulting 'dry' gas is then re-injected in order to maintain the reservoir pressure. An end stage of this technique is of course the production of the accumulated dry gas. The first project of this nature was started in 1938.

Water injection. Attention was drawn in 1880 to the possibility of increasing oil recovery by injection of water into the producing sands of the Pennsylvanian oilfields. However, it was in general, considered detrimental to allow water to invade a producing sand and in some states in the U.S.A. the practice was illegal until about 1920.

In the early 1920s secondary recovery by water flooding was confined almost exclusively to Pennsylvania. Initially water was injected into a central well and then into surrounding wells as the flood spread radially outwards. The next development was injection into a line of wells and production from parallel lines of wells. The five-spot pattern, in which there are alternately injection and producing wells, is now the most commonly used. In the early 1930s the advances in reservoir engineering led to the better understanding of the flooding mechanism and to the application of the method in depleted fields in many parts of the world.

Pressure maintenance by water injection early in the life of a reservoir was first tried experimentally in the East Texas field in 1936 and was extended to a full water-disposal programme by 1942. The added advantage of water disposal by the return of produced water to maintain pressure was appreciated and applied in other areas during the early 1940s, and by 1950 the maintenance of pressure in new reservoirs by water injection was an accepted engineering technique. The 1940s saw considerable improvements in the preparation of the injection wells and in particular in the treatment of the injection water to avoid bacterial growth and chemical precipitation. Considerable research and experimentation were also carried out to improve the efficiency of the flood by the addition of chemicals to the water and by other techniques.

XII. GAS AND OIL HANDLING

From the beginnings of the industry the operator has been confronted with a variety of problems in handling the produced oil and gas. In later decades important problems were the separation of the produced materials; the prevention of waste of natural gas and the lighter liquid hydrocarbons; the disposal of salt water; and the establishment of standard methods for measuring well and field productions, and for sampling and testing crude oil, natural gas, and natural gasoline.

Gas-oil separators. In early years various methods were improvised for the separation and handling of the produced gas and oil, sand, and water. The

FIG. 16.11. Multi-stage separation equipment at Nahorkatiya, India. The oil from the well passes through the separators, which are maintained at progressively lower pressures, allowing the gas to be removed, and retaining the valuable light hydrocarbons in the oil.

first units installed to collect gas for fuel were simple chambers in which the oil drained from the bottom and the gas flowed from the top. With the increase in gas pressures required for longer gas pipelines, separators capable of operating at higher working pressures were developed, and from the 1930s much work was done to improve the efficiency of separation of gas from liquids. In the 1920s multi-stage separation came into use for high-pressure flowing wells in order to retain a greater volume of the light hydrocarbons in the liquid phase by removing free gas from the flowing fluids in two or more stages, with controlled pressure drops to the stock tank (Fig. 16.11). Initially separators were designed primarily with vertical vessels, but in more recent years there has been considerable use of horizontal, high-pressure separators, particularly for the larger-flowing productions, as in the Middle East fields.

Tankage. In the beginning oil was run into all kinds of containers wooden, metal, earthen pits, etc. Bolted tanks appeared in 1913 and welded steel in

1926. In order to reduce evaporation losses the aluminium painting of tanks and installation of vapour recovery systems were introduced in the 1920s, as also was the floating roof for large storage tanks.

Oil treatment. Large volumes of water produced in some fields were separated at water traps or knockouts, and by settling. However, it was found that much water-in-oil emulsion was being formed during production, and three techniques were developed to break this down. These were respectively the application of heat, leading to the continuous-process heat-treater; chemical treatment (patented in 1914); and electrical treatment, initially introduced in 1909 on a batch basis, but later developed into a continuous flow process.

Natural gas. The use of gas as a fuel was initially a local development in certain regions, and because of the rapid decline in production in early oilfield operations preference was given to the more dependable supplies from non-associated gas accumulations. With the growing demands for gas as a fuel, and with more orderly oilfield development, the associated gas from a large number of oil wells was made to flow to collecting lines. Nevertheless there still continues to be a considerable wastage of oilfield gas in some regions.

From 1904 an increase in the need for gasoline, and later for propane, butane, etc., led to the treatment of the natural gas for the extraction of liquid hydrocarbons. Initially, single-stage compression was used, but by 1909 two-stage compression was introduced, followed in the 1920s by absorption processes. Various chemical treatments were also introduced to remove undesirable contents such as water, sulphur compounds, and carbon dioxide.

The first major application of natural gas was in 1883, when it was piped to Pittsburgh from the Pennsylvanian fields. By 1890, there were 500 miles of gas mains in Pittsburgh and over 27000 miles of piping for natural gas in the U.S.A. as a whole. Thereafter its utilization increased rapidly in North America, facilitated by the development of welded steel pipelines in the 1920s. Transmission over distances of many hundreds of miles to the major centres of population and industry became commonplace. In western Europe, by contrast, the contribution of natural gas was minimal until after the period of our present review. The French and Dutch gas fields were developed in the 1950s, followed by the fields beneath the southern part of the North Sea. Overstepping our limit only briefly, however, we may note that in 1951 the Union Stockyard and Transit Company in Chicago began to examine the possibility of transporting liquefied natural gas from Louisiana. In 1959, the first cargo of liquefied natural gas crossed the Atlantic to a specially constructed terminal at Canvey Island in Essex.

XIII. DOWNSTREAM OPERATIONS

The history of petroleum transportation, processing, and marketing operations in the first half of the twentieth century reveals general similarities to the progress of production operations: the simple, practical beginnings leading in the following decades to increases in size, scale, and complexity (particularly from the 1920s) with changing world production patterns and marketing requirements, and with the concurrent development of the relevant applied sciences.

In 1900 the few major American companies competed for European markets with the British companies supplying Russian products. In the east, markets were supplied from Russia, America, and Sumatra. With the passing of the years and with oil discoveries by many companies in an increasing number of countries, the world transporting, processing, and marketing pattern became much more complex. Oil products played a large part in the strategy and in the conduct of both World Wars.

Pipelines were first used for the movement of oil in the U.S.A. in the 1860s. The first pipeline from the Caspian oilfield was laid by the Nobel brothers in 1879, but most pipelines in 1900 were still of small diameter with screwed connections. In succeeding decades there were increases in diameter, improved steels, the introduction of welded joints, and a widespread change to electric or internal-combustion powered pumps and compressors to move crude oil, derived products, or natural gas.

Transportation by rail car, barge, and sea-going tankers also showed great developments in the fifty years.

Early refineries were basically small shell stills for batch separation of needed components of the differing crude oils. The increasing and changing demand for petroleum products, particularly the growing gasoline requirements, led through the years to increases in the number and capacities of refineries, to greatly improved processes and refinery technology, and to increasing investment in capital-intensive, labour-saving plants. Later decades also witnessed the development of petrochemistry, using both oil and natural gas components as basic raw materials for chemical manufacture.

BIBLIOGRAPHY

BRANTLY, J. E. *History of oil well drilling.* Gulf Publishing Co., Houston (1971).

DUNSTAN, A. E., NASH, A. E., BROOKS, B. T., and TIZARD, SIR HENRY (eds.) *The science of petroleum*, Vol. I. Oxford University Press, London (1938).

FORBES, R. J., and O'BEIRNE, D. R. *The technical development of the Royal Dutch/Shell Group 1890–1940.* E. J. Brill, Leiden (1957).

Golden Anniversary Number. *Oil and Gas Journal.* Petroleum Publishing Co., Tulsa (1951).

History of petroleum engineering. American Petroleum Institute, New York (1961).

MUSKAT, M. *Physical principles of oil production.* McGraw-Hill, New York (1949).

Petroleum panorama 1859–1959. *Oil and Gas Journal.* Petroleum Publishing Co., Tulsa (1959).

PIRSON, S. J. *Elements of oil reservoir engineering.* McGraw-Hill, New York (1950).

Proceedings of the World Petroleum Congress, London, 1933. World Petroleum Congress (1934).

REDWOOD, SIR BOVERTON *A treatise on petroleum*, Vol. II (4th edn.). C. Griffin, London (1922).

Secondary recovery of oil in the United States. American Petroleum Institute, New York (1950).

THOMPSON, A. BEEBY *Oil-field development and petroleum mining.* Crosby Lockwood, London (1916).

—— *Oil-field exploration and development.* Crosby Lockwood, London (1926).

UREN, L. C. *Petroleum production engineering—oil-field exploitation* (2nd edn.). McGraw-Hill, New York (1939).

WILLIAMSON, H. F., ANDREANO, R. L., DAUM, A. R., and KLOSE, G. C. *The American petroleum industry. The age of energy 1899–1959.* Northwestern University Press, Evanston (1963).

17

METAL MINING

JOHN TEMPLE

ONE of the most significant factors about metal mining in the first half of the twentieth century is that there was an enormous growth in the demand for metals and a consequent expansion in mining activities. Base metals such as iron, lead, zinc, copper, and tin became increasingly important as industrial societies became more complex and many new uses were found for them. For example, lead, because of its resistance to corrosion, had been widely used for water pipes as long ago as the Roman era. By the beginning of the twentieth century it was widely used in plumbing and to enclose underground telephone cables, but by 1940, with the development of the automobile industry, something like 40 per cent of all lead used for industrial purposes in the U.S.A. was used in batteries and for the manufacture of tetraethyl lead (anti-knock) in petrol. World production of tin in 1900 was 75000 long tons; by 1940 it had risen to 238000 long tons. This considerable increase was partly due to the extensive use of 'tin' cans as containers in the food packaging industry; for example, in 1924 nearly 10m. cans were made in Britain alone and by 1939 the figure had reached 400m. [1]. Tin was important in the nineteenth century in the manufacture of tinsmiths' and plumbers' solders, which are alloys of tin and lead, but the expansion of the radio and electrical industries in the first half of the twentieth century led to an enormous expansion in the use of tin as a constituent of solder. Some metals which were scarcely used at the beginning of the century became of prime importance because it was discovered they could be used to add toughness and strength to steel—in particular the group known as ferro-alloys, for example, chromium, manganese and nickel. Aluminium, for which a cheap reduction process of manufacture was discovered in the 1880s, became in the twentieth century a metal of enormous value because of its lightness, strength, electrical conductivity, and resistance to corrosion. Many new uses were found for it and its alloys, particularly in the aircraft industry.

It was not simply advanced industrial countries making greater demands for base metals that caused the expansion in mining. The emergence of the U.S.S.R. as an industrial power to rival countries such as the U.S.A. was based on a vigorous development of coal and mineral resources. From the mid-1920s onwards it was a primary Soviet task to increase the power of the state by exploitation of its own resources and to reduce its dependence on foreign supplies. Czarist Russia had imported considerable quantities of non-ferrous metals, but in the 1920s these became of basic importance for Soviet industrialization plans in electrical engineering, the chemical, automobile, and other industries. Thus, considerable expansion of mining took place and the emergence of a new industrial power must be reckoned as a factor in the extension of metal mining in the first half of the twentieth century.

Mining for precious metals continued to expand, but without any of the spectacular rushes such as those associated with California and the Klondike in the nineteenth century. Gold and silver continued to be used in considerable quantities for coinage, usually alloyed with other metals, and as the standard of many monetary systems. The advanced countries of the world accumulated gold in the belief that with considerable gold reserves their economies were sound. As in the case of base metals, Russia in the 1920s emerged as a major producer of precious metals, particularly gold, and in the 1930s she developed a number of new fields, emerging as a serious contender for the claim to be the world's leading gold producer.

I. OPENCAST MINING

Surface mining known as strip, opencast, or open-pit mining was being undertaken in a number of areas of the world by the beginning of this century and by the 1950s enormous advances were made in this technique. Initially, opencast mining was usually attempted because the ores to be mined were rich and lay near the surface. The technique was to remove the overburden with steam shovels, having first loosened it with preliminary blasting if necessary, to reveal the ore beneath. This was then scooped out with steam shovels and dumped into nearby railway trucks for transportation to reduction plants. As the ore was removed the track had to be constantly realigned in the open pit. This method was adopted by the United States Steel Corporation (founded in 1901) to mine rich iron ore deposits in the Mesabi ranges (Fig. 17.1) on the shores of Lake Superior. One of the features of such open pits was that the resulting ore was very rich, and the success of such methods in extracting rich ore bodies led to its adoption by other companies

FIG. 17.1. Missabe open-pit mine in the Mesabi Ranges in 1947. The immense scale of open-pit mining can be appreciated by comparing the size of buildings and trees on the sky line with the depth of the pit.

throughout the world, if they had the capital needed for such investment—for example, to mine bauxite to make aluminium.

A most significant step forward in opencast mining was made at the Bingham copper mine in Utah at the beginning of the century. Mining began there originally for gold and silver, but once these deposits were exhausted attention was turned to copper mining. The deposits were, however, of low grade, containing only 2 per cent copper, or less; this was considered to be unprofitable by the method then being used. Companies were used to working rich vein deposits such as those at Keweenaw in Michigan where 20 per cent copper was not unusual; most companies could not envisage making a profit with ores containing less than 10 per cent copper. In the circumstances, it was unthinkable that the Bingham deposits could be made to pay, since all that was left was vast quantities of low-grade disseminated ores. However, in 1899 two young engineers, Daniel C. Jackling and Robert C. Gemmell, wrote a report outlining a plan to mine and mill the ore by open-pit methods at the rate of 2000 tons per day. This took much courage for a number of reasons, not the least being that even a 500-ton mill was at that time considered huge; only a few existed, and those usually in more accessible areas. *The Engineering and Mining Journal* (an American publication) took a very poor view of this plan and in the issue of 27 May 1899 its editors wrote: 'It would be impossible to mine and treat ores carrying 2 per cent or less of copper at a profit, under existing conditions at Utah.' [2] However, the Jackling–Gemmell proposals were adopted, and through complex corporate and financial evolution the Bingham open-pit mine got under way by 1910. By 1913 the mill had a capacity of 4500 tons of ore per day. The most significant feature of this mine was that it proved that low-grade disseminated copper ores could be mined profitably if the scale of operations was sufficiently large. The Bingham engineers pioneered the large-scale exploitation of low-grade ores by open-pit methods and their techniques were to be copied in many parts of the world. Jackling and Gemmell's views were greeted with derision by many, but by the end of the Second World War ore containing as little as 0·75 per cent copper was regarded as profitable at many open-pit mines. The primary value of open-pit mining on this scale is that it can exploit poor ores; however, open-pit mining of good ores sometimes leads to underground mining once the most accessible deposits have been taken. This is what happened in the case of some copper mines in the Sudbury area of Canada. Open-pit methods were used until the 1920s and then underground mining began because the ore body dips very steeply at 45°.

In the early days of open-pit mining the companies stripped the ore with steam-driven shovels which dumped it into waiting railway wagons. Many mines, particularly copper mines in Chile and the U.S.A. continued to use this machinery except that electrically driven shovels replaced steam shovels, and they in turn were replaced by bulldozers powered by diesel engines. In such mines extensive systems of railway tracks were laid down, sometimes descending up to half-a-mile deep into the huge open pits that were created. With the invention of the internal combustion engine and the development of powerful diesel engines, many iron-ore mines changed their methods of transporting the ore and adopted diesel trucks, tractors, bulldozers, and belt-conveyor systems. This necessitated replacing tracks, locomotive repair shops, and locomotive engineers with roads, garages, and drivers. Such methods of transport were particularly suitable in the large, rather flat-bottomed iron-ore pits in the Lake Superior region of the U.S.A. By the early 1940s the change was well advanced in this area [3]. It should not, however, be assumed that all open-pit mining was conducted on this scale. The iron-ore deposits of Northamptonshire in England were discovered in 1852 and by the early twentieth-century steam shovels were being used to strip the overburden and mine the ore. This area never developed as did that around Lake Superior; for example, in 1933 there were some sixty open-pit mines, many of them employing no more than a dozen men. Even by the early 1950s the scale of operations in the most productive areas, around Corby and Scunthorpe (in Lincolnshire), was small. Walking dragline excavators were introduced about this date; they were used in American open-pit mines as early as the 1930s.

II. ALLUVIAL MINING

Surface mining was not confined to open-pit methods. Alluvial mining for tin and gold was common and could be divided into two main methods: hydraulic mining and dredging. By the beginning of the century dredging for gold was done with dredgers floating on man-made lakes dug from the gold-bearing gravels (Fig. 17.2). Powered by steam shovels, they dug out the gravel, washed out the gold, and deposited the debris or gangue behind, filling up the lake. This method was used in such widely separated places such as the western U.S.A., Australia, New Zealand, and Alaska. Dredging in the extreme climate of Alaska posed serious problems, since the ground was frozen for at least four to five months of the year. The working season was extended to some extent by using dredgers which were fitted with boilers

FIG. 17.2. A gold dredger working in Colorado in a man-made lake. Tin was mined in Malaya by a similar method.

and steam-spraying equipment. By the outbreak of the First World War many dredgers were electrically driven, particularly in the U.S.A. Dredging for tin was introduced into Malaya in 1912 by European companies and by 1940 accounted for 45 per cent of the total output. The earliest dredgers were steam powered, using wood as fuel but were later converted to burn coal. From 1926 onwards much larger dredgers were introduced which used electricity and by 1940 steam plants were rarely used in dredging operations. Conversion to electric power was expensive, and steam power thus maintained its popularity until the early 1930s.

Perhaps the most primitive methods of alluvial mining on a large scale were those used in the U.S.S.R., where the principal source of gold was the Lena goldfields in Siberia; these produced 83 per cent of Russian gold [4]. The gravels were dug out by pick and shovel and washed in primitive sluices, but from 1908 onwards European capital was obtained to instal dredgers and a hydro-electric plant. By 1910 it looked as if it could become the most profitable mining company in the world, but the uprising of 1913 and the subsequent revolution of 1917 changed everything. Once Stalin took over he initiated a major expansion programme in mineral development which was

designed, among other things, to ensure that Russian mines were state controlled. To develop gold resources he recruited American engineers who had
been working in Alaska, since they were familiar with geological and climatic
conditions similar to those of the far eastern provinces of the U.S.S.R. The
principal recruit was John D. Littlepage, who from 1928 to 1937 supervised
the machinery in some alluvial goldfields.

In 1931 a valuable source of alluvial gold was discovered on the Kolyma
river, east of the Lena goldfield. The whole area in which the Kolyma field
lay was virgin territory and was placed under the control of an organization
called Dal'stroy which developed the region. It is thought that the field soon
produced almost two-thirds of Russian output, but probably this was achieved
with the most primitive methods, using (it is alleged) convict labour estimated
to be as high as 5m.—the product of Stalin's purges in the 1930s [5]. Not
until the early 1950s did the size of the convict force decline and free labour
begin to be attracted to the area.

In some areas of the world where dredging operations on a considerable
scale were carried out it produced serious conflicts with agricultural interests,
who objected to the dumping of waste gravels once the gold had been extracted. Conflicts were common in California in the late nineteenth century and
early twentieth century, but the most serious problems were created by the
hydraulic method (Fig. 17.3). In this, powerful jets of water were used to
wash out the gravels; these were then re-washed, when the gold, being
heavier than the gravel, fell to the bottom of the washing boxes. The debris
remaining, which was usually suspended in water, found its way into near-by
rivers and streams and caused much damage to agricultural land further
downstream because water courses were blocked and flooding resulted. So
great was the outcry that in parts of California hydraulic mining was forbidden in 1884, and in this century it has been allowed only under licence,
such licences being granted only when due precautions had been taken to
prevent debris entering rivers [6]. Perhaps the most serious effects on a
landscape from hydraulic mining occurred in Malaya, where tin in considerable quantities was mined in this way. In the late nineteenth century the
Chinese extracted the most accessible tin ores but in the twentieth they concentrated their efforts into hydraulic mining; in the period 1924–8 almost 40
per cent of all Malayan tin was mined by this method. In hilly regions the
clean-weeding policy on estates caused soil erosion, which, with hydraulic
mining and the extensive use of water for ore-concentration purposes, resulted
in the choking of rivers with sand and clay. Consequently in some areas this

FIG. 17.3. Hydraulic mining. Gold bearing gravels were washed out using a powerful jet of water and carried down to sluice boxes where the gold is separated out. Such a method was used for tin mining too, and both caused serious environmental problems.

method was banned in 1933, although it was permitted near Kuala Lumpur with strict control over the disposal of tailings. The mines were required to build a tailings dam into which their muddy streams were finally directed, so that sand and clay were generally retained. However, hydraulic mining still left serious effects on the landscape, the areas of mining taking on the desolation of a lunar landscape [7].

III. UNDERGROUND MINING

Underground mining was carried out for most metals throughout the period under consideration but it lost ground to open-pit mining, although it must be recognized that some deposits could be worked only underground. For example, because of their depth the gold deposits of South Africa had to be extracted by underground mining. In some areas it has largely replaced open-pit mining; for example, the valuable iron-ore deposits of Kiruna in

FIG. 17.4. General view of the Kiruna iron ore mines in 1936; an open-pit mine where surface mining has ceased and been replaced by underground operations.

Sweden (Fig. 17.4), and in some of the copper mines of the Sudbury basin in Canada.

Mining methods underground were most varied, depending upon the configuration of the deposit, and it is impossible to provide an adequate picture of them all [8]. Perhaps one of the most common was the horizontal cut and fill method, which was used in many mines at Broken Hill in Australia. The ore body was reached from tunnels driven from stations in the shafts. These stations were at intervals of 100 to 150 ft and it was from the tunnels that mining was carried out. A long strip of ground was cut away until the stope was like a large rectangular hall 10 to 12 ft high, with stacks of timber to support the roof. Part of the bottom of the stope was filled with barren rock, leaving a space of only about 5 ft between floor and roof. From this platform of rock the miners drilled the roof bearing the ore, and as they raised its height by removing ore from it they packed more rock beneath them. Special care had to be taken when extracting the ore immediately below the level above. For mines in highly unstable rocks such a method was impossible, and in some Broken Hill mines the system of square-set timbering had to be used (Fig. 17.5), a method copied from the famous Comstock Lode in the U.S.A. Enormous quantities of timber were imported from Oregon to shore up the roof but the danger of fire was a constant hazard. In the Butte copper mine in the U.S.A. 40 to 50m. ft of board timber was installed annually using this square-set method of roof support [9].

G. 17.5. Broken Hill in the 1900s, showing the square set system of timbering the roof.

One of the features of open-pit mining was the removal of all the ore body in large quantities. The underground equivalent of such a method, developed in the early twentieth century, was block caving. Where this was first used is uncertain, but it was common in the copper mines of the Sudbury Basin in Canada by the 1930s and in copper mines opened up in the same period in Central Africa in what was formerly Northern Rhodesia. In the nineteenth century underground miners in metal mines in areas such as Cornwall followed the vein of ore very carefully until it was almost exhausted and then attempted to find it again, usually at another level. Block caving ignored such caution, removing all the ore body from the stope walls by undercutting and allowing the ore to fall down prepared cavities by gravity, shattering it in the process. The commercial success of such a method had already been demonstrated in many areas when, in 1947, the Anaconda Copper Mining Company embarked on one of the most ambitious and costly programmes up to that date, using the method to mine 180m. tons of low-grade ore at Butte. This mine had been exhausted of all high-grade deposits; only by block caving could the remaining low-grade copper be mined at a profit.

IV. EQUIPMENT

The advance in mining methods was made possible only by corresponding advances in equipment. One of the most time-consuming tasks in mining was to drill holes in the ore to shatter it with explosives. Early in this century holes were drilled by hitting a hard mining steel with a hammer, one man hitting while the other gave the steel a quarter turn after each blow. Pneumatic drills powered by compressed air (Fig. 17.6) were used in some mines but they were difficult to erect and dismantle since they weighed about 100 kg each. Further, they filled the workings with clouds of dust and thus were a serious health hazard to miners; one American rock drill was aptly named 'the widow maker'. Playing water on drills to lay the dust, and use of good ventilating systems were the answers to this problem, but many companies would not contemplate them because of the cost. In 1907 a drill called the water Leyner drill was introduced to some mines; this not only reduced the dust hazard but was lighter than other drills (around 60 kg) and could be operated by one man. The principle of one man per drill was resisted in some areas but, in general, miners were in no position to resist such innovations. One of the problems associated with improvement of drills was that of making bit heads tough enough to withstand the exceptionally tough rocks encountered in areas such as the Witwatersrand in South Africa. Much

FIG. 17.6. A South African gold mine in the early 1920s with a crew using a compressed air machine drill. These replaced hand drills but made mines dustier and noisier.

research went into the use of an abrasion-resistant material for the cutting edge of the drilling bit. Many materials were tried in the 1930s but the most successful proved to be tungsten carbide (Ch. 18, p. 454). Eventually a drill stem tipped with tungsten carbide was produced which required sharpening far less frequently than a forged steel bit. Just as the quality of drills improved, so did the quality of explosives; this reduced the hazard of blasting operations.

Many mines used horses underground to haul trucks along the main roadways; Broken Hill mines in the 1900s were a typical example. In the 1930s many mines began to experiment with diesel locomotives, although many more had electric locomotives, both trolley and battery types, by the 1930s. The latter required the installation of expensive equipment which was easily damaged. Diesel locomotives added to mine-ventilation problems, but by the early 1950s this had been solved by the use of what are known as exhaust scrubber units which removed harmful gases. Thereafter, diesel locomotives rapidly replaced electric since they cost less initially and were much more mobile. Some mines faced serious problems in ore removal if the dip of the strata did not permit the running of the mined ore into wagons by gravity. For many years the only solution was hand loading with shovels into trucks.

In the early 1930s a device known as a scraper winch was introduced into the East Rand mines in South Africa. It consisted of a double-drum winch, the ropes from which operated a hoe-type scraper which was dragged up and down the working face to pull the broken ore into a truck. So successful was this machine in South Africa that it was introduced in almost all mining operations where the broken ore did not run by gravity.

Considerable advances were made in mine ventilation which were made necessary by the increasing depth of mines, particularly in some areas, by the 1920s. Bratticing of shafts, i.e., separating compartments into upcast and downcast sections, was tried with varying success, but some mines still relied on natural ventilation. Some had shafts solely for ventilation purposes and equipped with fans to exhaust the foul hot air, thus drawing fresh cool air down the operating shafts. This method was very common in coal mines by the late nineteenth century. Attention was also directed to the efficient use of ventilation underground. Galvanized iron tubing, with small electric fans to force fresh air into, or to exhaust foul air from, the mining face began to take the place of earlier attempts to ventilate with brattice walls of brick or other material. These measures did not prove successful in all mines, particularly the deep gold mines of South Africa, where rock temperatures rose above 38 °C. Engineers began to consider artificial means of cooling. At first blocks of ice were sent down mines but this was soon discarded because it was not only expensive but gave little benefit. Then refrigeration plants to cool the ventilating air were installed on the surface; large quantities of cooled air were pumped down the shaft to the workings. This method had some success, but owing to the distance the air had to travel before reaching the working places much of the cooling effect was lost by heat absorption *en route*. To reduce this loss the plants were installed underground, but the choice of sites was limited because the plant had to be adjacent to the return airway leading to the upcast shaft into which the extracted air could be pumped without interfering with the general mine ventilation. The solution found for the South African gold mines in the 1940s was to use twin main headings, a practice common in coal-mines. It consisted of two parallel headings driven some fifty feet apart in place of one wide heading. The headings were connected every 500 ft; one was used as an intake airway connected to the downcast shaft and the other as a return airway connected to the upcast shaft. This was highly successful; large quantities of air could be circulated and the face of each heading was never more than 600 ft from the last through connection.

With the increasing depth of many mines, the design of hoisting equipment grew more complex, and a limit was being reached in the depth of wind possible with an economic load. This was because, with the increasing size of the winding ropes required, the weight of the rope itself would ultimately be greater than the maximum permitted breaking load. In mines deeper than 5000 ft consideration was given to a system of winding in two stages through a vertical and sub-vertical shaft system. Such a method was much too costly, and multi-rope winders were devised. The load was attached to a number of ropes of relatively small diameter and the weight was taken up by a compensating device, which allowed a much greater depth and heavier load to be attained than with a single rope.

V. ORE CONCENTRATION

Advances in mining technique in part depended upon the continuing development and improvement in the methods used to obtain concentrates of the minerals. If the ore contained comparatively little earthy matter, as in the case of iron ore, it was possible to remove the gangue during smelting. If the ore occurred in a simple geological structure the gangue could be removed by hand; thus some lead mines in the nineteenth century employed women and children to separate the ore from the gangue in preparation for concentration. Such a method was somewhat slow, expensive, and quite inappropriate to some mines. Minerals often occur in a complex form with three or more intermixed—for example, copper, lead, zinc. The separation of such ores for concentrating is a difficult problem and was never possible by hand picking. By the beginning of the century almost all mines used concentration mills which operated on the principle that various minerals have different specific gravities. In the case of lead the ore was crushed, mixed with water, and shaken on the rattling jigs or tables. The particles of galena (specific gravity 7·5) fell to the bottom but the ligher materials were swept away. In mills where the crushed material contained different ores, the heaviest was left but lighter minerals (for example, zinc, specific gravity 4) were lost. Mines with complex ore bodies lost a great deal of valuable ore; for example, one mine at Broken Hill in 1903 over a period of six months treated ore containing £380000-worth of metals but managed to extract only £90000-worth [10]. Such losses were not exceptional in mining areas where the ore bodies were complex. Where the material had magnetic properties the earliest attempt to counteract such losses was to build concentration mills with huge magnets, but neither this nor the gravity method was the answer to the problem.

The solution was found in the flotation process which was probably discovered independently, but simultaneously, by chemists in the U.S.A. and
Australia. One of the earliest inventions was a process known as skin or film
flotation. The crushed ore was covered in oil and dropped on to a tank of
water where the mineral particles floated and the barren particles sank. The
defect of this method was that it required a considerable surface area of
water to obtain a large output, but by 1910 one company was successfully
operating it at Broken Hill. The next step was to develop a flotation plant that
would select the minerals to be separated. By the end of the First World War
such a plant had been built—again at Broken Hill. These plants were vital to
mines that had complex ore bodies but were equally important to copper
mines, since copper ore has a low specific gravity and does not easily separate
from its gangue in gravity mills. Iron and copper are often found together in
copper mines; their specific gravities are similar and selective flotation was
vital to the continuing economic viability of copper mines with iron present.
The Anaconda Company in the U.S.A. increased its recovery of copper from
79 per cent to 95 per cent by changing from gravity mill to flotation mill in
1915; further success followed the adoption of selective flotation. After these
discoveries the flotation process was extended to tin, zircon, mica, and manganese, replacing fire and water as the dominant method of extracting the
world's minerals. 'In the last thousand years in metallurgy it stands with the
cyanide process, and the Bessemer process, as one of the three greatest
advances.' [11]

The cyanide process for the extraction of gold was discovered in 1889 by
two Glasgow physicians, Robert and William Forrest, and a chemist John
McArthur patented the process. Until then gold mined in vein deposits was
separated from its gangue by using mercury or chlorine gas, but neither was
suitable for ores with a low gold content. Cyaniding was first used with
success on the tailings on the Witwatersrand but was then used for all
extracted ore. Undoubtedly the continuing development and prosperity of
these mines was due to the application of this process; without it they would
have had to close. Since the beginning of the century the process has been
applied to all major gold and silver mines exploiting vein deposits.

The process consisted of treating the finely ground ore with a weak cyanide
solution as a solvent for the precious metal; separating the solids from the
clear solution; and recovering the metals in the form of a black precipitate,
by means of zinc shavings. Stamp batteries were used to crush the ore; early
accounts of gold mining around Johannesburg indicate that the town rever-

berated with the constant thud of these machines. An early improvement in crushing was the use of tube mills which consisted of a horizontal steel cylinder provided with a suitable liner (for example, pebbles) and made to revolve about its long axis. The grinding charge of pebbles rolled through and over the ore, grinding it into small pieces; stamp mills were no longer needed. In the 1920s the all-sliming process was introduced; water was fed into the tube mill, reducing the ore to a mixture of sand and slime. Ball milling was introduced in the 1930s; in this, the tube mill was charged with steel balls instead of pebbles as the agent to break down the ore. The final development in this part of the process was stage grinding; a first grinding circuit comprising ball mills, and a secondary or regrinding circuit of pebble mills. Once ground, the material was charged with cyanide in large circular vats. A regulated flow of zinc dust was added to the solution and the gold was precipitated as slime; most of the zinc was dissolved. The slime containing the gold was then purified by furnace methods.

By the early 1950s man had still barely scratched the surface of the earth in his quest for metals. However, the need to give careful attention to the depletion of resources and the quest for new sources was a feature of the technical literature of the period. Further sources were sought in deeper mines, in undeveloped countries in Africa and South America, and below the surface of the sea, although the problems of extraction from the sea were seen as a serious problem. Much of the literature of the period saw mineral needs only in the light of the requirements of western industrial societies. No serious consideration was given to the emergence of the 'Third World' and the possible changing demand for minerals. The total domination of western societies in the exploitation of minerals was drawing to a close, although few can have seen it at this time.

REFERENCES

[1] ALEXANDER, W., and STREET, A. *Metals in the service of man* (2nd edn.). Penguin Books, Harmondsworth (1945).
[2] *Engineering and Mining Journal-Press*, **167**, 80 (1966).
[3] *Minerals Yearbook, 1941*. U.S. Government, Washington (1943).
[4] CONOLLY, V. *Beyond the Urals*. Oxford University Press, London (1967).
[5] ARMSTRONG, TERENCE. *Russian settlement in the north*. Cambridge University Press, London (1965).
[6] JONES, S. J. The gold country of the Sierra Nevada in California. *Transactions of the Institute of British Geographers*, *15* (1951).
[7] KING, A. W. Changes in the tin mining industry of Malaya. *Geography*, **24** (1940).

[8] STOCES, B. *Introduction to mining*. Pergamon Press, London (1958). (A detailed examination of methods is provided in Vol. I, pp. 356–70.)

[9] *Engineering and Mining Journal-Press*, **167**, 82 (1966).

[10] BLAINEY, G. *The rush that never ended: A history of Australian mining*, p. 259. Cambridge University Press, London (1967).

[11] BLAINEY, G. *Op. cit.* [10], p. 271.

BIBLIOGRAPHY

ALEXANDER, W., and STREET, A. *Metals in the service of man* (2nd edn.). Penguin Books, Harmondsworth (1945).

BLAINEY, G. *The rush that never ended: a history of Australian mining*. Cambridge University Press, London (1963).

CARTWRIGHT, A. P. *Gold paved the way*. Macmillan, London (1967).

COGHILL, I. *Australia's mineral wealth*. Longman, London (1972).

CONOLLY, V. *Beyond the Urals*. Oxford University Press, London (1967).

JONES, W. R. *Minerals in industry* (4th edn.). Penguin Books, Harmondsworth (1963).

MacCONACHIE, H. Progress in Gold Mining over Fifty Years. *Optima*, September 1967.

RICHARDSON, J. B. *Metal mining*. Allan Lane, London (1974).

SHINKIN, DEMITRI B. *Minerals—A key to Soviet power*. Harvard University Press, Cambridge, Mass. (1953).

STOCES, B. *Introduction to mining*. Pergamon Press, London (1958).

WARREN, K. *Mineral resources*. David and Charles, Newton Abbot (1973).

WEBSTER SMITH, B. *The World's great copper mines*. Hutchinson, London (1962).

WILSON, M., and THOMPSON, L. (eds.) *The Oxford history of South Africa; Volume II, South Africa, 1870–1966*. Clarendon Press, Oxford (1971).

18

THE UTILIZATION OF METALS

W. O. ALEXANDER

DURING the nineteenth century, engineers and metallurgists had changed the making of iron and steel from an empirical art into a scientific process. They had found ways of using the mechanical properties of these metals for such diverse engineering purposes as the construction of bridges, power stations, steam locomotives, internal combustion engines, and ships. To make all such structures and mechanisms reliable, it became necessary to develop a working understanding of the mechanical properties of many metals, but particularly of steel, with its varying carbon content and alloys made by addition of other metals. These mechanical properties related to such aspects as response to tensile and compressive stresses; to the onset of brittleness; to the elementary understanding of the 'fatigue' properties of metals; and to a range of techniques for joining metal parts together, such as by riveting and bolting. This went hand in hand with great strides in the accuracy with which parts were machined and fitted together, and the development of metrology as a controlled science. Further, in the various machines and engines all the surfaces of metals were required to operate under rotating and oscillating conditions and it therefore became essential to improve the fit and smoothness of these sliding surfaces, and to improve the wear resistance and reduce friction to an absolute minimum. This led to the emergence of a branch of metallurgy where, by special heat-treatment procedures, the surface of the steel was given properties different from those in its bulk.

In the nineteenth century, the thickness of all steel and iron parts used was more than ample to allow for considerable corrosion, but even so various methods of surface protection were developed, partly to improve the appearance of the structure; in particular, red-lead paint and other paints based on linseed oil were used (Ch. 28). The engineers of those days kept all their machinery free from corrosion by frequent wiping with oily rags. However, during the twentieth century—partly because the sections decreased in thickness and partly because metals were required to operate in more severe environmental conditions than hitherto—the protection of metals by a range of coating and other techniques increased very considerably.

Although the techniques of joining by soldering, brazing, and hammer welding were well known from ancient times, other methods of autogeneous, or homogeneous welding, particularly of steels, were slowly developed during the period under review. Today, there are some thirty processes that can be used to weld varieties of steels and other metals, according to the different conditions of assembly and operation.

The techniques of casting molten metals, particularly into sand moulds, had reached a peak of perfection so far as accuracy and surface finish were concerned by the time of the Great Exhibition of 1851 [1]. Despite this, iron castings were regarded as relatively weak and brittle and not suitable for the major engineering applications. Cast iron had the advantage of conferring rigidity on any machine body and so was extensively used for the beds of lathes and reciprocating engines, but during the next fifty years it became clear that despite its advantages it had to be used much more efficiently, in thinner sections. This in turn, led to a whole range of improvements in casting procedures and techniques [2], [3].

By 1900, the metals which, in terms of tonnage, were most extensively used were iron and steel: iron in the form of cast iron, and steel in the form of low-, medium-, and high-carbon steels and a limited range of alloy steels such as Hadfield's manganese steels. There were also copper and copper alloys, including the brasses and bronzes; zinc, mainly in the form of zinc and galvanized sheet; lead, mainly in the form of lead sheet and pipe; and tin, as a coating for steel to make tinplate for canning, in the form of solders, and for some early collapsible tubes. None of the modern metals had yet emerged on the industrial and commercial scene. Although aluminium had been extracted commercially in the 1880s it was still costly, far from reliable, and subject to corrosion attack, particularly in sea-water or a marine atmosphere.

In the present context the further development of the traditional metals, and the utilization of the newer ones, may most appropriately be considered by following the concomitant engineering developments during the first half of this century. Many fields of engineering were created and expanded during this period, and many of these developments overlap and intertwine, particularly in respect of the metals and alloys used. The developments considered will be under the principal headings of: automobile engineering, aeronautical engineering; electrical engineering and power generation; the communications industry; the machine tool industry; transportation; and construction and assembly. All these fields are, of course, the subject of separate chapters elsewhere in this work.

I. AUTOMOBILE ENGINEERING

In the first decade of the present century, motor-cars were designed and constructed as one-off pieces of engineering (Ch. 30). During this initial period many improvements were made in the metals and alloys used for various components. One example is the automobile radiator, which was continuously modified to become an elaborate piece of constructional engineering. Even in the early days, use was made of copper or brass tubes. These were subsequently threaded through holes punched in strips of copper or brass. The whole assembly was completely soldered, giving a matrix through which water flowed in the vertical direction with air cross-flowing horizontally as the car moved forwards (Fig. 18.1). This construction called for

FIG. 18.1. (*Left*) A Sunbeam Mabley motor-car, showing the finned radiator made of copper with soldered brass header tanks, 1901. (*Right*) A Vauxhall, 1903, with a simpler copper tube radiator formed in a continuous coil. Note the paraffin lamps, good examples of brass metal work formed from sheet and strip, by cutting, blanking, pressing, spinning, and drawing. The single cylinder engine of the Sunbeam is made of cast iron, with fins integrally cast and a separate domed cylinder head.

accurate rolling of copper into strip and accurate drawing of thin-walled tubing; the latter was sometimes flattened in order to present a greater surface area for cooling purposes and less resistance to air-flow. There were also improvements in the solder alloys to give an appropriate melting-point, and in the fluxes to ensure proper wetting of the solder and copper or brass. By these means, capillary joints were made which were completely leakproof.

The internal combustion engine itself was by this time designed to use cast iron for the cylinder block and head, with a forged steel crankshaft and steel connecting rods. Cast iron was also used for the piston and piston-rings and the latter were continuously improved in design and manufacture to ensure good compression. It had already been found with the steam engine that cast iron rubbing on cast iron—with appropriate sealing rings and in the presence of a thin film of oil—had good wear resistance under conditions of high temperature and rapid oscillation. During this period a type of malleable cast iron was made by a special 'black heart' or 'white heart' annealing technique; this enabled cast iron to be used for the smaller components, such as steering linkages. In such applications, normal cast irons were brittle under shock loading; malleabilizing conferred ductility on the iron. This was in the days before mass production made it possible to produce such parts economically by drop-forging, and so a great number of parts were cast in jobbing foundries.

For the internal combustion engine a new range of parts had to be designed for the petrol carburation and electric ignition systems. Petrol tanks were made from pressed steel or brass by hand shaping and folding, the joints being sealed by soldering. Petrol was then conducted by way of appropriate brass stopcocks and copper piping into a carburettor, also made of copper and brass; the component parts were machined from tube, or in some cases, castings. Float chambers and needle valves were made of brass, and the induction chamber of either copper tube or cast iron. The spark for igniting the vapour in the compression chambers was generated either by magneto or, in some early cases, by hot flame. The design of the former necessitated the development of fairly finely braided cable to conduct electricity to each sparking plug. Special alloys had to be developed for the points of the plug, to generate the millions of reproducible sparks necessary for igniting the vapour. This was one of the uses of nickel, alloyed with small amounts of magnesium and silicon. The make-and-break points in the magneto did not have an adequate life when made from copper or silver; tungsten was found to give the best wear and spark resistance. In the earliest days, all lighting

FIG. 18.2. *Above*, the four-speed gearbox as fitted to the Rover Ten Special and Pilot 14 models. *Right*, a photomicrograph of case-hardened and carburized steel of the type used in early gearboxes. This shows increased carbon content at the top and a gradual decrease downwards. A uniform transition from a hardened microstructure to a softer one in the centre results. ($\times$ 50).

systems used acetylene gas and many cars were, therefore, without a battery, since enough electricity was generated from the magneto for ignition purposes.

The transmission from the engine at the front to the rear drive-wheels was by a steel shaft with a rather crude type of universal joint. The back axle embodied differential gears, the teeth of which had been case-hardened. The early gearbox (Fig. 18.2) was a continual source of trouble. The case was of cast iron but the shafts and gears were all carburized and case-hardened. The metallurgical control conditions for this last operation were not very well known and the carbon diffusion gradient from the tooth-face to its root was not good. The method of double-declutching had to be employed in order to get gears to mesh satisfactorily. There was no synchromesh and no ingoing relief of the gear teeth to facilitate a smoother change. Consequently, many gearboxes were wrecked by heavy handed gear-changing. The hard case flaked off from the working profile of the gear teeth, and as the flakes fell into the oil in the gearbox further damage was done.

During the twenty years between the wars, motor-car manufacture expanded rapidly and production-line manufacture became general practice. This meant that all parts to be fitted together had to be made accurately to within five thousandths of an inch; alternatively, fittings and fastenings had to be devised which could accept a greater tolerance. With the great increase in production during those years, metallurgical operations for the manufacture of strip and sheet for the body and chassis, and for the manufacture of castings and of many drop forgings underwent drastic improvements. Some were technical, others were aimed at better control and inspection; all sought to ensure cheap and reproducible properties in the metal parts. The jobbing iron foundry (Fig. 18.3) presented problems because varying qualities of cast iron were supplied to them, there was inadequate control of dimensions, and some variation in the surface finish of the finished castings.

During the 1920s and 1930s, every aspect of iron founding became subject to greater scientific control. So far as the metal was concerned, enough physical metallurgy and knowledge of the equilibrium structures of cast irons had been developed to ensure that—with appropriate control of composition and the subtle addition of certain elements in one form or another—a reproducible grey iron was obtained. This had good mechanical properties and a low surface friction, ensuring good wear resistance in the cylinder walls and piston rings. Apart from control of the quality of the iron, the development of moulding machines made it easier to ensure that the sand was compacted and rammed home uniformly. This gave a casting with good uniform surface appearance without any collapse of the facing sand of the mould, leading to surface markings in the metal. To ensure this reproducibility meant very accurate control of the particle size of the moulding sand, and of the nature of the additives necessary to ensure adequate bonding while the molten metal was being poured in and the subsequent liberation of a slight cushion of gas to protect the sand and prevent it from fritting on to the surface of the castings. To reduce the amount of man-handling involved in moving the heavy sand-moulding boxes when filled with molten metal, they were designed to be moved on roller conveyors. Casting machine circuit layouts were developed which consisted of a moulding machine; a conveyance to an appropriate casting section where the molten metal was poured in; a move to a cooling bay; and then transport to a knocking-out area in which the sand was removed

FIG. 18.3. The deep sand-moulding bay of a jobbing shop for cast iron. The man in the foreground on the right is ramming up the sand in the flask or moulding box, with a wooden pattern in position. Various flasks showing sand already moulded with appropriate cavities are also evident. The man in the background on the left is repairing a mould before casting, and the two in the centre background are sleeking or giving a smooth surface to the sand cavity. This photograph dates from *c.* 1950, but represents a survival of the process in its original form.

FIG. 18.4. A modern semi-automated foundry. In the foreground two men are pouring metal from a ladle using a hand shank. Moulds are on a roller conveyor; in the background are jolt squeeze machines for moulding the sand on to a pattern plate.

from around the casting and the moulding flasks were returned to the start for the manufacture of a new sand mould (Fig. 18.4).

Another significant development during the period was the improvement in the procedures, control techniques, and methods of manufacture of cast metal shapes by pouring molten metal into metal dies. This gave greater rates of output than could be obtained with sand moulds. It also gave more accurate dimensions, thus minimizing machining to prepare the finished surfaces. Such castings were known as die-castings and the development of die-steels which gave lives of the order of five hundred thousand, or even a million shots of molten metal was a very significant improvement. The achievement of lives of this order was also helped by devising special alloys which, while having good mechanical properties, did not attack the die steels during casting. In particular, three were developed in the aluminium industry, namely an aluminium alloy containing 11·5 per cent silicon and two other alloys which contained copper as well as silicon. In the 1920s, a zinc-based alloy would contain about 4·5 per cent of aluminium, which was added to prevent attack of the molten zinc on the steel of the dies. The aluminium content, however, led to severe corrosion problems, and zinc-based die-castings of that period often disintegrated in service into pieces of metal covered with white encrustations. The main reason for this rapid deterioration proved to be the presence of traces of lead, tin, and cadmium impurities in the zinc. By changing the process for distillation of zinc from horizontal batch production to integrated vertical distillation it became possible to manufacture zinc of 99·99 per cent purity, free from the harmful impurities. This development, coupled with the addition of about 0·5 per cent magnesium to the casting metal led to the introduction of a famous series of zinc die-casting alloys known as the Mazak of Zamac series (Fig. 18.5). All were based on 99·99 per cent pure zinc, and are used to the present day. These new alloys gave absolute dimensional stability and freedom from accelerated intercrystalline corrosion attack. Thus a new and expanding market for zinc was ensured and a high proportion of zinc-based die-castings were incorporated in all motor-cars of this period, particularly in the 1930s, for such parts as carburettor bodies (Fig. 18.6), door handles [4], interior and exterior trim, and some electric light fittings. In many of these applications a bright surface finish was obtained by chromium plating.

Aluminium die-castings in those days were relatively more expensive because the aluminium metal, although of lower density, cost more than zinc and was less pure. The development of an aluminium–silicon alloy was

FIG. 18.5. A pre-Second World War Madison Kipp pressure die-casting machine. The operator is withdrawing a finished casting made of Mazak.

FIG. 18.6. Die-cast part of a carburettor body showing the base plate and induction throat. Note the smooth surface finish, dimensional accuracy, the large number of holes, intricate re-entrant angles, and thin wall structures.

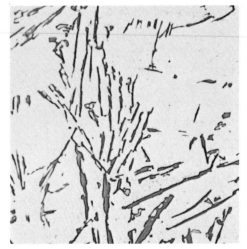

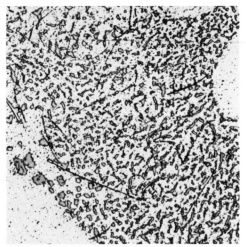

FIG. 18.7.(a). Microstructure of an alloy containing 88·5 per cent aluminium and 11·5 per cent silicon, in an unmodified condition. Needle-shapes of coarse silicon can be seen. (× 500).

FIG. 18.7.(b). The same alloy in a condition modified by the addition of 0·02 per cent sodium. The structure of the silicon is highly refined, and the mechanical properties and wear resistance thereby much improved. (× 500).

most important for the further improvement in the performance of the motor-car engine. Although cast iron pistons had been very satisfactory for many years—showing very low wear and hence adequate maintenance of compression—cast iron is relatively a very heavy metal to use for reciprocating parts. As mentioned later, the development of aluminium pistons for aeronautical engines became essential in the early years of the First World War, and it therefore became natural in the 1920s to utilize aluminium alloys for pistons in automobile engines. It was found that by adding about 11 per cent silicon and the merest trace of sodium (as low as 0·02 per cent) the structure of the silicon constituent in the aluminium alloy could be refined so that good mechanical properties and also excellent wear resistance were obtained (Fig. 18.7). The aluminium–silicon alloy became the standard alloy for pistons in most motor-car engines designed and manufactured during the remainder of the period. At about this time the more sophisticated and higher performance motor-car engines contained forged aluminium connecting rods, again borrowing from improvements made in aeronautical engines.

In the inter-war years, relatively heavy solid bearings in bronzes and Babbitt metals (so called after their nineteenth-century inventor Isaac Babbitt) used for the little and big ends of connecting rods, in gearboxes, and in the rear axles also gave way to much lighter bearings. Ultimate replacements were either steel shells with a thin coating of white metal, or a range of bronzes made by powder metallurgy [4]. These oil-less bearings were made from powdered bronze metal, compacted and sintered to give a coherent mass. Enough porosity remained for oil to be absorbed on quenching the bearing in an oil bath. The high-speed rotating parts of motor-cars which were not load-bearing tended to use roller or needle bearings. These in turn, had to be developed to a very high state of perfection both in surface finish, accuracy of dimensions, hardness, and freedom from distortion.

In the 1930s designs of gearboxes became more complex. Gears were required of much greater accuracy, much more reproducible in respect of dimensions, with far better surface finish and, above all, perfect diffusion gradient from the case-hardened surface through to the core of the gear. During this period, therefore, special alloy steels came into being which responded uniformly to carburizing, quenching, and tempering. Such heat treatments had to be devised to give the best combination of a high surface-hardness coupled with a softer core, but having the higher strength and higher ductility necessary to absorb the shocks imposed by gear-changing and the transmission of fluctuating loads from the engine to the rear axle.

One method developed was based on the principle of diffusing a certain amount of carbon into the surface layers by a pack-carburizing technique. The parts to be carburized are packed in a mixture containing a quantity of charcoal, together with bone-ash and certain accelerators such as barium carbonate. All is packed uniformly in a steel box and the whole is heated in a furnace to a temperature of about 900 °C for up to 48–50 hours. On cooling, after taking out of the furnace, the parts can be rapidly separated from the case-carburizing compound and quenched in water or oil or, alternatively, allowed to cool and re-subjected to heat treatment. A more elegant method was developed in the 1920s; in this nitrogen was diffused into the surface of certain steels containing specific alloying elements such as aluminium. Two processes were devised. One, using molten sodium cyanide salts, had the advantage that a very hard, though shallow, case-depth could be produced and distortion on subsequent heat-treatment was minimized. The other was developed by using 'cracked' ammonia, the nitrogen gas being activated to diffuse directly into the steel. This system, however, is rarely used now.

Apart from methods of increasing the surface hardness by diffusion techniques, use was also made of the controlled transformation which can be brought about by the addition of specific elements to plain carbon steels. In particular, a range of nickel–chromium steels was developed whereby use could be made of the phase transformations which occur in steels and are responsible for hardening and subsequently toughening them and increasing their ductility. This is normally carried out by a special double-quenching and tempering treatment. Such steels were obviously more expensive than the case-carburized carbon steels but they were used for gears, particularly for heavy duty gears, in the late 1920s and the 1930s. Such a steel would have the following typical composition: 3·7 per cent nickel, 0·8 per cent chromium, and 0·2 per cent carbon.

Further improvements came about by induction heating of metals. By accurate adjustment of the frequency and the size of the coil through which an alternating electric current was passed, eddy currents could be induced in the surface of the steel, thereby heating that portion to a higher temperature than the mass of the steel. On quenching, this gave a hardened surface with a softer and tougher core [5a], [12].

At the very end of this period, when it had become the practice to heat-treat shafts for large steam turbines and for the 200-MW generators then being manufactured, occasional failure by shattering of the shaft occurred in service. The results were catastrophic, since these shafts rotate at very high speeds. Such disintegration could be very dangerous, wrecking the whole machine and probably the building in which it was housed. This phenomenon was shown to be due to the presence of traces of hydrogen in the steel, which persisted throughout the melting and forging operations. This hazard was overcome by close attention to de-gassing and prolonged annealing procedures to enable the residual hydrogen to diffuse out of the forging.

II. AERONAUTICAL ENGINEERING

During the first half of this century, man developed from achievement of non-directional flight to the ability to carry about fifty passengers or up to 25 tons in freight in a single aircraft (Ch. 33). To obtain the necessary power with the minimum of weight, a series of sophisticated reciprocating petrol engines was developed with up to 24 cylinders disposed about the circumference of two circles or alternatively in banks of six, in two Vs. All these engines were superseded by the development of the gas turbine towards the end of the Second World War. Such engines were, however, not used in commercial aircraft until after 1950.

The fuselages of the earliest flying machines were almost entirely composed of non-metallic materials—mainly spruce and cane—together with fabric which was made taut and non-porous by painting with a cellulose lacquer dope. The structure as a whole was made rigid, though slightly adjustable, by the use of steel piano wire with torsion units which tightened the wire to a standard tension. This method of construction had the great advantage that it could be assembled by traditional woodworking and other simple techniques; some hundreds of thousands of machines were built in this way during the First World War.

In 1909, a revolutionary discovery concerning the hardening of aluminium was made by accident. This discovery arose as a result of some German research on the possible use of aluminium for coins. Aluminium metal in its pure form is relatively soft and it was required to harden it to make it more durable and wear-resistant. Alfred Wilm had prepared an aluminium alloy containing 3·5 per cent copper and 0·5 per cent magnesium together with traces of iron and silicon, which were always present as impurities in those days. This worked alloy was then quenched in water after annealing at about 500 °C. A hardness test was taken but was not found to be very significant. However, early the next week, further tests showed that the hardness and the strength were very much higher than the original values obtained. As a result, Wilm performed a series of experiments to study the effect of storing the alloy for different periods after heating and quenching in water. He found that the strength gradually increased to a maximum in four or five days. The phenomenon became known as age-hardening, but subsequently also by a variety of other titles, such as precipitation hardening, temper hardening, etc. In 1909, Wilm gave to the Durenner Metalwerke, at Duren, sole rights to work his patents; hence the name of the alloy became Duralumin. In the event, it was the Zeppelin and not the aeroplane which was destined to make the first use of these revolutionary properties in aluminium alloys. Intense efforts were made to overcome the difficulties of making the alloy into strips and sheets, and during the First World War large amounts of age-hardened duralumin were used to make girders, and rivets to join them, for the first Zeppelins. The alloy was only subsequently used for other types of aircraft (Fig. 18.8). Metallurgists all over the world attempted to explain this mechanism of age-hardening but it was not until some years after the original discovery that the problem was even partially solved. The investigation necessary to reveal the true cause of age-hardening was slow and laborious, because the only instruments then available, including the metallurgical microscope, were not sufficiently sensitive to follow the changes in the internal microstructure of the alloys [6].

At the National Physical Laboratory, Teddington, which was one of the main centres of metallurgical research in Britain, an investigation was commenced to see whether age-hardening could be induced in other aluminium alloys, and particularly in any alloys which would remain strong at the temperature at which an aircraft engine piston had to work. This led to the discovery of an alloy still known as Y alloy; this contained 4 per cent copper, 2 per cent nickel, and $1\frac{3}{4}$ per cent magnesium. The strength of this alloy can be increased by 50 per cent by age-hardening and it can be heat-treated either in the cast condition, as for a piston, or in the wrought condition, as for a connecting rod. It subsequently provided the basis of a number of special aluminium alloys, such as RR58, which contains 2 per cent copper, $1\frac{1}{4}$ per cent magnesium, $1\frac{1}{2}$ per cent iron, and 1 per cent nickel. This is used throughout the structure and surface cladding of some of the supersonic aircraft which are in operation today. It was only in the 1920s and 1930s that increasing quantities of aluminium alloys were used for aircraft fuselages. In the Second World War most aeroplanes consisted of a structure of girders and struts joined in various ways, with the main frame covered by duralumin sheet and strip which was fixed in place either by pop-riveting or projection

FIG. 18.8. A Hele–Shaw Beecham variable-pitch propeller mounted on the 11-cylinder Bristol–Jupiter Mark VI radial engine of a Gloster Grebe biplane. The blades are of forged duralumin, the nose cone and engine cowling of duralumin sheet metal.

welding. All these alloys had the drawback that they corroded very badly in sea-water and in marine atmospheres. A neat technique was developed for overcoming this problem by rolling pure aluminium on to each side of the sheet or strip, equivalent to about 5 per cent of the thickness of the sheet on each side. For the hulls and floats of flying boats and sea-planes, an aluminium alloy containing 5 per cent magnesium was produced. This was subsequently used extensively for making life-boat and power-boat hulls.

Magnesium alloys were also developed, particularly by the Germans, during the 1930s, largely because of an anticipated shortage of supplies of aluminium in the event of another war [7]. In many ways, the problems with magnesium were somewhat similar to those with aluminium, and extensive research work developed alloys that were much stronger and more resistant to corrosion. The commonest alloying metals were aluminium, zinc, and manganese, the latter helping to reduce corrosion. Surface treatments, for example, with chromates, were also developed to improve the corrosion-resistance of magnesium. The metal had one advantage over aluminium in that it has a higher modulus of elasticity and hence in the cast condition it can improve the rigidity of an engine. Magnesium alloys have at various times been used for crank-cases, oil sumps, and cylinder blocks. Nevertheless, although considerable increases in the production of magnesium occurred in the Second World War, its use decreased substantially immediately after the war.

From the point of view of mechanism, the engines of aircraft were similar in concept to automobile engines. There were however, a few radical differences, of which the most important are as follows: they had to be of the lightest weight possible for the power developed; many of them were designed to be air-cooled and only a few were water-cooled; they had to have an extremely low frontal area in order to reduce drag; they had to operate under overload conditions during take-off and then for extended periods; and finally, they had to be as reliable as possible, and fairly easy to maintain. All these factors required metals and alloys to work under extremely onerous conditions. In particular, the need to reduce weight meant that the factors of safety for strength and fatigue resistance were small—that is, of the order of two to three times the real strength.

One famous type of engine developed in the Second World War was the sleeve-valve engine. The sleeves of the cylinder were made of a steel which gave a hard wear-resisting surface after nitriding treatment. The connecting rods were of Y alloy, all forged and having polished surfaces to improve the fatigue resistance; that is, the ability to withstand continued reversals of stress. The crankshaft was an extremely important part of the engine since it had to transmit a great deal of power; bending, compression, torque stresses, and alternating stresses were imposed, and it was intermittently subjected to overload. Whether it was a long crankshaft, as required for in-line engines, or a shorter one, as required for radial engines, the steel normally used was a high-nickel chromium steel in the forged condition and subsequently heat-treated and tempered.

All these engines drove propellers. In the early days, these were made of laminated wood, but as the consistency and understanding of duralumin increased, the blades were made in that alloy. The final degree of sophistication was an arrangement to vary the pitch of the propeller by the use of an elaborate gear mechanism housed in the hub of the propeller (Fig. 18.8).

All other parts of aeroplane engines—notably the electrical generation and distribution system, and also the hydraulic system, particularly necessary for operating wing-flaps and retractable undercarriages—necessitated the special development of metals and materials which are too specialized to be described here. Likewise, the manufacture of fuel tanks and the design of fuel supply lines to the carburettors and provision for adequate mixing of fuel and air in the manifolds before being sucked into the cylinders needed special design and metallurgical requirements.

One very important aspect of the engine was the design and continual operation of the valves to control the supply of fuel and ensure adequate purging of the exhaust gases after firing. Such exhaust valves, which operated under conditions of red-heat, needed continual research work over the whole of the inter-war period; by the time the Second World War started, they had attained quite a sophisticated design. The mushroom-shaped cap and stem were made of a hot die-steel, and the rim of the mushroom was tipped with an alloy capable of resisting high temperatures, such as Stellite or Nichrome, which was welded into position. Likewise, there was an insert of Nichrome in the periphery of the valve cavity in the cylinder head. It was always difficult to achieve any significant cooling of these exhaust valves until the idea was conceived of making the stem hollow and filling it with sodium metal. The continual convectional movement of the molten sodium from the hot face of the valve head to the cooler part of the stem enabled the operating face to be kept somewhat cooler. As the horsepower per cylinder increased, it became more difficult to extract all the waste heat which was generated on combustion. With the water-cooled engines which Rolls Royce had developed, this was perhaps easier than with the air-cooled engines. Bristol Aero-Engines, with their sleeve valve operations, designed a cylinder head that was somewhat akin to an inverted top hat. The air-flow was guided down into the inside crown of the hat, but it was still somewhat difficult to extract the requisite amount of heat. The problem was not solved until the aluminium alloy was superseded by a copper–chromium alloy which had a nickel-plated protecting surface on the cylinder face.

During the late 1940s, the gas-turbine principle (Ch. 40) was used to develop a new generation of aircraft engines. These were developed principally in Germany and the United Kingdom but subsequently, and very quickly, in the U.S.A. Although many of the alloys which had been used in reciprocating engines could be used in totally different forms for the gas-turbine engine, there was, nevertheless, one radical problem to be solved. The basic concept required gases at about 800 °C to flow at very high velocity through impeller blades in a series of stages alternating between rotary blades and stators which guided the gases on to the next series of blades. These blades were subjected to creep, owing to the centrifugal force at high temperatures, and also to fatigue, owing to the combustion and fluctuation in load conditions. Today a whole series of different alloys based on combinations of nickel, chromium, and iron are used; the major breakthrough was in the development of a nickel chromium alloy known as Nimonic. This alloy was a development of an alloy which had been used for some years to make electric elements for fires and furnaces. By the addition of about 3 per cent of aluminium, and an equal amount of titanium, it was found possible to improve substantially the mechanical properties of an 80/20 nickel–chromium alloy at these operating temperatures. These alloys have been continuously developed since their initial introduction about 1940. The Nimonics form a very interesting series of alloys, exemplifying what can be achieved as a result of long and intensive laboratory and engineering research for one specific application only, namely the gas-turbine engine. The earlier alloy, Nimonic 80A, will withstand a force of 85 N mm^{-2} for 1000 hours at 850 °C, the later Nimonic 115 can withstand 240 N mm^{-2} for 1000 hours at 850 °C or 85 N mm^{-2} at 950 °C. Considerable quantities of these alloys are used in all the British gas-turbine engines [8], [9], [10].

III. ELECTRICAL ENGINEERING AND POWER GENERATION

An engineering term which had attained some significance in the late nineteenth century was 'prime mover'. This normally referred to machinery which converted thermal energy in a fossil fuel to reciprocating and, through a crank, to a circular motion. The most important prime mover was the steam engine, but engines were also designed and made to operate directly with certain grades of oils, and also with combustible gases from a variety of sources. Before 1900, the larger of these machines were being used to drive rotating generators to supply electricity. Before mid-century many of them had been superseded by the Parsons steam turbines (Ch. 41).

Almost invariably, the casings of the prime movers and the generators were made of metallurgically controlled cast iron with something like the following composition: 3·2–3·5 per cent carbon, 2·2–2·5 per cent silicon, 0·5–0·8 per cent manganese, 0·15 per cent sulphur, and 0·6–0·9 per cent phosphorus. The iron also contained special trace-alloying elements in order to secure as fine a microstructure and graphitic structure as possible. The development of more accurately controlled cast irons and alloy cast irons became a major metallurgical exercise. Cast iron had always been a very attractive alloy to use for the main structure of large engines because it was the cheapest form of iron, it was more rigid than steel, and had good wear resistance because of the graphite flakes present in the microstructure.

Faraday's discoveries contributed a great deal to the theory of electrical generation and the development of motors [4]. But it is not so well appreciated that one of his discoveries relating to the electrolysis of metals led quickly to a commercial process for producing pure copper by electrolysis. The first commercial extraction of pure electrolytic copper was made at Pembrey, South Wales, about 1885. Electrolytically refined copper had an appreciably higher and much more reproducible electrical conductivity than ordinary fire-refined copper. This means of purifying copper was rapidly adopted by most of the copper mines of the world and also in many of the refining units which had been established wherever there was a high consumption of copper. The technique consisted of smelting the copper ores in the usual way and obtaining a 'blister bar', containing about 98·0 per cent copper; fire refining then gave anodes ready for electrolysis. These anodes contained about 99·0 per cent copper, together with a variety of impurities. On electrolysis in a bath of copper sulphate, pure copper was deposited at the cathode as a thin copper sheet. The thin sheets were then remelted and cast into bars for rolling to rods and drawing to wire. Certain impurities remained in the electrolytic tanks. The sludge contained all the residual precious metals which were present in the ore and gave an important financial bonus and a source of many precious metals. These precious metals vary with the ore, but they include gold, silver, platinum, rhodium, and iridium. However, certain copper ores are relatively free from these precious metals, particularly those mined in Zambia.

The ready availability of pure copper had important consequences for the generation and distribution of electricity, for which large quantities of the metal were required because of its excellent conducting properties. It was fortunately discovered that with a residual trace of oxygen (about 0·03–0·05 per cent, existing as cuprous oxide) other trace elements were included in the oxide rather than the parent metal, thereby not interfering with the electrical conductivity of the pure copper. This development enabled a copper standard to be prepared; this was known as the International Annealed Copper Standard (I.A.C.S.) and all copper produced had to be 100 per cent I.A.C.S. Although there were developments and improvements in the techniques of smelting, refining, electrolysis, melting, casting, and working of copper for high-conductivity purposes, the basic principles did not change much during the first thirty years of this century.

However, for a variety of reasons, it was felt desirable to develop a method which eliminated all the residual oxygen from the copper. In the late 1930s, such a method was developed in America by the American Smelting and Refining Company; this was known as the OFHC (oxygen-free high conductivity) route. The production of this grade of metal overcame the one difficulty which was occasionally experienced with high-conductivity copper that contained oxygen. This was that under reducing gas-annealing conditions, or under prolonged usage at moderately low temperatures, some of the hydrogen reacted with the oxygen to form steam, which could cause blisters and eventual disintegration of the metal.

In parallel with the development of metals and alloys for the conduction of electricity, a range of alloys having specific electrical resistance properties was needed. One important range was that of alloys having a high resistance and an ability to retain that resistance even while operating at temperatures up to 1000 °C. Such alloys were invaluable for the development of electric furnaces, ovens, etc., in which the atmosphere could be kept relatively clean; electric heating overcame the operational problems of heating by the combustion of coal, coke, gas, oil, or tar. The most useful and widest range of these resistance alloys were those which were nickel-based with 20 per cent chromium. Increasingly tight controls were imposed on the manufacture of these alloys, and special trace elements were added to improve the mechanical properties at high temperatures, particularly to ensure that the oxidation-resistance and strength at high temperatures was maintained over a useful life period of around five to ten years. These were the basic alloys from which a new generation of alloys were developed for use as vanes and blades in gas-turbine engines.

It also became necessary to develop alloys having a fairly high electrical resistance and able to maintain that resistance constant irrespective of the temperature. Such alloys were required, for example, to ensure that generators and motors were driven at constant speeds. Two alloys were developed to meet these requirements; one, a 60 per cent copper, 40 per cent nickel alloy, known as Constantan; the other an 87 per cent copper, 13 per cent manganese alloy, known as Eureka or Manganin.

Since this fast-expanding field of generation, transmission, and usage of electricity depended upon electromagnetic phenomenon, a great deal of attention had to be paid also to the control of magnetic properties in the various metals and alloys used. It had been found in the design of generating equipment, in electric motors, and particularly in the manufacture of transformers, that considerable loss of electrical energy occurred if the magnetic losses were not minimized. One way of doing this was to concentrate the magnetic field by using cores of pure iron in the form of thin laminations cut from sheet. Such laminations still suffered from two defects; first, the direction of the magnetic field could not be controlled easily; secondly, the losses of energy due to hysteresis (alternations of magnetic field) were still quite considerable.

In the early 1900s, it was found that these hysteresis losses could be reduced if the iron contained up to 0·3 per cent silicon. Such an iron was not easy to make and it was slowly realized that the directional magnetic properties of such alloy sheets varied considerably according to the manufacturer's particular pattern of rolling reductions and annealing procedures. Empirical ways were found of augmenting the magnetic properties in certain directions and minimizing them in others. By assembling such sheets or strip materials with the appropriate directional properties, improved magnetic performance could be obtained and this became a regular feature of design specification. Almost invariably the empirical approach was ahead of the theoretical explanation, which began to emerge in the 1930s, but a close correlation between theory and practice was gradually established. The study of textures developed by specific rolling and annealing processes is particularly identified with the pioneer work of the American metallurgist N. Goss [5b].

The growth of the electrical industry in the first half of this century

Fig. 18.9(a). Stranding an aluminium overhead transmission cable with a steel core.

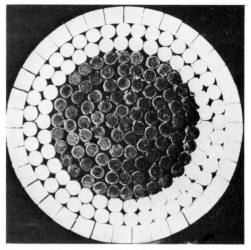

Fig. 18.9(b). Cross-section of a steel-cored aluminium overhead conductor for the River Thames crossing at Thurrock. Because of the great length of the span, the proportion of the steel wire is greater than usual.

imposed new demands on the metal industries for the provision of long-distance transmission cables. In Britain, for example, an eight-year programme to establish a National Grid was announced in 1927. By the end of 1935, there were 4600 km of primary transmission lines (132 000 V) and 1900 km of secondary transmission (66 000 and 33 000 V). This development was paralleled elsewhere in the world. As the transmission of electricity has been considered at length elsewhere (Ch. 45) we need here note only the salient points. Although copper was very satisfactory for electric cables it was expensive, especially after a price rise in the mid-1920s. Attention was therefore directed to aluminium—a good conductor, and lighter—but this was prone to severe corrosion in exposed places. Swiss metallurgists overcame this with the development of Aldrey, an alloy of aluminium containing 1 per cent magnesium and $\frac{1}{2}$ per cent silicon. Although stronger than pure aluminium it was still too weak for suspension between widely spaced pylons. To overcome this, the aluminium was wrapped round a core of high-tensile steel wire (Figs. 18.9(a) and (b)).

The very rapid expansion in the electrical industries necessitated also the development of a range of new techniques of assembly and manufacture. In order to ensure high conductivity at joints, various improvements on the old brazing procedures were introduced. Wide ranges of copper-based alloys, particularly ones containing silver and phosphorus, were developed so that this jointing could be done under workshop conditions with the minimum of oxidation and distortion.

One other field in this industry which called for special metallurgical work was the design of switches for making and breaking electrical circuits. This required that two metal surfaces of high conductivity be connected or disconnected in matters of milliseconds. Under such conditions great amounts of electrical energy have to be dissipated. This resulted invariably in a spark, which caused a severe local hot-spot on one or other of the metal surfaces. Successive makings and breakings of the circuit led to oxidation, pitting, and general surface unevenness which affected the subsequent making and breaking operations. This phenomenon was turned to practical advantage in the jointing technique of spot welding which rapidly superseded riveting, particularly for lighter engineering structures. It became the favourite method for assembling car parts, particularly the body and chassis, and indeed, even today it is probably responsible for 50 per cent of the jointing procedures used, worldwide, for steel and aluminium sheet and strip.

It had earlier become clear that pure copper was not ideal for making switches. Something much harder, having high conductivity and the ability to withstand oxidation during the high temperature generated during the make-and-break cycle, was essential. Various alloys to suit particular requirements were developed, for example by the addition of 0·5 per cent chromium to copper and also copper–silver–tungsten alloys were made by melting or powder metallurgy techniques.

IV. ELECTRIC ILLUMINATION AND COMMUNICATIONS

Another field of metallurgical development originated from a spur development, namely the widespread use of electricity for communication. For radio communication, there was an urgent need to improve the performance of thermionic valves, invented by J. A. Fleming at the turn of the century. To obtain reproducible performance and long life it was found necessary to use such unfamiliar metals as molybdenum and tantalum for various parts of the wire grid, plate, and support wires. It ultimately became possible to produce some of these metals, notably molybdenum, in cast ingot form, from which the metal could be readily rolled out accurately into strip, plate, and wire of small dimensions, as required for manufacture and assembly into radio valves.

Normally, such valves were evacuated before sealing-in all the components. But even so, undesirable traces of residual gases remained adsorbed on the surface of the glass and metal components. Eventually, it was found possible

FIG. 18.10 (a). The body of a prototype cavity magnetron, 10-cm wavelength. The twelve cylindrical holes, arranged peripherally, all had to be reproducible in size and have smooth surfaces accurate to ± 25 μm.

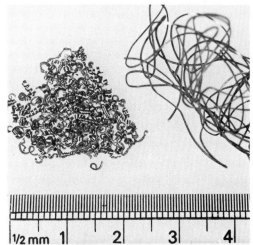

FIG. 18.10 (b). Machining copper. On the left are fine chip-like turnings obtained from a copper alloy containing tellurium. On the right are typical turnings obtained from tough pitch copper. Both specimens were machined on a lathe using the same cutting tool.

to eliminate these residual trace impurities by incorporating within the bulb a trace of barium or zirconium. These metals, known as 'getters', are highly reactive and on heating chemically combine with all the residual traces of nitrogen, oxygen, sodium, sulphur, etc.

In 1930, experiments began in Britain, under the general direction of Robert Watson-Watt, of the system of aircraft detection known as radar. From a metallurgical point of view, two of the main problems hinged round the development of the magnetron valve (Ch. 34) and the manufacture of waveguides. The former had as its base a very intricately shaped channelling and structural support device for the valve itself (Fig. 18.10(a)). This had to be made of copper and have a high thermal conductivity. However, pure copper is not an easy metal to machine (Fig. 18.10(b)), and ways had consequently to be found of overcoming this problem. This led first of all to the development of copper–selenium and copper–tellurium alloys which contained about 0·3 per cent of selenium or tellurium respectively. In more recent years, sulphur was often used instead. The early waveguides were made by very accurately shaping round copper tubes into the various rectangular forms which were essential for propagating the radar waves, but difficulties were encountered with manufacture of the necessary joints, elbows, and turns. Various techniques of electrodeposition and of the ancient lost-wax (*cire perdu*) process of casting

(Vol. I, p. 634) were invoked to produce these shapes on a commercial and reproducible scale.

V. CORROSION IN METALS

The most decisive development in the history of the steam engine was the introduction of the condenser by James Watt in 1765. His simple device became steadily more elaborate. By the early twentieth century the condenser consisted of a series of tubes arranged in bundles, nested together and connected at each end by plates to form inlet and outlet chambers. In the case of ships, the cooling water was always sea-water. With most power stations it was either river-water or, in many cases, estuarine water. In all condensers the cooling water ran through the tubes. Steam was led inside the shell of the condenser and condensed on the outside of the tubes; the condensed pure water was returned to the steam drums. By the beginning of the twentieth century, it had been found, by trial and error, that tubes made of brass containing 70 per cent copper and 30 per cent zinc were the best engineering compromise, but even these suffered corrosion from the cooling water; this was severe in sea-water and still more so in estuarine water. Bronzes had been found to be more resistant but were difficult to make and more expensive. The general practice was first to plug leaky tubes and finally to retube the condenser when leaks became intolerable. This operation had to be done frequently.

The state of the technology in condenser-tube manufacture is well exemplified in a famous British law case, which went to the House of Lords early in the present century. This case concerned the very limited life of a condenser which was fitted to a pleasure steamer, operating on the River Clyde, which had been re-tubed only six weeks before its first excursion sailing. The condenser tubes failed suddenly within a few miles of leaving port and the trip was cancelled. The current state of knowledge of corrosion was well-exemplified in this case, because the condenser tubes were found to have failed by a process known as plug-dezincification; in this, the zinc appears to be removed from the walls of the tube in the form of plugs, leaving only a relatively porous copper deposit. It was claimed that this was proof of the fact that the copper and zinc had never been properly mixed on melting. Nevertheless, it was argued, and finally accepted, that if that were the cause, then in the actual manufacture of the tube the copper and zinc would have to be extended 660 times along the length direction of the wall of the tube, not across it.

Plug dezincification was the most common cause of failure of condenser tubes and extended research work on this problem did not reach any conclusion until after the First World War. During the 1920s, intense efforts were made to improve the life of condenser tubes substantially. Addition of traces of arsenic (0·05 per cent) to the 70/30 brass immediately substantially reduced the incidence of plug-dezincification. Shortly after, adding 2 per cent of aluminium was found to lead to the development of such a coherent and corrosion-resistant oxide skin that this became the standard cheap alloy for condenser tubes for merchant vessels, and many power stations, throughout the world. The composition finally arrived at consisted of 76 per cent copper, 22 per cent zinc, 2 per cent aluminium, and about 0·05 per cent arsenic.

With the manufacture of pure nickel in 'tonnage' quantities, a considerable amount of work went into developing a copper–nickel alloy for condenser tubes. Eventually, a series was developed of which the most famous is the one containing 70 per cent copper, 30 per cent nickel. Although this was considerably more expensive, it was nevertheless much more durable and reliable and was therefore used in all naval vessels, luxury liners, and in certain power stations where the corrosion conditions of the cooling water were particularly severe. By the start of the Second World War, the condenser tube problem, or 'condenseritis' as it was known, had been virtually solved and reference to this was made by Mr Winston Churchill in presenting the Navy estimates to the House of Commons in 1940:

Our ships, great and small, have been at sea more continually than was ever done or dreamed of in any previous war since the introduction of steam. Their steaming capacity and the trustworthiness of their machinery are marvellous to me because last time I was here one always expected a regular stream of lame ducks from the Fleets to the dockyard with what is called condenseritis . . . and now they seem to steam on almost for ever.

In the First World War the average life of condenser tubes in naval vessels was three to six months; by the Second World War the average was seven years. Further minor improvements were later made in this field, more notably in the cupro-nickel series of alloys. First, the addition of about 2 per cent of iron and 2 per cent of manganese, with a lowering of the copper content, resulted in a condenser tube which is probably the most durable and reliable devised so far. It has also been found possible to cheapen these alloys by reducing the nickel content substantially down to about 90 per cent copper, 10 per cent nickel, and about 1 per cent iron. This alloy is now extensively used for the more general purpose of sea-water trunking and in pipes and ducts of large diameter to carry great quantities of water at fairly low velocities and which is now widely used for desalination equipment.

Steam condensers provide only one example of conditions in which metals are required to give service under severely corrosive conditions. For instance, we may consider the steelwork of a pier or of an oil rig immersed in sea-water. There one encounters the peculiar effect that intermittent immersion in sea-water, due to the rise and fall of the tide, causes even more intense corrosion than would occur if the steel were permanently immersed. Speed of corrosion is also affected by currents, salinity, water temperature, marine growths, and even by bacteria.

The methods of protection against corrosion are numerous, and vary considerably with the metal used and with the environmental conditions. By far the most common method is to coat the metal with some kind of protective coating; this can either be in the form of a paint (Ch. 23), a chemical treatment, or even an intensive oxidation of the surface of the metal. The use of paint has been developed extensively ever since the first cast iron and steel structures were erected. Whereas the old Forth Bridge needs continuous painting, year in, year out, many of the later steel bridges are so well protected by various pre-treatment techniques that repainting may be required only every ten years or so.

Finally, there is one other form of corrosion, stress corrosion, which although well known has been studied intensively only in recent years. This phenomenon also occurs in plastics. In almost all cases, a crack, usually inter-crystalline, appears to develop rather suddenly, owing to a combination of internal stress in the material coupled with the specific environmental factors. One of the earliest recorded cases occurred in brass cartridges. During the monsoon season in India in the early 1900s, it was found that many cartridge cases for .303 bullets unexpectedly cracked, jamming in the rifles. Further trouble was avoided simply by a low-temperature stress-relieving annealing of the cartridge cases after manufacture. Many simple copper and brass articles about the home develop cracks from time to time and these are examples of stress-corrosion cracking (Fig. 18.11). It occurs in cast iron vessels when caustic soda is boiled in them, and also in certain types of stainless steel used in chemical plant work or with hard domestic waters.

During the first half of this century the general trend in engineering was to require lighter structures to operate under more severe conditions. This has led to intensive research on corrosion all over the world [11]. For example, the sheet steel used for motor-car bodies had been reduced in thickness by

FIG. 18.11. A copper float from a cistern ballcock valve showing inter-crystalline cracking caused by stress corrosion.

50 per cent by the early 1950s. Meanwhile the use of salt on roads to combat frosty and icy conditions had very substantially increased. Salt solutions are highly corrosive to iron and steel, especially when this is hot, as in the exhaust system of a car. As a result, the motor-car industry has had to develop elaborate anti-corrosion treatments.

VI. MACHINE TOOLS

One area of increasing importance and technical difficulty during this period was that of shaping metals and alloys. Machine tools are the subject of a separate chapter (Ch. 42), and we need to consider here only those aspects which are of particular metallurgical significance. In practice, this means largely the cutting devices rather than the machines as a whole. In every case, the tool which removed the metal had to be designed to maintain a very sharp cutting edge in spite of the high heat and abrasive conditions in which it worked. The tool also had to be quite rigid in order to maintain the high accuracies required. During the nineteenth century, many medium- and high-carbon steel tools were used exclusively for this purpose, but they proved to be quite inadequate when mass-production became necessary in the twentieth century. The first breakthrough came in the first decade with the development of a high-speed steel by F. W. Taylor and M. White in America; this was achieved by adding about 18 per cent of tungsten. This

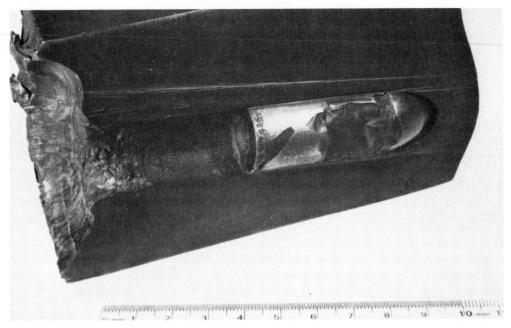

FIG. 18.12. A 20-mm tungsten carbide armour-piercing projectile embedded into 11-cm thick tank armour plate.

addition, with 0·7 per cent carbon and 5 per cent chromium, enabled tungsten carbide in minute particle form to be disseminated through the steel matrix and immediately enabled machining speeds to be increased from 10 metres per minute to 50 metres per minute. Although many modifications in alloy composition of these tool steels were developed by various companies and marketed under various brand names [12], [14], [16], little further progress was made until the idea was developed of using a much higher proportion of tungsten carbide (Ch. 42). It was found essential to manufacture this new type of tool, known as a carbide tool, by making tungsten carbide in fine granular or particle form, and mixing this intimately with finely divided cobalt metal in the proportion of 75 to 25. This powdered mixture was pressed to the shape of the machine tool tip, sintered to a biscuit consistency, and finally fired at high temperature in a hydrogen atmosphere. Such tool tips were exceptionally hard; in fact, the hardness is about 1000, compared with 700 for high-speed steel, and 500 for the ordinary steels used before 1900. Because the carbide tips were initially so expensive they were made as small pieces and brazed on to a steel stock. They had a limited life and so tended to be used only for extremely tough and difficult machining at high speed; such conditions arose in working the range of tough alloy steels which were increasingly used for engineering applications and with certain types of cast iron used for parts that had interrupted cutting requirements. Such tungsten carbide compositions and formulations were the subject of very considerable research effort, and from time to time carbides of other metals were incorporated in this basic composition in order to improve, in some cases markedly, the performance of such tips. In the Second World War, tungsten carbide tips of substantially the same composition as those used in the machine tool industry were used to tip small shells and bullets to facilitate the penetration of armour-plating (Fig. 18.12).

We may appropriately note at this point a range of alloy steels arising from a discovery made by Sir Robert Hadfield in 1882. He found that if steel containing about 1 per cent of carbon and 13 per cent of manganese is heated to 1000 °C and quenched in water, an extremely hard and shock-resisting metal is obtainable. As thus prepared, the steel is only moderately hard, but any attempt to cut or abrade the surface causes a change in structure, giving a product that is very useful for such heavy-duty applications as the jaws of rock-crushing machinery and railway crossings [13].

VII. CAST IRON

The question of increasing the rigidity of machine tools to maintain strict reproducibility under continuous operations was the province mainly of the engineer. The metal used for the beds of the vast majority of machine tools was cast iron. For a variety of reasons, all connected with developments in the engineering industry, this material underwent significant improvements throughout the period with which we are concerned. Cast iron contains about 2·4 to 4·0 per cent carbon, with silicon, sulphur, and phosphorus as impurities; manganese is added to combine with the sulphur. Variations in the proportions of these elements, particularly of carbon and silicon, result in different types of microstructure in the iron. In one, known as grey iron, an iron and iron cementite matrix known as pearlite is intimately mixed with graphite in the form of flakes (Fig. 18.13(a)). In white irons, iron and iron cementite exist mixed in complex shapes (Fig. 18.13(c)). White irons are extremely brittle but if this defect can be made unimportant in service, they are used, because of their extreme hardness, for parts of machinery where wear-resistance is a prerequisite; for example, as balls in grinding mills.

One of the most rewarding areas of metallurgical research occurred with grey iron. Control of the structure was exercised by maintaining strict chemical composition and by using additives to obtain a specific arrangement of

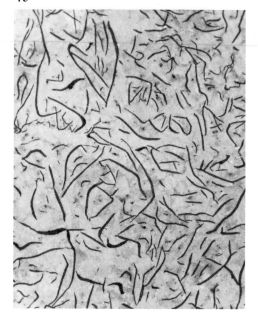

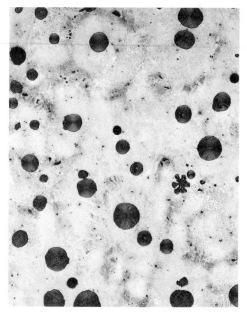

FIG. 18.13(a). Grey iron, showing graphite as black flakes in a grey pearlite matrix. ($\times$ 100).

(b). Cast iron of a similar composition to that in Fig. 18.13.(a), but modified to spheroidize the graphite into nodules. ($\times$ 100).

the various constituents. During the Second World War, it was discovered simultaneously in the U.S.A. and Britain that the addition of small amounts of magnesium caused the graphite to solidify in a spherical instead of a snow-flake form (Fig. 18.13(b) and (d)), which always led to brittleness. The toughness was very significantly increased, without too much interference with machin-ability and other properties. Only since the 1950s has the merit of this out-standing discovery been fully appreciated.

VIII. HIGH-TEMPERATURE OPERATIONS

Outstanding developments were made during the first half of the twentieth century in the design and construction of furnaces for melting and heat-treating metals. The main developments were undoubtedly in engineering, in design, and in improving the performance of ceramics or refractories (Ch. 25). At the same time, the need for metal parts to withstand continuous operation at high temperatures and under various atmospheric conditions demanded extensive improvements in the properties of steels and of nickel–chromium alloys. In addition to being worked to shape, they could also be cast to form the various trays, chain-links, support pieces, and elements of some of the more complex furnaces, many of which were electric.

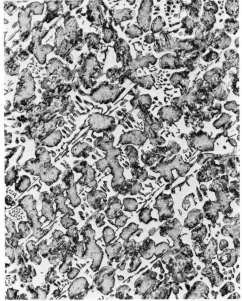

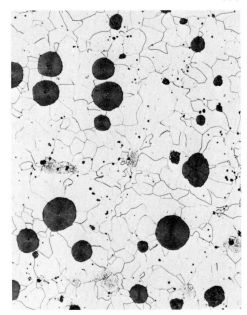

(c). White iron showing cementite (white) in a matrix of pearlite. ($\times$ 100).

(d). Nodular graphite in a ferrite matrix. ($\times$ 100).

Similar improvements in the high-temperature strength and creep-resistance properties of steels and other metals were also necessary for the extrusion industry and for drop-forging and hot-stamping. Dies had to be made which would stand up to hundreds of thousands of pressings, forgings, and stampings. A similar performance was also needed for the die-casting industry. For all these uses, alloys had to be developed to withstand the alternate heating and cooling which was liable to cause cracking because of the fluctuating thermal stresses [12], [13], [14].

IX. MANUFACTURE AND ASSEMBLY

During the second quarter of this century, the manufacturing procedures established in the first quarter were very greatly developed. In particular, output increased enormously and continuous processes were developed for rolling and other manufacturing operations. The rolling mills which were in use in the early years of the twentieth century were larger versions of the two-high rolls used from about 1700. For strip and sheet, according to the coil weight for strip and the length and width for plate and sheet, the barrel of the working part of the roll would vary in length from 3 inches to 12 ft while the diameter varied from 2 inches to about $2\frac{1}{2}$ ft. Sheets were frequently extended

FIG. 18.14(a). Cold sheet rolling of aluminium at the Alcoa works, New Kensington, Pennsylvania, U.S.A., 1912. The metal was passed to and fro through two-high rolls and occasionally turned at right angles to widen the sheet.

FIG. 18.14(b). A multi-stand mill for rolling metal continuously at La Providence, Réhon, France. It has six stands, each housing a four-high mill.

by changing the rolling direction at right angles—this was known as cross rolling.

Owing to the heavy bending forces on the rolls, breakages were frequent. This was overcome by having substantial backing rolls to support the work rolls—these were known as four-high mills.

As the engineering and knowledge of the stresses encountered in rolling various metals, both hot and cold, increased, so the motor horsepower and the reductions per pass also increased. Eventually it became possible to roll strip continuously 4 ft wide and cut it up into sheets for motor-cars and many other applications. Fig. 18.14(a) is a good example of a shop rolling aluminium in two-high sheet rolls, while Fig. 18.14(b) shows six-stand four-high mills arranged in series so that large reductions of strip thickness at high speeds can be produced.

Apart from this, many ancillary processes and industries developed remarkably. For example, initially most steel structures were assembled by riveting. Hot riveting was used for large structures, and cold riveting for small ones. Welding was a process familiar to the ancient metal workers, but it was not until the second quarter of the present century—when control of the microstructure of the weld metal became possible—that welding procedures and machines were systematically developed. Mention has already been made of spot welding; this enabled riveting of thin sheet steels to be superseded and it is the jointing technique most widely used throughout the world today. Nevertheless, such joints do not have the same physical strength as the parent or matrix metals. A very wide range of other welding techniques was consequently developed which give the joint virtually the same strength as the basis metals [15], [17]. Most of these techniques are based on melting a wire of continuous filler metal, originally by oxy-gas methods, but now almost invariably by using an electric arc. In addition to applying the heat at source, filling the molten metal drops into a V-notch between the two pieces to be joined, it is necessary also to obtain a good surface tension condition for the weld bead and to protect it from being oxidized. Various ways of doing this were developed by using a slag and, later, by developing gas protection using the inert gases helium or argon.

Throughout this period, the making of new alloys was principally the result of trial and error; in the case of some alloys, the position is still little better today. This method has one advantage, in that samples are immediately available for the engineer to test. On the other hand, a vast number of alloys frequently have to be synthesized, from which selections are made.

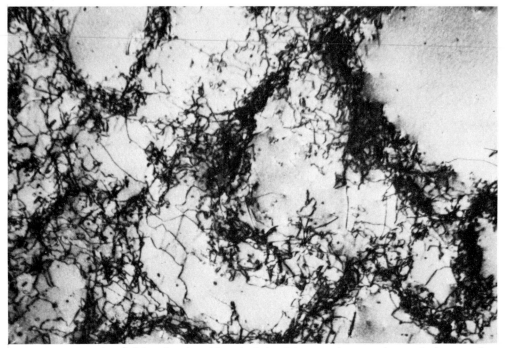

Fig. 18.15. Tangles of dislocations in cold-worked aluminium, an electron micrograph. (× 50000)

This encouraged theoretical studies designed to give a more basic understanding of the way in which the structure and properties of alloys are determined; among the pioneers was W. Hume Rothery at Oxford.

Towards the middle of the century the development of the electron microscope revealed a new phenomenon in metals. This was that their crystalline structure was not perfect; there were always discontinuities or disarrays of atomic arrangements (Fig. 18.15). These gaps or holes were called dislocations and their investigation led to the development of an extensive range of dislocation theories. Such theories have given metallurgists a much more thorough and more unified concept of the mechanisms by which metals are deformed, and of their failure through creep or fatigue. Although such concepts were revolutionary, they have not yet resulted in any outstanding improvement in metals and alloys for practical use.

Another area which became very important during this period was the development of non-destructive testing. With the vast quantities of metals being fabricated, and the increasingly onerous duties which engineers were placing on them, it became important to ensure that the quality of the metal remained constant. It was obviously impossible to destroy large quantities of products simply to test them before they were put into service, and so various physical techniques were developed to ensure that the metal did not contain any gross discontinuities such as blow-holes, inclusions of slag, or cracks which might have developed on heat-treatment. Such techniques exploited the use of X-ray, magnetic, electrical resistance, and eddy-current methods. A simple technique commonly employed for the detection of surface cracking was to soak the part in paraffin, dry it, and determine whether any paraffin had seeped into cracks from which it could be seen to ooze out.

Towards the end of the Second World War, a new technique depending upon the progagation of ultrasonic waves through metals was developed. This was similar in principle to the echo-sounding or asdic method that had been developed for locating submarines. In searching for defects in metals, however, it became necessary basically to alter the apparatus, the method of propagating the waves, and the reading of the signals which were returned. This procedure is now widely used.

REFERENCES

[1] Literature at Ironbridge Trust, Blists Hill, and Abraham Darby Museum, Coalbrookdale.
[2] AITCHISON, L. *A history of metals.* Macdonald and Evans, London (1960).
[3] DENNIS, W. H. *A hundred years of metallurgy.* Duckworth, London (1963).
[4] STREET, A. C., and ALEXANDER, W. O. *Metals in the service of man.* Penguin Books, Harmondsworth (1972).
[5] Metallurgical Society Conferences, American Institute of Mining and Metallurgical Engineers, Vol. 27. *The Sorby Centennial Symposium on the history of metallurgy.* Gordon and Breach, 1965.
  (a) E. C. Bain
  (b) N. Goss
[6] Edwards, J. D., Frary, F. C., and Jeffries, Z. *The aluminium industry.* McGraw-Hill, New York (1930).
[7] BECK, A. *The technology of magnesium and its alloys.* F. G. Hughes, London (1941).
[8] TAYLOR, T. A. *Metallurgia,* **10**, 316 (1946).
[9] GADD, E. R. *British Steel Maker,* **20**, 386 (1954).
[10] New Engineering Materials. *Production Engineering,* **6** (1953).
[11] EVANS, U. R. *Metalic corrosion, passivity and protection.* Edward Arnold, London (1937).
[12] BENBOW, W. E. *Steels in modern industry.* Iliffe, London, (1951).
[13] *The making, shaping and treating of steel.* Carnegie Steel Company (1925).
[14] Metallurgical progress. *Iron and Steel.* Iliffe, London (1957).
[15] TWEEDALE, J. G. *Metallurgical principles for engineers.* Iliffe, London (1960).
[16] TEICHERT, E. J. Metallography and heat treatment. *Ferrous metallurgy,* Vol. III. McGraw-Hill, New York (1944).
[17] SEFERIAN, D. *Metallurgy of welding.* Chapman and Hall, London (1962).

19

IRON AND STEEL

M. L. PEARL

ETWEEN 1900 and 1950 world production of pig iron increased from
about 40m. tons a year to just under 200m. tons, a fivefold increase.
In the same period steel production went up rather more, from about
28m. tons of ingots to 208m. tons, an increase of over seven times. This was
not a steady upward curve confined to the major producers at the beginning
of the period: the U.S.A., Britain, and Germany. The two sets of figures do
not reveal times of deep depression, followed by spells in which new heights
of production were achieved. In the early years of the First World War; in
the brief slump of the 1920s; and in the longer one of the thirties, world
production was lower than it had been in 1913. The statistics do not reveal
other important changes which had taken place at the close of the period:
the rise of new producers, principally the U.S.S.R. and Japan, whose proportions of total world production rose at a much greater rate than those of the
older industrialized countries; the development of different processes within
each country; and the changing relative position of the world's leaders in
iron and steel production.

The nineteenth century closed with two great technical changes in steelmaking which had been developed in its last four decades: the Bessemer
converter which, following Gilchrist Thomas's invention of a basic dolomite
lining, produced steel from phosphoric pig iron (hence the name Thomas
furnace used on the Continent) and the open-hearth furnace invented independently by William Siemens and the Martin brothers to which a basic
lining was soon applied, and which, while it was a much slower process than
the Bessemer converter, enjoyed the advantage that it could consume large
quantities of steel scrap. These two processes dominated the production of
steel in the first half of the twentieth century but by the end of the period, the
open-hearth method had become the predominent process throughout most
of the world, a transition which had begun earliest in Britain in 1894 when

Fig. 19.1. Newport Ironworks, 1900.

the open-hearth process first produced more steel than the Bessemer process (Vol. V, Ch. 3). It was not until 1908 that the same thing happened in the U.S.A. and not until 1925 that it took place in Germany, but that was a transition influenced by the loss of the highly phosphoric *minette* ores after the First World War.

The increase in the production of iron in the blast-furnace was due to no such striking inventions as had taken place in steelmaking but to increases in the scale of operation; improvements of larger furnaces and their equipment; a much higher driving rate than hitherto; developments in the manufacture of coke; the utilization of waste gas and greater fuel economy; and the discovery, exploitation, and transportation of more plentiful, richer, and cheaper ores.

I. IRON ORE SUPPLY

The most remarkable changes took place in the U.S.A. In 1892, the major opening up by the Merritt brothers of the Mesabi Range of the Lake Superior Region (which had been mined commercially since 1852) provided a vast source of cheap ore, easily mined in opencast pits, rich in iron content, but soft and fine in texture—the last, a problem which would plague blast-furnace operators for another twenty years. Meanwhile, a novel system of mechanization was swiftly organized, aided by the control which the Carnegie Steel Company came to have over the orefields in alliance with the Rockefeller railroad and steamship lines. Giant steam shovels, said to cost $10 000 each which had been developed in 1885, were in operation in 1898. Particularly suited to scooping up the soft soil-like ore, they lifted 5 tons of material at a time, taking $2\frac{1}{2}$ minutes to fill one of the ten or twelve railway wagons which were combined in a train alongside the open pit. To the benefits of cheap mining costs was added the bonus of a specially reduced railway tariff to the lake shore docks.

A transportation and handling system thus arose far in advance of anything known elsewhere in the world. To transport the ore some 800 miles to the lower lakes, transport shipping facilities were rapidly expanded. From six ore-carrying vessels in operation on the lakes in 1886, the number grew to 296 in 1899, almost all steam driven; many mining companies acquired their own fleets. In the 1890s, the 'whaleback' design of McDougall made an important but brief contribution. European observers were impressed by the specially designed ore ships, each with ten or more hatches (Ch. 31). By 1901 the largest vessel carried about 7500 tons but this was soon eclipsed by the *Augustus B. Wolvin*, launched in 1904, with a capacity of nearly 11 000 tons.

Loading and unloading also required novel features of mechanization and organization. To facilitate loading into the boats, railway wagons ran out on a high trestle beside the docked vessel to allow the ore to flow into bins, each containing about 150 tons, mounted above the boat hatches. In the mid-1890s, 3000 tons of ore could be loaded into a ship in one hour. Similar improvements were introduced for unloading the ore. In 1894, 300 tons an hour were unloaded from the boats. The mechanical unloader, a massive structure with huge buckets moving down into the hold of a ship, invented by George H. Hulett in 1898, shortened this time drastically. By 1904, a rate of 2500 tons an hour had been achieved. Other limitations on the movement of the ore were removed or reduced. The size of lake steamers was limited by the size of the locks at Saulte Ste. Marie (completed in 1855); by 1896 a new lock had been constructed some 800 ft long but it could still take only one vessel at a time. Improvements in ships, docks, and waterways were accompanied by improvement of railway facilities. Earlier, in the mid-1880s, the Minnesota Iron Company, exploiting another part of the Lake Superior ore deposits, the Vermillion Range, had pioneered the use of wagons of 24 tons in trains of 500 tons. In the 1890s, the Carnegie Steel Company built a railway equipped with 45-ton wagons to run in trains of 1500 tons from Lake Erie to the blast furnaces in Pittsburgh. The extent of this American advance may be seen when it is compared with European practice. From 1890 onwards, 15-ton wagons were being increasingly used in Germany. By 1906 just over half were of this capacity, but 20-ton wagons made of steel and designed for mechanical unloading were just being introduced. In Britain 7-ton wagons were commonly used although a 20-ton steel self-discharging wagon had been introduced by the North Eastern Railway in 1900.

In 1900 the Lake Superior region supplied three-quarters of the iron ore requirements of the iron and steel industry of the United States. The Mesabi Range, by far the largest deposit, produced 9m. tons in 1901, reaching a peak of 42·5m. tons in 1916. One commentator wrote in 1903, 'It has become possible to transport ore a thousand miles and make pig-iron for less than half a cent a pound.'

II. COKE CONSUMPTION AND PRODUCTION

In 1900, for every ton of pig iron made in British blast-furnaces about 25 cwt of coke was required. This was inferior to the best American and German practice where around 21·5 cwt of coke was then consumed, but coke consumption varied according to the chemical and physical nature of the ores and the coke, the size and driving rate of the blast-furnace, and the quantity of special irons made, such as those for forge and foundry. International comparisons tend to be misleading for the early years of the century because of these highly complex factors and because total fuel consumption was not consistently recorded. The extent of the possible variation appears from figures of coke consumption studied by Sir Lowthian Bell at his Clarence Works in 1872, when only 22·4 cwt of coke was consumed for every ton of iron made from Cleveland ores. In the 1880s, Bell improved this figure exceptionally to 20·5 cwt, a better achievement than the consumption reported in 1916 when Cleveland ore was said to require 23·3 cwt of coke to make a ton of iron. The effect of the type of ore used is vividly illustrated by another extreme figure of 1916: Lincolnshire ore required 33 cwt of coke to make a ton of iron, but for Britain as a whole from 1900 to 1930 the average hovered around 25 cwt. Improvements in ore and coke quality, and increases in the size, driving rate, and control of the blast-furnace helped to reduce consumption everywhere by the end of the half-century. In 1950, the leading blast-

FIG. 19.2. Beehive coke ovens in operation in Britain, 1920.

FIG. 19.3. Coke ovens in the 1950s. The oven in the foreground is being 'pushed' to unload the coke. Hot coke will next be transported to the quenching tower in the background.

furnace operators required only 13 cwt of coke for every ton of iron made, presaging the 10 cwt achieved in the 1970s.

III. COKE

In 1900, coke was mainly produced by the beehive oven (Fig. 91.2), a dome-shaped circular brick structure with a diameter of about 12 ft and a height of 7 ft; rows of such ovens were constructed in long single or double lines. The gases and volatile matter were allowed to escape into the atmosphere, and when the process was over, part of the brick structure was broken open, the coke was 'watered' and then drawn out by labourers using long-handled scrapers. It was heavy and unpleasant work and it took about $1\frac{1}{2}$ tons of coal to make a ton of coke in 48 hours. The alternative process, the by-product oven or the 'retort' oven, as it was sometimes called, had been known since at least the 1860s, but only in Germany, France, and Belgium had it become widely established in the last decades of the century. From 1880 to 1900 there was rapid progress in by-product recovery in these countries, particularly in Germany, where the Otto Company, developing a design of the Belgian non-recovery Coppée system (which went back to 1861), built 5000 by-product and 7000 non-by-product ovens. Britain lagged behind. In 1901 there were less than 1000 by-product ovens and over 26 000 beehive ovens. The U.S.A. was also backward: in 1900 over 95 per cent of its coke was produced by beehive ovens.

Worldwide progress can be charted by the dates at which the production of beehive coke fell below 10 per cent in the leading industrial countries: Germany, 1914; France, 1920; Britain, 1923; the U.S.A., 1928. A combination of factors delayed for a long period the introduction of the by-product process which was known to be much more efficient in many respects, although the advantages if offered were uneven and favoured those producers who, like the Germans, could organize the supply of surplus gas to town consumers after they had consumed what they needed in open-hearth furnaces. The by-products—tar, benzole, ammonia, and naphtha—were of little commercial value in the late nineteenth and early twentieth centuries: the beehive oven was simple and cheap to build compared with the by-product oven (one American estimate calculated an average price of $750 as against $15 000); a prejudice developed against 'retort' coke in some places because it had been made from inferior coal (contrarily, in the U.S.A. much beehive coke was made from the Connelsville coal of West Pennsylvania, an ideal coking coal). Other old prejudices took a long time to die. One such, based on a limited knowledge of chemistry, claimed that the recovery of by-products lowered the quality of coke, another claimed that the dull appearance of retort coke proved its inferiority.

By 1906, practice in Europe had stabilized with a by-product oven of about 8 tons capacity for which the coking time was 28 hours. A big surge of development took place during the First World War. The U.S.A. made up for its late start and led the world in installing higher, narrower ovens and in the use of silica brick. These improvements in higher thermal conductivity and load-bearing qualities produced a considerable increase in coke-oven throughput and gave a much reduced coking time. The much greater demand for by-products after 1914, above all for benzole, and the huge increase in their prices (coal-tar derivatives had risen over four times in value from 1915 to 1916) provided an inducement for heavy investment in by-product ovens. Ovens with a capacity of 15 tons and over were by then being installed in America.

In 1923, when the average daily output in both Germany and Britain was 4 to 5 tons of coke a day, the American average was some two to three times higher, between 11 to 14 tons a day. But Germany, unlike Britain where too many small plants continued unproductively, was making rapid progress and by 1926 one-third of her plants had been rebuilt. In 1924, the Columbus Steel Company of Provo, Utah, erected an oven 13 ft high, widely regarded as a remarkable innovation. Three years later, the Carnegie Steel Company put in four batteries of ovens 14 ft high. Such ovens, each about 42 ft long but only 18 inches wide, had capacities of over 25 tons. Technical developments from the 1930s continued with the use of greater capacities and centralized facilities based on integrated iron and steel works. Batteries in the U.S.A. and Germany had capacities of an average of 500 000 tons a year. At Magnitogorsk in the U.S.S.R. as part of a large iron and steel complex a coking plant was constructed with a capacity of over 2m. tons a year. Progress continued to the Second World War, again principally in the U.S.A. and Germany, with the widespread introduction of regenerators to use blast-furnace waste gas (permitting the richer coke-oven gas to be used in the steel plant or fed into a national system); the decrease in pressure differentials in oven heaters; the provision of uniform heating in all parts of the ovens by an underjet which burned blast-furnace gas; the improved ovens of the Koppers Company; coal washing (which eliminated ash and sulphur); and the Sulzer method of dry coke quenching. The increase in the scale of operation of coke ovens during the first half of the twentieth century is shown by comparing Figs. 19.2 and 19.3.

IV. IRONMAKING

In the first few years of the twentieth century America's largest and most up-to-date blast-furnace—the Edgar Thomson (of the United States Steel Company) built in 1902, and then the largest in the world—produced about 500 tons of pig-iron a day. It stood nearly 89 ft high and had a hearth diameter of 15 ft 6 in. At the end of the half-century the best blast-furnaces in the U.S.A. were producing nearly 1400 tons a day, the height had risen to 108 ft, and the diameter of the hearth to 31 ft. This near-trebling of output per unit of production in fifty years was due to a whole range of improvements and not to size alone, but none of the developments in ironmaking can be compared with the great technical innovations in steelmaking which had taken place in the previous century. Much was due to progress in mechanical engineering and to improved methods of ironmaking on principles or practices already fairly well known since the 1870s. If a young man who had begun work then had survived until the 1940s he would have found little changed in the essentials of the blast-furnace operation. A towering structure with massive pipework at the top forming an O- or Y-shaped silhouette; much mechanical equipment, most notably the skip hoist conveying the burden to the top of the furnace where it would be automatically discharged instead of being tipped in by hand barrows; rows of coke ovens and a minor chemical works beside them; and hot-metal transporters would all be new and distinctive. But much of the rest, at least on the surface of things, would be familiar to a veteran, for the process was essentially the same.

One of the most important American achievements, and one which would come to affect the increase in size and the driving rate of blast-furnaces everywhere, was the long struggle to smelt satisfactorily the ores from the newly opened-up Mesabi range. Because of their very fine texture these ores presented many difficulties that were not fully overcome until the end of the First World War—about twenty years after their large-scale introduction. High iron content and good reduceability of the Mesabi ores permitted much faster driving rates but their fine texture caused choking in the furnace. The upward-moving gases tended to create channels in this mass instead of reducing it completely as it descended at the rate of about 500 tons an hour. This, in turn, caused sudden slips of the unequally reduced burden with interruptions of the process and an excessive consumption of coke. The fine ores also produced a great increase in flue dust (over five times that of other ores in some cases) and complications in the heating stoves by clogging the brick-

FIG. 19.4. A view of four blast-furnaces named after British Queens at Appleby Frodingham. The combined capacity in the mid-1950s of 23 500 tons per week was increased later to 30 000 tons per week by improved blowing techniques.

work. The problems were solved by altering the design of the furnace and by much more thorough cleaning of the gases. An effective change in design was to increase the furnace height in relation to the hearth diameter, thus producing slightly more tapered walls (typically, 1894 furnaces tapered inwards from the hearth to the top at an angle of $86°$ compared with $83°$ in 1918). At the same time, the bosh, the area immediately above the hearth, was lowered and given a reduced taper (in the bosh, the walls tapered inwards and downwards to the tuyeres); in one typical example the angles were steepened from $74·6°$ in 1894 to $82·2°$ in 1918. The bigger American furnace with its new lines and its much higher output came to be the model adopted by the advanced countries. By the 1930s, heights of furnaces in the U.S.A.

had reached 90 to 103 ft, hearth diameters 25 to 28 ft, and production by the leading operators was over 1000 tons of pig-iron a day.

The problems caused by the excessive dust produced by Mesabi ores were gradually overcome by improved methods of gas cleaning. The earlier dry type of dust-catcher was replaced by a wet process using a gas washer developed by B. F. Mullen in 1903. It was widely used until it was succeeded in 1914 by the Brassert–Bacon tower washer which combined multiple bands of water sprays with devices to ensure an even cleaning of the gases. Dry cleaning of gases by bag filters continued to be improved, however; a widely used system introduced in 1914 was the German Halberg–Beth plant. In 1903, the introduction of German-designed blowing engines driven by blast-furnace gas had presented a challenge to the older reciprocating steam engine, and had given an additional impetus to the search for more effective means of gas cleaning. A Thiessen drum-type washer had been developed in Germany and it was introduced as a secondary cleaning system in America in 1907. Turbo-blowers, first applied in the blowing of American blast furnaces in 1910 were to raise blowing rates from the old level of 45 000 ft^3/min attained by the reciprocating steam engine to volumes of 60 000 ft^3/min, and finally to over double this rate by the late 1940s. They also impelled further improvements in secondary cleaning. This was provided by electrostatic systems in which a difference of potential between two electrodes in the gas caused electrically charged dust particles to travel to the collecting electrode and adhere to it. The phenomenon, known since the 1820s, had been rediscovered by Sir Oliver Lodge in the 1880s, but it was not until the development of alternating current and rectifiers which could produce high voltages that it could be applied in industry. F. G. Cottrell of the U.S. Bureau of Mines had worked on the problem from the early 1900s and his system was widely adopted after the First World War. The first full-scale installations for blast-furnace gas cleaning were in 1915 at Bethlehem, and in 1920 in Britain at Skinningrove. Improvements were continuous. On the eve of the Second World War, cleaning had become 99 per cent effective and blast-furnace gas was widely used for underfiring the by-product coke ovens and blended with the richer coke-oven gas for use in the steel plant and elsewhere.

V. MECHANICAL HANDLING

The higher driving rate of the blast-furnace brought about a great increase in the mechanical handling and stocking of raw materials from the 1890s. An intensive period of American mechanization described by a visiting British delegate as applying 'engineering common sense which almost amounts to genius' included the introduction of the inclined skip hoist. The first such installation was at the Lucy furnace in 1883 and improved versions followed at other furnaces in the next few years. Skip buckets were filled from bunkers at ground level, weighed, carried to the top of the blast-furnace, and emptied by a double hopper. It was a system which would become universal, eventually displacing the vertical hoist which raised wheelbarrows from ground level, their contents being manually dumped into the furnace (Fig. 19.8, p. 486). It took over twenty years, however, for hand charging to disappear in most works. An American report of 1914 showed that out of 42 blast-furnaces just over one-third were still being charged by hand. Better ore handling provided the next example of highly organized American engineering progress. In 1895, construction of the Duquesne blast-furnace near Pittsburgh included the building of an ore yard with novel handling and stocking facilities: an ore bridge system invented by A. E. Brown, and stockhouse bins permitting gravity charging of large moving buckets. These innovations, sufficiently advanced to be termed the 'Duquesne revolution' some twenty years later, were widely adopted at all big plants built in the next few years. From the 1930s, skip buckets of over 250 ft^3 capacity were common and with them large ore bridges, many equipped with 15-ton scoops. The interwar years also saw the introduction of 60-ton 'gondola' cars and improved systems of coke conveyors, ensuring a continuous flow of materials to the blast-furnace.

VI. ORE TREATMENT

American experience was also influential in various measures to improve iron ore by pre-treatment. Grinding lumpy ore by roll or core crushers was one such improvement. Another was washing at the mine to remove silica, which was reduced from 30 per cent at one works in 1907 to less than 10 per cent. Because of this beneficiation less limestone was needed as a flux to remove impurities. At this time, Britain, which relied heavily on foreign ore, mining about two tons for every ton imported, was also making progress in ore treatment. In 1908, the Iron and Steel Institute was told that as a result of installing a mechanical ore-cleaning belt 13 per cent of impurities were removed and mineral consumption was cut by nearly 10 cwt a ton in the blast-furnace. Imported ores were being similarly pre-treated at the time. Progress was also achieved in the United States by the improvement of imported ore, although such ores accounted for only a small part of America's total needs, usually less than 5 per cent for most years of the first two decades of the twentieth century. Of this relatively small quantity, the largest part came from Cuba, where the Mayari low-phosphorus ore had been discovered in 1904. The iron content of these ores was raised from 38 per cent to over 56 per cent when they were dried out. In 1910 they were further improved by nodulizing, a process of heating the ore in kilns.

As had become evident with the attempt to use the Mesabi material, the ore charged to a blast-furnace should be in the form of evenly sized lumps to ensure good permeability and efficient reduction. This requires screening out 'fines' and agglomerating them into lumps. Processes designed to agglomerate finely powdered iron ore by heat treatment and to upgrade the iron content began before the end of the nineteenth century.

Agglomeration of flue dust was practised as early as 1896 at the South Works of the Illinois Steel Company, where a plant was erected for briquetting this material with a mixture of limestone as part of the blast-furnace burden. G. Grondal, a Swedish engineer and a pioneer in magnetic separation, while at work in Finland in 1899 developed a briquetting process in which ore fines were mixed with water, compacted in moulds under pressure, and fired in a tunnel kiln using heat regenerators. The main emphasis, however, was to lie in sintering, a process largely influenced by American developments in which powdery fines are mixed with coke breeze, conveyed on a moving grate, and subjected to a heat treatment which causes them to form into lumps for charging into the blast furnace. Sintering owed much to pioneering efforts in non-ferrous smelting, particularly of lead, by T. Huntington and F. Heberlein, who, working in Italy in 1896, desulphurized partially roasted ore in pots, a batch process requiring a great deal of manual labour. Two American engineers, A. S. Dwight and R. L. Lloyd, while working on a lead smelter in Mexico developed in 1906 a method of continuous sintering on a moving grate using a downdraught for the first (Fig. 19.5). time. Their earliest commercial machine was 30 ft long, and used an endless conveyor which travelled at a speed of 1 ft/min and had a capacity of 50 tons a day. The process was soon applied to iron ore, although combustion problems were very different from non-ferrous sintering where the sulphur content of the ore provided a fuel. Early applications in America between 1910 and 1920 were mainly confined to the sintering of flue dust using the Dwight–Lloyd machine, as first introduced at Birdsboro in 1911, or the J. E. Greenawalt stationary pan type. During the interwar years Sweden developed sintering at Domnarvet, 90 per cent sinter burdens being used in 1950.

In Britain, a little progress was made and some 14 per cent of the burden

FIG. 19.5. Richard Lloyd, superintendent of the Cananea Consolidated Copper Co., Mexico, shown charging the first continuous sintering machine which he and Arthur S. Dwight developed in 1906 to convert copper sulphide ore fines to a material suitable for charging a blast furnace. The Dwight–Lloyd was first applied to the iron and steel industry at Birdsboro, U.S.A. in 1911.

charged to blast-furnaces was sintered by 1950, when plant had increased considerably in scale and output and was capable of producing 3000 tons a day on machines 6 ft wide and 100 ft long. But the main change to a high-sinter burden came in the next two decades. The earlier machines had burned coal but they were superseded by others burning oil, coke, or blast-furnace gas or mixtures to provide process heat; coke breeze and mill scrap were also added. Cooling of the red-hot sinter had always been a problem because of the rubber belts used to convey the material to the blast-furnace bunkers. For this purpose, water quenching was also being superseded by crushing the larger lumps to a smaller size and then air cooling them in a shallow bed, a British innovation in the late 1940s by Dorman Long.

Pelletizing of fine ores, the next agglomeration process to be applied, requires fines to be rolled in a rotary drum to form small pellets which are later hardened by heat treatment. In 1913, in Sweden, A. G. Andersson invented a process for the balling of fine moistened ores subsequently dried and heat treated, but the method was not widely applied. Another unsuccessful attempt was made by C. A. Brackelsberg in Germany, who constructed a pelletizing plant at Rheinhausen in which sodium silicate was used as a binding agent, the balls being hardened on a Dwight–Lloyd sintering machine. The most successful advance in pelletizing was achieved in the United States in work by E. W. Davis sponsored by the U.S. Bureau of Mines on the Mesabi taconite ores between 1934 and 1936. The process involved grinding these hard ores of low iron content to a high degree of fineness so that they could be magnetically concentrated. In the late 1940s the Allis Chalmers and A. G. McKee Companies developed a heat-hardening machine that promised the greatest success yet for the pelletizing process. Unlike the sintering process it gave a relatively long heat treatment at a closely controlled temperature.

VII. CHARGING

A device to prevent gases escaping from the top of the blast-furnace every time the furnace was charged with raw materials had been introduced by G. Parry at the Ebbw Vale Ironworks in Wales as early as 1850. This was a cone-shaped hopper, the bottom of which was closed by raising a cup, or 'bell' as it was called, thus providing a partial seal against the escape of gases. Lowering the bell caused the charge to fall quickly into the furnace, minimizing the loss of gases during charging. The system continued with little change until the mid-1920s when it was greatly improved by making the hopper of a revolving type, giving important advantages in the even distribution of the charge, and by the addition of a second, larger bell lowered into the top of the furnace stack providing an efficient gas seal for the first time. The new system was widely adopted, particularly in America, where it formed part of the drive to automate the charging of the furnace. By the end of the Second World War nearly half of all blast-furnaces in the U.S.A. were equipped with automatic charging in which scale cars had their contents automatically weighed and recorded, and the operations of the skip hoist, the distributive hopper, and the small and large bells were all governed by automatic correlated devices controlled from a central panel.

VIII. HEATING STOVES

The heating of the air blow by blast-furnace gas using Cowper stoves (Vol. V, p. 58) lined with refractory brickwork had been universal since the 1870s. These were large cylindrical structures about 100 ft high, looming as high as the blast-furnace itself, with diameters of 20–4 ft, the interior brickwork laid in a chequer pattern. In the first two decades of the twentieth century it was usual to have four stoves to each blast-furnace, each stove alternating in use with another. With later improvements, three stoves were common, their diameters being increased to about 26 ft. The great amount of brickwork involved in these stoves (in 1900 some 40 000 bricks were required for each) led to research for more efficient working. Moreover, the constant problem of plugged chequer openings clogged by dust from the blast-furnace gas gave, as we have seen, further incentives for better gas cleaning. The need to operate furnaces at higher temperatures also encouraged efforts to improve the heaters. In 1890, a two-pass stove invented by Julian Kennedy for the Edgar Thomson works showed a notable advance. It was found that smaller chequer openings in the brickwork combined with more efficient gas cleaning provided a much higher surface ratio as well as greatly reduced clogging by dust and impurities. From the beginning of the First World War varieties of brickwork design were introduced with special shapes and patterns and much smaller openings. The improvements were dramatic. By 1950, a stove that had provided about 80 000 ft^2 of heating surface was relined to yield a surface of more than 250 000 ft^2.

IX. FURNACE LININGS

Ironmakers have always been concerned to prolong the life of the blast-furnace before it has to be blown out for repair. To obtain the longest 'campaign' (the traditional name for such periods) while driving the furnace to its maximum 24 hours a day is a measure of efficient operation. In the modern period the biggest part of such repair is relining with refractory bricks, a task that can take three months to complete, although in the 1950s British blast-furnacemen at Appleby Frodingham set a world record of less than a month (see also Fig. 19.7).

Lost production during a shutdown and the cost of the operation itself prompted efforts to extend furnace life. The insertion in furnace linings of cooling and protective devices was the main measure adopted. Previously, blast-furnaces were constructed with very thick linings as if in imitation of the early stone structures, but even as early as 1872, at Glendon in Pennsylvania, modern constructional methods were being introduced. In that year the blast-furnace was protected by cast iron or steel plates built into the brickwork. In 1876, horizontal metal plates made up of iron castings containing a double coil of cooling pipe were used in the Dunbar furnace in Pennsylvania. Around 1883, Julian Kennedy was probably the first to introduce a fixed bronze bosh plate on one of the Lucy furnaces. This was soon followed, at the same works by James Scott and at the Edgar Thomson works by James Gayley, by a removable bosh plate provided with a high-pressure water supply. Exterior shower cooling was introduced but furnace cooling systems remained basically unaltered in principle until the late 1940s, when world attention was drawn to the invention in the Soviet Union of evaporative stave cooling in which specially designed cooling elements around the inner periphery of the furnace shell utilized the latent heat of vaporized water to prevent overheating: an ideal heat absorption system which began to find application in the following decades, first in the open hearth and then in the blast-furnace. The progress achieved in linings and refractories in the half-century is indicated by the changes in blast-furnace life. From 1900 to 1920 it was common to produce about 700 000 tons of pig-iron in a campaign before relining. By 1950, the life of a typical blast-furnace (operating at a much higher rate of output) had increased to 2m. tons a year; in a few cases 3m. tons were being produced.

We have seen that the last decades of the nineteenth century were notable for establishing the main lines of mechanical development in ironmaking. In contrast, the early years of the twentieth were marked by a turn to a discussion of the scientific principles involved in blast-furnace operation. Developments which stimulated such a debate and were themselves stimulated by it came fast in this period: such as the refrigerating equipment for drying the air blast by James Gayley at the Isabella furnace in 1904, the reported benefits stimulating a sharp controversy, and the invention of the stockline recorder by David Baker at South Works in 1901, giving operators a real insight for the first time into the flow of materials in the blast-furnace. In 1904 also, American standard specifications were prepared for pig iron based on chemical analysis rather than grading by the visual appearance of fracture, the common practice until then.

Procedures for opening and closing the tap-hole, previously time-consuming and dangerous occupations, also underwent marked technological improvement. The pneumatic drill was first used to open the tap-hole at the Sparrows Point works of the Maryland Steel Company in 1890. The next advance in tapping occurred in 1906, when an oxygen lance was first applied for the purpose, a practice which was to become the common method everywhere in subsequent years. A clay gun for stopping the tap-hole was invented by Samuel W. Vaughen in 1896 and installed in the Cambrian Iron Works in Johnstown, Pennsylvania. Widely used, it was replaced by a rotary gun in 1928.

X. HOT METAL HANDLING

Hot metal handling, which was to influence the development of mixers and greater and faster production of iron and steel, also advanced rapidly. Open-top ladles handling 12 tons were superseded by the first mixer-type ladles of 75 tons capacity installed in America in 1915. Capacities at the largest works rose sharply to 125 tons in 1922 and to 150 tons in 1925. An enclosed ladle on rails found application where metal needed to be conveyed some distance, as in 1899 when up to 800 tons of hot metal was railed 5 miles from the Duquesne blast-furnaces to Homestead. A later version with a capacity of 90 tons, called a torpedo or a submarine car from its shape, was developed in 1915. Later developments were linked with the increased output of integrated works; as in 1928 at the Middletown, Ohio, works of Armco where 150 tons of molten pig-iron was brought to the open hearths from Hamilton, 8 miles away over a specially built railroad, and in 1939 at Colvilles in Britain where a development plan to integrate the Clydebridge steelworks and the Clyde ironworks separated by the river necessitated the building of a bridge over the Clyde for the transport of liquid pig-iron. Capacities continued to increase, and by the late 1940s submarine cars in America were handling up to 210 tons of iron at a time. Slag handling was also improved in the early years of the century. In 1909, M. Killeen at the Edgar Thomson works invented a device for skimming slag from hot metal. Molten slag was poured directly into flat-bottomed cars from slag pots, which in 1920 held about 15 tons, and slag-disposal plant was developed, sometimes operated by remote control.

XI. MIXERS

Improvements in handling hot metal from the blast-furnace accompanied the invention of the hot metal mixer by William Jones at the Edgar Thomson works of the Carnegie Steel Company in 1889. The mixer, quickly adopted in the U.S.A., Germany, and Britain (where the first installation was at Barrow in 1890) was a large brick-lined storage tank in its earliest form, into which molten iron from the blast-furnace was poured and maintained in a liquid state until it was required for the Bessemer converters. The effect of this constant drawing off and topping up was to even out any variations in the composition of different blast-furnace casts. The next step was to use the mixer as an adjunct to the steelmaking process, and to treat the hot metal with some further chemical refinement. 'Active' mixers, as they were called, came into use as a result of the work of J. Massanez at Hoerde in Germany, who reported in 1891 on the desulphurizing effect of manganese in the mixer (having removed up to 70 per cent of sulphur from the hot metal), and of E. H. Saniter in Britain, who in the following year added lime and calcium chloride for the same purpose.

A major invention which followed the same line of development was made by Benjamin Talbot (Vol. V, p. 60), an Englishman who worked in both England and America. His Talbot process, a continuous method of steel-making, was developed at the Southern Iron and Steel Company, Chatta-nooga, Tennessee, where he carried out experiments for the Tennessee Iron and Coal Company of Birmingham, Alabama, who were troubled by their highly siliceous pig-iron. Talbot conceived the idea of desiliconizing in a continuous, basic-lined, gas-heated, tilting furnace by means of a basic slag, developing work already done in America by H. H. Campbell in 1889 and S. T. Wellman in 1895 on open-hearth tilting furnaces. In 1899, as superin-tendent of the Pencoyd Steel Works, Pennsylvania, he introduced his process, the main feature of which was the very rapid decarburization of pig-iron when mixed with up to five times its volume of molten mild steel. The Talbot furnace was continuously charged and tapped, the chemical reaction between highly oxidizing slag and molten iron speeding the process. The heavy slag produced by highly phosphoric pig-irons was poured off by tilting the furnace, the steel being retained in the bath. The tilting furnace, besides its other economic advantage of having a large capacity, could thus make use of a lower quality of pig-iron than could be refined in a fixed type of furnace. It was quickly adopted in America but more slowly in Europe, although in 1901 Talbot's permanent return to England to become managing director of the Cargo Fleet Company in Middlesbrough stimulated its introduction there, as we shall see later in describing steelmaking developments. Large cylindrical mixers of similar design to the Talbot furnace, but of greater capacity, became common and were sometimes used for either purpose. By the First World War, several British firms had mixers of up to 300 tons capacity; at Ebbw Vale there was one of 750 tons, and in the United States mixers with capaci-ties of about 1100 tons were in use.

The invention of the hot-metal mixer abolished the need for a cupola furnace in the steelworks to remelt iron which had been cast into pigs and allowed to solidify. Pigs were otherwise broken up for use in the Bessemer converter or sold to the merchant iron trade. But hot metal held in a molten state could be used only where steelmaking was operated alongside ironmaking in an integrated plant, and thus was much less common in Britain than in the United States, where in 1906 about 80 per cent of pig-iron was made at such plants. In Britain in 1902 only about 25 per cent of open-hearth steel was made by firms with their own adjacent blast-furnaces. Improvements in pig casting continued to be made, however. In 1895, E. A. Ueling invented a pig-casting machine which was in operation at the Lucy furnace in the following year: iron was poured into a number of shallow moulds in buckets on a conveyor system and loaded on to rail cars. The system was installed in Britain at Palmer's works in Jarrow in 1900 and at the new Dowlais Cardiff works of Guest Keen in 1907, but pig-casting and pig-breaking machines were generally much less used in Britain, even in the 1930s, than on the Continent or in America.

XII. BESSEMER CONVERTER STEELMAKING

At the close of the nineteenth century, the acid Bessemer process was still the predominant form of steel production in the U.S.A., accounting for about 7·5m. tons of steel, nearly two and a half times the amount produced by the open-hearth process in that year. Germany then produced nearly 4·5m. tons of Bessemer steel (almost all basic) which, like America's total, was about two and a half times the amount of her open-hearth steel. In Britain, at this period, the relative positions of the two processes were reversed, 1·825m. tons of Bessemer steel (mainly acid) being produced against just over 3m. tons of open-hearth steel.

The American industry had switched to larger converters in the last decade of the century and a two-converter plant had become standard practice, enabling a continuous flow of production to be achieved. It required the coordination of a number of short-time operations, the provision of mechanical equipment, and a carefully planned layout to prevent the obstruction of one operation by another. The process entailed charging the molten pig-iron, usually held in a mixer, into the converter, sometimes with a small amount of scrap; blowing to convert the iron into steel (a blow of only 8 minutes was required (Fig. 19.6)); pouring the metal into a ladle; casting it into moulds and allowing it to solidify; and the frequent repair of the bottom of the converter. Two vessels, made it possible for one to be blown while the other received metal or discharged it. A heat of steel, up to about 15 tons in 1900 and up to 35 tons in the 1930s, required a complete cycle of about 12 minutes from start to finish; an American innovation by A. L. Holley introduced frequent changes of converter bottoms, replaced after about 25 heats, and carried out in 15

minutes. The American Bessemer industry enjoyed the advantages of a supply of iron low in silicon, sulphur, and phosphorus, enabling it to operate a
predominantly acid process and to achieve, by means of larger, continuously
worked converters, a massive superiority in productivity and the production
of good-quality steel mainly used for rails. In Britain, where such favourable
conditions did not apply, about the same number of units produced less than
one-third of the American output in 1901, and the steel was not considered
suitable for many purposes. Long after the worldwide decline of the Bessemer
process and the rise of the open-hearth, American Bessemer production was
still considerable and highly efficient: visiting British steelmen in 1932
(at the height of the Depression) commented that the most striking thing
was 'the enormous production of a comparatively small shop'. In 1930, over
5·6m. tons of Bessemer steel was produced in the U.S.A., about one-eighth
of total steel production. Thereafter, the proportion declined rapidly each
year and although a post-Second World War peak of 4·9m. tons was
reached in 1951, this was then about one-twentieth of total steel production.

Fig. 19.6. A 25-ton Bessemer converter
'blowing' (1950).

By the late 1960s Bessemer production in America had become negligible.

The basic Bessemer process was eminently suitable for application in Germany, making it possible for her to exploit the large deposits of phosphoric ore in Europe, particularly those taken from France after the Franco–Prussian War. The German industry adopted the most advanced practice and increasingly used molten pig iron direct from the blast-furnace, after the introduction of the mixer in 1889. As in America, it was a development made easier by integration with adjacent blast-furnaces. A British delegation to Germany and Belgium in 1895 was greatly impressed by the progress achieved by the German Bessemer industry. It had arrived in the midst of a major development of the industry which was to more than double the average output of German Bessemer plants in fifteen years, raising it in 1905 to 225 000 tons of steel a plant. Basic Bessemer steelmaking in Germany reached its peak in 1913, when over 10·5m. tons were produced, about 56 per cent of total steel production, open-hearth steel accounting for around 39 per cent. Thereafter, Bessemer steelmaking started to decline, a transition slow at first but accelerated by the loss, after the First World War, of the Lorraine ore-fields and about one-third of steelworks capacity. By 1925 the relative positions were reversed, basic Bessemer steel accounting for 42 per cent of production and basic open-hearth steel for 53 per cent. The proportions were not substantially different some thirty-three years later in the late 1950s, although by then Germany had almost doubled her total steel production, having recovered from the devastating defeat in the Second World War and the loss of her eastern territory.

Basic Bessemer steelmaking in Britain declined rapidly after 1894, although a small acid Bessemer industry survived, based on the haematite ores of Cumberland. A number of complex factors, partly technical and partly economic, worked against the setting-up of a new basic steelmaking centre, for which the north-east coast, where the process was first successfully applied, seemed at one time to be the most favourable site. But the locally mined ores were excessively high in silicon and sulphur, and ironmakers found it difficult to get rid of both impurities at the same time. Steel thus produced tended to be brittle: it had a bad reputation in Britain and this, too, hindered its development.

This outline of Bessemer steelmaking by the three major producing countries has left out of account one of the most important ways in which the converter was utilized: the coupling together of two or more processes in series to take advantage of the technical advantage of each, a mutual dependence which has been likened to the analogy of the roughing and smoothing planes plied by a carpenter. The Bessemer converter was the obvious choice for the first stage of the process: a heavy metallurgical load was turned rapidly into molten steel, and the open-hearth furnace completed slowly and effectively the second stage, the final refining and conditioning of the metal. A combination of Bessemer and open-hearth duplexing was tried as early as 1872 in Neuberg, Styria, but as we have seen its first wide application was carried out, at the beginning of the century, in the U.S.A. after the invention of the tilting furnace, although fixed furnaces were also so employed.

American steelmakers enjoyed the advantages of plentiful supplies of rich iron ores of fairly low phosphorus content, while the low availability of steel scrap in the main iron- and steelmaking districts provided conditions eminently suitable for duplexing by acid Bessemer and basic open-hearth processes. Modifications of the duplexing techniques took place in the 1930s, and 1940s in the U.S.A., principally in the cutting down of the amount of metal remaining in the tilting furnace from one-third of the contents to the point of completely emptying the vessel. There were variations also for using Alabama high-phosphorus metal, but in essentials the techniques remained the same for the whole period of American Bessemer production.

German duplexing developed under very different conditions. The industry had been built up on the principle that phosphoric irons were handled in the Thomas converter and only low-phosphorus materials were charged into the basic open-hearth. An essential ingredient, unlike American practice, was molten scrap and the process was again unlike American in being not 'all liquid' metal, small proportions of iron and scrap entering the open hearth. Thus, the converter functioned essentially as a producer of molten synthetic scrap of low phosphorus content.

In Britain the decline of the Bessemer converter made the development of duplexing techniques less necessary, although experiments had taken place with two open-hearth furnaces, one primary in which molten iron was treated with iron oxide and lime and then refined carefully, excluding the primary slag in a secondary furnace. In 1889 the arrival of the tilting furnace, invented as we have seen by Benjamin Talbot, provided a satisfactory alternative, especially advantageous for use with the highly phosphoric iron made in Britain. The first tilting open-hearth furnace in Europe was built at Frodingham, Lincolnshire, in 1902 with a capacity of 110 tons. It was very successful, and by the end of the year tilting furnaces of almost double its capacity were

FIG. 19.7. Bricking the bottom of a 360-ton tilting furnace at the Lackenby works of Dorman Long in 1930.

under construction at two other works. Britain continued to make a greater proportion of its steel in large tilting furnaces than western Europe or the U.S.A., mainly because of its greater usage of phosphoric iron for open-hearth steelmaking. The interior of a tilting furnace is shown in Fig. 19.7.

XIII. THE SIDE-BLOWN ACID CONVERTER

In 1891, Alexandre Tropenas in France produced a small side-blown acid converter, improving on designs by others developed in the 1880s in America and Britain, which had been extremely wasteful of metal. The Tropenas converter proved to be of much value to foundries in supplying small quantities of steel. By the beginning of the century, over thirty Tropenas converters of 1–2 tons capacity were in operation in Europe. The process remained basically unchanged for the half-century, although the furnace became much more common and capacities grew larger, usually to 3 tons. The operation differs essentially from the Bessemer converter in that the blast of air impinges on the surface of the bath of metal, forming a complex cover of slag mainly composed of iron oxide. Thus, the reactions are between the slag and metal and not between air and metal as in the Bessemer converter, where air is blown from the bottom through the vessel. Another essential difference from the Bessemer process is that considerably more heat per unit of carbon is generated because of more efficient burning of carbon monoxide within the vessel.

Experiments with oxygen enrichment of air in the side-blown converter were carried out in the 1940s, and other experiments were then undertaken with oxygen in the blast in Bessemer practice and in the open-hearth furnace. These were to point to a revolution in technology in the late 1950s and 1960s: namely oxygen steelmaking, destined to replace both the Bessemer converter and the open-hearth furnace as the chief production method of the future.

XIV. THE OPEN-HEARTH PROCESS

In the closing years of the nineteenth century, Britain made around 3m. tons of steel by the open-hearth process, more than any other country although America would catch up and surpass her in the first year of the next century. In 1909, a year after American Bessemer production started to fall behind open-hearth output, the U.S.A. was producing 14·5m. tons by the open-hearth process, over three and a half times British production of just over 4m. tons. A year later, in 1910, Germany overtook British output for the first time, a lead narrowly maintained for most of the succeeding years. Open-hearth steel constituted a much larger proportion of total steel production in Britain than in most other countries (nearly two-thirds in 1900 against about one-third elsewhere, and over 90 per cent in the 1930s against about one-half in Germany and about one-quarter in France, although by then the U.S.A. had also raised her proportion of open-hearth steel to over 80 per cent of the whole). The main developments in the process were larger furnaces; improved design; better refractories; mechanical charging of cold materials; the charging of hot metal and the associated duplex working; the replacement of producer gas by other fuels; and advances in instrumentation and control.

Mechanical charging of cold materials was developed early in the U.S.A. replacing the former hand-charging (Fig. 19.9). Wellman introduced a hydraulically-operated machine for charging cold pig, ore, and scrap into the open-hearth furnace at the Otis Steel Company in 1888. The saving in time was considerable and it was calculated that one additional heat a week for each furnace could be obtained. Excessive cooling of the furnace roof was avoided and working conditions improved. In 1894, Wellman replaced his hydraulic machine by an electric charger, thereby halving his labour costs. The older works in Britain, many of fairly small capacity or with cramped layout, were slow to change. Despite the labour

FIG. 19.8. Hand-charging a blast-furnace, Carnegie Steel Company, Pittsburgh (1900). (See p. 472.)

economy of mechanical charging, British operators found considerable advantage in slower charging by hand although it could take $3\frac{1}{2}$ hours to fill a 40-ton furnace: successive layers of scrap, lime, and pig iron were heated more quickly than the heaped lumps mechanically charged, which tended to pile up in front of the furnace doors. Hand-charging continued in Britain until about 1910, even when existing shops were extended. Thereafter, the mechanical charger was widely adopted everywhere (Fig. 19.9), not least because of the increasing difficulty of getting men to do the particularly hot and arduous work.

Open-hearth furnace design was originally based on the type of reverberatory furnaces used for puddling. The fuel was producer gas and the problem was to ensure efficient combustion of this lean gas to give a hot flame directed downward on to the charge to be melted. A variety of furnace contours was tried, as were variations in the number and size of gas and air ports. As the

FIG. 19.9. An electric charging machine for the open-hearth furnace at the Cleveland Iron and Steel Works in 1920.

principles of efficient combustion were gradually understood, the advantages of a single gas port admitting gas at a relatively high velocity, with free access of air admitted through a full width air port, became evident and such arrangements became general after the First World War.

At the beginning of the century, gas for the open-hearth was made at steelworks in a steel shell producer some 12 ft high and 10 ft in diameter; this was lined with firebrick, closed at the top, hand-fed with coal nuts, fired, and subjected to an air-steam blast. Such a producer would gasify about 6 cwt of coal an hour, and two would be needed for a 20-ton open-hearth furnace, a fairly common size in the 1890s. It was an inefficient process. The producers had to be cleared of clinker by hand several times a day, a particularly hard and unpleasant job, which also lowered the quality of gas while it was being done. Improvements were gradually made in mechanization and design. At the beginning of the century an ash and clinker extraction machine was introduced. Another innovation, from 1910, was an automatic coal-charging machine. From 1924 the development of small steam turbines directly coupled to an air fan led to a design in which a battery of producers, in line, fed their gas into a large bricklined main connected to each open-hearth furnace. By 1926, the system had been improved by controlling the gas pressure in the collecting main to a predetermined level. These various innovations raised the gasification rate of a 10 ft diameter producer from the 6 cwt an hour of the 1890s to 32 cwt an hour in the 1940s, saving fuel and manpower and supplying a better product. But producer gas was still deficient in use when compared with other fuels then being widely used in the U.S.A.

Probably the outstanding development in the open-hearth process occurred in the early 1930s in the U.S.A. when heavy fuel oil residues, a product of the petroleum industry, began to be used in the open-hearth. The new fuel, sometimes used in conjunction with natural gas, coke-oven gas, or tar, gave immediate advantages over producer gas. It was comparatively cheap, of high calorific value, clean and, because it entered the furnace at high velocity, it could be directed at the metal bath, saving wear of the linings and roof and ultimately even eliminating the need for long water-cooled ports. Indeed, most port design problems in the U.S.A. in the 1930s and in Britain after the Second World War, until then a major concern in open-hearth practice, were solved by the introduction of these fuels.

Perhaps no aspect of the open-hearth has had more attention paid to it than the refractories used in the process (Ch. 25). Magnesite bricks were first made by the Fayette Company in Pennsylvania in 1895 and chrome bricks by the same company in 1896. Magnesite was installed in 1913 at the Phoenix Iron Works by N. E. McCullum, but silica bricks formed the main component of refractories until 1930. In 1936 Austrian brickmakers introduced chrome magnesite bricks for roofs and ports. They enjoyed a considerable success on the Continent with smaller furnaces, but they were more than three times as expensive as silica and were less successful in larger furnaces. They also required a suspended roof and were more costly to install. The Second World War interrupted both supplies and application in Britain and the U.S.A. It was not until the late 1940s that trials were resumed.

A notable advance was achieved in Britain by the development of a process for producing magnesia from sea-water but the chrome magnesite brick did not become widely used in either Britain or the U.S.A. until the late 1950s when intensive bath-lancing with oxygen was introduced. At that time, too, basic roofs which required suspension from hangers were introduced in America.

XV. OXYGEN STEELMAKING

Oxygen steelmaking, the most important development of the whole half-century, lies largely outside our period. It was introduced in the late 1940s and in two or three decades it replaced the open-hearth process in one country after another. The idea of using oxygen was not new, however. As early as 1856 Bessemer had taken out patents for the use of oxygen-enriched air in his converter. From the 1930s the work of C. von Linde and M. Fränkl in producing commercially pure oxygen in bulk at a low price ('tonnage oxygen') had led to increasing interest and experimentation in both open-hearth and electric furnaces. Trials from 1937 by R. Durrer and C. V. Schwarz in Germany in the use of oxygen as a sole refining agent and further work after the war by Durrer and H. Hellbrügge in Switzerland led to the construction of a pilot plant at Linz, Austria, in 1949. In it hot metal and scrap were charged into a barrel-shaped furnace with a solid bottom, and high-purity oxygen was blown downward into the bath through a water-cooled lance inserted through the mouth of the vessel. Results were favourable and commercial plants began production in 1950 at Linz and Donawitz, giving the process the name L–D.

XVI. ELECTRIC FURNACE SMELTING AND STEELMAKING

In 1879, Sir William Stevens built the first experimental electric arc furnace for steelmaking based on principles which have not altered in modern design. He demonstrated its effectiveness at his Charlton works in 1881 to visiting members of the Iron and Steel Institute, before whom he melted 5 lb of iron in 20 minutes. Other types of electric furnaces were soon being built on the Continent and in the U.S.A. C. A. Faure patented a 'resistance' furnace in 1883 which used the heat generated by the passage of an electric current through solid conducting rods, an invention which was developed by the brothers E. H. and A. H. Cowles in America in 1884. A different type of furnace, using induction heating, was invented in 1887 in Britain by S. Z. de Ferranti, who had worked for Siemens, but the process found no commercial application until it was developed successively in Sweden by F. A. Kjellin in the early years of the century and by E. Northrup in the U.S.A. in the 1920s. Yet another line of development was initiated in 1898 in Italy by E. Stassano, who attempted the smelting of iron in a reduction furnace resembling a blast-furnace, the tuyères being replaced by electrodes. His invention was followed by other experiments in the direction of direct reduction, the most successful combining upper and lower furnaces, notably those of C. Keller in France between 1901 and 1905, and of E. A. A. Grönwall, A. R. Lindblad, and O. Stalhane in Sweden, where rich iron ore and cheap hydro-electric power were both abundant. The Swedish experimenters developed an 'Elektrometall' furnace with a high shaft in which the upper part, similar to a blast-furnace, used charcoal as a reducing agent and the lower chamber contained up to eight electrodes. By 1918, electrically smelted iron in Sweden produced by a number of Elektrometall furnaces had reached a total of 100 000 tons a year, one-eighth of total pig-iron production. The process was superseded after 1928 by a Norwegian low-shaft electric smelting furnace, called the Tysland–Hole after its inventors, G. Tysland and I. Hole. This used coke, coke breeze, and sometimes anthracite coal and even lignite as a reducing agent. It became the most widely used electric iron smelting furnace in the world, the largest furnaces in the 1950s having a capacity of about 250 tons a day.

The main application of electricity was in a different direction, the one indicated by Siemens in the nineteenth century. The first commercially successful electric arc furnace was constructed by P. Héroult in France. Following experiments with the non-ferrous metals in the 1880s, he produced his first heat of molten steel using the electric arc in 1900. In 1906, a Héroult furnace was installed by the Halcomb Company in Syracuse, New York. It was basic-lined, powered by a single-phase 500-kW generator with two electrodes, and had a capacity of 3 tons. At first it was cold charged but soon duplex working was adopted in combination with the open-hearth furnace. Progress was slow up to the First World War. The greatest advance took place in Germany, where electric steel production was nearly 20 000 tons in 1908 when American production was under 7000 tons. On the outbreak of the First World War, German production had risen to just under 90 000 tons, still about three times the American tonnage, and furnaces had been introduced of 30 tons capacity for duplex working with the basic Bessemer converters. During the war there was a surge of interest in the process and the U.S.A. became the largest producer after 1916, first producing more than 500 000 tons in 1918. This was well in excess of crucible steel production, which was similarly displaced all over the world as the economies of the process became apparent.

The development and application of electric furnace steelmaking was hastened by the provision of cheaper and more plentiful power and by improvements in electrodes. The flexibility of the process, which was well adapted to duplex working, gave it another advantage. Its rapid growth in the U.S.A. during and after the First World War was associated with the increased availability of scrap and also with the increasing demand for special steels, free of impurities, for the growing automobile, aviation, and engineering industries. What was then the largest electric furnace in the world was installed by the American Bridge Company in 1927, having a capacity of over 100 tons; in 1950 American electric steel production had passed 6m. tons and 80 to 100 furnaces were in use. The biggest technical changes had taken place in the size of furnaces; the development of cylindrical tilting shells; three-phase equipment; the method of charging; and in electrode and power control. The greatest impetus for change came from top charging. In the earlier stages of development many features of design followed open-hearth practice, including the practice of door charging by machine. By 1920, in the U.S.A. the Snyder 'coffee-pot lid' furnace, which was designed to cant the electrodes and roof upwards and backwards, marked a decisive change in charging. It was followed by the Swindell–Dressler furnace in 1924 which maintained the horizontal position of the roof while it was raised and moved aside. In Europe a different method was widely employed, the roof being lifted while the shell was transferred for charging; by the mid-1930s top charging had been universally introduced with considerable advantages in the charging time and operation.

The main electrical problem facing early operators was the need to equalize inevitable fluctuations in voltage during melting, heavy current being required with full kilovolt amperage in the first stages of melting with lower voltages at reduced amperages later. A further problem was that any break in the arc, or the direct contact of the electrodes with the charge, caused a reduction in the power input. The resistance of the arc, dependent on its length, made it necessary to provide voltage controls to maintain a constant energy input. Early regulators utilized handwheels or cranks which increased the length of the arc by raising the electrodes and decreased it by lowering them. Motor-driven winches to move electrodes soon appeared, raising or lowering being done by switches; electric lamps were connected between the electrodes and earth to indicate voltage by the light they emitted. These were followed in the 1930s in Europe, and in 1940 in the U.S.A., by the first rotary regulators based on the Ward–Leonard principle, in which ultimately there was instantaneous response to conditions in the furnace, motor-generator units being used to move electrodes rapidly. Oil circuit-breakers designed to switch power on and off were also superseded, being replaced by air circuit brakes towards the end of the Second World War, enabling much more frequent and efficient switching to take place.

XVII. ALLOY STEELS

Electric furnaces of both the induction and the arc type came to be used mainly for special or alloy steels but early production of such steels was in the crucible furnace: in America in gas-fired crucibles although in Britain the traditional cementation and coke-fired crucible process was used, almost as in Huntsman's day, until the First World War. High-speed machine tools (Ch. 42) had been improved in the last decade of the nineteenth century with the introduction in 1890 of steels containing 18 per cent tungsten and 4 per cent chromium.

A considerable advance was made in 1895 by F. Taylor and M. White of the Bethlehem Iron Company, who invented a method of heat-treating tool steels almost to melting-point and then cooling them rapidly. The new steels could cut for hours at 150 ft/min, some five times faster than the speed of carbon steels previously used. Speeds of 500 ft a min were soon attained, with a far-reaching impact on the machine tool industry.

Sir Robert Hadfield developed manganese steel in the 1890s at his family steelworks in Sheffield, producing his famous 'Hadfield Steel'. This had the valuable quality of abrasion resistance and found worldwide application as a material subjected to intensive wear, as in the teeth of power shovels which dug most of the Panama Canal before 1914. Hadfield also developed silicon steel, also called 'electrical steel'. This turned out to be an essential component for electrical transformers and generators because of its excellent magnetic properties and high electrical resistance (and thus low energy losses), although Hadfield had first marketed it as a tool steel with exceptional qualities of hardening after quenching. Vanadium steel, towards the development of which the experiments and writings of L. Guillet in France and J. O. Arnold in Britain in the 1890s had contributed much, found application in the early years of the twentieth century in the manufacture of motor-cars, its properties of resistance to shock and fatigue making it particularly suitable for crankshafts, gears, and springs. In the last decade of the nineteenth century, work on nickel–chromium steels, used orginally for armaments, led to the discovery of stainless steel, in which chromium was the main alloying element. The credit for its discovery, attributed to H. Brearley in Britain, and to B. Strauss and E. Maurer in Germany between 1912 and 1915, must be shared with others, for the alloy had been investigated by L. Guillet,

FIG. 19.10. Manipulating a handrolling sheet mill in the 1940s. It took about eight man-hours to produce one ton of steel.

A. M. Portevin, and P. Monnartz in France and by W. Giesen in England between 1903 and 1910. Brearley was the first person to discover its commercial utility. He did this accidentally in 1912 while testing the suitability of high-chromium steels for rifle barrels. Noticing their exceptional resistance to corrosion, he suggested their use for cutlery, to the eventual advantage of one of Sheffield's leading trades.

XVIII. ROLLING MILLS

At the beginning of the twentieth century America led the world in rolling practice. The size and rapid growth of home demand had enabled her to achieve a degree of specialization impossible under European conditions where handrolling survived until mid-century (Fig. 19.10). In speed of rolling, in continuous mills, in the provision of electric drives, and in maintenance she was also far in advance of her competitors. To produce billets, rails, plates, and sections, American rolling mills had separate roughing mills attached to each specialized finishing mill. In Britain, a single cogging mill served all the finishing mills (Fig. 19.11). The continuous mill, a development

FIG. 19.11. Cogging mill in 1900 at the Cleveland Iron and Steel Works Cogging is the first stage of the rolling processes, in which the ingot, shown on the right, is given its first substantial reduction.

FIG. 19.12. The first modern semi-continuous hot strip mill installed at Ashland, Kentucky in 1924.

of a mill designed by J. P. Bedson, a British engineer, in the 1860s for wire-rod rolling, was introduced in the 1890s in the U.S.A., but it was not until 1907 that the first semi-continuous Morgan mill was installed in Britain, mainly for rolling thin strip. The output of the American mills at the beginning of the century was reported to be three times that of the best British mills. Britain was in the forefront, however, in heavy rolling. In 1902, Dorman Long produced the largest section girders in the world and British armour plate mills were outstanding (Fig. 19.13). By 1914 Colvilles had installed what was claimed to be the largest slabbing mill in existence, capable of rolling a 30-ton ingot into heavy plate slabs. But it was surpassed four years later in the U.S.A., when an even larger plate mill was installed in the Coatesville plant of the Lukens Company, starting a new generation of four-high reversing mills. The most significant advance in rolling took place in 1924, when the first semi-continuous hot strip mill capable of rolling sheets 58 inches wide and up to 30 ft long was in operation at Ashland, Kentucky (Fig. 19.12). Earlier efforts to make continuously rolled strip had been made at Teplitz in the later Czechoslovakia between 1892 and 1907 but without much success.

FIG. 19.13. Armour plate being rolled in 1940 at the English Steel Corporation's works (River Don) driven by 12 000-hp engines.

J. B. Tytus made important discoveries at the American Rolling Company (later Armco) on the effects of composition and temperature in the rolled material. He also found that slightly convex rolls, followed by rolls of lesser convexity, enabled him to roll strip of much greater width in proportion to thickness than had been thought possible previously. The Ashland mill, which had been used as a secret experimental plant since 1922, had a number of defects and it was not a continuous mill. In 1926, A. J. Townsend and H. M. Naugle constructed a mill for the Columbia Steel Company at Butler, Pennsylvania, in which these handicaps were overcome. It produced strip up to 48 inches wide on the continuous principle and, unlike the Ashland mill, used no intermediate heating. In a short time the two complementary processes were brought together through the purchase of the Butler mill by Armco. Two major developments were incorporated in the new strip mill: the application of the four-high mill which gave the necessary rigidity and the correlation of roll speeds in successive mill stands.

The growth of the automobile industry in the U.S.A., where the all-steel body made rapid progress in the 1920s, created a demand for better physical properties, particularly in surface finish and in deep-drawing qualities. The new continuous strip mill was able to provide both the greater quantities and the higher qualities now called for. The next important step was taken by the Wheeling Steel Corporation in the U.S.A. in 1928. Their introduction of a cold-reduction tandem mill showed that suitable material could be produced for tinning from coils from hot strip mills. The replacement of hand-mill stands in the U.S.A. went on apace. By the outbreak of the Second World War 28 strip mills had been built at a cost of $500m. It was calculated that in American conditions in the mid-1930s 400–600 tons of material could be rolled in a day in the new mills by a workforce no bigger than had been needed to produce 25 tons on a hand mill. The first British hot-rolled wide strip mill was built at Ebbw Vale in 1938, rolling sheets of 56 inches width. It was followed in 1939 by the John Summers mill at Shotton. By the beginning of the Second World War nearly two-thirds of British sheet or tinplate capacity was still being produced outside the strip mills. It was not until the first post-war Development Plan in Britain had been completed that the situation changed radically, the Port Talbot strip mill, on a new site, beginning production in 1951.

XIX. CONTINUOUS CASTING

Continuous casting of steel, a process which offers considerable economic advantages by its elimination of a number of stages in conventional steel-making from the pouring of metal into ingot moulds to primary rolling in the mills, began commercial production in the 1950s and thus lies largely outside our period. Sir Henry Bessemer had filed patents in 1856 for a machine to cast steel sheets continuously between rolls but he failed to establish the method as a production process and for the rest of the century little progress was made, although some inventions suggested future lines of development. These included the conveyor-type mould machines patented in Britain in 1872 by W. Wilkinson and G. E. Taylor, and in the U.S.A. by A. Matthes and H. W. Lash in 1885 and G. Mellen in 1913. Bessemer's idea of casting continuously between rolls was revived by F. W. Wood in Britain in 1897. Ideas for vertical water-cooled moulds were also being developed, as shown notably by B. Atha in Britain in 1886; in Germany by R. M. Daelen; and in America by J. Illingworth in 1897 and J. O. E. Trotz in 1898. The introduction of continuous casting of non-ferrous metals in the 1930s gave valuable experience and aroused fresh interest in its application to steel, a much more difficult and complex problem, however, because of the physical properties of iron which has a relatively high melting-point and low thermal conductivity. In the U.S.S.R. much experimental work was done from 1931 and a Bessemer-type direct roll casting machine was operational from 1952 producing iron sheet. S. Junghans in Germany, using a machine developed in 1933 for continuously casting copper alloys, began trials in steel casting in 1939. His work, held up by the war, led to success in 1943 when steel from a small Bessemer converter was continuously cast on a vertical mould machine at Scharndorf. In the U.S.A. experiments were also conducted at the Republic Steel Co. during and after the war. As a result, an agreement with the Babcock and Wilcox Co. led to a pilot plant being started up at Beaver Falls in 1948. In the following year, Allegheny Ludlum installed a continuous casting machine at Watervliet, New York, using a reciprocating copper mould developed by I. Rossi, an improvement which led to the Junghans–Rossi machine. In Britain, the United Steel Companies installed a Junghans–Rossi pilot plant in 1952 at Barrow, where important discoveries were made in methods of fast casting, the mould moving downwards slightly faster than the hot metal. Steelmakers everywhere paid much attention to the new process in the early 1950s but many technical problems remained to be solved. After the war, Soviet research was intensified under I. P. Bardin. The first industrial-scale vertical machine, using a single-piece vertical mould, was installed at the Krasny Octyabr works in 1951. In 1955, the then largest unit in the world began production at Krasnoe Sormovo, with a capacity of 50 tons of steel slabs an hour. In the West applications were slower at first and in the U.S.A. the earliest major commercial unit did not begin operations until 1962, although the Atlas Steel Co. in Canada had commenced production ten years before.

BIBLIOGRAPHY

ARCHIBALD, W. A. History of the all-basic open hearth furnace. In *The all-basic open hearth furnace* (Iron and Steel Institute Special Report 46), London (1952).

BURN, D. L. *The economic history of steelmaking.* Cambridge University Press (1940).

——. *The steel industry 1939–1959.* Cambridge University Press (1961).

BURNHAM, T. H., and HOSKINS, G. O. *Iron and steel in Britain. 1870–1930.* Allen and Unwin, London (1943).

CARR, J. C., and TAPLIN, W. *History of the British steel Industry.* Blackwell, Oxford (1962).

DAVIS, E. W. *Pioneering with taconite.* Minnesota Historical Society, St Paul, Minn. (1964).

DURRER, R. Elektrische Ausscheidung von festen und flussigen Teilchen aus Gasen, *Stahl und Eisen*, 1377–85, 1423–30, 1511–18, 1546–54 (1919).

——. *History of iron and steelmaking in the United States.* American Institute of Mining, Metallurgical and Petroleum Engineers, New York (1961).

HOGAN, W. T. *Economic history of the iron and steel industry of the United States.* 5 Vols. D. C. Heath, Lexington, Mass. (1971).

KING, C. D. Seventy-five years of progress in iron and steel, in *Seventy-five years of progress in the mineral industry 1871–1946* (ed. A. B. Parsons), pp. 162–98. American Institute of Mining, Metallurgical and Petroleum Engineers, New York (1947).

JOHANNSEN, O. *Geschichte des eisens* (3rd edn.). Verlag Stahleisen, Dusseldorf (1953).

LECKIE, A. H. Faber fabrum adjuvet 1869–1973. Primary iron and steelmaking. *Journal of the Iron and Steel Institute*, 823–34 (December, 1973).

McCLOSKEY, D. N. *Economic maturity and entrepreneurial decline. British iron and steel. 1870–1913.* Harvard University Press, Cambridge, Mass. (1973).

PETIT, D. A Century of Cowper Stoves. *Journal of the Iron and Steel Institute*, 501–9, April (1957).

POUNDS, N. G., and PARKER, W. N. *Coal and steel in western Europe.* Faber and Faber, London (1957).

SCHMIDT, G. Hundert Jahre Roheisenmischer, *Stahl und Eisen*, 733–9, 2 August (1973).

SHEDDEN, C. T. (ed.) Diamond Jubilee Number. A Record of Sixty Years Progress in the Coal, Iron and Steel Industries, 1867–1927. *Iron & Coal Trades Review.* (1927).

TALBOT, B. Presidential Address, *Journal of the Iron and Steel Institute*, **1**, 33–49 (1928).

WARREN, K. *The American steel industry 1850–1970. A geographical interpretation.* Clarendon Press, Oxford (1973).

——. *The British iron and steel sheet industry since 1840.* G. Bell, London (1970).

Grateful acknowledgement is made to Drs. A. H. Leckie and E. W. Voice for much helpful information.

20

THE CHEMICAL INDUSTRY:
A GENERAL SURVEY

L. F. HABER

I T has long been a commonplace that the chemical industry is a growth
industry. Was this also true in the first half of the twentieth century and,
if so, what characteristics distinguished the industry from others that were
equally new and just as technologically advanced? But before these questions
can be answered, another must first be dealt with: what is meant by the
chemical industry? Various definitions based on identifiable product groups
or distinct processes were adopted by different countries, and were changed
from time to time thereby causing confusion and preventing comparisons.
The usual classification comprised the manufacture of industrial organic and
inorganic compounds, fertilizers and other agricultural chemicals, plastics,
soaps, paints, and pharmaceuticals. In addition related, though quite distinct,
branches were often included, among them the carbonization of wood and
coal; animal-, vegetable-, and mineral-oil refining; artificial fibres; and synthetic rubber. The problems of definition remained unresolved until the
1950s and the term 'chemical industry' as used here is unavoidably imprecise.
This leads to further difficulties in charting the industry's growth. In the
first place, the scope of its numerous branches altered importantly. And
secondly, the indicators which, in default of systematic data collection, were
employed to measure the production of entire sectors, themselves altered:
thus sulphuric acid, soda, superphosphate, and dyes would have served well
enough to indicate the activity of the entire industry in 1900, but a generation
later they would have been incomplete and at mid-century positively mis-
leading.

In the principal industrial countries, chemical manufacture grew fast, both
in absolute terms and also relative to other industries. The annual production
increases before 1914 were of the order of 4 per cent in Britain, and 7 per
cent in Germany and the U.S.A. During the 1920s, the recovery from the
1921 trough to the 1929 peak was marked by even higher rates. Then came

the Depression and rearmament: in Britain and the U.S.A. the industry averaged 3–3½ per cent a year, in Germany about 6½ per cent. Finally, from the late 1940s to the mid 1950s, fast growth was resumed, in the first two countries at 5–6 per cent annually, in the third, where the base was extremely low, at about 17 per cent. In all three, chemical production developed faster than industrial production as a whole, the ratio being usually 3:2, occasionally even 2:1 [1]. These numbers support the widely held opinion of the progressive nature of chemicals, but other modern industries have grown at much higher rates. For example, the index of car output regularly surpassed that of the American chemical industry except in the 1930s, and elsewhere electrical engineering has expanded more rapidly.

The precise cause-and-effect relationship which forms the mechanism of growth cannot usually be determined. Often the result of expansion in one sector was the origin of development in another. There were also advances in pure chemistry which had unpredictable repercussions, often after a considerable interval, on chemical technology. In the 1890s and 1900s the mainspring of expansion was the increasing demand for well-known products embodying chemicals: soap, glass, paper, textiles. The demand for chemicals was indirect and derived from population growth and the expectation of a continuing improvement in the standard of living. Essentially the consumers asked, not for new products, but more of the old. For example soap production in Britain rose from 260 000 tons in 1891 to 353 000 tons in 1907 [2]. The outcome was reflected in a larger demand for alkalis. Besides the increase in the traditional business there were, even then, signs of that diversification which, in later years, became such an important characteristic of the entire industry. At this early stage explosives and agricultural chemicals were emerging as important branches. The former made possible vast civil engineering enterprizes and led to the establishment of numerous factories for the manufacture of dynamite and similar compounds. The latter, quantitatively much more important, was the industrial response to the needs of agriculture and viticulture. Manuring with superphosphate and sulphate of ammonia had been followed by the setting-up of many works to prepare or compound fertilizers. The size of the industry is indicated by the 4·6m. tons of phosphatic fertilizers made in 1900; the volume would double within the next 10 years [3]. The principal customers were the cotton and tobacco growers in the U.S.A.; European sugar beet cultivators generally; and German wheat farmers. Viticulture was another big market. The vineyards having been replanted after the phylloxera epidemic henceforward needed constant protection by spraying with copper sulphate solution. The supply of this simple compound formed one of the pillars of French chemical manufacture.

The development of the chemical industry between the wars reflected a more diverse array of factors. Synthesis and substitution, pushed on by wartime needs, were unquestionably the most conspicuous and direct-acting. Nitrogen fixation exemplified both and became of outstanding economic importance. An instance of substitution was the adoption of novel routes to obtain solvents for surface coatings and for degreasing and dry-cleaning operations: products made by wood distillation were replaced by supplies from fermentation processes and from acetylene via acetaldehyde. The commercial significance lay not only in the removal of supply constraints, but in their coincidence with the boom in consumer durables in the U.S.A. The demand for cars appeared insatiable: registrations rose from 9·2m. in 1920 to 26m. ten years later and to 32m. in 1940. Cars embodied chemicals in diverse ways, notably the fast-drying cellulose lacquers without which the body panels would have clogged the assembly lines, and rubber chemicals for tyres. Once his vehicle was on the road the motorist called for methanol or ethylene glycol anti-freeze in winter, and, in all seasons from the mid-1920s, for petrol containing tetraethyl lead to improve its quality. Thus even when car production fell in the 1930s, those on the roads still provided an outlet for chemicals. Similarly with electrical goods, notably radios. Even Europe shared in that boom, which provided the first impetus for phenolic resins. An even greater outlet for these novel materials was found when enamelled metal tubing for telephone handsets was replaced in the 1930s by the plastic telephone housing.

While these and other branches of chemical manufacture, among them man-made fibres, pharmaceuticals, photochemicals, and safety film were prospering, fertilizer sales declined. Nitrogen, produced chiefly by capital-intensive synthetic methods, fell from 2·1m. tons in 1929 to 1·6m. tons in 1932 and did not pass the 2m. tons mark again for another two years [4]. These arid years continued until protection and various forms of support led to a gradual recovery of agricultural purchasing power. After 1939 excess capacity disappeared and throughout the 1940s there was a fertilizer nitrogen shortage.

The post-war years witnessed yet another pattern of growth which was much broader and more sustained than any previously encountered. Polymer chemistry came to fruition slowly, but was extraordinarily versatile in its applications. Five main types of polymers—nylon, polythene, polyvinyl-chloride, polystyrene, and styrene-butadiene rubber—known before the war but barely developed commercially, grew very rapidly after 1940, and during the 1950s were joined by polyester and acrylic fibres. These, together with nitrogen, proved to be the pacemakers of chemical manufacture, and since these more recent developments have been fully documented, the scale is known. In 1954, the first year when reasonably dependable statistics of world plastics were first collated, it was found that sales amounted to 2·1m. tons, compared with 2·4m. tons for aluminium and 2·7m. tons for copper. The comparison does not measure the substitution of the new materials for the non-ferrous metals, though it was already beginning, and plastic pipes, sheets, and panels were even then being introduced.

Substitution on its own, accounted for much of the post-war growth. There was, owing to the absence of traditional materials, a good deal of restocking after 1945 with plastics and synthetics, at first called pejoratively *Ersatz* but soon found to be every bit as good as the stuff they replaced. Later, the consumer goods economy provided new openings for plastics in packaging. During the 1950s they joined paper and board, glass, and tin cans in a complementary system of subtle complexity until it appeared that supply created its own demand. But the product improvements, though necessary, were not a sufficient condition for the expansion. This was supplied by process improvements, so that with rising volume, unit costs fell sharply and the consequent price reductions opened wider markets, which were buoyed by rising disposable incomes. Such a combination represented optimum conditions for chemical sales and is the explanation for the high growth rates since 1945.

In chemical production, as in other science-based industries, one can often identify the origins of those major innovations which have given a particular sector an exceptional impetus. In addition the First World War, rearmament, and the Second World War acted as general accelerators for many branches of the industry. As regards the former, research was (and remains) a key factor. But the occurrence of a significant new development depends on there being a direct causal link between research, discovery, development, and commercial utilization. Experience has shown that each of the four has its own tempo, that chance plays a big part, and that chemical inventiveness is unpredictable. But there can be no doubt that the systematic support of research and development in some branches of the industry provided the right kind of environment for the transfer of discoveries to marketable products [5]. The work in the large research departments of the German dyestuffs companies led to a stream of patents, of which the most promising were taken to the pilot plant stage, and a few became outstanding commercial successes. The establishment of fruitful industrial research took years and the sponsoring firm needed to be of a minimum size to finance the necessary laboratories and experimental-scale plants. But once past this threshold, the companies concerned developed an innovative thrust which placed them at an advantage over others which did not possess the same characteristic. Patents awarded for plastics will serve to illustrate the process and the statistics show not only the gradual disappearance of the individual inventor, but also the important positions occupied by a handful of large concerns.

The German firms, represented by I.G. Farben easily led the field until the end of the war, when they were overtaken by Du Pont, the largest American chemical business. Other American manufacturers, as well as I.C.I. in Britain and CIBA in Switzerland became significant patentees in plastics after 1945.

The Germans set the pace generally in research and development until 1914, and in some areas until the 1930s. Why were other companies not able to imitate them? The answer was that laboratory research and bench trials

TABLE 20.1

Patents awarded for plastics 1791–1955 [6]

	1791–1930		1931–1945		1946–1955	
	Number	Per cent	Number	Per cent	Number	Per cent
Patents awarded to individuals	1803	43	791	15	489	8
Patents awarded to firms	2436	57	4341	85	5749	92
Total	4239	100	5132	100	6238	100
Principal firms awarded patents:						
I.G. Farben*	564	13	978	19	325	5
Du Pont	78	2	321	6	637	10
Monsanto Chemical	—	—	37	1	283	5
American Cyanamid	—	—	60	1	266	4
Imperial Chemical Industries (I.C.I.)	25	1	90	2	253	4
Dow Chemical	—	—	115	2	187	3
Eastman Kodak†	169	4	235	5	187	3
CIBA	42	1	56	1	101	2
Total of above eight firms	878	21	1892	37	2239	36

Note: percentages have been rounded.

* Up to 1925 and after 1946 predecessor and successor companies of I. G. Farben, including Wacker and Chemische Werke Hüls.

† Including associated companies outside the U.S.A.

were only two among many items of expenditure. Scaling up the work entailed costs of a different order of magnitude and that was followed by commercial development. Large resources were required. In addition, eventual success called for two further ingredients: chemical engineering 'know-how' and 'feedback' from customers. Neither was patentable, but the First World War demonstrated that both were essential. Patents could often be licensed or there were compulsory-working clauses in the legislation, but 'know-how' was carefully guarded, for it enhanced the value of the patent and unless paid for, or otherwise obtained, had to be acquired through the licensee's own experience, and that took time. For example, in Britain after the First World War, the Ministry of Munitions, and Brunner, Mond and Co. needed about eight years to build a large nitrogen fixation plant, a task on which Badische Anilin- und Soda-Fabrik (B.A.S.F.), the original developers, had spent four. Finally, beyond chemical technology there was the market for the goods. The Germans and later the Americans excelled at customer relations. British and French manufacturers, with few exceptions, were very bad at it. The differences were traditional and had often been noted, but their continuing failure to strengthen the two-way flow of practical information between makers and users was damaging because it delayed improvements in the quality of the products.

Research needed constant support in good times as well as bad. In combination with development, research could occasionally yield very large returns if it was able to break into new fields. Most big companies at one time or another were able to score some notable achievements when they extended their dyestuffs research into other areas of organic chemistry. The Germans took the decision in the 1900s; Du Pont and a few other American companies in the 1920s; and I.C.I. in the 1930s. In the case of I.C.I., the existence of the dyestuffs cartel actually stimulated 'competitive research' which led to technical and quality improvements among related product groups not affected by cartel restrictions, such as rubber chemicals, paint resins, and intermediates [7]. Once a change in direction had taken place, systematic diversification generated a powerful momentum of its own and so accelerated the industry's overall development.

Between the wars, and especially after 1945, technical and commercial development became so expensive that attitudes changed and cross-licensing of patents and even the exchange of know-how became acceptable because it saved money. Another step in the diffusion of chemical technology was the emergence, notably in the U.S.A., of consultants and later of specialist chemical engineering companies which, for a fee, put up a complete plant. They were barely noticed in the 1920s; gained ground in the 1930s; and from the 1940s onwards built scores of works.

Both world wars produced major changes in the relative importance of the different product groups and in the industry's relative position among the belligerents and neutrals. The outbreak of the First World War was followed by two distinct crises: in Germany a shortage of nitrogen owing to the cutting off of imported nitrates and ammonium sulphate; among the Allies a lack of dyestuffs and many organic compounds needed for explosives, textiles, and pharmaceuticals. Both problems were solved by expensive improvisation. In 1918, various processes yielded Germany 185 000 tons of nitrogen against 225 000 tons, mostly imported, consumed in 1913. The pattern of use was, of course, wholly distorted: before the war over four-fifths had gone to agriculture, but in 1918 the bulk was diverted to the supply of munitions. As for the Allies, they were initially forced to scrape around and even trade with the enemy. The small Swiss firms filled many gaps and, supported by swap deals of coal-tar distillates for dyes, soon increased their production. Thus the British, French, and Americans gained time to build their own manufacturing plants. The effect of all this activity was a major shift in the sources of supply. In 1913, Germany provided roughly 85 per cent of the volume of dyes; in 1924, the first year when trade was normal after the war and inflation, she supplied 45 per cent. In terms of value the reduction was not so great, but in the other countries experience and knowledge was accumulating and the quality difference virtually disappeared in the 1930s.

The chemical companies would not have been able to accomplish these changes unaided. In this respect the First World War marked a turning-point because Government assistance and even direct intervention became necessary and were accepted. British Dyes Ltd. was backed by the state and it was Government pressure (since the users were not sufficiently interested) that led to the creation of the British Dyestuffs Corporation in 1918–19. In Germany, B.A.S.F. asked for and obtained help to build a big nitrogen-fixation plant at Leuna. The state also supported the construction of nitrogen plants in France and the U.S.A. and of chlorine works in Britain, France, and Russia.

In all countries governments lent a willing ear to demands for the protection of the 'key' chemical industry. In Britain, France, and the U.S.A. tariffs on organic chemicals became prohibitive and in Britain an elaborate and, on the whole successful, import licensing system was introduced. The Germans on the other hand sought a remedy in trading agreements. But everywhere in the 1920s and early 1930s dyestuffs-making capacity exceeded demand. Nevertheless, behind the shelter of extreme protectionism the new chemical manufacturers could develop enough diversity to strengthen their country's position in war. The pursuit of chemical self-sufficiency took different forms in different countries, but it was always a powerful expansionary force. Germany and Russia endeavoured to safeguard their supplies of rubber, and the former also sought to replace imported oil with domestic production from coal or lignite hydrogenation. Both achieved a considerable measure of success in the 1940s. Similar difficulties did not arise in the West, where Britain and the U.S.A. were plentifully supplied with the scarce materials of the previous war: liquid chlorine, ammonia, dyes, acetone, and acetic acid. What they had not reckoned on was the sudden cessation of rubber supplies when Japan overran south-east Asia in 1942. That event proved to be another turning-point in the history of the industry: the synthetic rubber programme was based on general-purpose butadiene-styrene rubber (GR-S). The initial work had been done in Germany, but its adaptation to American conditions was wholly American and was achieved in less than three years (1941–4) at a total cost to the Government of $700m. The following figures show the speed and scale of the programme and the extent to which substitution continued after the war when natural rubber again became available.

The butadiene capacity for synthetic rubber was about two-thirds petrochemical and this single statistic illustrates forcefully the sudden arrival of a new raw material. Although petroleum and natural gas had been used for chemical synthesis in the 1920s, economic conditions were not suitable for their progress. Even at fuel value it was difficult for aliphatics made by petrochemical routes to compete in the solvents business with products made by the fermentation of molasses. The Second World War transformed the

TABLE 20.2

U.S. rubber consumption excluding reclaimed rubber (000 long tons) [8]

	1941	1943	1944	1947	1949
Natural rubber and latex	775	318	144	563	575
GR-S	—	132	496	449	321
Other synthetic rubber	6	39	71	111	93
Total consumption	781	489	711	1123	989
Per cent synthetic	1	35	80	50	42

situation. Corn and molasses found other uses and their prices rose fast. The adoption of hitherto under-used materials created entirely novel opportunities for chemical manufacture. In 1941, organic and inorganic compounds made from oil or gas were insignificant; four years later the volume was estimated at 4m. tons and in 1955 at almost 15m. tons [9]. Europe, not having these resources at the time, remained tied to coal and acetylene chemistry until well into the 1950s. But even in Europe the trend to petrochemistry was irresistible, the more so as prices fell relative to other materials. The change in chemical feedstock was another turning-point in the history of the industry: supply, costs, location, and structure were profoundly affected by it for the next fifteen to twenty years.

In Britain, the Leblanc alkali factories required salt, coal, and access to navigable waterways (for their raw materials were bulky), and railways. These factors determined the location of the soda trade, first in south Lancashire, later in north Cheshire. There were lesser centres on Clydeside and Tyneside. The fertilizer business was well represented in the Eastern Counties and the Thames Estuary. A similar pattern obtained on the Continent for the French and German soda works, while access to the Rhine and the Main was essential to the dyestuffs factories. Fertilizer plants were scattered across northern France, Belgium, and Germany and were built in ports or alongside inland waterways.

Around 1900 there were some noteworthy developments in several other areas. The chemical enterprizes of Basle specialized in dyestuffs, and later pharmaceuticals, and so overcame the locational drawbacks. In Russia there were several growth-points, especially in the southern Ukraine where foreign capital and technicians contributed to the opening of coal-mines, brine pits, and soda works. The turn of the century also witnessed the early stages in the utilization of electrochemical processes and plants to make carbide, and electrolytic caustic and chlorine were established in remote locations near hydro-electric power stations. The American pattern, however, was different: the alkali business developed later than in Europe and was characterized by the construction of well-sited factories, more spaciously built than their English counterparts, in the salt-bearing areas of Syracuse (New York State), Wyandotte (near Detroit), and Saltville (West Virginia). The rest of the chemical industries were to be found mainly on the eastern seaboard, especially around New York and Philadelphia.

The First World War upset the locational traditions and widely dispersed the industry. In northern and eastern France and parts of Russia, factories were destroyed or occupied by the Germans. The immediate response was to replace the lost works and build additional plants to make essential chemicals elsewhere, so that in post-war France Lyon, Grenoble, and even places in the Pyrenees emerged as centres of chemical manufacture. In central Germany large factories were built in 1915–17 because lignite and land were cheap, and in the U.S.A. the Allies paid for the construction of vast munitions plants. Du Pont, for instance built Hopewell (Virginia), Carney's Point, Haskell, and Parlin (all in New Jersey), and the last three plus the dye works at Deepwater (New Jersey) became foci for the company's post-war growth. There were other locational repercussions of the war: countries abruptly cut off from their chemical supplies were forced to start manufacture, with the result that chemical production, formerly negligible or non-existent, suddenly became significant in Scandinavia, Holland, Spain, Italy, Canada, and, most notably, in Japan. Yet further shifts arose from the redrawing of maps in 1919–20 which affected production in eastern Europe.

In the late 1930s and early 1940s further locational changes occurred in central Europe. Germany dispersed her new oil and synthetic rubber works beyond the 1919 frontiers to put them out of reach of hostile bombers. Many factories were nevertheless badly damaged during the Second World War, but were rebuilt and even enlarged after 1945 to become major chemical enterprises in Poland, Austria, and Czechoslovakia. The Soviet Union also benefited from the extension of her frontiers and from dismantling, and its chemical industry became more widely diffused. In the U.S.A. the changes were even more far-reaching than in Europe. The choice of petroleum derivatives for the synthetic rubber plants necessarily led to the selection of southern and eastern Texas and Louisiana as centres of the new industry. The region has abundant energy resources, salt, and sulphur which, conjointly with petrochemical feedstock, acted like a powerful magnet: the rubber factories were soon followed by the branch plants of chemical manufacturers and the sale of the GR-S installations to private enterprise in the mid-1950s brought many oil and rubber companies into this region of diversified chemical activity. By 1958 petroleum-derived chemicals represented 55 per cent by value of chemical production in the U.S.A., and the bulk was manufactured in the Texas Gulf region [9].

The outcome of these geographical movements, superimposed on technological changes, was to emphasize the major role of the U.S.A. in the chemical industry: it accounted for an estimated 34 per cent of world production in 1913 and 43 per cent in 1951. Germany's share fell from 24 to 6 per cent (reflecting new borders and extensive destruction), Britain's from 11 to 9 per cent. Japan and Russia, which between them represented 4 to 5 per cent of production before 1914, saw their shares raised to 4 and 11 per cent respectively [10]. Since then Germany and Japan have gained much ground.

Chemical manufacture has long been characterized by large companies. They used to dominate the alkali and dyestuffs business in the 1900s; nowadays they are typical of the plastics and synthetic fibres sector. Technical considerations and returns to scale favoured vertical integration and so contributed to the emergence of large concerns. In the U.S.A. bigness did not invariably lead to monopoly and the producers competed with each other for markets, which, except in the Depression of the 1930s, were generally buoyant. In Europe, by contrast, concepts of 'orderly marketing' were deeply ingrained. Hence the paradoxical situation that while production overall was rising, individual firms tended to combine to fix prices and conditions of sale. Many agreements extended beyond frontiers (some American firms even joined them between the wars) and lasted for years. Such cartels cemented business relations and established demarcation lines between companies which remained noticeable long after the courts had ruled against them or wars had put an end to them.

The market-sharing devices were nevertheless unable to prop up indefinitely firms that depended on obsolescent processes: thus, companies relying on Leblanc soda, wood distillation, and cyanamide had to adopt new technologies or go under. But in the newer branches where the spillover effects of dyestuffs research and development were particularly valuable, associations were often followed by mergers to strengthen the operational base of progressive companies and to afford them better opportunities to invest where the returns at the margin were most promising. The German dye makers early realized the full range of benefits from the concentration of economic power. In the 1900s they grouped themselves into two loose federations and in 1916 established closer links to avoid overlap and also to support each other after the war. Nine years later they merged into I.G. Farbenindustrie which had a monopoly of a few products and was the leading producer of many others—not all of them chemicals. I.G.'s technological competence, together with its frequently unscrupulous marketing techniques, enabled it to dominate the chemical world until 1939.

There were many other mergers: the formation of I.C.I. in 1926 was the British response to the I.G., but for many years the new concern was financially over-extended. In the U.S.A., Allied Chemical and Union Carbide and

Carbon resulted from the amalgamation of previously autonomous companies, and there were numerous mergers among small chemical firms. Du Pont, however, grew through acquisitions and by the purchase of companies in the leathercloth, paint, and general chemicals business widened its product range. The rationalization movement in France led to the emergence of a few large enterprizes, surrounded by tributary small undertakings which they held in common.

Neither cartels nor mergers were able to insulate chemical manufacturers from the slump. I.G. Farben and I.C.I., with their large interests in nitrogenous fertilizers, suffered severely between 1930 and 1934, and the dyestuffs business, despite elaborate regulations, also went into decline. The war abrogated cartels and they were not reformed afterwards.

The major post-war structural change was the dissolution of I.G. Farben and the creation of about half-a-dozen successor companies, a process which took several years. The new firms once again carried the pre-1925 names and, released from the restraints of the I.G. bureaucracy and those of the Allied Control Commission, they developed extremely fast. Their diversity was their strength: there was little overlap between them, and thus little competition, and each had certain lucrative chemical specialities. Elsewhere in continental Europe the pre-war structural pattern remained unchanged until the 1960s. But in Britain and the U.S.A. the rise of petrochemicals and the termination of the long-standing I.C.I.–Du Pont links in 1952 produced a new situation. In the first place, firms in the newer branches of applied chemistry expanded rapidly. Secondly, the petroleum companies began to play a much more active part in chemical manufacture. Thirdly, the relative importance of many older firms, which had grown by merger between the wars but lacked adaptability, declined. Finally, I.C.I. and Du Pont ceased to respect each other's home ground; by creating transatlantic subsidiaries they set a precedent which was followed by many others and which emphasized the changes in the structural pattern that took place from the late 1940s onwards.

At first chemical production was not capital-intensive. Equipment was simple, batch operation the norm, and manhandling unavoidable. Fixed costs were relatively small and variable costs fluctuated widely according to the price of raw materials and the work-load; the fertilizer trade was largely seasonal, and in the Alps and Russia climatic conditions frequently caused interruptions. Where cartels operated effectively and there was no substitute, manufacturers could charge what the traffic would bear, and it bore a lot. As a consequence, satisfactory profits were earned for many years, except where the business was tied to obsolescent technologies or was saddled with an unrealistically high item for goodwill.

The picture changed after the First World War with the introduction of high-pressure processes. The synthetic ammonia plants created many financial difficulties: capital input per unit of output was much higher than anything previously encountered. The equipment was not only expensive, but useless without ancillary plant such as power stations, storage silos, and tanks. Overdrafts for working capital and loans for buildings and machinery became very large. The debt burden was easily managed while the works operated continuously near capacity levels, but once sales fell returns dropped steeply because demand failed to respond to price cuts and capital charges became a millstone. This happened to I.G., I.C.I., and others with investments in the fertilizer business and they survived during the early 1930s only by shutting some plants and relying on those sectors which reacted positively to lower prices.

Statistics of the return on net worth for the American chemical industry since 1925 reflect wide changes in profitability. The peak was reached in 1929 with 18 per cent (manufacturing industry averaged only about 13 per cent): the trough was 5 per cent in 1932. The rate of return then recovered to 15 per cent four years later, only to drop once more, though still well above the level of industry generally [11]. These fluctuations reacted on investment and rendered it essential to optimize capital expenditure. In retrospect, it is easy to identify the mistakes that were then made in allocating resources. In Britain plastics were neglected; in France ageing equipment was retained in works too small to yield economies of scale; in Germany uneconomic fertilizer factories continued to operate. The financial difficulties were only in part avoidable. The main problem was the failure to develop an effective management system for the large chemical concern. It was not enough to vest ultimate financial control in the head office: the top management had also to allow for sufficient flexibility at the periphery to encourage responsiveness to technical advances and the needs of the market.

A practical solution to the division of power was not found, but the problems receded after 1939. For many years afterwards profits were excellent for those firms that had shifted the emphasis to plastics, fibres, and pharmaceuticals. Such were the boom conditions that cash flow was ample and the rewards of new investment spectacular. For a quarter of a century after the war the combination of sustained growth, product diversification, economies

of scale, and price reductions gave the chemical industry a generally prosperous character.

What, looking back, were the most notable achievements of the chemical
industry in the first half of the century? Unquestionably the technical advances
have been very great, but to acknowledge them is no longer a sufficient cri-
terion of progress. One needs to consider other significant aspects of the
industry's growth; only a few can be singled out here.

Chemical manufacturers were bad environmentalists, and until recently
neglected smoke abatement and water pollution. Their waste heaps and
residues testify, even now, to earlier, extensive damage before legislation and
its enforcement put a stop to it. By contrast, the internal operating conditions
have greatly altered so that by-product utilization, fuel efficiency, and good
working conditions have become accepted practice. Chemical work, once
dangerous, dirty, and very heavy, has been lightened. Good labour relations,
sometimes inspired by the paternalism of the employer or the remoteness of
the location, helped by steady employment (except in the 1930s) freed the
industry from many of the tensions and frictions encountered in the engineering workshops, though earnings were often substantially lower.

The scope of these and other desirable attributes has been widened by the
pervasive effect of investment in new technologies. There have been two
sides to the progress of applied chemistry. On the one hand, competition for
most of the period was deliberately restrained, though never entirely eliminated. On the other, research and development were fostered and sufficient
resources found to create more jobs. Looking beyond the industry to its
customers it could be said that 'Better things for better living through
chemistry' [12] was no idle boast: new products appeared and they made life
easier, healthier, and safer in many ways that were not anticipated fifty years
earlier.

REFERENCES

[1] FEINSTEIN, C. H. *National income, expenditure and output of the U.K. 1855–1965*, Table
   51. Cambridge University Press (1972).
   HOFFMANN, W. G. *Das Wachstum der Deutschen Wirtschaft seit der Mitte des 19. Jahr-
   hundert*, pp. 359–63. Springer, Berlin (1965).
   U.S. DEPT. OF COMMERCE. *Historical statistics of the U.S. to 1957*, pp. 413–4. Government
   Printing Office, Washington (1960).
[2] WILSON, C. *The history of Unilever*, Vol. 1, p. 116. Cassell, London (1954).
[3] LAMBERT, E., and LAMBERT, M. *Annuaire statistique*, p. 298. Librairie Agricole, Paris
   (1912).

[4] STOCKING, G. W., and WATKINS, M. W. *Cartels in action*, p. 126. Twentieth Century Fund, New York (1946).

[5] There is now a large literature on the subject; cf. FREEMAN, C. *The economics of industrial innovation.* Penguin Books, Harmondsworth (1974).

[6] FREEMAN, C. The plastics industry: A comparative study of research and innovation. *National Institute Economic Review*, 36–7, November (1963).

[7] READER, W. J. *I.C.I.: A history*, Vol. 2, p. 330 ff. Oxford University Press, London (1975).

[8] *Industrial and Engineering Chemistry*, **42**, 997 (1950). S. T. CROSSLAND. *Report on the rubber program 1940–1945*, p. 58. Rubber Reserve Co., Washington (1950).

[9] BATEMAN, R. L. Petrochemicals on the move. *Oil and Gas Journal*, 126–7, 1 September (1958).

[10] *Chemische Industrie Düsseldorf*, **4**, 890 (1952). The proportions are based on values, corrected for exchange rate changes; they are orders of magnitude rather than precise statistics.

[11] BACKMAN, J. *The economics of the chemical industry*, pp. 222, 350. Manufacturing Chemists Association, Washington (1970).

[12] The slogan was coined by Du Pont in 1935.

BIBLIOGRAPHY

BAUD, P. *L'Industrie chimique en France.* Masson, Paris (1932).

HABER, L. F. *The chemical industry 1900–1930.* Clarendon Press, Oxford (1971).

HAYNES, WILLIAMS. *American chemical industry.* 6 vols. D. van Nostrand, New York (1945–54).

21

THE CHEMICAL INDUSTRY

FRANK GREENAWAY, R. G. W. ANDERSON, SUSAN E. MESSHAM,
ANN M. NEWMARK, AND D. A. ROBINSON

PART I: HEAVY INORGANIC CHEMICALS

THERE were substantial changes in the manufacture, output, use, and type of heavy inorganic chemicals produced during the first half of the twentieth century. One of the most obvious changes was that at the beginning of the century the industry was largely confined to western Europe and the U.S.A. (Vol. V, Ch. 11). By 1950, plant for the manufacture of heavy inorganic chemicals was much more widely spread throughout the world, with major industries having arisen in the U.S.S.R. and Japan. In 1900, Britain had the world's largest chemical industry, with Germany and the U.S.A. not far behind. By the middle of the century, America had, for most heavy inorganic chemicals, the largest output and had increased her world's share. Britain, Germany, and France were producing relatively much less. A traditional indicator of the state of the chemical industry was production of sulphuric acid. In 1900, Russia and Japan produced 11 per cent and 4 per cent as much sulphuric acid as Britain. By 1958, comparative figures were 215 per cent and 167 per cent. The world share for the U.S.A. in these two years was 23 per cent and 29 per cent respectively.

The first quarter of the century was of greater importance so far as innovation was concerned. During this period, the basis of the nitrogen fixation industry and the electrochemical industry was laid, much of the pioneering work being carried out by German chemists. Another new industry, that of fertilizer production, was growing during this period, being dependent on nitrogen fixation for ready supplies of ammonia and on discoveries of new mineral deposits: of sulphur in the southern states of America for sulphuric acid and of phosphates in north Africa, from which superphosphate was manufactured. The method of production of traditional heavy inorganic chemicals was changing, as was their use. Chlorine was made, for most of the nineteenth century, by the oxidation of hydrogen chloride, but supplies became available from electrolytic processes as soon as cells were developed

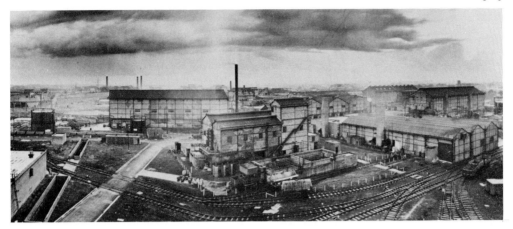

FIG. 21.1. A typical British pre-war 'shadow' factory—in this case for plastics manufacture—built by the Ministry of Supply and subsequently taken over by I.C.I. (1947).

for the production of alkali metals. In the early part of the twentieth century, most chlorine produced was used for bleaching; by the middle of the century, most was used in the manufacture of organic solvents.

I. SULPHURIC ACID

World production of sulphuric acid increased from 4·05m. tonnes (100 per cent H_2SO_4) in 1900, to 9·49m. tonnes in 1920, and to 21·3m. tonnes in 1948. The two processes of manufacture, the lead chamber and the contact process (Vol. V, pp. 244–8), were both operated throughout the period. The U.S.A. produced three to four times more sulphuric acid in 1952 than the second greatest producer, the U.S.S.R.; in the U.S.A., about one-third of total production was from lead chambers in 1920 and in 1950, though for a short period in the mid-1930s quantities produced by the two techniques were similar [1].

For both processes, the starting material is sulphur dioxide. Before the twentieth century, the main sources of this gas were from roasted native sulphur (mined mainly in Sicily), from roasted iron pyrites, and from oxidized hydrogen sulphide, obtained either from calcium sulphide ('alkali waste', a by-product of the Leblanc soda process) or from 'spent oxide', produced in the manufacture of coal-gas. Britain and Germany had their own sources of alkali waste and spent oxide at the beginning of the century; the U.S.A. was forced to import most of her sulphur supplies: in 1901, 96 per cent of elemental sulphur and 66 per cent of iron pyrites used in sulphuric acid production was imported.

Fig. 21.2. Apart from technological changes, chemical plant assumed a more sophisticated appearance. These pictures contrast two sulphuric acid plants of the 1930s (*left*) and the 1970s (*right*).

This situation was soon to change radically. From 1894 to 1897 Herman Frasch evolved a cheap method of mining the vast deposits of native sulphur which were associated with the salt domes of the Gulf Coast of Texas and Louisiana. This consisted of inserting three coaxial tubes into the sulphur layer. Water under pressure at 160 °C was pumped through the outer tube to melt the sulphur at the base of the well. Compressed air was pumped through the centre tube and the molten sulphur was forced up through the middle tube, and then allowed to solidify in vats. The product was of very high purity (99·5 per cent) [2]. By 1950, more than half of total world sulphur production (10·4m. tons annually) was mined by the Frasch process. A decade later, France was the second largest producer of sulphur, extracting three-quarters of a million tons of sulphur by purification of natural gas; four years later, output from this new source had doubled.

In the lead chamber process, a solution of sulphur dioxide was oxidized by nitrogen peroxide to sulphuric acid in a Glover tower. Recovery of the nitrogen gases took place in Gay-Lussac towers. In 1930, this traditional method was further developed by Pierre Kachkaroff in France, the new process allowing simultaneous production of sulphuric and nitric acids. The difference was that sulphur dioxide and air were passed into packed towers through which a concentrated solution of nitrogen oxides in sulphuric acid was circulated. The theory of the method was that sulphuric acid was formed in the liquid phase and that the rate of formation was significantly increased at high concentrations of nitrogen oxides [3]. The efficiency of the process was high. Nitric acid was produced by denitrating sulphuric acid with high nitrogen oxide content by injection of steam, which liberates the oxides to form nitric acid containing only traces of water. By the early 1950s the process was being operated in France, Italy, and Britain.

A major advantage which the contact process held over the lead chamber process was that oleum (a solution of sulphur trioxide in sulphuric acid) could be produced by the former but not by the latter method unless expensive concentration processes were resorted to. Oleum was used increasingly mixed with nitric acid for nitration in the manufacture of explosives, to produce nitrocellulose for plastics, and in the dyestuffs industry, as well as in its pure form in petroleum refining. The process entailed the catalytic conversion of sulphur dioxide to the trioxide, followed by absorption of the gas in strong solutions of sulphuric acid. Though there were still technical difficulties to be overcome, the process was employed in Germany, and to a lesser extent in Britain, at the beginning of the twentieth century. The chief problem was catalyst poisoning: platinum—a very expensive metal—had been used from 1831 onwards but gaseous impurities were found to 'poison' it, and it rapidly lost its effectiveness (arsenic oxides were particularly detrimental). In 1852, iron oxides were used and were found to be less susceptible, though they were also much less efficient in converting dioxide to trioxide. In Germany, the Badische Anilin-und Soda-Fabrik (B.A.S.F.) conducted extensive researches into the contact process, and in a lecture to the Deutsche Chemischen Gesellschaft in 1901 R. Knietch revealed some of the results of this work. The behaviour of platinum and other catalysts had been studied under varied conditions of temperature and rate of flow of reactants, as well as problems of catalyst poisoning by pyrites-derived sulphur dioxide. The widely accepted premise made by Clemens Winkler in 1875, that sulphur dioxide and oxygen had to be mixed in stoichiometric ratio for maximum yield, was shown to be false and it was revealed that contact vessels required cooling rather than heating for optimum results. Another German firm, Verein Chemischer Fabriken in Mannheim, had meanwhile been investigating iron oxide catalysts and in a patent taken out in 1898 [4] described the technique (the Mannheim Process) by which oxidation took place in two stages using a double catalyst of ferric oxide and platinum. Purification of the gases before reaching the catalyst took place by filtration through porous, granular, or fibrous substances. Other processes devised at the start of the century were the Schröder and Grillo process (using platinum-impregnated magnesium sulphate) [5] operated in Germany and the U.S.A., and that used by the Tentelov Chemical Company (of St Petersburg) in which the sulphur dioxide was preheated and the catalyst was arranged in the converter in several sections [6]. By 1911, 24 Tentelov-type plants were in operation. In Britain, the contact process was not widely operated before the First World War: prior to 1915, the plant capacity for 20 per cent oleum was about 25 000 tons per annum. By 1918, it had risen to 450 000 tons.

Vanadium was first patented as a contact process catalyst by the German Ede Haen in 1900 [7]. However, it proved to be of low activity and it was not until 1915 that B.A.S.F. used a catalyst of vanadium pentoxide and potash with powdered pumice or kieselguhr on a commercial scale [8]. By 1928 it had ousted all other catalysts used by that company, though it was not until 1927 that it was used for the first time in the U.S.A. For the remainder of the period under review, it persisted as the most widely employed material.

II. THE ELECTROCHEMICAL INDUSTRY

Although Sir Humphry Davy prepared sodium by electrolysis in 1807, electrolytic methods of preparation of chemicals were of little importance until the end of the nineteenth century because of the undeveloped state of the electric power industry. Supplies of cheap electricity did eventually become available, and in 1890 an American working in Britain, Hamilton Castner, developed a method for producing sodium that was commercially viable, in which molten caustic soda in an iron pot was electrolysed using an iron anode and a nickel cathode. Castner's work was directed towards the production of sodium for reducing aluminium chloride to the metal (he had a few years previously devised a novel method of producing sodium by the reduction of caustic soda with charcoal weighted with iron). In 1886, however, C. M. Hall and P. L. T. Héroult had independently patented their electrolytic method of making aluminium, and this rendered Castner's cheap sodium redundant. There was nevertheless an increasing demand for alkali metal cyanides which were needed for gold and silver extraction and Castner's Aluminium Company Ltd. commenced the production of sodium cyanide by passing ammonia over molten sodium to form sodamide, which was reduced to the cyanide with red-hot charcoal. This industry expanded dramatically with increasing gold production: up to 1892, world consumption (excluding the U.S.A.) did not exceed 100 tons per year; by 1899 annual production was 6500 tons and by 1915, 22 000–24 000 tons [9]. In 1891, the Aluminium Company agreed to supply the German refining company

Deutsche Gold- und Silber Scheide Anstalt with sodium for fusing with potassium ferrocyanide to produce potassium cyanide.

Castner then developed an electrolytic cell for the production of caustic soda of purity high enough to allow sodium to be produced by a continuous process. The cell incorporated a mercury cathode and carbon anodes. Brine was electrolysed, forming a sodium amalgam which, by rocking the cell, came into contact with water in a central compartment with which it reacted to form caustic soda. A number of similar mercury cells were developed, including one by Carl Kellner which was used by the Belgian firm Solvay et Cie. The Castner and Kellner patents were cross-licensed and thereafter the cell was called the Castner–Kellner cell.

A development on slightly different lines was the diaphragm cell. From 1884 the German company Chemische Fabrik Griesheim experimented with Breuer's cell, which consisted of an outer iron chamber (which also acted as the cathode) containing porous cement boxes which acted as diaphragms in which anodes of magnetite were dipped [10]. The company produced caustic potash and chlorine which was used to make bleaching powder; by 1900, these products were being made by the same method under licence in France and Russia as well as by other German firms. The cell underwent various modifications. One of these, containing an asbestos diaphragm, was patented by Ernest LeSueur in the U.S.A. [11], who set up the first commercially successful American plant at Rumford Falls, Maine, in 1892. His firm, Electro-Chemical Company, managed to weather the price-cutting operation on imported bleach and the American alkali-bleach industry was thus established. In 1903 the Development and Funding Company (later to be renamed the Hooker Electrochemical Company) set up a works at Niagara to exploit a cell patented by C. P. Townsend and many more companies and cells followed. By 1944 about 250 American patents on diaphragm cells had been taken out [12], of which 32 became commercially proven, being operated in 47 plants. By 1952, the U.S.A. produced 2·366m. tonnes of chlorine, eight times more than Britain, the next largest producer.

The Castner cell underwent various slight modifications during the first quarter of the century, but in 1924 the American J. C. Downs patented a cell for the production of sodium from fused sodium chloride [13]. This consisted of a steel tank lined with firebrick containing a massive cylindrical graphite anode projecting through the base, surrounded coaxially by a cathode of iron gauze. By adding calcium chloride to the sodium chloride, the melting-point of the electrolyte is reduced from 800 °C to 505 °C. The energy effi-

FIG. 21.3. Plant for production of magnesium compounds from sea water.

FIG. 21.4(a). The original rocking cells with mercury cathode set up at Oldbury, England, to demonstrate the Castner–Kellner electrochemical process for caustic soda manufacture.

FIG. 21.4(b). Later horizontal cells for the manufacture of chlorine and hydrogen peroxide.

ciency of the Downs cell process from salt to sodium is about three times greater than that of the composite process of first producing sodium chloride in a mercury cell, followed by further electrolysis in a Castner cell. However, boht processes were in operation in 1950. The price of sodium in the U.S.A. dropped from $2.00 per pound in 1890 to $0.15 in 1946 [14].

Other heavy chemicals were manufactured in electrolytic cells first developed at the end of the nineteenth century. A cell for the production of sodium chlorate was devised in 1887 by H. Gall and A. de Montlaur [15]; a 25 per cent solution of brine was electrolysed in a cell in which a diaphragm separated a platinum–iridium anode from a platinum or nickel anode. Five years, later, a diaphragmless cell was patented in which a solution of magnesium or calcium chloride and potassium chromate was electrolysed. By 1900, the firm of Corbin et Cie in the French Alps, was producing 4500 tons of chlorates per annum. The industry spread to the U.S.A. and Germany, and cells employing graphite or (in Europe especially) magnetite anodes were developed [16]. By 1950, about half the annual world production of 20000 tons of sodium chlorate was used as a herbicide, the other main use being for conversion to potassium chlorate. Cells for the production of solutions of sodium hypochlorite were introduced at about the same time as chlorate cells [17]. Such solutions were prepared by the electrolysis of brine, the product being high in sodium chloride content. However, this method proved to be only about half as economical as that of hypochlorite manufacture by the absorption of electrolytically prepared chlorine in caustic soda solution. Hypochlorite solutions were used extensively as a bleaching agent in the textile and paper industries. Production in the U.S.A. of a 4–6 per cent solution rose from 15000 tons in 1925 to 75000 tons in 1937; estimated production in 1948 was 100000 tons.

III. NITROGEN FIXATION

At the beginning of the twentieth century there were two major sources of 'fixed' nitrogen (that is, nitrogen in a chemically combined state), namely sodium nitrate and ammonium sulphate. Demand for nitrogenous fertilizers was rapidly increasing and it was widely realized that new sources were required [18]. Sodium nitrate was being exported in large and increasing quantities from Chile, where it occurred in large natural deposits. In the second half of the nineteenth century, this accounted for about 70 per cent of the world supply of nitrogen fertilizer; in 1902, 1·3m. tons was exported. Fifty years later, production was 1·4m. tons (though in 1913 it had been

2·8m. tons, declining to 433 000 tons in 1933). However, with increased demand and efficient techniques of fixing nitrogen, Chilean nitrate accounted for only 1–2 per cent of the world supply of nitrogen fertilizers in the mid-1960s. Ammonium sulphate was obtained as a by-product from gas works and coke ovens, destructive distillation of coal producing about 1·5 per cent free ammonia. The 'ammoniacal liquors' were concentrated and absorbed in sulphuric acid. In 1913, 550 000 tons were produced by Germany, 440 000 tons by Britain, and 166 000 tons by the U.S.A. out of a world production of approximately 1 300 00 tons [19].

Fixed nitrogen was required not solely as a fertilizer: it was of prime national importance in the production of explosives (p. 547) for which nitric acid was needed and efforts were concentrated to devise a method for chemically combining atmospheric nitrogen. It had been known from the end of the eighteenth century that when an electric spark was passed through a mixture of nitrogen and oxygen, nitric oxide was formed. As early as 1859, a method of preparing nitric acid based on this reaction was patented by a Frenchwoman, Madame L. J. P. B. Lefêbvre [20]. Though the method was chemically feasible, it was highly inefficient. The reaction is reversible and the combined gases have to be cooled rapidly; only 3 per cent of electrical energy is converted into chemical energy for the combination of the reactants. Cheap electricity was thus essential for the process to be economically worth while. Technical problems included designing electrical equipment which would provide a stable arc and absorbers which would separate efficiently the small proportion of the product. The first industrial plant was operated by the Atmospheric Products Company of Jersey City, New York, in 1902 using a process patented by C. S. Bradley and R. Lovejoy [21]. However, the nitric acid always contained nitrous acid and nitrates and these impurities could not be removed economically; the company ceased to operate the method in 1904.

The first commercial success was the plant devised by the Norwegians Christian Birkeland and Samuel Eyde [22]. By applying a magnetic field to an arc formed by an alternating current, the arc was distorted to an oscillating disc form. The electrodes were formed from copper tubes internally cooled by water. In 1903 the Norwegische Elektrische Aktiengesellschaft constructed small furnaces at Ankerlökken, near Oslo. In 1905, full-scale operation commenced at Notodden [23]. Meanwhile, Walther Nernst and Fritz Haber were studying the thermodynamic equilibrium of the oxides of nitrogen in Germany. The results interested B.A.S.F. They had employed O. Schönherr to study the reaction in 1897. In 1904 he patented a method of producing a steadier arc than had been attained by Birkeland and Eyde [24]. In 1906, the German and Norwegian firms jointly started production as Norsk Hydro at Notodden using Schönherr furnaces, but five years later B.A.S.F. pulled out because potentially more efficient methods of nitrogen fixation were under investigation by Haber. However, production of calcium nitrate from Norwegian arc process nitric acid increased fairly steadily from 1600 tons in 1907 to 109 000 tons in 1919 [25]. With the wide introduction of the Haber process for ammonia in the 1920s, the arc process went into decline: the Nitrogen Products Committee, set up in Britain by the Ministry of Munitions in 1916, reported in 1920 that 100 000 kW of continuous power could produce 119 00 tons of nitrogen as nitric acid or 230 000 tons of nitrogen as ammonia. by the arc and Haber processes respectively.

Although the Haber Process was to prove ultimately of greatest significance, another important method of fixing nitrogen was developed at the turn of the twentieth century. This took as its starting material calcium carbide, which had first been prepared by Friedrich Wöhler in 1862 [26]. This reacts with water to produce acetylene. A commercial preparation was developed in 1892 in the U.S.A. by T. L. Willson, and in France by H. Moissan, by heating a mixture of lime and coke at temperatures of 2000–2200 °C in an electric arc furnace. Initially production was used nearly entirely for lighting purposes; special burners were devised in which water dripped on to the carbide, the gas being lit at a jet. This form of illumination did not develop as rapidly as had been predicted, but a new use for carbide was soon discovered. In Germany, in 1898, Adolph Frank and Nikodem Caro found that when barium carbide was heated in a nitrogen atmosphere a mixture of barium cyanide and barium cyanamide was formed [27]. The object of this investigation was to discover a novel method of preparation of sodium cyanamide, Castner's method being under patent. It was found that calcium carbide could also be nitrogenated; calcium cyanamide only was produced, which could be converted to sodium cyanide with soda ash [28]. In 1900, Frank discovered that ammonia could be produced by hydrolysing calcium cyanamide with superheated steam [29], which led to the suggestion by him (and also slightly later, but independently, by Hermann Freudenberg) that calcium cyanamide should be used as a fertilizer [30]. A year later, F. E. Polzenius found that by adding calcium chloride to the carbide, the temperature at which nitrogenation occurred dropped from 1100 °C to 700–800 °C. This important discovery alleviated technical problems concerning furnace linings.

In 1904 the first commercial plant was set up at Westeregeln in Germany, but it was not an economic success. In 1905, Frank and Caro's Cyanidgesellschaft started operating a plant at Piano d'Orto in Italy, but this was abandoned after a year because of the short life of the externally fired retorts. These were replaced by electric ovens, developed by Frank in 1906, which incorporated a carbon electrode. By 1910, there were plants in Germany, Italy, Canada, France, and Japan producing about 20 000 tons of cyanamide a year. This output was increased tenfold by 1913; by 1918, wartime nitrogen requirements for fertilizer and munitions had forced production up to 600 000 tons. In 1934, a new type of kiln was installed at Knapsack, Germany. It could produce cyanamide continuously instead of in batches, and consisted of a carbon-steel cylinder with dished ends, lined with refractory brick. While the kiln slowly rotated, carbide was introduced through a gas-tight seal and was subjected to jets of nitrogen at 10 atmospheres pressure. This furnace was found to be suitable only for granules, and tunnel ovens were introduced to deal with fine carbide in which the carbide moved along the kiln on flat trucks on rails against a current of nitrogen. The low yield and high cost of these furnaces restricted their use [31].

By 1947 there were 49 cyanamide plants in 17 countries, with an annual production of 750 000 tons (about half maximum capacity). Most was used for agricultural purposes, though polymerization at high temperatures of dicyandiamide (prepared by acidifying calcium cyanamide and dimerizing the product) produces the plastic melamine (p. 555), which was made in increasing quantities from 1939. Although in the earlier part of the century most calcium carbide was used for cyanamide production, acetylene became increasingly important as a basic material in the production of organic chemicals. During the Second World War, production rose in Germany to nearly 3m. tons per annum (in 1937 it had been 1m. tons). In 1962, world production of carbide was over 9m. tons, Germany, Japan, and the U.S.A. being the principal producers.

The third, and in perspective by far the most important, method of fixing nitrogen was the synthesis of ammonia from its elements, nitrogen and hydrogen. This method was also developed in the early part of the twentieth century, though Johann Döbereiner discovered as early as 1823 that nitrogen and hydrogen combined in the presence of a catalyst [32]. The reaction was investigated spasmodically for the next eighty years and it was realized that the reaction was reversible and would proceed favourably at high pressures. In 1903 Haber was asked to study the reaction, and with G. van Oordt it was

21.5. Compressors for ammonia designed to operate at 250 atmospheres.

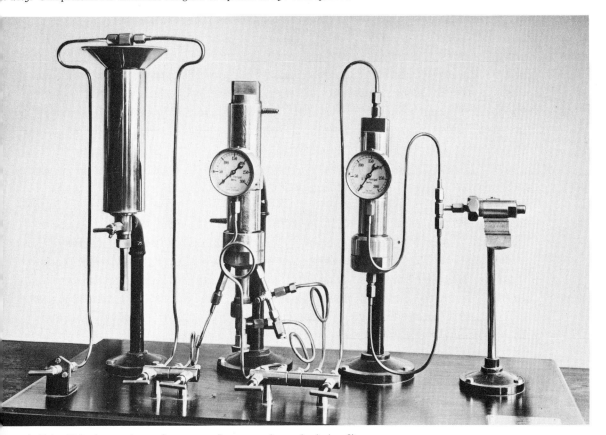

21.6. Fritz Haber's experimental apparatus for ammonia synthesis (1908).

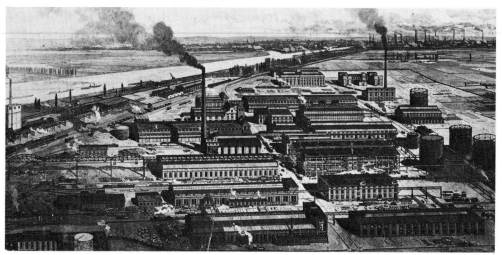

FIG. 21.7. The first full-scale plant for ammonia synthesis by the Haber process, Oppau (1914).

shown in a series of experiments that ammonia was formed from the elements using iron and other metal catalysts at temperatures of 1000 °C; elevated pressures were not, however, used (it was thought that this would produce great technical difficulties) and the yield was small [33]. In 1906, Nernst investigated the equilibrium at enhanced pressures and a temperature of 700 °C, but the yield was even smaller. Haber agreed to re-examine the equilibrium and between 1907 and 1910 he successfully developed the basis of a commercial technique using pressures of about 200 atmospheres. B.A.S.F. became interested in the work and extensive research was conducted to find the most efficient catalyst. A larger-scale test was performed at Ludwigshafen in 1910 and it was realized that, although the engineering problems were challenging, a commercially successful synthetic method had been devised. Increasingly large converters were built and by 1913 the first full-scale plant, employing a converter of $8\frac{1}{2}$ tons, was constructed at Oppau.

The vessel in which the synthesis occurred was a high-quality steel bomb. Because hydrogen at high pressure attacks steel, making it brittle, an inner lining of low-carbon steel was introduced which absorbed the hydrogen; small holes in the lining ensured that the pressure was maintained at the same level on either side. The lining was wrapped with asbestos paper, round which the heating wire was wound. A quartz tube contained the catalyst, and the reacting gases were circulated, the product being withdrawn either by liquefying or by dissolving it. A further investigation tackled at the same time was directed towards finding the most economic source of the reactants. Initially nitrogen was obtained by liquefaction of air, and hydrogen by liquefaction of water gas (a mixture of hydrogen and carbon monoxide, obtained by treatment of coke with steam). By 1915, a method of obtaining a mixture of the gases required had been developed: producer gas (a mixture of nitrogen and carbon monoxide obtained by passing air over coke), water gas, and steam were mixed; the hydrogen was temporarily removed and the carbon monoxide was reacted with the steam to yield more hydrogen and carbon dioxide. The latter was removed, the gases were washed to remove impurities, and the correct mixture of hydrogen and nitrogen was thus obtained [34]. Other methods (but all utilizing water gas) were devised and many patents were taken out. Production of ammonia in 1914 at Oppau was 6000 tons (in terms of nitrogen). A year later this had risen to 12 000 tons; by 1916, it was 43 000 tons. In 1916, B.A.S.F. opened a second plant at Leuna, near Leipzig; by 1918, the combined output was 95 000 tons, representing about half Germany's output of nitrogen compounds.

The Haber process was not adopted by other countries until the 1920s, though an unsuccessful works was erected by the General Chemical Company at Sheffield, Alabama, in 1918. Several modifications were developed utilizing the same basic method but varying pressures, gas velocities, temperatures, and catalysts (the original patents were controlled by B.A.S.F. and the technical details were closely guarded). The most widely adopted of these were the Casale and Claude processes, using pressures of 600 and 900–1000 atmospheres respectively, and temperatures of 500 °C and 500–650 °C. By 1928–9, 75 per cent, 11 per cent, and 5 per cent of world synthetic ammonia was being produced by the Haber, Casale, and Claude processes. The basic Haber process did not undergo fundamental modification, though many variations of catalyst were tried. The most effective to emerge was finely divided iron in a mixture of an alkali metal oxide and an amphoteric oxide such as alumina, silica, or titanium oxide. The oxides act as 'promoters', enhancing the action of the catalyst. By 1930, about half the world production of fixed nitrogen was by ammonia synthesis. Just before the Second World War this had risen to two-thirds; and by 1950, to nearly four-fifths.

An important corollary to the Haber process was the development of an industrial method for oxidizing ammonia to nitric acid. At the beginning of the century the oxidation on a laboratory scale had been known for some time [35] but research was revived in 1900 by W. Ostwald, who found that platinum promoted the reaction to form nitric oxide, higher oxides, and nitric acid, and that a second reaction took place simultaneously which produced free nitrogen [36]. The first works to operate this process on an industrial scale were at Gerthe, near Bochum, in 1908, which used treated coke-oven ammonia. In 1912 plants were built at Vilvorde, Belgium, and at Dagenham, England, though both soon ceased production. A major problem of exploiting the method was that the traditional source of nitric acid (treatment of sodium nitrate with sulphuric acid) was too competitive. The decreasing cost of ammonia from the Haber process altered the situation, and the demand for nitric acid increased during the First World War: the Gerthe installation was enlarged. The constructional details developed in Germany at this time did not significantly change up to the middle of the century, though oxidation plants operating at enhanced pressures were introduced during the 1920s. The reaction took place in a conical chrome-steel converter, across which were interposed horizontal layers of platinum–rhodium alloy gauze. The reaction temperature rises to 900 °C, and cooled nitrogen peroxide is absorbed in a weak solution of nitric acid. By 1950, this method of industrial production had ousted all others. The largest producer of nitric acid in 1954 was the U.S.A. (2 760 000 tons of 100 per cent acid) followed by France, Italy, and Britain. In 1937, 65 per cent of all nitric acid produced in the U.S.A. was used in the manufacture of explosives; a decade later, this had dropped to 18 per cent.

IV. FERTILIZERS

In addition to carbon, hydrogen, and oxygen, thirteen elements are essential for plant nutrition. Of these, nitrogen, phosphorus, and potassium are needed in relatively large quantities, and during the first half of the twentieth century a chemical industry developed steadily to supply compounds containing these three elements. In 1900, 2·2m. tons of artificial fertilizer was consumed in the U.S.A.; by 1950 this had increased eightfold to nearly 18m. tons. In 1965, the figure had reached 31m. tons [37]. World figures show a similar rate of increase, approximately fifteenfold over the period 1900–1965. Two trends have become apparent since the beginning of the century: the use of mixed fertilizers (those containing two or all three of the vital elements) increased, and the proportion by weight of these elements in the applied fertilizer grew larger. In 1900, mixed fertilizers in the United States contained an average of 13·9 per cent $N+P_2O_5+K_2O$ (by weight). This had risen to 22·6 per cent by 1949.

The most important source of nitrogen at the beginning of the century was from natural organic materials, but the relative importance declined as

FIG. 21.8. The storage of products in seasonal demand presents problems for the chemical, as for other industries. This picture shows the interior of an enormous silo for the storage of synthetic fertilizers.

ammonium sulphate, ammonium nitrate, and ammonia itself were applied directly to the soil. Both the sulphate and nitrate were manufactured by the reaction of ammonia with solutions of the appropriate acid. By the mid-1960s, about 85 per cent of the world production of ammonia was used for fertilizers.

Up to the end of the nineteenth century the largest source of phosphate fertilizer was ground bones or bone meal and guano (Vol. V, p. 254). However, enormous deposits of calcium phosphate were discovered in north Africa and from 1889 Algerian mines started production. After the First World War further sources were discovered in Russia and on the island of Nauru in the Pacific. By 1938 annual world production had reached 13m. tons, the U.S.A. accounting for nearly one-third, the other major producers being the U.S.S.R., Tunisia, Morocco, and Nauru. Calcium phosphate is not very soluble in water, and most was converted to 'superphosphate' (which had first been prepared on a large scale by John Bennet Lawes in the nineteenth century [38]). This was manufactured by treating mineral phosphate with sulphuric acid to form a solid mass consisting of a mixture of monocalcium phosphate, dicalcium phosphate, and calcium phosphate which was then ground to the desired size. Originally the reactants were mixed in a container (a 'den'), allowed to react, removed, and left for three weeks or so to 'cure'. Batch mechanical dens were developed, two of the most widely used being the Beskov (in which the batch was contained in a cast-iron vessel which ran on wheels towards an excavator which dug out the super-phosphate [39]) and the Sturtevant, first used in the U.S.A. in 1921 (in which the excavator moves into a fixed den [40]). A continuous process, the Broad-field den, was introduced in 1936 [41].

A more concentrated form of phosphate fertilizer, 'triple superphosphate', first produced in 1872, was of increasing importance in the middle of the twentieth century, especially in the U.S.A. It accounted for 12 per cent of world production by 1956–7, but in the U.S.A. in 1964, approximately equal amounts of super- and triple-phosphate were used. Triple superphosphate is monocalcium phosphate, made by treating mineral phosphate with 75 per cent phosphoric acid in plant similar to that used for continuous process superphosphate. It has the advantage of not being diluted with calcium sulphate. A further source of phosphorus as fertilizer was the slag of steel-works in which phosphoric ores were smelted. This was first sold in Germany as a fertilizer in 1885, and in 1956–7 it still accounted for 16 per cent of the world fertilizer market in phosphates.

Potassium fertilizers ('potash'), in contrast to fixed nitrogen and phos-phates, were simply applied as they were mined, mainly as potassium chloride. Potash was a German monopoly until the First World War, deposits having been discovered at Stassfurt in Saxony in the late 1850s. In 1952, Germany (West and East combined) produced 2·8m. tons (as K_2O), twice as much as their next rival, the U.S.A. (1·48m. tons).

Mixed fertilizers gradually increased in importance from the beginning of the century. Until the late 1920s, multinutrient fertilizers were simply mixed mechanically. After the advent of cheap Haber process ammonia, the ammo-niation of superphosphate to form a mixture of dicalcium phosphate, ammo-nium phosphate, and ammonium sulphate became common.

REFERENCES

[1] *Chemical economics handbook*. Stanford Research Institute, Menlo Park, Calif. (1951–).
[2] HAYNES, W. '*Brimstone: The stone that burns*', pp. 40–62. Van Nostrand, Princeton (1959).
[3] SNELLING, F. C. *Chemistry and industry*, p. 300 (1958).
[4] British Patent No. 17255 (1898).
[5] British Patent No. 10412 (1901).

[6] British Patent No. 11 969 (1902).

[7] British Patent No. 8545 (1901).

[8] British Patent Nos. 23 541 (1913); 8462 (1914).

[9] MUHLERT, F, F. '*Die Industrie der Ammoniak und Cyanverbindungen*', Vol. 2, pp. 171–6, 223–7, 267. O. H. O. Spamer, Leipzig (1915). WILLIAMS, H. E. 'Cyanogen Compounds', p. 98. Arnold, London 1948; *Fifty years of progress, 1895–1945: The story of the Castner–Kellner Alkali Company*, p. 20. I.C.I., London (1945).

[10] HALE, A. J. *The applications of electrolysis in chemical industry*, p. 92. Longmans, London (1918).

[11] United States Patent 723 398 (1903); HALE, *op. cit.* [10], p. 101.

[12] VORCE, L. D. *Transactions of the Electrochemical Society*, **86**, 69 (1945).

[13] United States Patent No. 1 501 756 (1924).

[14] *Kirk–Othmer encyclopaedia of chemical technology*, Vol. 1, p. 442. Interscience, New York (1947).

[15] HALE, *op. cit.* [10], p. 123; British Patent No. 4686 (1887).

[16] WHITE, N. C. *Transactions of the Electrochemical Society*, **92**, 15 (1947); JAMES, M. *Ibid.*, **92**, 23 (1947).

[17] ESCARD, J. *Les industries electrochimiques*, p. 176. Béranger, Paris (1907).

[18] CROOKES, W. Report for British Association for the Advancement of Science, pp. 3–38 (1898).

[19] PARTINGTON, J. R., and PARKER, L. H. *The nitrogen industry*, p. 124. Constable, London (1922).

[20] British Patent No. 1045 (1859).

[21] British Patent No. 8230 (1901).

[22] British Patent No. 20003 (1904).

[23] WAESER, B. *The atmospheric nitrogen industry*, Vol. 2, pp. 565f. Churchill, London (1926).

[24] British Patent No. 26602 (1904).

[25] PARTINGTON, and PARKER, *op. cit.* [19], p. 240.

[26] WOHLER, F. *Annalen der Chemie und Pharmacie*, **124**, 226 (1862).

[27] WAESER, B. *op. cit.* [23], Vol. 1, p. 26.

[28] British Patent No. 25475 (1898).

[29] WAESER, B. *op. cit.* [23], Vol. 1, p. 27; German Patent No. 134 289

[30] WAESER, B. *op. cit.* [23], Vol. 1, p. 26.

[31] KASTENS, M. L., and McBURNEY, W. G. *Industrial and Engineering Chemistry*, **43**, 1020 (1951).

[32] DOBEREINER, J. W. *Journal für Chemie und Physik*, **38**, 321 (1823).

[33] HABER, F., and VAN OORDT, G. *Zeitschrift für Anorganische Chemie*, **43**, 111 (1905).

[34] WAESER, B. *op. cit.* [23], Vol. 2, p. 524.

[35] MITTASCH, A. '*Salpetersäure aus Ammoniak*', p. 17. Verlag Chemie, Weinheim (1953).

[36] British Patents Nos. 698 and 8300 (1902).

[37] *Kirk-Othmer encyclopaedia of chemical technology*, Vol. 9, p. 37. Interscience, New York (1966).

[38] British Patent No. 9353 (1842).

[39] SCHUCHT, L. *Die Fabrikation des Superphosphats*, p. 155. Friedrich Vieweg & Sohn, Braunschweig (1926).

[40] WAGGAMAN, W. H. *Phosphoric acid, phosphates and phosphoric fertilizers*, p. 260. Reinhold, New York (1952).

[41] WAGGAMAN, *op. cit.* p. 263.

BIBLIOGRAPHY

HABER, L. F. *The chemical industry during the nineteenth century*. Clarendon Press, Oxford (1958).
——. *The chemical industry 1900–1930*. Clarendon Press, Oxford (1971).
HARDIE, D. W. F., and PRATT, J. Davidson. *A history of the modern British chemical industry*. Pergamon Press, Oxford (1966).
HAYNES, W. *American chemical industry*, 6 vols. Van Nostrand, New York (1945–54).
KIRK, R. E., and OTHMER, D. F. *Encyclopaedia of chemical technology*, 15 vols and 2 supplements. Interscience Encyclopaedia, New York (1947–60).
——, and ——. *Encyclopaedia of chemical technology* (2nd edn.), 22 vols and 2 supplements, Interscience Publishers, New York (1963–72).
Materials and technology. 9 vols. Longmans and J. H. de Bussy, London and Amsterdam (1968–75).
METZNER, A. *Die chemische Industrie der welt*, 2 vols. Econ-Verlag, Dusseldorf (1955).
READER, W. J. *Imperial Chemical Industries: a history*, 2 vols. Oxford University Press, London, 1970 and 1975.
WILLIAMS, T. I. *The chemical industry past and present*. Penguin Books, Harmondsworth (1953).

PART II: RAW MATERIALS FOR ORGANIC CHEMICALS (INCLUDING EXPLOSIVES)

I. GENERAL CONSIDERATIONS

From its early development the organic chemical industry has had two main functions. The first comprizes the extraction and purification of feedstock chemicals and suitable derivatives from a variety of natural sources. The second comprizes processes using these chemicals in the manufacture of sophisticated products, such as dyestuffs, explosives, and plastics. This part of the chapter deals with the production of feedstock chemicals from raw materials and, in addition, studies the development of explosives during the period 1900–50.

The first half of the twentieth century saw dramatic changes in the availability and utilization of raw materials in the production of organic chemicals. By 1900 coal-tar was the chief source of organic compounds, largely aromatic, produced commercially. Coal-mining was a flourishing concern, providing coal both for the production of steel (as coke) and for the coal-gas industry. In Britain, coal-gas remained of primary importance until the discovery and exploitation of North Sea gas in the 1960s. However, with the advent of large natural gas reserves, the use of coal-gas declined in many countries during the second quarter of the twentieth century. After the Second World War natural gas gradually superseded coal-gas in several European countries (Italy, France, Holland, and West Germany). By 1970 natural gas—which had been exploited since the beginning of the century—accounted for 95 per cent of the fuel gas in the U.S.A. Canada was a substantial exporter to the U.S.A. In Europe, the U.S.S.R. became a substantial exporter to eastern European countries. By 1950, therefore, the main source of coal-tar was changing from fuel-gas production to the production of metallurgical coke and was thus tied to the demand for coke by the steel industry. Although the steel industry has, on the whole, expanded fairly steadily during the twentieth century, supplies of coal-tar have not increased at the same rate. Owing to improved technology in steel manufacture (Ch. 19), the amount of coke required to produce 1 ton of pig iron fell from about 2200 lb in 1913 to 1850 lb in 1950. However, coal-tar remained a useful source of aromatic compounds, especially in Europe, providing a significant revenue to help offset the cost of coke manufacture.

At the turn of the century the rapid developments in the production of coal-tar chemicals were dominated by Germany: the onset of the First World War exposed serious deficiencies in the chemical industries of Britain and the U.S.A. The shortage both of products such as drugs, dyestuffs, and explosives, and of the raw materials required for their manufacture, stimulated these countries, especially the U.S.A., to strengthen their industries. Not only did coal-tar chemicals begin to be used in America, but also preliminary research into the production of petrochemicals was initiated. America emerged at the end of the war with the leading chemical industry, which, not least because of its isolation from European economic troubles in the decade after the war, expanded rapidly during the 1920s.

Although some benzene and toluene were produced from Borneo crude oil during the First World War for the manufacture of explosives in Britain and France, the petrochemical industry had its roots in the United States. In the early twentieth century, mass production of the motor-car created a vast commercial outlet for petrol (gasoline). Methods of modifying the chemical constitution of petroleum and its frequent companion, natural gas (itself used for much industrial and domestic heating), were introduced to improve the yield and quality of the petrol produced. These techniques were later adapted specifically for the production of organic chemicals. In 1918 isopropanol (a useful solvent) was prepared from propylene (present in some refinery gas streams) by the Carleton–Ellis process (treatment with sulphuric acid and subsequent hydrolysis). By 1920 the Standard Oil Company, New Jersey, was producing substantial volumes of petroleum-derived isopropanol, from which was produced acetone or other solvents. In 1925, a year after its formation, the Carbide and Carbon Chemicals Corporation was producing ethylene glycol, important as an antifreeze, and in 1926 the Sharples Company commenced production of amyl alcohol, a valuable solvent, from natural gas. Thus the 1920s witnessed the birth of the major rival of coal-tar chemicals, the petrochemical industry which, at that time, was based on the extraction and utilization of simple olefins from petroleum. The second phase of the petrochemical industry started after the Second World War. The First World War was responsible for the creation of the industry; the Second promoted its widespread expansion. Better techniques were developed and a wider variety of hydrocarbon feedstocks were used to meet the demands of the rapidly developing polymer industries (p. 551), which were able to use the aliphatic compounds of which the more volatile fraction of petroleum is mainly composed. In addition, methods of extracting the less volatile aromatic compounds from petroleum were introduced. The rapid expansion of the industry in the U.S.A. is illustrated by the following figures for the production of oil-based chemicals. In 1925 100 tons were manufactured; this increased to 45 000 tons by 1930, 1m. tons by 1940, about 5m. by 1950, and 49m. by 1967.

Outside America, virtually no petrochemicals were produced until the Second World War, even in the U.S.S.R. where development might have been expected. During the 1940s, the western European petrochemical industry developed along the lines of the early United States industry, restricting production to the lower olefins and their derivatives. By 1950, in Britain, Shell and Petrochemicals Ltd. were manufacturing petroleum-derived chemicals, while an I.C.I. plant was under construction at Wilton for the same purpose, and B.P. were on the verge of entering the field. Oil-based chemicals in western Europe amounted to 100 000 tons a year in 1950. The giant Italian company of Montecatini–Edison, which controlled the pre-war Italian fertilizer and dyestuffs industries, expanded into the manufacture of plastics from petrochemicals very rapidly in the post-war period. In 1955, the Japanese companies of Sumitomo Chemical and Mitsui and Co. began making Ziegler polyethylene, and the Mitsubishi Petrochemical Co. was formed to produce mainly plastics and organic chemicals. The U.S.S.R. had only a small chemical industry, which remained based on coal-tar until the late 1950s because Stalin made no attempt to encourage the development of petrochemicals.

Rather, he wished to encourage the coal-mining industry, a stronghold of the proletariat. After his death, policies changed slowly until the mid-1960s, since when growth has been very rapid, reaching 15 per cent a year.

There are several reasons for the time-lag in the development of the petrochemical industry outside the U.S.A. First, the growth of the car industry in the U.S.A. was unequalled anywhere else in the world: by 1950 the United States had over 270 cars per 1000 population, whereas the corresponding European figure was only 23. In addition to the large quantities of petrol required, the car industry had opened up a market for antifreeze compounds (ethylene glycol was marketed for this purpose first by the Carbide and Carbon Chemicals Corporation in 1927); fast-drying paints and lacquers; and anti-knock compounds (petrol additives permitting high compression ratios in engines). Secondly, the American petrochemical industry was based on the readily available natural gas as well as the catalytic cracking of gases produced from petroleum. In Europe, however, no suitable natural gas reserves were known in the early part of this century. In addition, there were relatively few catalytic cracking plants; consequently the production of olefins was based largely on the heavier naphtha feedstocks. In Europe, too, coal remained the major source of energy until after the Second World War, when prices began to rise, partly owing to industrial disputes. The availability of coal chemicals made petrochemicals uneconomic until the development of the new polymer industries created the need for aliphatic chemicals.

The petrochemical industry grew rapidly: by 1950 about a third of all organic chemicals in the U.S.A. were petroleum-derived: in 1968 the figure had risen to 90 per cent by weight; the corresponding U.K. figure was 85 per cent. Nevertheless, petrochemicals have not ousted coal chemicals. Instead, the expansion has followed that of the new polymer industries. It was not until the 1940s that techniques were developed in America for the manufacture of aromatic chemicals from petroleum. By that time as we have, noted, production of coal-tar was largely dependent on the steel industry, and was inadequate for the growing demands for aromatic chemicals.

II. COAL-TAR CHEMICALS

Peat is formed by anaerobic decomposition of plant debris. It is then slowly metamorphosed by heat under the pressure of accumulated younger sediments, first to lignite or brown coal, and finally to bituminous coal. When heated above 400 °C in the absence of air, coal softens and becomes plastic. Volatile materials are driven off and the remaining involatile matter coalesces, swells, and solidifies to form coke. The four major components into which the volatile material can be separated are coal-gas, coal-tar, crude benzole or light oil, and ammoniacal liquor. The last-named product is removed from the gas stream by absorption into sulphuric acid and is sold to producers of ammonium sulphate, while coal-tar and crude benzole are used in the production of road and roofing materials, creosote, etc, and as organic chemical feedstocks. Coal-tar was first produced in the eighteenth century for treating wood. It then had a period as an unwanted by-product of the manufacture of coal-gas by carbonization for city gas-works at the beginning of the nineteenth century, first in Britain and then more widely in Europe and elsewhere. Its potential value as a rich source of organic chemicals was not exploited until 1845, when A. W. Hofmann extracted benzene—first isolated by Michael Faraday in 1825—from tar. His subsequent conversion of benzene, via nitrobenzene, to aniline, one of the mainstays of the nineteenth-century dyestuffs industry, ensured a dominant position for coal-tar in the chemical industry during the earlier years of the twentieth century.

The relative economic importance of the products varied widely as the century progressed. At first, medium-temperature carbonization gave a fuel

FIG. 21.9. Until the middle of the century coal-tar was the principal raw material in Europe for the manufacture of a wide range of organic chemicals. The picture shows a tar distillation plant at the Beckton Works of the Gas Light and Coke Company.

gas which was the principal source of profit, with the coke and tar of secondary importance. Later, with the development of the organic chemicals industry, particularly those areas which started from aromatic compounds, the coal-tar distillation products became increasingly profitable. With the advent of gas from sources other than coal, and then with the wider exploitation of natural gas, medium-temperature carbonization died out. The steel industry dictated coke production with coal-tar as a by-product which must be made to yield a maximum profit. Fortunately for the steel industry, a number of aromatics such as the cresols, naphthalene and other polycyclic compounds, and pyridine remained most readily obtainable from coal-tar.

Before the advent of natural gas, coal carbonization was the chief source of gas for domestic and industrial heating. In Britain before 1880, gas was produced in horizontal retorts; a quarter of Britain's gas was still obtained by this method in the 1950s. The inclined retort enjoyed a brief vogue between 1890 and 1930 and the vertical retort was first successfully applied in Germany in 1902. All these were batch processes: the more efficient continuous vertical retort was introduced commercially after 1920. This became the most important process in Britain; about 60 per cent of British gas industry plants used the continuous method by the mid-1950s. Outside Britain, gas industries made more use of the coke-oven, from which metallurgical coke is obtained. However, all this was dramatically changed, when, after a brief period of generating gas from petroleum, the North Sea gas deposits were extensively exploited.

Metallurgical coke for the steel industry was originally produced in beehive ovens from which all by-products were lost to the atmosphere. As the extraction of chemicals from coal-tar produced by the gas industry increased in importance, special chemical recovery ovens were developed. Several Knab–Carvès ovens were set up in France before A. Hüssener introduced similar plants in Germany in 1881. In Britain, H. Simon of Manchester erected three Simon–Carvès plants. However, prejudice against coke from any but beehive ovens prevented further development. In the U.S.A. the first recovery ovens were not installed until 1892; these were not specifically for the production of coal-tar, but to provide metallurgical coke and ammonia for the Solvay process. At the beginning of the twentieth century Germany dominated the coal-tar industry, as may be inferred from Table 21.1. During and after the First World War, the American coal-tar industry was widely developed so that by the 1940s the U.S.A. led the field in the production of coal-tar chemicals, even though this industry took second place to the American

TABLE 21.1

Metallurgical coke produced in chemical recovery ovens as a percentage of total metallurgical coke produced

	1900	1909	1914
Germany	30	82	100
Britain	10	18	—*
U.S.A.	5	16	30

* Figures not available.

petrochemical industry. Russia first introduced by-product ovens in 1910 and the development of coal-tar products was closely linked to the use of these ovens in the Ukrainian steel industry.

The temperature at which coal is carbonized is dictated by the products required and it directly affects both the quantity and chemical composition of the coal tar produced. Coal is carbonized at high temperatures (about 900–1200 °C) in metallurgical coke-ovens and in retorts for the gas industry. By 1950, almost all coal-tar was obtained from coke-ovens, the exception being Britain where the production of town gas in retorts was still important, though town gas was to be almost completely replaced by natural gas within the next twenty years. In America a medium-temperature process (700–900 °C) provides domestic coke. Low-temperature carbonization (below 700 °C) yields smokeless fuels (such as coalite from the Coalite and Chemical Co., Britain). This process was developed in Europe and was established in Britain in 1927, becoming of increased importance with the growing demand for smokeless fuels. During the Second World War low-temperature carbonization, which may provide three to five times as much tar as the high temperature process, was of paramount importance for the production of aviation fuel in Germany and Japan, where access to petroleum was limited. In the U.S.A., even by 1950, development of the low-temperature process was minimal. Between 1920 and 1930 various commercial ventures had failed, costing an estimated $50m.

The chief use for crude coal-tar in the early twentieth century was in road-surfacing. Changes in the character of road traffic led to its being tried out for laying dust, first on a small scale in 1907 in Britain. It was later used on a larger scale for binding macadam. In recent times new road-surfacing materials based on asphalt or petroleum pitch have replaced tar. It was also distilled to give creosote for wood preservation and creosote pitch fuel (for the steelworks). In the first half of the century, tar distillation was an important industry for the manufacture of chemicals. Coal-tar from which most of the water had been decanted was distilled to give a pitch residue and five chief fractions: light oil (b.p. below 170 °C), carbolic oil (170–200 °C), naphthalene oil (200–230 °C), creosote oil (230–270 °C), and anthracene oil (270–350 °C). The crude benzole obtained during coal carbonization was combined with the low-boiling light oil fraction from coal-tar and further separation and purification of all the fractions was carried out to give the products outlined in Table 21.2. Light oil was subjected to simple fractional distillation to separate the aromatic hydrocarbons. Phenols were recovered from the carbolic and naphthalene oils by extraction with sodium hydroxide. Subsequent extraction with sulphuric acid removed the pyridine bases, leaving solvent naphtha. Naphthalene crystallized out of the naphthalene oil on cooling, leaving creosote as the residue.

Similarly, anthracene and phenanthrene were obtained from the anthracene fraction. The method of carbonization affected the composition of the tar produced: coke-oven tar contains a higher proportion of aromatic hydrocarbons, whereas gas-works tar yields higher percentages of tar acids (phenols). Low-temperature carbonization yielded a tar with a significant percentage (about 30 per cent) of paraffinic and olefinic hydrocarbons.

TABLE 21.2

The distillation of coal tar and subsequent products obtained

	Boiling points (approx.)	Fraction	Products
COAL-TAR →	Below 170 °C →	Light oil+Crude benzol ————→	Benzene / Toluene / Xylenes / Naphtha
	170–200 °C →	Carbolic oil →	Phenols, cresols, and xylenols / Pyridines / Solvent naphtha
	200–230 °C →	Naphthalene oil →	Naphthalene / Phenols / Creosote
	230–270 °C →	Creosote oil →	Used as obtained
	270–350 °C →	Anthracene oil →	Anthracene / Phenanthrene / Creosote

The chemicals produced from tar were used for the synthesis of a wide variety of compounds for many industries. All the basic reactions were known by 1900, but here we can only outline some of the basic synthetic techniques applied to the feedstocks. Oxidation, usually effected by heating in the presence of a catalyst, converts toluene to benzaldehyde, naphthalene to phthalic acid, and anthracene to anthraquinone. Nitration, using nitric acid and oleum, converts benzene to nitrobenzene, which may be reduced to aniline; similarly toluene gives toluidine via nitrotoluene, so providing a starting-point for many reactions yielding dyestuffs. Sulphonation of aromatic hydrocarbons, followed by hydrolysis, provides a route to phenols. Thus, naphthalene gives β-naphthol, which on further sulphonation yields the naphthol sulphonic acids required by the dyestuffs industry. Research had also shown that a number of compounds were high explosives. The first of these to be developed and widely used was trinitrophenol (picric acid). Its use for this purpose was patented by E. Turpin in 1885 (French Patent 167 512) and it was adopted by the French Government under the name Melinite. Trinitrotoluene (TNT) was known by 1900, and began to be used by the Germans in 1904, but it was only widely developed during the First World War, when it was required in quantities which exceeded the capacity of the coal-gas industry to supply toluene. This encouraged petroleum distillation in those installations capable of yielding toluene. The antiseptic action of phenol was first appreciated by Joseph Lister in 1867, but the phenols themselves have been superseded by other germicides. In the present concept the chloro-derivatives are important, particularly since 1927, the date of the first German patent for a chlorinated xylenol antiseptic. Another chloro-derivative is benzene hexachloride the λ-isomer of which was developed in 1942 as the insecticide Gammexane. Research continues in these fields, but the distinction between research in coal-tar derivatives and petroleum derivatives is no longer meaningful.

III. ALIPHATIC CHEMICALS

As previously observed, coal-tar chemicals are aromatic, but aliphatic compounds may be produced from coal in several ways. Introduced after 1875 as a fuel, water-gas is generated by the reaction between steam and hot coke and is composed chiefly of carbon monoxide and hydrogen. In 1913 Badische Anilin- und Soda-Fabrik AG (B.A.S.F.) began research on the synthesis of methanol from water-gas and they were producing synthetic methanol by 1924. During the 1920s Franz Fischer and Hans Tropsch found that when water-gas was passed over suitable metal oxide catalysts at about 200 °C under moderate pressure hydrocarbons suitable for petrol resulted. P. E. M. Berthelot used nascent hydrogen to convert coal into oil in 1890, but it was F. Bergius in 1913 who hydrogenated coal catalytically using molecular hydrogen. B.A.S.F. developed the process, which became of great importance to Germany during the Second World War.

Acetylene is manufactured from coke and lime, via calcium carbide. The process was developed commercially in 1892 simultaneously by T. L. Willson in the U.S.A. and H. Moissan in France (p. 523). The high temperature required for reaction (about 2000 °C) is achieved by the passage of a large electric current through the mixture, and the economics of the process is dominated by the cost of the electricity used. Acetylene was initially developed for gas lighting, but after 1904 it found a vast commercial outlet in oxy-acetylene welding (p. 459) and later as an important synthetic starting material. The work of Walter Reppe in Germany provided the major impetus for industrial acetylene derivatives, produced in significant quantities during the Second World War. Acetylene production reached its peak in 1943–4 in Germany when 440m. cubic metres were manufactured, of which 86 per cent was from calcium carbide. Europe was largely dependent on acetylene for supplies of ethylene and other chemicals demanded by the newly developed polymer industries at this time. By mid-century, however, petroleum-derived ethylene rapidly superseded acetylene as a feedstock, especially in countries such as Britain where electricity was expensive: in Norway, Sweden, and Japan, where hydroelectric power was cheap, acetylene remained an important synthetic source.

In addition to the above-mentioned raw materials, certain agricultural and forestry products were developed as sources for particular organic compounds. One of the most important aliphatic chemicals has always been ethanol, ever since distilled spirits produced in the fourteenth century were recognized to be valuable as a solvent. Especially in the early part of the twentieth century, ethanol was an important starting material for the synthesis of many chemicals, such as ethylene, acetaldehyde, etc. The production of alcoholic beverages by fermentation is one of the oldest industries known, and in 1900 the manufacture of ethanol was still carried out almost exclusively by the fermentation of carbohydrate products, followed by distillation of the dilute aqueous alcohol so obtained. Molasses, the syrupy residue obtained from the extraction of crystalline sugar from sugar-cane or beet, was the principal source, although in Germany, for example, potatoes were the preferred material, and grain or wood could also be used; the last three had to be hydrolysed before fermentation. Although in many countries where there is a surplus of cheap fermentable material ethanol is still produced mainly by fermentation, most industrial alcohol in countries like the U.S.A. and Britain with a petrochemical industry is now synthesized from petroleum-derived ethylene (p. 535).

'Pyroligneous acid' (a complex mixture containing methanol and acetic acid), a by-product of wood carbonization for the production of charcoal, has long been a route to certain chemicals (methanol, acetic acid) and thence to products such as acetone. In the United States, the peak of wood utilization was reached early in the century but gradually declined, partly owing to depletion of forests and partly to replacement of sources. Charcoal continued to be made since some was still required for use as an adsorbent—with a special requirement for gas masks—but the amount of 'wood alcohol' produced from wood itself in 1950 was only a quarter of that produced in 1920. Synthetic methanol from water gas was exported to the United States by B.A.S.F. in competition with the American wood alcohol manufacturers in the 1920s. In retaliation, the duty on methanol was raised by half (to 18 cents per U.S. gallon), effectively shutting out German methanol from America, but American exports of methanol were halved between 1923 and 1929. By 1950 petroleum-derived synthesis gas (carbon monoxide and hydrogen) had replaced water-gas as the major source of methanol.

The principal use of acetone before the First World War was in the manufacture of smokeless powder for explosives. Until then most acetone was produced via calcium acetate, itself derived from acetic acid present in pyroligneous acid. At the outbreak of war, demand for explosives by the combatants soared and supplies of acetone were totally inadequate, even using acetic acid derived from fermentation (alcoholic and acetic) of sugars. Attempts were made in the U.S.A. and Britain to copy a German process in which acetylene was converted catalytically to acetic acid, but German patents, seized in America, proved to be singularly uninformative on the initial hydration of acetylene to acetaldehyde. The acetone problem was eventually solved in quite a different way. During his researches on the production of butadiene and derivatives for the manufacture of synthetic rubber, Chaim Weizmann isolated a bacterium (*Clostridium acetobutylicum*) which fermented grain to butyl alcohol (butanol) and acetone in a 2:1 ratio. In early 1915, the British Admiralty Powder Department became interested in the process and by 1916 significant quantities of acetone were available for cordite manufacture in Britain. The full history of this work has an important political element, since Weizmann chose as his reward an agreement by the British Government to support the principle of a Jewish national home in Palestine. He subsequently became the first President of Israel.

During the First World War, interest in synthetic rubber waned and butyl alcohol piled up as a comparatively useless by-product of Weizmann fermentation. After the war, demand for acetone fell dramatically, but within a few years butyl alcohol and its derivatives found a new outlet. In the 1920s, butyl acetate proved to be an extremely valuable solvent for quick-drying nitrocellulose lacquers in the car industry (p. 601): previously the conventional finishes for car bodywork had caused a bottle-neck in production. In 1923 the recently formed American Commercial Solvents Corporation manufactured about 2000 tons of butanol and by 1924 this had risen to nearly 6500 tons. By 1969, when fermentation processes had been overtaken by production from petroleum, U.S. production was over 100 000 tons per annum.

IV. PETROCHEMICALS

The term 'petrochemical' includes chemicals manufactured from both petroleum and natural gas, which may occur in association with petroleum or, more often, in isolation. Believed to be of biological origin, crude petroleum is a rank, green or brown liquid with a high viscosity and is composed largely of aliphatic hydrocarbons. Natural gas, found in association with petroleum ('wet' gas) contains appreciable quantities of easily liquefiable hydrocarbons (with up to seven carbon atoms per molecule) in addition to methane, whereas most non-associated ('dry') natural gas consists almost entirely of methane.

Commercial drilling for petroleum was first carried out in America in 1859; paraffin for lamps (kerosene) was the product chiefly desired from distillation. In the later years of the nineteenth century the increase in the number of machines in use everywhere and in their speed of operation outstripped the supply of animal and vegetable oils formerly used for lubrication. Mineral oils took their place and became important petroleum products.

With the blossoming of the new car industry, petrol became valuable and oil prospecting boomed. Since then the major products from petroleum have been fuels (petrol, paraffin, diesel fuel, etc.). Petrochemicals were developed subsequently as a valuable by-product; today, not more than 0·7 per cent of the total consumption of natural gas and petroleum in the United States is

TABLE 21.3

The distillation of crude oil and the uses of the resulting fractions

	Boiling points approx.	Fraction	Products
	Below 40 °C	Liquid petroleum gas (C$_1$–C$_5$ paraffins) →	Fuel Chemical feedstock
	40–180 °C	Light naphtha (straight-run gasoline) →	Gasoline Chemical feedstock
	130–220 °C	Heavy naphtha →	Solvents
CRUDE OIL →	160–250 °C	Paraffin (kerosene) →	Domestic heating fuel
	220–350 °C	Gas oil →	Diesel oil Industrial fuel oils
	—	Residual oil →	Lubricating oils Waxes Asphalt

used as a chemical feedstock, even though by 1950 petroleum-derived chemicals accounted for a third of the organic chemical industry.

Initially, petrol was obtained by straight distillation (Table 21.3) but the voracious demands of the motor-car industry prompted the development of processes to increase the amount of petrol produced. Thermal cracking, by which long-chain hydrocarbon fractions are broken down to produce smaller molecules suitable for petrol engines, was developed in the liquid phase by the Standard Oil Company, Indiana in 1913 (the Burton process) and Carbon Petroleum introduced the Dubbs process. Vapour-phase cracking was not commercially viable until 1925. Later techniques included hydrogenation (1927); reforming (by which naphtha is upgraded to higher octane gasoline); and catalytic cracking (on alumina and silica). These methods were later modified for the production of chemical feedstocks by treatment of the two most volatile fractions from petroleum (Table 21.3). Heavy naphtha provided a source for organic solvents.

Before the First World War, the petrochemical industry was largely concerned with the manufacture of simple olefins and their derivatives. Ethylene was prepared by thermal cracking of liquid petroleum gases. Propylene and butylene were extracted later, either by direct cracking or as refinery by-products from reforming or catalytic cracking. Separation was achieved by

FIG. 21.10. Typical mid-century chemical plant. This picture of I.C.I.'s Wilton Works shows an olefine plant, two polyolefine plants, and a power station (1956).

FIG. 21.11. By the middle of the twentieth century the chemical industry was capital-intensive rather than labour-intensive. Batch processes were increasingly replaced by highly automated continuous ones. This picure shows the control panel for I.C.I.'s No. 3 olefines plant at Wilton (1959).

fractionation at low temperatures under pressure. At first, most petro-chemicals were derived from ethylene by one of two chief reactions. Hydration to ethanol, by absorption in concentrated sulphuric acid followed by hydrolysis of the ethyl sulphate so formed, provided an alternative route to fermentation (p. 541). In the United States about 90 per cent of industrial alcohol was made by fermentation in 1930. By 1939, 48m. U.S. gallons, 24 per cent of the total, was obtained via ethyl sulphate, and in 1949 the figure reached 165m. U.S. gallons (44 per cent). Direct hydration of ethylene at elevated pressures became commercially successful after the Second World War, when new catalysts became available. In Britain production of ethanol via ethyl sulphate commenced in 1942 and direct hydration was introduced in 1951. Ethylene-derived ethanol provided a synthetic route to many aliphatic chemicals, for which industry had previously been dependent on other raw materials.

Conversion to ethylene oxide by reaction with hypochlorous acid led to a completely new range of chemicals. Ethylene glycol, made by hydrolysis of ethylene oxide, probably accounts for over half the ethylene manufactured. Initially used for preventing accidental dynamite explosions in winter (caused by crystallization of liquid components), it was sold as the first permanent antifreeze for car radiators by Union Carbide in 1927. Glycol ethers, obtained by hydrolysis using alcohols, became useful as high-boiling solvents and in surface coatings. The number of useful derivatives from ethylene increased rapidly; in 1926 Carbide and Carbon Chemicals manufactured about five different derivatives; this rose to 41 in 1939. The corresponding numbers for propylene were two in 1926, and 27 in 1939. Butylene derivatives were introduced after 1930.

From 1940 the industry expanded: the number of hydrocarbons used as raw materials increased and the original, rather limited outlets for products diversified. Methods of separation were improved. While the utilization of simple olefins continued to increase, aromatic compounds, dienes, and acetylenes were also obtained, leading to further derivatives. Several important new olefin reactions were introduced in the post-war period, some of which had been discovered earlier. Direct oxidation of the olefin and direct hydration to the alcohol became commercially viable. Reaction with carbon monoxide and hydrogen, discovered in Germany in 1938, gave primary alcohols containing one extra carbon atom; for example, reaction with ethylene yields n-propanol. High-temperature substitutive chlorination led to synthetic glycerol and to new intermediates for the plastics industry.

Methane, derived from natural gas, became widely used as a source of petrochemicals through the possible conversion with water (methane–steam) to synthesis gas (carbon monoxide and hydrogen), or with oxygen (methane–oxygen) to hydrogen after removal of carbon dioxide. Thus, in America, methane largely replaced coal in the production of synthetic methanol. Towards the end of the first half of this century acetylene derived from petroleum was manufactured from methane, by reaction at temperatures above 1200 °C. In Europe, this process was developed using methane derived from coal.

A range of oxygenated compounds (formaldehyde, methanol, acetaldehyde, etc.) were made available by oxidation of the lower paraffins, especially propane and butane. Toluene and xylene were prepared from petroleum naphthenes during the Second World War, and benzene was added to the range of aromatic chemicals soon after, supplementing coal-tar derived benzene, which was no longer sufficient to meet the demands of the American chemical industry. During this period the growth of the polymer industries was creating a vast market for petrochemicals. Research on the development of synthetic rubber was directed towards the development of a specific route for the production of butadiene from petroleum-derived butane and the butenes, and styrene from benzene and ethylene. Nylon opened up a market for cyclohexane and the investigation which led to the formulation of Terylene entailed the isolation of *p*-xylene. Nitrile fibres required the synthesis of acrylonitrile from ethylene or acetylene. Plastics depend on a wide variety of chemicals: styrene, vinyl chloride, and polyethylene from ethylene, and formaldehyde from methanol, to name but a few. In addition detergents, which by 1968 exceeded soap in demand in western countries, were petroleum-based from the outset. The order of magnitude of world sales in 1968 was: soap 7m. tons, surfactants 15m. tons.

V. EXPLOSIVES

Most explosives contain oxygen–nitrogen bonds as in nitrates or nitro-compounds, and may be divided into two categories. High explosives are insensitive to normal shocks, requiring an initiating explosion from a detonator; low explosives burn rather than explode, evolving large volumes of gas. The foundations of the twentieth-century explosives industry were laid in the nineteenth century with the development of nitroglycerine and nitro-cellulose high explosives (dynamite, blasting gelatine, cordite, etc.) and by

1900 picric acid explosives, such as Lyddite (British), Melinite (French), and Shimosite (Japanese), were supplanting black gunpowder for military purposes (Vol. V, Ch. 13). In Germany, B.A.S.F. tested many nitrated organic compounds, finally deciding in favour of trinitrotoluene (TNT). Research was also directed to the development of detonators superior to the original mercury fulminate type. Lead azide, first prepared by T. Curtius in 1890–1, gradually replaced fulminates from the beginning of the century and diazodinitrophenol was patented by W. M. Dehn in America in 1922. Electric blasting was introduced in the early twentieth century. The period 1940–50 also saw the dawn of a new era of explosives: the atomic bomb (Ch. 11).

The outbreak of the First World War had a profound effect on the explosives industry, TNT and picric acid becoming widely used. In Germany, the flourishing coal-tar industry provided the aromatic raw materials while the Haber process supplied the all-important nitric acid. The French and British allies were initially unprepared and suffered a shortage of both feedstocks. Supplies of phenol and toluene from the revived coal-tar industry were supplemented by aromatic derivatives from Borneo crude oil. CIBA-owned Clayton Aniline Co. in Manchester raised their production of TNT from a pre-war level of 200 tons per annum to 3000 tons per annum by 1916: 700 000 tons of explosives were produced in total in Britain between October 1915 and December 1918. Nobel's Explosives Ltd., the largest British manufacturer, formed part of the Nobel Dynamite Trust Co. which dominated the industry in Europe. In America, Du Pont had a monopoly in military explosives before 1913, when the Hercules Powder Co. and the Atlas Powder Co.

FIG. 21.12(b). Cotton linters being loaded into nitrators at the same plant; note the protective mask.

came into being, after a court decree. Du Pont diversified its interests by buying up firms specializing in nitrocellulose-containing products, such as celluloid, metal lacquers, and 'leather cloth'; these changes began reaping their rewards in the 1920s and 1930s.

Commercial explosives were very important also in mining, road-building, etc. In Canada, for example, the mining boom led to large-scale manufacture of explosives. Large explosives works were in operation in South Africa before the First World War and, like their British parents, the Nobel and Kynoch firms were combined in 1918. This new firm merged with De Beers in 1924, forming the African Explosives and Chemicals Industries Ltd. Blasting gelatine (92 per cent nitroglycerine and 8 per cent nitrocellulose) is still the strongest commercial explosive, although explosives based on ammonium nitrate (AN) increased in importance over the period with which we are concerned.

The oxygen balance (that is, the number of grams of oxygen lacking or in excess for the complete combustion of 100 g of explosive, where all carbon and hydrogen present is converted to carbon dioxide and water) is of great importance in mining. Nitroaromatics, for example, have a negative balance, causing production of carbon monoxide in place of the dioxide. For mining purposes, where this could cause accidental explosions, an oxygen-carrier (chlorate, nitrate, etc.) must be added. France was first to introduce safety regulations in 1890. British introduced 'permitted' explosives in 1897, and the American 'permissible' list came soon after. The permitted list was redrawn in 1912, and was regularly updated thereafter.

At the start of the Second World War, TNT and picric acid still remained the preferred military explosives. Hexamethylenetetramine (RDX or cyclonite) had been known from 1899 but, because of its sensitivity and high cost, was not generally in use. The British developed production techniques when more powerful explosives were demanded, and large-scale manufacture from formaldehyde and ammonium nitrate was initiated by the Tennessee Eastman Co. Pentaerythritol nitrate (PETN) is almost as powerful as RDX, and was manufactured by the nitration of pentaerythritol. Both RDX and PETN were commonly used in conjunction with TNT in bombs, mines, and torpedoes. The Second World War led to enormous production of explosives: in the U.S.A., 73 new plants were built between 1939 and 1944. During the Second World War, 30m. tons of military explosives were manufactured in the United States. Commercial explosives produced between 1912 and 1951 amounted to about 9m. tons.

BIBLIOGRAPHY

AYRES, E. Raw materials for organic chemicals. *Chemical and Engineering News*, **32**, 2876–82 (1954).

COLES, K. F., KERNER, H., and PORGES, J. W. Chemicals from petroleum. *Review of Petroleum Technology*, **12**, 337–53 (1952).

COOKE, M. A. *The science of high explosives*. American Chemical Society Monograph (1958).

GOLDSTEIN, R. F. History of the petroleum chemicals industry in *Literature resources for chemical process industries*. American Chemical Society (1954).

HABER, L. F. *The chemical industry 1900–1930*. Clarendon Press, Oxford (1971).

HIBBEN, J. H. Organic chemicals, productions and value. *Industrial and Engineering Chemistry*, **42**, 990–7 (1950).

HOIBERG, A. J. *Bituminous materials: asphalts, tars and pitches*. Vol. III. Interscience, New York (1966).

IHDE, A. J. *The development of modern chemistry*. Harper and Row, New York (1964).

KIRK, R. E., and OTHMER, D. F. *Encyclopaedia of Chemical Technology* (1st edn.), Vols. 1–15. Interscience, New York (1947–56).

KLAR, M. *The technology of wood distillation*. Chapman and Hall, London (1925).

LUNGE, G. *Coal-tar and ammonia* (5th edn.). Gurney and Jackson, London (1916).

McADAM, R., and WESTWATER, R. *Mining explosives*. Oliver and Boyd, Edinburgh and London (1958).

MARSHALL, A. *Explosives* (2nd edn.). Churchill, London (1917).

Petroleum: 25 years retrospect 1910–1935. Institute of Petroleum Technologists, London (1935).

REUBEN, B. G., and BURSTALL, M. L. *The chemical economy*. Longman, London (1973).

URBANSKI, T. *Chemistry and technology of explosives*, Vol. III. Pergamon Press, Oxford (1967).

WADDAMS, A. L. *Chemicals from petroleum* (3rd edn.). John Murray, London (1972).

WILLIAMS, T. I. *The chemical industry: past and present*. Penguin Books, Harmondsworth (1953).

WILSON, P. J., and WELLS, J. H. *Coal, coke and coal chemicals*. McGraw-Hill, New York (1950).

PART III: POLYMERS, DYES, AND PIGMENTS

Some of the greatest changes in personal life and industrial techniques experienced in the twentieth century have been brought about by changes in the materials available for fabrication. This is true of nearly all materials, including those known throughout civilization, such as metals or glass, of which new forms have multiplied as the result of advancing chemical knowledge of their nature. But the greatest change has been in the introduction of materials of a totally novel character, loosely described as polymers, having in common a particular type of complex chemical constitution. Some naturally occurring materials have this kind of structure, produced by the union of many small identical or similar components. These include the building materials of living organisms (cellulose, protein) and some of the earliest of this family of new materials to be introduced were, in fact, modifications of natural products. The greatest advances, however, have been made since the chemistry of polymerization, the building up of large molecules by sequential linking of many small molecules, was investigated from the 1920s.

To estimate the changes brought about by polymers, one could compare the conditions in a domestic kitchen in, say, 1860 with those in, say, 1960. In 1860, working surfaces, wall and floor coverings, containers, cooking vessels, and implements would all have been made of materials known since very early times. In 1960 hardly anything in use in the kitchen would not have had, as at least part of its constitution, substances discovered or invented in the chemical laboratory and having properties of cleanliness, colour, resistance to heat, transparency, and much else not attainable by any traditional material. The comparisons in the world of industry would be even more detailed and striking, except where sheer size and load-bearing capacity were the important characteristics. The development of these new synthetic materials is described below. It is followed by an account of dyes and pigments—not unrelated because new polymers, especially those used as textiles, posed new dyeing problems.

I. POLYMERS BASED ON NATURAL MATERIALS

Cellulose nitrate (*CN*). The origins of the plastics industry lie in the nineteenth century and it is proper to mention them here since they were not considered in the previous volume of this work. It was Christian Frederick

Schönbein, a professor at the University of Basle, who early in 1846 first isolated from the reaction of paper and a mixture of concentrated nitric and sulphuric acids a substance (cellulose nitrate) which was capable of being shaped into attractive vessels. The mention of this work by Schönbein's patent agent, John Taylor, to an English metallurgist, Alexander Parkes, led to the proper formulation of the first new plastics material. Parkes recognized the need of the nascent electrical industry for an insulating material to replace shellac, horn, and gutta percha, and experimented with a wide range of compounds as potential plasticizers or solvents for the cellulose nitrate in the hope that a correctly formulated new material might fulfil that role. He exhibited his 'Parkesine' at the Great Exhibition of 1862 and unveiled his plasticized cellulose nitrate (using camphor as plasticizer) at a meeting of the Royal Society of Arts in December 1865. This was not, however, to be the birth of the industry; Parkes's original Parkesine Company was wound up in 1868, leaving his assistant, Daniel Spill, to continue the work first as the Xylonite Company, which failed in 1872, and thereafter alone using the trade name 'Ivoride'. Meanwhile, in the U.S.A. a prize of $10 000 had been offered for an alternative to ivory, there being a severe shortage of this material for billiard-ball manufacture. John Wesley Hyatt, in seeking this prize, experimented with a variety of materials and came to examine cellulose nitrate as described in a series of patents from the middle of 1869. It was Hyatt who, after painstaking researches, made the all-important discovery that a solution of camphor in ethanol was not only the ideal plasticizer but also the prime solvent which enabled the cellulose nitrate—or 'Celluloid' as his brother Isaiah named it—to be processed into marketable products. Camphor still played the same vital role in the 1970s, no superior substitute having been found. The Hyatt brothers exploited their discoveries commercially, and celluloid was soon used in the U.S.A. not only for billiard balls and for detachable collars and cuffs for shirts, but as an alternative to rubber in artificial dentures, and in a variety of decorative wares. Hyatt patented an injection-moulding technique in 1878 for the covering of buckles with celluloid. In Britain, Daniel Spill joined forces with L. P. Merriam and Amasa Mason, a representative of the Hyatts' Celluloid Manufacturing Company, and in 1877 formed the British Xylonite Company (which still trades a century later within the British Industrial Plastics organization). The fate of the Hyatts' own organization was to become, many years later, the Plastics Division of the Celanese Corporation of America.

Cellulose nitrate was the first of the thermoplastics, that is, a material which under the application of heat and pressure can be constrained into a particular shape or form which it retains after the constraints have been removed, the whole cycle of events being capable of indefinite repetition. The availability of cellulose nitrate in the form of film with outstanding mechanical properties made feasible the development of cinematography (Ch. 53) in the latter part of the nineteenth century. The heyday of cellulose nitrate was the 1920s, when its use as a film base covered the whole range of photography and cinematography. British production at that time reached a peak of 40 000 tons a year. Production in Britain had declined to only half that figure by 1963, and this fall typified a worldwide trend after the introduction of the much less flammable materials cellulose acetate butyrate (CAB), cellulose acetate propionate (CAP), and cellulose triacetate (CTA), as film base stock. Only CTA, available after the Second World War and widely adopted for professional motion picture films, matched cellulose nitrate in its outstanding dimensional stability and ousted the older material from its remaining uses in precision aerial photography and photogrammetry.

Casein (CS). A second semi-synthetic material came on the scene at the turn of the century. In response to a German demand for a white 'blackboard', W. Krische and A. Spitteler took out German and U.S. patents in 1899 and 1900 respectively for a process wherein a hard, white, water- and acid-resistant material could be coated on cardboard by first treating the cardboard with a solution of casein (the chief protein in skimmed cows' milk) followed by reaction with formaldehyde. The International Galalith Gesellschaft Hoff and Company made casein products in Germany from before the First World War. In Britain early manufacture was associated with the Syrolit Company, still trading as Erinoid Ltd. in the 1960s. Various manufacturers have been associated with the product in the U.S.A., where the trade names 'Aladdinite', 'Kyloid', and 'Ameroid' have been used at various times. Casein's main uses have always been in the buttons and decorative arts trade, where it has held its own against newer synthetic materials. Its attractions have been its manufacture from waste natural products, its fabrication by machining techniques, and its ability to accept an almost infinite range of colours and shades.

Cellulose acetates (CA and CTA). The complete acetylation of cellulose gives the triacetate (CTA), and it was this material which was the product of the early experimenters. It was first made under most vigorous conditions by P. Schutzenberger in 1865, but it was the work of C. F. Cross and E. J.

Bevan in 1889–94 and their patent of 1894—employing the milder conditions of precipitation of cellulose hydrate from zinc chloride or cuprammonium solutions, followed by heating with acetyl chloride in the presence of crystalline zinc acetate—which laid the foundations for the industrial production of cellulose triacetate. However, the triacetate is soluble only in relatively expensive and toxic solvents like chloroform and carbon tetrachloride, and commercial production of this polymer, for both film and fibre, began only in 1954, at the end of the period now under discussion, when methylene dichloride became available cheaply and on a large scale as an alternative solvent. In 1905 G. W. Miles discovered that Cross and Bevan's primary product could be partially deacetylated under mild aqueous conditions to give a second polymer, commonly referred to as cellulose acetate (CA), which was soluble in acetone. This material could be processed as fibres or gave films of reasonable clarity. By 1910 the Swiss brothers Henri and Camille Dreyfus were producing nonflammable cellulose acetate film. The mechanical properties of this diacetate film were, however, inferior to those of cellulose nitrate, and the latter was preferred as the cinematographic film base until the later introduction of the film forms of the mixed cellulose esters and cellulose triacetate, already referred to. The First World War saw cellulose acetate playing a vital role as a 'dope' for aircraft wing fabrics. The overcapacity which resulted at the end of the war was accommodated by the development of the fibres market, the Dreyfus brothers forming the British Celanese Company (U.K.). Acetate rayon (as the fibre form was called) was available in Britain in 1921, and in America three years later. Its future was further ensured by the discovery by Alexander Clevel, in 1922, of an entirely new class of dyes, the disperse dyes, specifically suited to this new fibre. In 1921, Arthur Eichengrün designed the prototype of the modern injection-moulding machine, and cellulose acetate became the principal thermoplastic moulding material until after the Second World War, when it was superseded by polystyrene and polyethylene.

Thermosetting materials from formaldehyde and phenol, urea, and melamine. Adolf von Baeyer first described the resinous products from the reactions of phenols with aldehydes in the *Berichte der Deutschen chemischen Gesellschaft* in 1872. Leo Hendrik Baekeland, a Belgian working in the U.S.A., who by his exploits became one of the giants of polymer history, having already made a fortune for himself by the sale of his 'Velox' photographic printing paper rights to Eastman Kodak in 1899, began to concern himself with these resinous products, and with the need for a substitute for such materials as shellac, ebonite, rubber, and asphalt. He carefully assessed all the previous work and his own systematic and brilliant experiments resulted in over 100 patents (of which the most famous is the 'heat and pressure' patent of 1907) concerned with regulated manufacture from phenol and formaldehyde of an entirely new, and the first completely synthetic, material, Bakelite. With this material, in contrast to the thermoplastics, the application of heat and pressure confers upon the polymer molecules final structural details which are manifest by the material emerging from the mould as a hard solid incapable of further changes in shape or form: hence the designation thermosetting.

James (later Sir James) Swinburne, an electrical engineer, was concerned to find an improved material for cable insulation, and a phenol-formaldehyde resin made by Adolf Luft was brought to his notice in 1902. Unaware of the work going on in the U.S.A., he improved the method of preparation of this thermosetting material, only to file his patent application in 1907 just one day after Baekeland. The General Bakelite Company was formed in the U.S.A. in 1910 and soon established a market for its products in the electrical and motor vehicle industries. Commercial success came more slowly in Britain. After the war Swinburne and Baekeland came to an agreement. Swinburne's evocatively named Damard Lacquer Company (of Birmingham) and the Bakelite Corporation of Great Britain, together with other smaller concerns, merged their interests to form Bakelite Ltd. in 1926–7. By the time of Baekeland's death in 1944, world production of phenolic resins was 125 000 tons annually, their range of applications covering electrical and household fittings fabricated from thermosetting moulding powders, laminates, adhesives, binders, and surface coatings.

The disadvantage of phenol-formaldehyde (PF) mouldings lay in their dark, drab colours, though cast PF resin could be obtained colour-free and pigmented to yield more attractive products. Nevertheless, the desire for brighter products spurred on research with analogous materials. The urea-formaldehyde–thio-formaldehyde moulding powders, first marketed as 'Beetle' in 1928 by the British Cyanides Company, were one such group of materials, though they possessed the drawback of water absorption. A second class of resins having improved water- and heat-resistance were those derived from melamine and formaldehyde developed by Henkel in Germany in 1935 and soon afterwards by Ciba in Switzerland. Not least important to their success were the developments about the same time of practical routes to melamine discovered by Ciba, and in the U.S.A., by Monsanto and American

Cyanamid. The range of applications of these two groups of resins today follows that of phenol-formaldehyde polymers.

Poly(vinyl chloride) *(PVC)* *and poly(vinyl acetate)* *(PVAC).* The monomers vinyl acetate and vinyl chloride were synthesized from acetylene by F. Klatte in 1912, and polymerization studies of the former compound soon followed. By 1920 the firm of Wacker in Burghausen were making poly(vinyl acetate) for lacquers, but its copolymerization with vinyl chloride to yield new materials which could be processed without decomposition, as described in the independent patents—filed in 1928, of the Carbide and Carbon Chemical Corporation, and E. I. Du Pont de Nemours, both of the U.S.A., and the German I. G. Farbenindustrie—represents its most significant contribution to the development of plastics. Poly(vinyl acetate) cannot be used for shaped articles and its major outlets are in adhesives, emulsion paints, and in the preparation of related polymers, e.g. poly(vinyl alcohol).

E. Baumann in 1872 observed that powdery solids resulted from the action of sunlight on vinyl chloride, an olefinic compound first reported by H. V. Regnault in 1835. I. Ostromislensky (British Patent 6299, 1912) described forms of polymerized vinyl chloride, but these could be processed only in the melt (with some concomitant decomposition) giving rigid substitutes for ebonite, gutta percha, and celluloid applications. The copolymers with vinyl acetate paved the way to the all-important discovery in 1930 by W. L. Semon of the B. F. Goodrich Company in the U.S.A. that poly(vinyl chloride) (PVC) could be plasticized with high-boiling liquids like tritolyl phosphate to give a rubber-like mass. The subsequent processing of the polymer at 160 °C was, for plastics, the first break with traditional rubber processing. Commercial manufacture began in the U.S.A. and in Germany at Ludwigshafen by 1933, giving Germany the advantage of having a substitute for rubber in many of its applications before the start of the Second World War. British efforts in this area were accelerated by the needs of the war, and by 1943 the Distillers Company Ltd. were producing paste polymer, while I.C.I. were operating an emulsion polymerization plant and delivering granular poly(vinyl chloride) for electrical requirements. The civilian market for PVC was expanded rapidly (and in some cases inappropriately) after the war; cable insulation, leathercloth, chemical plant, and packaging were among its applications. At the time of writing it is, with polyethylene and polystyrene, one of the three most important plastics, assessed in terms of tonnage.

Polystyrene *(PS)* *and some copolymers.* The discovery of polystyrene is attributed to E. Simon, a German apothecary, in 1839. It was another German, Hermann P. Staudinger, who questioned the colloidal (T. Graham, 1861) and micellar (C. D. Harries, *c*. 1910) theories of the constitution of rubber. As a result of his experiments with rubber, polystyrene (which he renamed), and polyformaldehyde, he propounded in the 1920s the macromolecular theory of polymers. This embodied the concept of the distribution of molecular weights within a batch of polymer, and (in 1935) the chain reaction mechanism for the formation of addition polymers from unsaturated precursors. For these monumental contributions to the understanding of the nature of polymers Staudinger received the Nobel Prize for chemistry in 1953. Polystyrene was a clear glass-like material and a very good electrical insulator, and these properties, becoming apparent at a time of growing interest in synthetic rubbers, were sufficient to bring about its commercial development by I. G. Farbenindustrie in Germany from 1930. The last impediment, a satisfactory large-scale manufacture of the monomer (output of which was 100 000 tons per annum towards the end of the war), had recently been overcome by B.A.S.F. Limited. Production of polystyrene by the Dow Chemical Company took place in the U.S.A. before the war, and the first full-scale commercial production plant in Britain at Carrington (formerly Petrochemicals Ltd., now Shell) started up in October 1950. Postwar capacity again had to be directed to civilian uses, and by 1962 polystyrene, measured by consumption, was the third most important plastic. Its range of applications in, for example, domestic appliances, food containers, toys, and packaging, and, in the expanded form, in thermal insulation, has been further extended into fields demanding even more durable properties. This was effected by the introduction in 1948 (in the U.S.A.) of toughened (or high-impact) polystyrene which incorporates styrene-butadiene rubber in the polymer matrix, and in the 1950s by the availability of the acrylonitrile-butadiene-styrene (ABS) copolymers.

Other copolymers had elastomeric (or rubber-like) properties. Between the wars B.A.S.F. (latterly as part of the I. G. Farbenindustrie complex) conducted extensive research into synthetic rubbers, and two new products known as Buna-N and Buna-S were developed commercially. The latter was a copolymer of butadiene and styrene whose synthesis was catalysed by sodium (hence the acronym Buna-S), and German output of this general-purpose synthetic rubber ran at 5000 tons per annum by 1938. The Japanese occupation of the rubber plantations in 1941 led to hectic efforts to develop similar much-needed materials in the U.S.A., and output of GR-S (Government Rubber-Styrene, later named SBR, styrene-butadiene rubber) rose from a few thousand tons in 1939 to 820 000 tons in 1945.

Poly(methyl methacrylate) *(PMMA)* *and polyacrylates.* R. Fittig in 1877, and G. W. A. Kahlbaum three years later, both described glass-like polymers of acrylic acid esters, whilst in 1901 Otto Rohm presented a doctoral thesis on the subject of acrylate polymers, soft materials with potential uses as lacquers or safety-glass interlayers. In 1927 Rohm and Haas A. G. in Germany produced poly(methyl acrylate), the first commercial acrylic polymer. In the early 1930s, W. Bauer in Germany, and Rowland Hill of I.C.I. in Britain, began to study the esters of methacrylic acid and their polymers. In 1931 Hill found that the polymer of the methyl ester of this acid could be cast as a clear glass-like substance which was light, unbreakable, and weatherproof. Its tremendous potential could not be realized, however, until a second I.C.I. chemist, John W. C. Crawford, developed in 1932–3 a cheap, commercial synthesis of the monomer from acetone, hydrogen cyanide, methanol, and sulphuric acid, a route virtually unmodified some 44 years later. Commercial production of I.C.I. poly(methyl methacrylate), Perspex, began in 1934, sheet output being consumed almost exclusively until 1945 by the Royal Air Force for aircraft window glazing. Since that time the polymer's optical properties have been exploited in display signs and light fittings; other outlets have been in dental prostheses and in automotive finishes.

Polyethylenes *(PE)* *and polypropylene* *(PP).* Interests in the Haber process and a consultancy held with I.C.I. by A. M. J. F. Michels of Amsterdam, an expert in the field of high-pressure chemistry, motivated the programme of research into the effect of high pressures on chemical reactions begun at the I.C.I. Alkali Division laboratories at Winnington in 1932. E. W. Fawcett and R. O. Gibson—hoping to achieve addition reactions with ethylene in the presence of benzaldehyde, aniline, and benzene—noticed that after one reaction on 27 March 1933, in the presence of benzaldehyde at 170 °C and 1000–2000 atmospheres, a white waxy solid coated the walls of the reaction vessel; analysis revealed the solid to be a polymer of ethylene. It was not until December 1935, when ethylene was compressed alone, that a further eight grams of polyethylene were obtained by M. W. Perrin J. G. Paton, and E. G. Williams. The polymer had outstanding properties: it was a very good electrical insulator, possessed good chemical resistance, and could be moulded or made into film or threads. After further development work the first ton of polyethylene was available by

FIG. 21.13. Although its full impact was not felt until the second half of the century, polythene was one of the great chemical discoveries of the 1930s. These pictures show, from left to right, an early high-pressure reactor vessel (1937–8); a mercury-sealed automatic gas compressor (Michels 1937–8); and a very early (pre-war) sample of polythene.

the end of 1938 for evaluation in submarine cable technology. Full-scale production began on the day that Germany invaded Poland (31 August 1939). Sir Robert Watson Watt, the inventor of radar (p. 844), has stated that the availability of polyethylene as a dielectric material transformed the tasks of the Allies in this field from the impossible to the comfortably manageable. During the war I.C.I. allowed production rights in the U.S.A., and by 1941 Du Pont and Union Carbide and Carbon Corporation had established manufacturing plants. After the war polyethylene became one of the major commercial polymers, with diverse applications ranging from domestic goods to chemical plant, from packaging film to toys.

A second polyethylene, high density polyethylene (HDPE) emerged in 1953–4 when three distinct low-pressure processes for making new variants of the polymer were described almost simultaneously, by Karl Ziegler, working at the Max Planck Institut für Kohlenforschung at Mülheim in West Germany, and by the Phillips Petroleum Company and Standard Oil of Indiana in the U.S.A. Ziegler, who with Giulio Natta (see below) received the Nobel Prize for Chemistry in 1963, had a long-standing interest in organometallic chemistry. Studying the polymerization of ethylene catalysed by aluminium alkyls, Ziegler undertook a systematic search for an impurity which he believed was inhibiting the growth of the polymer chain beyond about 100 monomer units. In his search, he discovered the contrary effect of titanium tetrachloride, which accelerated the polymerization and could yield virtually unbranched polymers with molecular weights in excess of 100 000.

The process had no requirement for elevated pressures and the higher-density, stiffer polyethylene had uses which complemented those of the high-pressure variety. Höchst A. G. in Germany produced Ziegler polyethylene from 1955; Phillips and Standard Oil plants came on stream in 1956 and 1961 respectively.

Propylene had not been polymerized usefully until, in 1954, Giulio Natta and his colleagues at the Polytechnic Institute in Milan succeeded in polymerizing propylene to solid polymers of high molecular weight by employing Ziegler-type catalysts. But Natta's contribution was to discover that, by varying the catalysts selected, various stereospecific forms of polypropylene could be produced with corresponding changes in their properties, successive monomer units being added to the growing polymer chain at the catalyst surface. Commercial production was begun by Montecatini in Italy in 1957; polypropylene (PP) was used for fibres, films, and in the injection moulding of diverse products. In 1962 world production of both HDPE and PP was about 250 000 tons.

Polyamides (*PA*). Wallace Hume Carothers joined the staff of E. I. Du Pont de Nemours at Wilmington, Delaware, in 1928 from Harvard University. He embarked on a programme of research directed at the synthesis of substances of high molecular weight, and was particularly interested to produce polymers which could yield fibres. He introduced the concept of condensation (as opposed to addition) polymerization and by 1929 described 18 microcrystalline polyesters—that is, potential new fibre materials—from the reactions of dibasic acids and diols. His understanding of the correlation between the repeating structural unit of a polymer and its physical and chemical properties was paramount in his work. In 1931 he discovered that chloroprene (2-chloro-buta-1,3-diene) could be polymerized 100 times more rapidly than isoprene, the monomer of natural rubber, to give a superior synthetic rubber, polychloroprene or 'Neoprene' as it was better known. Commercial manufacture began in 1932 and chloroprene polymers have since maintained their position as speciality rubbers. Carothers maintained a prodigious output of publications on a wide range of novel condensation polymers. Included in his patent filed on 2 January 1935, claiming various new polyamides, was the polyamide from adipic acid and hexamethylene diamine. Production of the polyamide, Nylon 66 (Ch. 27), began on 28 October 1938 after almost four years' development work, involving 230 chemists and engineers and costing Du Pont $27m. Nylon stockings appeared

FIG. 21.14. The advent of man-made fibres was one of the most important developments in the chemical industry. This picture shows the first sample (manufactured by Du Pont) of a polyamide capable of being successfully knitted. The successor to this polyamide swept the world as 'nylon' first as a nearly ideal material for stockings and then for many other uses.

in 1939 and 64m. pairs were sold in the first year. By 1941 Nylon moulding powders followed the earlier fibre outlet, but were not well known before 1950. By 1962 four polyamides were of commercial significance: Nylon 66, 610, 6, and 11.

Poly(ethylene terephthalate) (*PETP*). Carothers and Hill had described in 1931 aliphatic polyesters which were capable of being extruded and subsequently cold-drawn into strong fibres, but which had proved unrewarding commercially, having low melting-points and being too readily hydrolysed. During 1939–41, two British research workers, J. R. Whinfield and J. T. Dickson, working in the laboratories of the Calico Printers Association Ltd, investigated the polyesters more fully, and the foundations for the strong fibre and lustrous film that is recognized today as poly(ethylene terephthalate), an aromatic polyester, were laid. After further investigation by the then Ministry of Supply in Britain, the new synthetic fibre was evaluated by I.C.I. in 1943 and 'Terylene', as it was called, was seen to rank in the same class as nylon. Large-scale manufacture by I.C.I. at Wilton began in January 1955, Du Pont in the U.S.A. producing their own 'Dacron' about a year earlier. PETP is now extensively used in fibre form by clothing manufacturers, and the extremely strong clear film form of this polymer is now also manufactured on a large scale.

Polyurethanes (*PU*). Further evidence of the stimulus of Carothers's work was provided by the findings of Otto Bayer (of I. G. Farbenindustrie and later Farbenfabriken Bayer, Germany) during 1937–9 that another group of bifunctional compounds, the di-isocyanates, reacted with glycols to give polyurethanes (PU) with properties making them of interest as plastics and fibres. Later research showed that their usefulness extended to applications as adhesives, surface coatings, and rigid foams. Commercial production began in 1941 in Germany, Bayer pressing further ahead after the war with the development of elastomers in 1950 and flexible foams in 1952. Though known through Intelligence reports during the war, work on polyurethane foams in the U.S.A. dates from 1946. Production in Britain from home-produced (I.C.I.) raw materials waited a further decade. Mass-production of polyester-based foams was established in most industrial countries by 1955, and the application of foams was further extended by the introduction of the polyether-based versions by several American companies in 1957.

Polytetrafluoroethylene (PTFE). We have seen many instances of accelerated progress in the development of recently known polymers because of the demands of war. One patent however, published in the U.S.A. in 1941 in the name of R. J. Plunkett of Kinetic Chemicals Ltd, made the first claims for an entirely new material. The Germans O. Ruff and O. Bretschneider synthesized tetrafluoroethylene in 1933, and fluorine chemistry was an active field of research in the Second World War because of its involvement with the separations of the isotopes of uranium (Ch. 10, Pt II). Going to a cylinder of tetrafluoroethylene which he knew to be full, Plunkett found that on opening the valve none of the gas remained, although no leak was suspected. The cylinder was duly cut open and the walls were found to be coated with a white solid. This accidental discovery of PTFE in 1938 revealed a new polymer with astonishing properties, inert to almost every chemical agent, uncharred by electric arcs, completely resistant to sunlight and moisture, and possessing an extraordinarily low coefficient of friction. So heat-resistant is this polymer that its processing resembles more the techniques of powder metallurgy than conventional plastics practice. Its very low power factor over a wide frequency range, together with its chemical resistance, indicated potential use in critical electrical equipment under highly corrosive conditions. Du Pont had a pilot plant operating by 1943, and began large-scale manufacture in 1950. Regular production in Britain began with I.C.I. in 1947. The original cost of the I.C.I. product was £5 per pound, falling to one-third of that price by 1962. Nevertheless, PTFE is still a specialist and relatively expensive polymer.

Other polymers. In the early 1930s the Thiokol Corporation of the U.S.A. announced a polysulphide rubber which they called 'Thiokol'. It has remained a low tonnage product and is most familiar in its role as flexible hoses for petrol pumps, possessing as it does exceptional solvent resistance. Another synthetic rubber, called butyl, was first developed in 1937 by the Standard Oil Development Company (now Exxon). The most successful copolymer blend was isobutylene-isoprene, the small amount of diene ensuring a product capable of being vulcanized. Production did not begin until 1941 and was completely absorbed by military needs until after the war, when the very low permeability of the polymer to gases ensured its outstanding success as the material for automotive tyre tubes. A further advance, namely the tubeless tyre, standard on Detroit's 1955 models, caused butyl to lose 40 per cent of its market within two years. Its unique range (for a rubber) of desirable properties nevertheless ensured its commercial survival.

The potential of polyacrylonitrile as a textile fibre was recognized in the 1930s, but development was thwarted by the fact that the polymer would neither dissolve nor melt. In 1938 I. G. Farbenindustrie spun the first acrylic fibres but their solvent was not commercially satisfactory. E. I. Du Pont de Nemours succeeded in dry-spinning polyacrylonitrile from dimethylformamide in 1942, and soon afterwards it was recognized that copolymers of acrylonitrile with, for example, vinyl acetate or vinyl chloride were most suited for commercial development. The modacrylic fibres, containing 35–85 per cent acrylonitrile in the monomer mix were introduced by the Union Carbide Corporation in 1949; Du Pont introduced acrylic fibres, in which the acrylonitrile content exceeds 85 per cent, in 1950. Apart from its textile applications, polyacrylonitrile is noteworthy as the precursor for carbon fibre.

The silicones have their origins in the work of Frederick S. Kipping between 1899 and 1944 at Nottingham University, and in the researches in the 1940s of J. F. Hyde of the Corning Glass Company (U.S.A.). Hyde, who knew Kipping's work, sought to prepare new materials intermediate in properties between the organic and wholly inorganic polymers. Industrial production of organosilicon polymers—with their wide range of temperature tolerance, water repellance, and anti-adhesive properties—was first begun by the Dow Corning Corporation in 1943. Albright and Wilson operated the Dow process in Britain from 1954. Organosilicon polymers remain an expensive, speciality product; world production in 1962 amounted to 20 000 tons.

II. DYES AND PIGMENTS

Chemistry has contributed vastly to the ease of life, but it has also contributed increasingly to its intensity. Nowhere is this more true than in the visual field. The twentieth century must be the most vivid epoch in history. The foundations for this chromatic revolution were laid in the mid-nineteenth century. The discoveries in dye chemistry made by the organic chemists who followed the example of W. H. Perkin have already been described (Vol. V, Ch. 12). The intricate structure and development of the industrial organizations devoted to dyestuffs and related chemical manufactures, particularly in Germany, Britain, and the U.S.A., in the first half of the twentieth century is expounded elsewhere (Ch. 20). It is proposed therefore to do no more here than highlight certain areas of progress or innovation. Increasingly stringent requirements of fastness, especially light-fastness, have been one continuing motivation for research in the laboratories of the dye manufacturers.

The onset of the First World War in a world almost wholly dependent upon Germany for its supply of dyestuffs provided a stimulus in Britain, and later elsewhere, to learn how to manufacture these existing colours, and thereafter how once more to become significant innovators. A textile manufacturing concern, Morton Sundour Fabrics Ltd,. set up the Solway Dyes Company at Carlisle, England only a few weeks after the outbreak of war, and, with no previous experience of dyestuffs manufacture, had produced small quantities of Indanthrene Yellow G (Caledon Yellow) by November 1914: bulk production began in February 1915 and by August 1917 eight indanthrene vat dyes and two alizarine colours were being manufactured in Britain by this company alone. Other British companies already engaged in dyestuffs manufacture before the war achieved notable, if less spectacular, results under duress, while in British India the moribund indigo plantations (Vol. V, pp. 261–3) were revived; over 600 000 acres were planted in 1916–17 nearly four times the figure for 1913–14. In the U.S.A. during 1915–19 $466m. was invested in the establishment of a dye-manufacturing industry and the nation went from being the largest importer of dyes to being a significant exporter by 1920. France by 1919 manufactured 70 per cent of her own requirements, having made but 20 per cent of these needs in 1913. Ciba-Geigy and Sandoz in Switzerland achieved remarkable progress during the war years and were limited only by the availability of the essential raw material, coal, which reached Basle via the Rhine.

The third major area of invention has been the development of new classes of dye and techniques of application to provide the means to colour the wide range of new man-made textile fibres which have become available. Significant advances have come about in understanding the mechanics of dyeing, and dyestuffs and pigments now find not only the traditional applications but also uses

FIG. 21.15. The rise of the dyestuffs
industry was one of the major
developments of the late nineteenth
and early twentieth centuries. This
picture shows the dyehouse of
Levinstein's works at Blackley,
Manchester, *c.* 1912.

FIG. 21.16. Although the chemical
industry was going through a period
of rapid modernization between the
wars, primitive conditions were slow
to disappear altogether. This
wartime picture shows women
washing casks at I.C.I.'s Dyestuffs
Works, Blackley (1943).

in the colouring of plastics and rubbers, of foodstuffs, in printing, display and duplication processes, and in colour photography.

Fastness. The appreciation of light-fastness as a desirable quality of a particular dye dates at least as far back as the first century A.D., being referred to in Plutarch's *Life of Alexander.* By the end of the nineteenth century not only were many of the new synthetic dyes recognized as having a sadly fugitive nature, but the influence of the type of fibre being dyed on the fastness of the dyeing was also appreciated. Around the turn of the century several companies developed and published fastness-evaluation criteria and procedures, notably Friedrich Bayer and Company (1898) and the Berlin Aniline Company (1904), who concluded: 'Absolutely fast dyes do not exist, sunshine and rain finally bleach them all'. Details of systematic testing and calibration of dyes on a numerical scale of fastness date from about 1911. Shocked by the rapid fading of the dyed fabrics his company (Morton Sundour of Carlisle) produced when these were displayed in the shop windows of his customers, James Morton began in 1902–3 a systematic evaluation of the light-fastness of dyes then available. He resolved, with pioneering spirit, to use only those which, after his assessment, he was prepared to market with a guarantee. Hardly any claims were ever made against the Sundour unfadeable fabrics after their introduction in 1906, though we have described already the resourcefulness required of Mortons to ensure supplies of the selected vat dyes after the onset of war. Committees were appointed nationally (in Germany in 1911, in the U.S.A. from 1922, in Britain in 1927) to standardize the methods of fastness testing and scales of fastness. Tests for fastness to washing, perspiration, and to a great many other challenges were devised in addition to the original evaluation for light-fastness. Dyeings of the new fibres, the acetates, nylons, polyesters, and acrylics, had also to be assessed, and by 1956 over 20 colour-fastness tests had been standardized internationally.

Vat dyes. One group of dyes characterized by their outstanding light-fastness is the vat dyes. The first synthetic vat dye was described by René Bohn and introduced as Indanthrene Blue in 1901. This is an anthraquinone derivative, and the other chemical class which goes to make up the vat dyes is the indigoids, of which the natural dyes indigo and Tyrian purple are examples. Vat dyes are characterized by their ability to exist in two forms: a reduced, colourless, soluble form, taken up by the textile from the solution in the dyeing vat, and an oxidized, coloured, insoluble state which is reformed by subsequent exposure of the textile to the air, and is firmly held within the fibre. About a hundred vat dyes were known by 1910, but additional valuable colours were added to the series in the period 1920–30, notably the first really fast green dyestuff, Caledon Jade Green. This was announced by James Morton of Scottish Dyes Ltd. in 1920. Solubilized vat dyes that did not require the alkaline conditions of the conventional vat dye for their application, and therefore allowed animal fibres to be so dyed, were introduced in 1924; the Soledon series was a further achievement of the Scottish Dyes chemists.

Dyeing man-made fibres. Until the introduction of the new synthetic fibres, dyeing had been based almost exclusively on water-soluble colours, which dye by a process of absorption in which fibre and dye show an affinity for each other. The chemical nature of this affinity entailed either salt formation between dye ions and ionized fibre, as in the case of wool, or intermolecular association (for example, hydrogen bonds), as in the case of cotton. The hydrophobic nature of the man-made fibres rendered the conventional classes of dyestuffs ineffective. The commercial advance of the first of these fibres, cellulose acetate, was very considerably delayed by the difficulty of dyeing it. It was, however, found that certain insoluble amino-azo dyes would dye acetate fibre if used in a finely divided state. The British Dyestuffs Corporation in 1922 introduced the first of these so-called disperse dyes specifically developed for the new textile. These were the Ionamines, soluble but unstable compounds which decompose in the dyebath liberating the insoluble dye in a form readily taken up by the fibre. A year later a simultaneous discovery by the British Celanese Company and the British Dyestuffs Corporation provided a further type of dye for acetate fibres. The process involved the use of certain anthraquinone dyes in combination with a dispersing agent and was especially important by virtue of the range and fastness of colours it provided. The techniques for dyeing polyamide (nylon), polyacrylonitrile, and polyester fibres with dispersed anthraquinone colours have evolved by adaptation of these original findings. It is now understood that these fibres are dyed by a reversible process involving solution of the dye in the fibre. Since dyeing is a competition between the fibre and the dye-bath solution, a good dye is one whose solubility in the fibre is high compared with its aqueous solubility, and selection for these properties is more important than selection for the chromophore (the colour-producing unit of the molecule). An additional approach to the dyeing of acrylic fibres has been to devise ways of inserting specific cationic sites in the polymer matrix. Two methods have proved effective: first, copolymerization of the acrylonitrile with a little vinyl pyridine to give basic groups capable of first accepting a proton and then interacting with the dye molecule; second, treatment of the polyacrylonitrile fibre with cuprous ions, which results in almost unlimited capacity for anionic dyes.

Phthalocyanines: an entirely new chromophore. H. de Diesbach and E. von der Weid first reported in 1927 a strongly blue-coloured organic compound of copper as a bothersome by-product of their researches. Subsequent elucidation of this compound showed that they had in their paper disclosed copper phthalocyanine, thereby preventing its exclusive exploitation through patents. Meanwhile, in 1928, complaints of a bluish contamination of phthalimide being manufactured by Scottish Dyes (by then merged with I.C.I.) were investigated. The process entailed reaction of phthalic anhydride and ammonia in a glass-lined kettle, a fault in which had led to the formation of an intensely blue-coloured iron-containing compound. Analogues containing copper and nickel were systematically synthesized. Brilliant studies by R. P. Linstead and his co-workers at the Imperial College of Science and Technology, London, reported in a series of papers during the period 1934–8, led to a full practical understanding of the manufacture, properties, and structures of a number of metal phthalocyanine derivatives. I.C.I. began limited production of the copper compound during 1935–7 under the market name Monastral Fast Blue B.S. The metal phthalocyanines are almost perfect pigments, their brilliance, stability, and fastness surpassing all others. The Second World War delayed the expansion of the market, but by 1956 two phthalocyanine pigments were of great commercial significance: the original blue copper compound and its hexadecachloro derivative, which is green. World sales during this period reached a maximum of about $7·5m. in 1953, almost ten times the 1945 figure.

It was obviously desirable to attempt the conversion of phthalocyanines to forms suitable for dyeing textiles. The first successful example was a turquoise direct dye for cotton, which was a disulphonated copper phthalocyanine claimed in 1938 (U.S. patent 2 135 633). Later successes included quaternized copper compounds which were soluble and had a high affinity for cellulose (I.C.I. Dyestuffs Division 1947) and a cobalt phthalocyanine which was applied under conditions resembling those used for the vat dyes (U.S. patent 2 613 128, 1952).

BIBLIOGRAPHY

ABRAHART, E. N. *Dyes and their intermediates*. Pergamon Press, Oxford (1968).

AMERICAN CHEMICAL SOCIETY. *Chemistry in the economy*. Washington, D.C. (1973).

BAWN, C. E. H. *Plastics: a centenary and an outlook*. Sir Jesse Boot Foundation lecture, University of Nottingham (1962).

BEER, J. J. *The emergence of the German dye industry*. University of Illinois Press (1959).

BRADLEY, W. *Recent progress in the chemistry of dyes and pigments*. Royal Institute of Chemistry, London (1958).

BRITISH DYESTUFFS CORPORATION. *The British dyestuffs industry 1856–1924*. Manchester (c. 1924).

CAMPBELL, W. A. *The chemical industry*. Longman, London (1971).

FORRESTER, S. D. The history of the development of the light fastness testing of dyed fabrics up to 1902. *Textile History*, **6**, 52–88 (1975).

GREENHALGH, C. W. Aspects of anthraquinone dyestuff chemistry. *Endeavour*, **35**, 134–40 (1976).

HABER, L. F. *The chemical industry 1900–1930*. Clarendon Press, Oxford (1971).

HARDIE, D. W. F., and DAVIDSON PRATT, J. *A history of the modern British chemical industry*. Pergamon Press, Oxford (1966).

I.C.I. MOND DIVISION. *A hundred years of alkali in Cheshire*, by W. F. L. Dick, Kynoch Press, Birmingham (1973).

I.C.I. PLASTICS DIVISION. *Landmarks of the plastics industry*. Kynoch Press, Birmingham (1962).

I.C.I. PUBLICATIONS FOR SCHOOLS. *Colour chemistry*. Kynoch Press, Birmingham (1972).

KAUFMAN, M. *The first century of plastics—celluloid and its sequel*. The Plastics Institute (c. 1962).

MERRIAM, J. *Pioneering in plastics—the story of Xylonite*. East Anglian Magazine (1976).

MIALL, S. *A history of the British chemical industry*. Ernest Benn Ltd, London (1931).

MORTON, J. *Dyes and textiles in Britain: 1930*. Lecture to the British Association Meeting, Bristol, September 1930, published privately.

——. *A history of the development of fast dyeing and dyes*. Lecture to the Royal Society of Arts, 20 February 1929, published privately.

MORTON, J. W. F. *Three generations in a family textile firm*. Routledge & Kegan Paul, London (1971).

REUBEN, B. G., and BURSTALL, M. L. *The chemical economy*. Longman, London (1973).

ROBINSON, S. *A history of dyed textiles*. Studio Vista, London (1969).

SAUNDERS, K. J. *Organic polymer chemistry*. Chapman and Hall, London (1973).

SCHOOLS COUNCIL, PROJECT TECHNOLOGY HANDBOOK. *Design with plastics*. Heinemann, London (1974).

SCHOOLS COUNCIL, PROJECT TECHNOLOGY. *Fibres in chemistry*. English Universities Press, London (1974).

SHERWOOD TAYLOR, F. *A history of industrial chemistry*. Heinemann, London (1957).

The Focal Encyclopaedia of Photography, Vol. 2. Focal Press, London (1965).

WHITE, H. J., JUNIOR (ed.). *Proceedings of the Perkin Centennial 1856–1950*, American Association of Textile Chemists and Colorists. (c. 1956).

WILLIAMS, T. I. *The chemical industry*. Penguin Books, Harmondsworth (1953).

22

GLASS MANUFACTURE

R. W. DOUGLAS

T HE development of the glass industry in the second half of the nine-teenth century has been discussed elsewhere in this work (Vol. V, Ch. 28). By the end of that period, as a result of the growth of chemical science, the mixtures of raw materials for glass-making (batches) had begun to look very similar to those in use today. The regenerative tank furnace was established, but the forming of glass articles still involved a great deal of handwork by skilled craftsmen. In 1903 the Lubbers cylinder process for sheet glass was brought into production and the Owens bottle-making machine had started its rapid advance. By 1927 a new bottle machine, the I.S., was introduced which in 20 years almost completely replaced the Owens. By 1930 flat-drawn sheet glass and the continuous rolling, grinding, and polish-ing processes for plate glass were well started, only to be made obsolescent in the decade 1960–70 by the development of the new float process. These rapid changes were prompted by economic and social demands and facilitated by the availability of better fuels and refractories (Ch. 25) and by advances in physical science and technology, particularly in increasing the efficiency and control of batch mixing and melting, so that continuous supplies of good glass were available to feed the machines.

Although the development of new glass batches for optical glasses had led to some dozens of new glasses being available commercially, soda–lime–silica for bottles and windows and potash–lead–silica for high-class tableware were still the only glasses produced in large quantities at the end of the nineteenth century.

In 1915 a new glass containing significant amounts of boric oxide and alumina was put on the market; this glass has a low thermal expansion and was soon utilized for chemical apparatus and domestic ovenware. The new electrical industry created demands for special glasses for thermionic valves and the new metal-vapour discharge lamps. The camera and motion-picture industries and aerial photography lead to markets for new optical glasses of better quality. The growing automobile industry increased the demand for high-quality flat glass. By 1950 the glass industry had become a modern mechanized industry based on engineering and chemical technology. In the space available here it will be possible to illustrate these tremendous changes by describing only a few particular examples; the bibliography at the end of the chapter gives a guide to more information.

I. GLASS BATCHES AND RAW MATERIALS

Glass for chemical apparatus and ovenware. In the first few years of the century more and more glass companies began to employ chemists, particularly in the U.S.A. From a programme of work on new compositions the Corning Glass Company produced commercially, in 1915, a low-expansion borosilicate glass which had a coefficient of thermal expansion of about 3×10^{-6} per °C compared with 9×10^{-6} for typical soda–lime–silica glasses. Approximate analyses of this glass, and of several others to be mentioned, are given in Table 22.1.

The ability to resist sudden changes of temperature increases inversely as the thermal expansion; if, for example, an article made in soda–lime–silica glass could withstand immersion in water at 20 °C immediately after standing in water at 80 °C, a vessel of the new glass of the same size and shape could withstand immersion at 20 °C from an initial temperature of 200 °C; this is equivalent to a moderately hot domestic oven. Although the size of the ware has not been specified, it is clear that such ware could withstand the usual practice of being left to cool in air after removal from the oven. Borosilicate glass became very important for chemical apparatus.

Glasses for the electrical industry. The widespread and rapid growth of the electrical industry led to demands for new glasses. In the electric lamp, the current is led to the filament through wires which are sealed into a glass tube by heating the glass until it is soft enough to be 'pinched' on to the wires, thus making a vacuum-tight glass-to-metal seal. Soda–lime–silica glass is the cheapest but it was found that when this glass was used for the 'pinch' discoloration, bubbling, and cracking occurred around the wires. It was soon established that this was due to the passage of a very small current through the glass between the wires. The conduction is electrolytic, by sodium ions moving under the action of the electric field as they do when in solution in water. It was found that the potash–lead glass, used since the end of the seventeenth century in Britain for 'lead crystal' tableware, had a much higher

TABLE 22.1
Approximate compositions of some commercial glasses* (wt. %)

	SiO$_2$	Al$_2$O$_3$	CaO	Na$_2$O	MgO	PbO	K$_2$O	B$_2$O$_3$	BaO	ZnO	F	La$_2$O$_3$	ThO$_2$
Container glass													
1925	73	0·5	9	17									
1935	73	1·0	10	16									
1950	73	2·0	10	15									
1970	74	2·0	10	14									
Sheet glass													
19th century	71·5	1·5	13	14									
Machine cylinder	72·5	1·5	13	13									
Fourcault process	72	0·5	10	14	3·0								
Lead crystal	56	1·5		4		30·0	7						
Ovenware borosilicate	80	3		4				12					
Sodium-vapour resisting	23	23	10	6				37					
Mercury-vapour resisting and 'E' glass	55	22	14					7	3				
Opal lighting ware	60	10	5	8·5			2·5	1·5		9	4		
Optical glasses													
Special barium crown	10·5							24·5	27·0	4		18	13·5
Special barium flint	15					9		15	32	2		13	11
Ordinary light flint	53			5		35	8						
Barium flint	43	10		1·0		33	7·5		10	5·0			

* Compositions at about 1950 unless stated otherwise.

electrical resistance than the soda–lime glass and 'pinches' made from this glass were free from electrolytic breakdown.

Fortunately the thermal expansion coefficients of the lead and soda glasses are not very different and soda-glass bulbs could be fused satisfactorily to the lead-glass tubes carrying the wires. Although initially platinum was used for the wires sealed in the glass it was quickly replaced by a nickel–iron wire with a thick coating of copper; this gave a net radial thermal expansion matching that of the glass. When high-powered lamps and thermionic valves were made, borosilicate glasses had to be used for the envelopes. Glasses were then designed with thermal expansion to match that of molybdenum and tungsten.

The metal-vapour discharge lamps developed around 1930, that is the mercury and sodium vapour lamps now widely used for street lighting, presented new problems. The hot metal vapours are extremely reactive. Glasses containing very little silica were found to be suitable for sodium lamps; other glasses containing no sodium were necessary for mercury lamps. The glass that would resist sodium vapour was so much less viscous than ordinary glasses that it had to be supported as a thin layer on the inside of soda-lime glass tubes. This coating was accomplished by a traditional craft process known as 'flashing', in which some of the soft glass is gathered on the iron and then, while still hot, it is coated with support glass; the composite mass is then drawn into tubing. The sodium-free glass was found to be particularly resistant to attack by atmospheric moisture and a glass of similar composition is now used for much of the glass fibre manufactured.

Optical glass. To meet the demands for lenses of better quality, the designer needed a range of glasses with different refractive indices and in which the refractive index varied in different ways with wavelength. (This variation of refractive index with wavelength is known as 'dispersion'.) Existing glasses tended to have the same general relationship between refractive index and dispersion, but newer glasses, particularly some containing quite large amounts of the rare-earth compound lanthanum oxide, were found to have the desired properties.

Container or window glass. The compositions of the ordinary soda–lime–silica glasses have changed as furnaces have improved and mechanical methods of forming have been introduced. For container-making, reduction in the soda content and addition of a little alumina increased the durability, that is the resistance to attack by moisture. These changes also resulted in a quicker-setting glass so that rates of production could be increased. Improvements in refractories and furnaces which led to higher melting temperatures enabled the soda content to be reduced still more. In sheet glass manufacture it was found that the best operating temperature could coincide with the temperature at which the glass devitrified most rapidly. (Devitrification means the growth of crystals in the glass.) By replacing part of the lime by magnesia this trouble was overcome.

II. BATCH MATERIALS

Magnesia could be conveniently introduced into sheet glass by using dolomite in place of limestone. Many other new raw materials have been used to aid melting. All materials have to be relatively free from iron if glass of good colour is to be obtained. Glass-making sand has to be specially sought; some deposits in Europe need only water-washing to bring them to the required standard of purity. In Britain it is economic to remove heavy minerals by using flotation techniques; crushing, milling, and sieving may also be required.

Materials are now stored so that transfer to automatic batch weighing and mixing plant is convenient, thus ensuring a constant supply of accurately weighed and properly mixed batch. This was a great change from the shovel and the pile of raw materials on the floor, which could still be seen occasionally as late as 1940.

Soda ash and limestone. By 1900 the Solvay process for the production of sodium carbonate was well established and the industry had available to it a flux of high purity and constant composition. Limestone sufficiently free from iron is available in most countries. Some figures relating to the growth of the production of sodium carbonate in the U.S.A. give an interesting commentary on the growth of the glass industry. In 1867 the *Scientific American* reported that not a pound of soda ash was made in the U.S.A. In 1869 a company was set up to manufacture the material, apparently using the old Leblanc process, and in 1881 the Solvay process was introduced. The Columbia Chemical Company was formed in 1899 through the efforts of John Pitcairn, one of the founders of the Pittsburgh Plate Glass Company, which in 1920 absorbed Columbia Chemical as one of its manufacturing divisions. By 1935, 600 000 tonnes of soda ash were used by the glass industry in the U.S.A., manufactured by five organizations; the consumption had risen to 1·1m. tonnes in 1949.

III. FURNACES AND REFRACTORIES

The regenerative principle invented by Siemens (Vol. V, p. 58) started the development of the continuous melting of glass in tank furnaces. These furnaces consist essentially of a rectangular refractory container with an arched roof to provide space for the combustion of the fuel above the glass. The largest tanks feeding sheet glass machines could contain up to 2000 tonnes; large container plants would hold about 500 tonnes. The maximum depth of glass in a tank furnace is about 1·5 m and the 500-tonne tank would have a glass surface area of about 220 m², say 11 × 20 m.

The mixture of raw materials is fed into the furnace at one end and is withdrawn to feed the machines at the far end. This flow of the glass through the furnace, the 'pull' on the furnace, as it is called, together with currents due to thermal gradients helps in homogenizing the glass. Any lack of uniformity of composition which inevitably occurs during the melting of the batch can be removed only by diffusion and this can be effective only over very small distances. The inhomogeneities have to be stretched and so made small. This is achieved in optical glass manufacture by stirring, and by the convection currents in a normal tank furnace. Currents and diffusion both increase as the furnace temperature is raised. The maximum temperature attainable increased when oil replaced producer gas and as the height and volume of regenerators increased. Glass melts are, however, very reactive at high temperatures and the operating temperatures could be increased only as better refractories became available (Ch. 25). The operating temperature of container tanks increased from 1300 °C in 1929 to about 1470 ° in 1950; by the 1970s, it reached about 1550 °C.

In 1920 fireclay blocks or natural sandstone were used for the walls and bottom of the tank. It was known from chemical analysis that fireclays were clays of high alumina content, and naturally occurring sillimanite, kyanite, and andalusite (all having the composition $SiO_2.Al_2O_3$) were introduced as raw materials for refractories. When fired, these materials break down into mullite ($3Al_2O_3.2SiO_2$) and cristabolite (SiO_2). Studies in the Geophysical Laboratories in Washington had established that mullite was the compound of silica and alumina with the highest melting-point, and in 1925 G. S. Fulcher of the Corning Glass Works patented processes for the casting of refractory blocks from a melt of appropriate composition. Although mullite was expected to result from the freezing of the melt, it was found that corundum (Al_2O_3) was first precipitated as a metastable phase, leaving a more siliceous liquid from which mullite separated in large amounts as the block cooled. The blocks thus contained corundum, mullite, and a glassy material which made it possible to relieve cooling stresses and so to cool the blocks without their cracking. In attempts to eliminate cracks, zirconium oxide (ZrO_2) was added to the melt and in 1926 Fulcher was granted a patent covering the addition of 10–60 per cent ZrO_2. By 1942 refractories containing 31–42 per cent ZrO_2 were in use and these effectively eliminated refractory problems. For the crown, or roof, of the tank silica bricks are used. Here again the studies in the equilibria of various binary silica mixtures led to an understanding of the best compositions and cements to be used.

In the U.S.A. natural gas became available in about 1885 and the industry tended at first to concentrate around the natural gas fields. As these became exhausted plants used producer gas, but eventually more natural gas was piped from Texas. Rising transport costs for the finished products have now resulted in a spread of the industry throughout the U.S.A.

In Europe, producer gas was used at first; each weekend the plant had to be shut down while the flues were freed from soot by burning. This was a smoky, messy process which was followed by a rush to increase the furnace temperature again so that production could restart on the Monday morning. When fuel oil became available it provided welcome relief from the problems associated with producer gas and efficiency of operation was increased.

The development of the tank furnace can be illustrated from Table 22.2, which presents some figures relating to U.K. furnace practice. These show the changes in terms of regenerator height and furnace temperature with the efficiency of the furnace in terms of the area required to pull a tonne of molten glass from the furnace in a period of 24 hours. The life of the furnace in years is also given, it being defined as the duration of one period of continuous operation before a shut down for a major repair. It was a second generation of bottle-blowing machines, which required glass to fall into the

TABLE 22.2
Development of the tank furnace, 1929–70

	Furnace temperature (°C)	Regenerator height (m)	Furnace efficiency (m²/tonne)	Life (months)
1929	1300	3	2	8
1950	1470	10	1	24
1970	1550	10*	0·5	48

* But oil-fired.

FIG. 22.1. A view of an electric melting furnace showing the device for spreading batch on the surface of the melt.

mould from above that first prompted the increase of the height of the regenerators. Later, the need for efficient transfer of heat to the incoming combustion air from the exhaust gases lead to the design of better regenerators.

In container manufacture it became increasingly common to provide means of introducing electrical energy into the melt through molybdenum or graphite electrodes inserted through the walls or bottom of the tank. This increased energy, or boost, enabled greater quantities of glass to be pulled from the tank. The additional circulation produced by the electric heating increased the homogeneity, so that an increased quantity of good glass was obtained relatively cheaply.

All-electric furnaces were also introduced. In Switzerland cheap hydro-electric power during the summer made such furnaces commercially profitable; during the winter, the same furnaces could be heated by oil. For the production of low-expansion borosilicate glass an interesting furnace was introduced in which the batch was fed on to the top of the melt. No crown was necessary; the batch crust was sufficiently cool to retain the heat. Furnaces built on this general principle are likely to be used more and more in the future, particularly as concern for the environment increases the care which has to be exercised to control all forms of industrial effluents.

IV. FLAT GLASS

The Lubber cylinder process previously described (Vol. V, p. 678) continued in use until 1933 when it was replaced by sheet glass drawn flat, and continuously, from a tank furnace. The process was first introduced in a patent of William Clark of Pittsburgh in 1857 and stages in its development into a commercial process are mentioned in Vol. V. Useful commercial production did not begin, however, until 1913.

Sheet glass is drawn from the furnaces with surfaces which have been formed while the glass is soft and in contact only with air, a condition often described as 'fire polished'. The continuous sheet emerging from the machine passes through the annealing chamber and is then cut off in appropriate lengths and the glass is ready for use. Plate glass by contrast, is made by causing molten glass from the furnace to flow between rollers to produce a continuous flat 'ribbon' which may be a few metres in width. The rolling action gives a rough surface to the glass, which has subsequently to be ground and polished. The final product, plate glass, is of a higher quality than sheet glass. Sheet glass is used for windows; plate glass for motor-cars, mirrors, and shop windows.

In 1900 plate glass was still made by a discontinuous process first invented in France in 1687. Molten glass, which had been melted in a large pot, was cast and rolled on to an iron table to provide a flat sheet for subsequent grinding and polishing. The growth of the motor-car industry provided an increased demand which stimulated the development of the continuous process. In 1920 the continuous production of the ribbon was invented by the Ford Motor Company; it was brought into commercial use by cooperation with Pilkington Bros. of St Helens, England. This long-established glass-making company was able to contribute a vast experience of tank furnace operation, and the invention was turned into a commercial success. At first, the rough ribbon was cut into pieces of appropriate size and fixed on to large flat beds for grinding and polishing, each piece having to be turned over and reset after the first side had been ground and polished. A continuous grinding and polishing process was invented by Pilkingtons in 1925, and by 1937 the process had been developed into one in which the polished plate emerged continuously from a machine in which the rolled ribbon passed between grinding heads which ground both sides simultaneously. The continuous grinder plant was 652·5 m long; the length of the new twin grinder and polisher was 406 m. Both machines are now obsolescent, having been almost completely replaced by the new float process invented by Pilkingtons in 1952; the length of this equipment is 197 m.

In 1913 the output per man-hour of the cast-glass process was 0·093 m²; the early continuous grinder and polisher had in 1923 increased the output to 0·42 m²; and by 1956 the twin grinder and polisher had increased the output per man hour to 1 m². In 1970 the output from a float glass plant had more than doubled the productivity per man. As these increases in productivity were occurring, the demand for the output increased so that more and more jobs were created. In Pilkingtons the number of employees grew from 1350 in 1854 to 10000 in 1923; it reached 19000 in 1953.

By 1970 licences to manufacture float glass had been granted to firms in many countries. It is such an important innovation that it seems appropriate to describe it briefly here, although its commercial success was not made public until the beginning of 1959, a little beyond the period of our immediate concern.

A continuous ribbon of glass flows out of the melting furnace and floats along the surface of molten tin contained in a gas-filled enclosure to prevent oxidation. One surface of the glass is thus formed while the glass is still fluid enough to flow and assume the perfectly plane surface of the liquid metal. The upper surface of the glass also takes up a plane fire-polished surface under the heat of the enclosure. The density of the glass and the forces of surface tension combine to give a natural thickness of about 7 mm to the floated ribbon. However, by pulling the glass lengthways and constraining it sideways, it is now possible to manufacture float glass in thicknesses from 2 mm to 7 mm. Fig. 22.2 gives a diagrammatic representation of the float

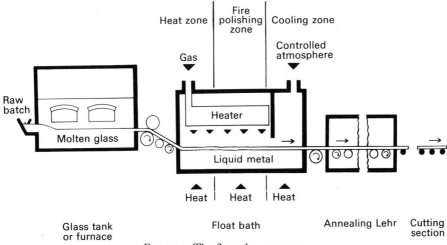

FIG. 22.2. The float glass process.

glass process. Coupled with this process is a computer-controlled cutting operation which cuts the ribbon into optimum sizes to fulfil orders with a minimum of waste glass.

V. CONTAINERS

The Owens machine. The Ashley semi-automatic machine (Vol. V, p. 674), had established the principle of forming first the mouth and neck of the bottle, a procedure used by all subsequent bottle-making machines. Nevertheless, because in the manual process the mouth had always been formed last, the neck and mouth part is still known as the 'finish' (Fig. 22.3). The machine holds the finish while the remainder of the glass is formed into the final shape of the bottle. All machines that make containers need to be fed with lumps of molten glass of the same weight as the finished bottle; Owens first solved this problem with his suction machine in which arms were arranged to advance the moulds so that appropriate amounts of glass could be sucked from a revolving pot arranged at the mouth of the furnace. The arms of moulds moved over the pot at a point diametrically opposite that at which glass flowed into it from the furnace. The machines were about 4·5–5·5 m in diameter, and the outer part of the machine rotated at 2–7 revolutions per minute. The finish is formed first, during the sucking of the glass into the mould which also forms the 'parison', that is, the partly-formed article. The parison is then moved to be blown to size in the blow-mould. A machine making 340-g beer-bottles might have 10 heads, each head making two bottles simultaneously; it would produce 60 such bottles per minute.

An experimental machine built by M. J. Owens in the United States in 1898 commenced successful production in 1904. An Owens machine (Fig. 22.4) was installed and set in operation in a model factory near Manchester, England in April 1907. The glass manufacturers of Europe were invited to inspect the machine and were offered an opportunity of buying the European patent rights; if this offer had not been accepted the Owens Company would have proceeded with plans for rapid development in Europe. The European manufacturers were compelled to joint action and raised £600 000 to purchase the patent rights. By 1913, 164 Owens machines were operating: 24 in Germany, 6 in Canada, 6 in Austria, 5 in Britain, and 1 each in France, Holland, Mexico, and Sweden.

Feeder-fed machines: The I.S. machine. A rival method soon appeared in which 'gobs' (lumps of glass of the correct weight and shape) were produced

FIG. 22.3. The 'finish' of a cod bottle in the hand making process.

FIG. 22.4. An Owens bottle-making machine; the diameter is about 16 ft, the blow moulds can be seen towards the bottom of the machine.

from a feeder. This was a long channel (about 0·5 m × 0·5 m × 3 m) into which glass flowed from the melting furnace and out of which the glass was ejected through an orifice by gravity, assisted by a plunger (Fig. 22.5). The forming of a gob into a bottle is shown diagrammatically in Fig. 22.6.

Several different machines were designed to form the bottles; some were suction machines but most were feeder-fed. The individual section I.S. machine, first introduced in 1925, is now used in the production of over 90 per cent of the glass bottles that are made. It departs from the rotating turntable principle introduced by Owens, and instead consists of a number

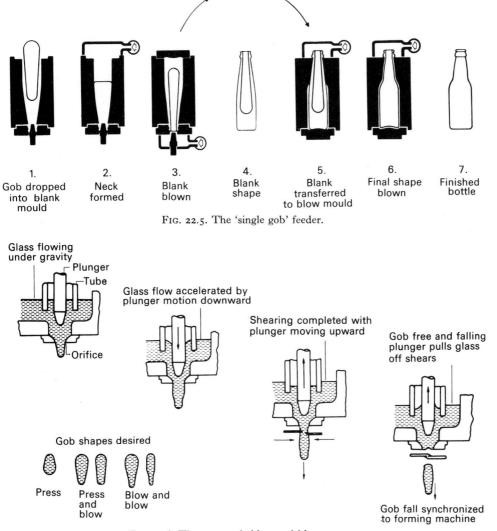

1.	2.	3.	4.	5.	6.	7.
Gob dropped into blank mould	Neck formed	Blank blown	Blank shape	Blank transferred to blow mould	Final shape blown	Finished bottle

FIG. 22.5. The 'single gob' feeder.

Glass flowing under gravity

Plunger
Tube
Orifice

Glass flow accelerated by plunger motion downward

Shearing completed with plunger moving upward

Gob free and falling plunger pulls glass off shears

Gob shapes desired

Press

Press and blow

Blow and blow

Gob fall synchronized to forming machine

FIG. 22.6. The automatic blow-and-blow process.

of independent stations in line which are fed with gobs, in turn, through a system of chutes. This machine was a great step forward, because it avoided the necessity of moving the heavy turntables of the earlier machines and had the further advantage that each station is in effect a complete machine unit, which facilitates maintenance. The parison is transferred to the blow-mould and the neck ring which forms the finish is immediately returned to the parison mould, thus permitting the machine to be blowing one bottle to its final shape and at the same time to be forming the next parison. There is thus very little idle mould time, unlike the moulds on a rotating table which remain empty during the greater part of the cycle. The overlapping cycle of the I.S. machine increases the rate of production per mould and gives a higher output per mould than the rotating machine. Machines are now produced with as many as eight stations in line and they may be fed by a feeder which makes up to three gobs simultaneously. When two or three gobs are fed, they are done so to moulds with two or three cavities respectively. Bottles weighing 300 g could be produced with double-gob feed on an eight-section I.S. machine at the rate of 84 per minute.

By 1922 the feeder as at present in use—embodying the orifice, plunger, and shears shown in Fig. 22.5, had been developed by the Hartford Fairmount Company of the U.S.A. and was in production. In 1920, 3000m. bottles were produced on Owens suction machines and 2000m. from gob feeders; in 1927 approximately 3000m. pieces each were being produced by the two types of machine. Thereafter, Owens machines declined while gob fed machines increased. In 1950 15000m. units were made in the U.S.A. and about 2500m. in Britain. In 1905, the total production in the U.S.A. by hand, semi-automatic, and machine processes was about 1700m.

The development of new markets accompanied the growth of the container industry. The milk bottle is one example. In 1918 milk was not delivered in bottles; even in London it was served from churns which were pushed round the streets on hand carts; but in 1950 over 200m. milk bottles were made in Britain. Many more foods, beverages, medical and toilet preparations were packed in glass containers. A comparison of the numbers of glass containers for various products produced in Britain in the years 1918 and 1950 is given in Table 22.3.

Semi-automatic production for small quantities of special containers could be seen in a few places, even in the mid-1970s, but in significant numbers it

TABLE 22.3
Numbers of glass containers manufactured

	Containers (millions)		1970
	1918	1950	
Beer, cider, etc.	88	288	
Wines, spirits, minerals	88	417	
Jams, fruit	63	504	
Chemical, medical, and toilet	174	911	
Potted meats, fish pastes	22	81	
Others	126	958	
Total	561	3159	6000

had practically disappeared by 1950. The container made in the machine is reproduced with constant weight and dimensions. For example, the one-pint milk bottle was supplied in 1960 to the following dimensional tolerances: height $21 \cdot 336 \pm 0 \cdot 114$ cm; diameter $7 \cdot 62 \pm 0 \cdot 101$ cm. The contents of machine-made bottles are thus closely controlled, individual bottles falling between the limits of about $\pm 0 \cdot 2$ per cent. The reproducibility of the lip or mouth of the containers greatly facilitates vacuum sealing in various forms and the constant size, shape, and volume make it possible to design very fast filling machines; bottles may be filled with free-flowing liquids such as milk and whisky at around 300 per minute on one machine.

The shape of glass containers. The consistent dimensions of bottles produced on automatic machinery made it possible to pay much more attention to the design of a bottle to obtain maximum strength. For example, when a bottle is subjected to pressure or a blow from the outside, magnification of stresses occurs at the corner where the base merges with the wall. By increasing the radius of curvature, taking out the sharp bend, the strength is increased; by altering the shape, so that external blows are less likely to occur at this point, a considerable increase in strength could thus be obtained.

The weight of a bottle for a specific purpose has been progressively reduced The British pint milk bottle, which weighed 560 g in 1920, was reduced to 400 g in 1960; the U.S. quart bottle from 700 g in 1932 to 500 g by 1940.

Attention has also been given to the protection of the surface of bottles because it has been found that the strength of glass is determined by the state of the surface, the extent to which it has been scratched and bruised having a considerable effect on the strength. Bottles are therefore coated with tin oxide before entering the annealing oven and, at the cold end, with various organic materials which increase the ease with which one surface slides on another; in this way considerable strengthening has been effected. The hot-end process has become known in some places as titanizing, because at first compounds of titanium and tin were used. Today, it is mostly stannic chloride which is used; this breaks down on heating to give a coating of tin oxide on the surface.

Industrial relations. Many of the changes in methods of manufacture were at first resisted by the unions (Ch. 5); loss of employment was feared, but eventually the contrary occurred.

In the U.S.A., in the period 1903–9, the union concerned agreed to a three-shift system in place of the traditional two-shift system because it encouraged the employment of more blowers. Higher piece rates were demanded for machine operators, to try to prolong the use of semi-automatic machines, the introduction of which it had at first opposed. Finally, the union continually reduced the agreed piece rate in hand factories in order to compete with output from the new machines.

Earlier, there had been opposition to the tank furnace. However, in 1948, the President of the Glass Blowers Association of America reported that when automatic machines were first introduced there were 10 000 members of the Association, but now that the industry was fully mechanized the membership had reached 38 000 and was still growing. In the United Kingdom from 1920 to 1950 production increased sixfold, while the number employed on the industry rose by about 50 per cent. Since 1950 the production and the efficiency of the industry has increased considerably; in 1970, 6000m. containers were made in Britain and 38 000m. in the U.S.A., nearly one per day per person.

VI. THE ELECTRICAL INDUSTRY

Westlake and Corning ribbon machines. In the electrical industry, glass was required in the form of bulbs for electric lamps and thermionic vacuum tubes, and in the form of tubing for the lead-in pinches (Ch. 18, p. 448). This industry, new at the beginning of the twentieth century, rapidly expanded and by 1950 bulbs were being produced by one machine at the rate of 1800 per minute. These machines were developed in the U.S.A. by the Corning Company; in 1950 a machine was installed in Britain near Doncaster, where coke-oven gas was available as fuel. This one machine, at the time that it was installed, could produce enough bulbs for the lamp-making industry of the whole of Europe.

The first bulb-making machine, the Westlake, was suction fed; it was a mechanization of the action of the hand blower in having attached to the turntable of the machine a series of blow-pipes which were fed by two arms, mounted at the top, which went into the furnace and sucked up a gob of glass; this was dropped on to the blow-pipe and held there mechanically. The blow-pipe then rotated, blew the glass, and swung into a vertically downward position. A mould then closed about the parison that had been formed and this was then blown into a bulb. The finished bulbs were dropped on to a conveyor belt on which they moved to another machine; there the piece of glass by which they had been held at the end of the blow-pipe was burnt off by a ring of flame. Developments of this machine shortened the blow-pipes and made the relation to the man with the blowing iron less immediately apparent. The Corning ribbon machine was based on an entirely new principle; instead of gathering gobs by suction a continuous ribbon of glass about 2 inches wide issues from the furnace. This ribbon passes first between moulds which impress upon it depressions of an appropriate size, rather like a series of egg poachers mounted on the ribbon. The ribbon, with these depressions formed in it, then passes between two caterpillar conveyors, the bottom one carrying the moulds and the top one the blowing apparatus, so that the depressions could be formed and blown into bulbs. At the end of the caterpillars, after a sufficient space for cooling, the bulbs are knocked from the ribbon, the remains of the ribbon being returned to the furnace for remelting as 'cullet', waste glass added to the batch. These machines required very little supervision; a team of four or five men was sufficient to supervise the operation of a group of two or three. This was a vast change from the time when bulbs were blown by hand. A visitor would then be amazed at the apparent danger of standing on a blowing platform surrounding a five-pot furnace, where perhaps as many as four men would be working out of each pot blowing bulbs at the rate of several a minute each, the red-hot glass swinging on the end of each man's blowing iron.

Tube-making machines. The other product required for the lamp bulb industry is tubing. At the beginning of the century this, too, was made in a traditional way. The blower gathered a large blob of molten glass on the end of a blowing iron, blew, and formed it into the shape of a very thick-walled large bottle. A punty (a preheated tool) was attached to the bottom end of the glass and then one man held the punty and walked away while the other held the blowing iron and occasionally puffed into it to maintain the correct diameter. In this way lengths of tubing of up to 50 ft could be made at a time but each time much glass remained on the blowing iron and the punty; less than 25 per cent of the gather would become useful tubing. The Danner tubing machine made this process continuous by allowing glass, in the form of a ribbon about 2 inches wide, to issue from the furnace on to a heated hollow refractory mandril. Air could be blown down this so that the ribbon, in effect, formed a thick-walled bottle covering the mandril. The glass was drawn from the end of the mandril and the tubing so formed passed through caterpillars; the ribbon renewed the 'bottle' and the caterpillars drew the tubing continuously. In other developments, the glass is fed on to the inside rather than the outside of the mandril, as in the Vello vertical draw processes.

Glass for television tubes. The television industry has made special demands on the glass industry for large pieces of glass of quality equal to that of optical glass and of special composition. This is necessary both from the point of view of sealing to metals, and also to prevent the passage of X-rays generated by the impact of the electron beam on the fluorescent screen and its supporting glass. An entirely new glass-making process was developed for making the bodies of television tubes. Large gobs of glass were dropped into a mould which was then spun violently so that the body was formed by centrifugal force. The large screens are pressed and the screen is joined to the body on a 'glass lathe'; this is a machine with two rotatable heads on the same axis so that the body and screen may be 'offered up' to one another and the joint sealed by rotating the two parts together in a ring of burners. To aid melting, electricity may be introduced through the flames, the additional energy greatly facilitating the fusion process.

VII. GENERAL

From 1900 to 1950 the manufacture of glassware changed from dependence on hand-work, some of it very skilled, to a highly mechanized industry. The changes which have been described were possible, in part, because the ability to control the steps in this high-temperature technology was greatly increased by the growing availability of scientific and control instruments. The traditional process of annealing was given a firm mathematical interpretation in 1920 by L. H. Adams and E. D. Williamson, although Maxwell had shown the way in the middle of the nineteenth century. Thermocouples, radiation pyrometers, and devices to indicate the level of the glass in the tank are examples of the sensors used for control purposes. By 1950 the mechanized industry had spread all over the world, into more than twenty countries.

Handworking shops survive but have also changed; furnaces and batch materials are much improved, giving better glass consistently. The products of the craftsman find ready markets and there has been growth in this branch of the industry in many countries. The art schools have supplied designers; in Sweden, for example, there were in 1950 about forty small hand factories, most of them with a designer specially employed.

Two major U.S. glass companies, Owens-Illinois and Corning, sought to expand their markets with new products; out of these programmes eventually grew the Owens-Corning Fibre-glass Corporation, formed in November 1938. A new industry was established, and in many countries other companies have entered the field and contributed to its growth. The annual sales of fibreglass in the U.S.A. in 1939 were valued at $4m.; by 1950 the value was $82m.

Textile fibres are usually made from marbles of 'E' glass (see Table 22.1). The marbles are made from glass melted in a small tank furnace, and are remelted in a platinum bushing heated by an electric current passing through the platinum. The length of the bushing may be some 150 mm or more and it may have some 200 orifices about 1 mm in diameter. Through these orifices continuous filaments are drawn. They are then twisted together and drawn by winding on rapidly revolving drums. In another process, the glass leaving the bushing is blown into very fine streams which are collected on a large drum and then twisted together and wound as staple fibre yarns. Both products can then be made into textiles by any standard weaving process. For thermal insulation, glass wool can be formed by collecting the blown fibres on a travelling belt.

Other special products are toughened and laminated glass for automobiles. Some machines have been adapted, others designed, for the mechanical production of such widely used items as tumblers and wine glasses.

As with many other branches of technology the development of the glass industry led to the formation of appropriate scientific societies in many countries. Among these may be mentioned the American Ceramic Society (1899) and the Society of Glass Technology in Britain (1916). The publications of these societies include abstracts of the world literature.

BIBLIOGRAPHY

BUSBY, T. *Tank blocks for glass furnaces*. Society of Glass Technology, Sheffield (1951).

DOUGLAS, R. W., and FRANK, S. *A history of glass making*. Foulis, Henley-on-Thames (1972).

GARSTANG, A. Fifty years of furnace building. *Glass Technology*, **12**, 1 (1971).

MALONEY, F. J. TERENCE. *Glass in the modern world*. Aldus Books, London (1967).

MEIGH, E. The automatic glass bottle machine. *Glass Technology*, **1**, 25 (1960).

——, and GOODING, E. J. *Glass and W. E. S. Turner*. Society of Glass Technology, Sheffield (1951).

MOODY, B. E. *Packaging in glass*. Hutchinson, London (1963).

NORTON, L. E. Some furnace developments, 1928–68. *Glass Technology*, , **101** (1969).

SCHOLES, S., and GREENE, C. H. *Modern glass practice*. Cahners Books, Boston, Mass. (1974).

TAYLOR, W. C. The effect on glass of half-a-century of technical development. *Bulletin of the American Ceramic Society*, **34**, 328 (1951).

TOOLEY, F. *Handbook of glass manufacture*. Ogden Publishing Co., New York (1960).

TURNER, W. E. S. Twentyone years. A professor looks out on the glass industry. *Journal of the Society of Glass Technology*, **42**, 99 (1938).

23

PAINT

HENRY BRUNNER

WHILE paint and varnish have been known from earliest times, it was not until the twentieth century that paint became sophisticated—both in its process of manufacture and in its usage and method of application. Not only did coating compositions come to be required for industrial purposes, but industry itself provided an additional impetus with the development of highly efficient paint-making equipment. The expansion of world trade, coupled with the greater availability of the raw materials used in paint and varnish manufacture, helped considerably. An even more significant factor has been the evolution of the science of organic chemistry, which in turn facilitated the creation of the plastics industry (Ch. 21). Many of the chemicals, synthetic resins, and polymers utilized in plastics are substances basic to modern paint and varnish manufacture. Varnish is included here with paint since the technology of both is so closely linked. Except for a few minor technicalities, varnish is virtually unpigmented paint, and provides the medium wherein the pigment is dispersed and whereby the paint is able to dry as a coherent film or coating.

Paint technology made most rapid advances in those countries where industrialization developed fastest, that is, the United States and western Europe. Russia presented an entirely different story; paint technology there starting virtually from scratch at the time of the October 1917 revolution and continuing apace despite relatively little communication with the West. The influence of both the United States and Britain on developments in paint technology became more pronounced with the cessation of hostilities in 1945. It was not surprising to find—since paint raw materials such as resin intermediates, cellulose derivatives, and solvents are all products of the chemical industry—that a number of leading chemical concerns such as Du Pont and I.C.I. diversified into paint manufacture. Furthermore, since the larger companies had numerous associate companies and licensees throughout the world, new paint compositions and novel methods of paint application soon became available on a world-wide basis.

I. THE MANUFACTURE OF PAINT

For centuries it was common practice for painters to prepare their own products from pigments and oils (mainly linseed oil). Manufacturers of 'ready-mixed' paints made only slight progress until just before the beginning of the twentieth century when 'factories were established in such great numbers that within a decade or two their proprietors were forced to reach out for wider markets. Thus was ushered in an era of extraordinarily keen competition' [1]. Three factors play a leading role in paint manufacture:

(i) The pigments, which determine the colour and opacity characteristics of the paint.
(ii) The medium (vehicle), which governs the nature or type of the paint and whether it will dry (cure) by exposure to the air, by heating (stoving), by cold-curing (catalysis), or merely by solvent evaporation.
(iii) The efficient dispersion of the pigments in the medium so that large agglomerates are broken down and also thoroughly wetted by the medium. This is to ensure that the paint dispersion is stable as well as capable of yielding a smooth film of uniform appearance. Thus dispersion regulates most of all the quality of the paint.

Pigments. The war years, the growth of the motor-car industry, and the increasing awareness of the toxic hazards of white lead all played their part in the development of pigments. This is well illustrated by the selection of three pigments from the many new ones which became available to the paint manufacturer in the early part of the twentieth century.

Zinc chromate (zinc yellow), little used before 1914, assumed a dominating role in the field of rust-inhibiting pigments during the Second World War. By then many millions of pounds of this pigment were being used annually in paints for the protection of metal of all kinds of war equipment.

Also of outstanding importance was I.C.I.'s discovery in the 1930s of phthalocyanine blue (Monastral Blue) and later its halogenated derivatives, the Monastral Greens. R. P. Linstead [2] identified a coloured impurity in phthalimide manufacture as iron phthalocyanine, which led to the deep blue copper phthalocyanines. These pigments were found to be of exceptional light-fastness, unaffected by acids or alkalis, and, in addition, 'non-bleeding' in oils and solvents.

Perhaps the most significant contribution of all was that of titanium dioxide. Introduced shortly after 1918, this met considerable resistance because of its high cost. Nevertheless because of its chemical inertness, extreme white-ness, excellent covering power (the highest of all the white pigments), and the added advantage of being virtually non-toxic, titanium dioxide was soon dominant in the manufacture of white paint. White lead's share of the white pigment usage in paint fell during the period 1900–45 from nearly 100 per cent to less than 10 per cent. The share of lithopone, a co-precipitate of zinc sulphide and barium sulphate, introduced before the First World War, rose to 60 per cent by about 1928 but fell to 15 per cent by 1945. The most spectacular and consistent rise of all was that of titanium dioxide, which by 1945 had captured 80 per cent of the white pigment market. An important consequence was a dramatic fall in the incidence of lead poisoning in the paint and allied industries. In Britain, fatalities reported to the Factory Inspectorate fell from 38 in 1910 to nil in 1950.

Pigment dispersion. Before 1900 most paint plants were built when labour costs were not of primary concern and it took many more man-hours than now to produce a gallon of paint. Subsequently, particularly during the twenty-five years following the mid-1920s, progress in paint manufacture was marked by a steady decrease in labour costs of mixing, 'grinding', and filling, resulting from the use of better equipment and mechanization (Fig. 23.1).

The importance of pigment dispersion has already been stressed; the wet-ting of the pigment by the medium is fundamental to good paint manufacture. Adequate dispersion ensures that layers of moisture or moist gases on the surface of the pigment are displaced by the liquid medium. In paint, 'grind-ing' is a misnomer for 'dispersion'. Optimum dispersion generally entailed a two-stage process, using first a simple mixer, followed by dispersion of the pre-mix in one of a variety of types of mills. The latter, in effect, squeeze and rub together the pigment particles with the molecules of the medium.

Various types of mills were introduced, each with its own characteristics and designed for different paints. Thus, new mills supplemented rather than replaced existing ones, contributing to a wider range of dispersion equipment available to the paint manufacturer. Such machinery ranged from pug mills for stiff pastes to edge-runner or pan mills (Fig. 23.2) and single-roll, two-roll, three-roll, and even five-roll mills. American practice tended to make use of three- to five-roll mills, whereas European mills commonly used only one roll and a scraper. In the 'Uniroll' mill, paint is forced between the single roller and a vane bar which is held against the roller by accurately controlled hydraulic pressure.

FIG. 23.1. Early twentieth-century paint manufacture; mixing, 'grinding', and filling operations all under one roof.

FIG. 23.2. Edge-runner or pan mills, in use prior to the Second World War.

Not least in importance was the pebble- or ball-mill: essentially a rotating cylinder containing pebbles or steel balls which occupied about 45 per cent of the volume of the cylinder. Pebble mills, which were lined with tiles, were employed for white and light tints; darker-coloured paints were made in steel-lined mills and made use of steel balls. Ball-mills came into favour because they dispensed with a pre-mix stage and because for their operation only a minimum of supervision was required. In addition there were no losses of solvent by evaporation, with a consequent reduction in fire hazards. Hence ball-mills were ideal for the manufacture of cellulose lacquers.

II. CLASSIFICATION OF PAINTS

There are two main purposes for paint; basically, they are decorative and protective. A few other reasons exist; for example, the colouring of pipes and conduits for identification, and the use of temperature-indicating paints.

Decorative paints found their main outlets for use in homes and buildings. They underwent considerable technical improvement during the first half of the twentieth century. During that same period, however, the creation of specially formulated paints for industry was little short of revolutionary. Although decorative paints have a secondary importance in providing protection to woodwork, stonework, and plaster, protection rather than decoration is usually paramount for the vast array of metal-based products of industry. For instance, the utilization of paint and allied products for the protection of iron and steel has saved millions of tons of iron and steel alone from corroding to rust (ferric oxide). It is perhaps appropriate to mention that various grades of naturally occurring ferric oxide provide useful yellow, red, and brown pigments.

Another classification of paints derives not so much from the purpose for which they are intended, but from the nature of their formulation. There are several types, including:

   (i) conventional air-drying paints and varnishes,
  (ii) stoving finishes,
 (iii) lacquers,
  (iv) water-borne paints, and
  (v) cold-curing (two-pack) compositions.

Air-drying paints. These are based principally on so-called drying oils. They dry partly as the result of the evaporation of the solvent which is used to thin them to brushing or spraying consistency, and partly because oxida-

Fig. 23.3. A varnish kitchen or room set aside for varnish manufacture, at the Cornwall Road Works of Nobles and Hoare Ltd., North Lambeth, London. The factory, destroyed by enemy action in 1940, was subsequently re-erected at Leatherhead. Note the wheeled cradles for moving the gum-running and varnish pots. In the foreground are oil cans.

tion of the oils by the air converts the liquid paint to a solid film. To accelerate this oxidation process, driers are added to the paint. These are mostly solutions of soaps of certain metals, notably cobalt, lead, and manganese. Research on driers has led to progress, particularly over the selection of compositions for optimum performance; too much drier can in fact retard drying. A greater understanding of the mechanism of film formation by oxygen uptake has also been achieved. From early times various natural resins had been used to reinforce linseed oil and other drying oils, since paints based merely on pigment and oil will yield only very soft films. One such resin was kauri, a fossil resin from the largest forest tree of New Zealand (*Agatha australis*), but by the turn of the century its use had diminished, partly owing to dwindling supplies and partly because of the availability of a more useful fossil resin from east Africa known as Congo copal. The latter was a much harder resin but required prolonged heat treatment, which came to be known as 'gum running' (Fig. 23.3). Another useful resin was rosin (colophony); it is a solid residue obtained in the manufacture of turpentine from the oleo-resin of pine trees. For technical reasons, rosin itself requires improving and this was achieved by causing it to react with glycerine (glycerol) to yield what is known as ester gum.

The rapid growth of the chemical industry made an impact on varnish technology, and soon completely synthetic resins were produced. Of the more important types of oil-soluble synthetic resin, the earliest were the phenolics, that is, phenol-formaldehyde (Bakelite) condensation products. Normally, when phenol (carbolic acid) is reacted with formaldehyde an infusible, oil-soluble resin is obtained, but by various chemical devices, which were the subject of numerous patent applications, phenolics completely soluble in oil were made available. These devices included replacing ordinary phenol by higher phenols such as cresylic acid (a coal-tar product), butyl phenol, and octyl phenol (the last two are synthetic), using different catalysts, or heating together the phenol and formaldehyde in the presence of the fatty acids which make up the drying-oil molecule. By far the most important oil-soluble phenolic was that obtained by carrying out the phenol–formaldehyde reaction in the presence of rosin. In the forefront of this type of resin were the German 'Albertols'; Kurt Albert was one of the original inventors. American resins comparable to the Albertols were known as Amberols, which was aptly descriptive of the appearance of these resins. Albertols and Amberols were later to be known as 'reduced phenolics' to distinguish them from the '100 per cent phenolics'. The latter proved of little value to the manufacturer of air-drying products. Although derived paints had useful properties, such as improved water resistance, they were deficient in drying performance, virtually of little value for the manufacture of white or light-tinted paints, and tended to discolour on exposure. All these drawbacks were attributable to the

TABLE 23.1

Resins as finally used in protective coatings* in the U.S.A. (millions of lb)

	Alkyd	Rosin	Ester gum	100% phenolic	Reduced phenolic	Natural resin
1926	—	40·1	42·7	—	11·4	41·0
1936	46·7	18·6	18·8	1·4	72·1	27·7
1941	131·7	42·5	22·5	33·7	97·7	43·0

* Protective coatings would be expected to cover not only air-drying finishes but the stoving types as well as lacquers. This explains the increased usage of 100 per cent phenolics (stoving) and continuing usage of natural resin (lacquers).

Throughout the period under review, as well as for centuries previously, linseed was the principal oil employed; in paint manufacture it far exceeded the usage of all other drying oils together (Table 23.2).

Source: Reference [3].

presence of unreacted phenol(s) in the synthetic resin. Far more important were the reduced phenolics which for many years were to remain the vital resin component of oleo-resinous varnishes and orthodox paints derived from them.

What might be termed the first breakthrough in modern paint technology was the advent of the alkyd resin, a type of polyester. The alkyd most frequently met with was a product of the chemical reaction of phthalic anhydride and glycerol with certain vegetable oils (or their corresponding fatty acids). The development of alkyd resins was greatly stimulated, around 1918, by the manufacture of cheap phthalic anhydride from the coal-tar product naphthalene. Oil modified alkyds, which were being used in the U.S.A. by 1928, enabled the paint manufacturer to formulate products with good drying performance, excellent colour characteristics, and good durability. In the second quarter of the twentieth century, alkyd-based paints were destined to overtake paints based on oleo-resinous vehicles although the latter were still preferred for certain compositions such as undercoats, where the advantage of lower cost of the vehicle was not outweighed by less satisfactory technical performance.

Partly because of the upsurge in output of the paint industry, partly because linseed crops were subject annually to variable climatic conditions, and partly because the early twentieth century saw two world wars, alternative oils to linseed were actively sought.

One of the earliest of these was tung oil (Chinese wood oil). Its drying characteristics were excellent except for the fact that derived coating com-

TABLE 23.2

Consumption of drying oils by the paint, varnish, and lacquer industry in the U.S.A. (millions of lb)

	Linseed	Tung	Perilla	Fish	Soya bean	Castor*	Oiticica	Total excluding linseed	Total drying oils
1904	357·1	9·2	—	—	—	—	—	9·2	366·3
1914	410·8	27·0	—	—	—	—	—	27·0	437·8
1919	385·8	46·2	4·7	—	20·0	—	—	70·9	456·7
1931	411·6	81·7	10·5	12·1	6·3	1·8	0·3	112·7	524·3
1935	409·0	116·7	49·8	18·3	13·0	3·5	1·9	203·2	612·2
1941	650·5	64·9	7·0	40·7	41·6	44·2	35·2	234·2	884·7

 * Castor oil, not being a drying oil, is an anomaly in this table. The figure would refer either to its subsequent conversion to dehydrated castor oil, or to its use as a plasticizer in lacquers.
 Source: Reference [3].

positions had a tendency to wrinkle on drying. Because of this, tung oil was, more often than not, blended with linseed oil. Oiticica oil from Brazil and perilla oil from Asia were other vegetable oils that were considered in place of linseed, and even fish oils were used. Fish oils, however, particularly in their early stages of development, were prone to yield finishes in which the distinctly fishy odour was all too obvious. Castor, a non-drying oil, became the subject of much research, leading to many patents, all aimed at modifying it to render it air-drying. In the main, this was achieved by catalytic dehydration. Dehydrated castor oil became an important drying oil in paint and varnish technology. So too was another oil, namely, soya bean; the use of this as a paint ingredient developed more slowly however. This was partly because, unlike linseed and castor oil, it was also an edible oil, and the Second World War and its aftermath called for concentration on its use for food. Secondly, although soya bean oil had an advantage over linseed in yielding paints and varnishes with a reduced tendency to yellow on ageing, its air-drying characteristics were not as good as those of linseed. As a result, soya bean oil came into its own, initially, with a growing demand for white stoving finishes (see later) and subsequently as an alternative to linseed in oil-modified alkyds and paints based on them.

An interesting development arising from the immediate post-war shortage of vegetable oils for paint manufacture was their extended use by modification with the hydrocarbon styrene. On the termination of the war in 1945 the United States had large stocks of (and plant capacity for) this chemical, which had been a vital component of styrene–butadiene synthetic rubber. A number of leading American companies, notably Archer–Daniels–Midland, American Cyanamid, Alkydol Industries, and Spencer–Kellogg soon made use of styrene in resins and paint vehicles. Lewis Berger and Sons Ltd., in 1947 introduced in Britain their Bergermaster enamels. These were eventually ousted, when drying oils became freer again, by the somewhat more durable conventional oil-modified alkyds but styrene came to stay in the form of styrenated alkyds, on which were based useful stoving finishes.

Stoving finishes. The second important class of paints are those which require stoving (baking is the American term). By 1929 stoving finishes were being applied to metal articles such as iron bedsteads, and the 1930s saw their wider usage in the industrial equipment market and the motor industry.

Stoving is effected by hot-air convection in ovens, either by a low-bake system, or a high-bake process, the latter usually requiring much shorter times in the oven. The low-bake system was first used by the Ford Motor Co. in the U.S.A.; Briggs Motor Bodies Ltd. were the first to use this system in Britain in 1936–7. The Ford Motor Co. had also, in 1935, tried using infrared stoving, that is, heating by radiation. For this purpose they employed carbon filament lamps in reflectors. Later refinements included the use of batteries of tungsten-filament lamps in quartz tubes.

The chemical structure of both phenolic and alkyd synthetic resins enabled them to be utilized in stoving finishes; stoving times ranged from a few minutes to an hour or more, according to the temperature employed. Stoving schedules were geared to cater for those industries relying on mass-production methods, particularly for such articles as car bodies, electrical equipment, and domestic appliances constructed of metal.

The phenolics were soon widely used, either 'straight' (100 per cent phenolic) or modified (that is, plasticized to render them more flexible). Despite poor colour characteristics, phenolics were attractive because of their relatively low cost and because derived finishes were found to possess good resistance to chemicals, acids, petrol, lubricating oils, and water. During the Second World War stoving phenolics were used for the coating of propellers, aircraft engines, ammunition boxes, petrol containers, and portable water tanks. After the war they continued to be utilized where colour was not of great importance, as for instance in wire enamels and insulating varnishes. Much better than the stoving phenolics in initial colour and lightfastness were the stoving alkyds. However they did not give films of the hardness and resistance often required by industry.

Sometimes natural resins, such as rosin, or synthetic resins, like the reduced phenolics, were added to improve the hardness characteristics but it was the introduction of the 'nitrogen resins'—urea-formaldehyde and melamine-formaldehyde—which gave stoving finishes the boost they needed.

Synthetic urea, manufactured cheaply by the reaction under pressure of carbon dioxide with ammonia, arrived in the early 1930s. Unmodified urea-formaldehyde provided colourless resins which stoved, however, to hard but rather brittle films. They proved ideal for hardening oil-modified alkyds, with which they were fortunately found to be compatible; alternatively, it could be said that the alkyds were excellent plasticizers for the urea-formaldehyde resins. Alkyd urea-formaldehyde finishes were quickly adopted where hardness and flexibility, coupled with good colour characteristics were required, for example, for the car industry and for white domestic appliance finishes. Shortly after the Second World War the majority of British car manufacturers changed over from cellulose-based finishes to stoving enamels formulated on alkyd urea-formaldehyde.

In about 1939–40, melamine became commercially available. Like urea, it reacts with formaldehyde to yield colourless resins, valuable for hardening oil-modified alkyds. Although melamine was more expensive than urea, films of melamine-formaldehyde were harder than those of urea-formaldehyde and were able to withstand heating to much higher temperatures without discolouration.

Lacquers. A third, and no less important, category of paint compositions are the cellulose lacquers. Originally, 'lacquer' was used to describe varnishes based on solutions of natural gums and resins, such as shellac. Later, it was used to include not only varnishes but coloured compositions containing dyes or pigments. Since lacquers dry merely by solvent evaporation, they are easily removed by re-dissolving in paint thinners.

Sometimes lacquers, and even the more conventional paints, are referred to as enamels. The word 'enamel' crept into the paint industry for the description of coating compositions of great hardness and high gloss, both of which are characteristics of the vitreous finishes of the ceramic industry.

The colour of the lacquer was often derived from that of the component natural resin; the principal volatile solvent was alcohol. Thus shellac (Vol, II, p. 362), Manila, and Sandarac gave yellowish lacquers, and those from Dragon's Blood, a resin from the rattan tree (*Calamus draco*), were red. In about 1900, attempts were made to introduce oriental lacquers into Britain but these were abandoned because of their skin-irritant properties. These lacquers were made from the sap of a Chinese tree, *Rhus vernicifera* (the varnish tree), and were developed by the Japanese. The thick milky emulsion from the tree, when purified and applied in thin films, dries to dark, even black, coatings of high gloss, hardness, and durability.

The coming of nitrocellulose (known in America as nitro-cotton), transformed the lacquer scene and made available products which were initially colourless, but could be pigmented to a wide range of light-fast colours. Nitrocellulose was a product of the explosives industry (Vol. V, Ch. 13). As far back as 1855, Alexander Parkes had taken out a patent [4] for nitrocellulose protective coatings, but at the turn of the century nitrocellulose lacquers had not assumed much importance. Initially, their development was hampered by the lack of suitable solvents, but the introduction of ester-type solvents, notably butyl acetate, came to the rescue. Another important factor in the growth of the nitrocellulose lacquer industry in the mid-1920s was the search for uses for nitrocellulose other than for explosives after the First World War had ended.

Coinciding with all this was the rise of the car industry (Ch. 30). In America, Du Pont took a substantial interest in the capital of the General Motors Corporation, partly for general financial reasons but also to ensure an outlet for nitrocellulose lacquers. The British company Nobel Industries (to be merged later in I.C.I.) also invested in General Motors to secure the advantage of their world-wide organization. While in America, following an anti-trust suit, Du Pont had to contend with the Hercules Powder Company, and in Germany Dynamit AG had to compete with such companies as Köln-Rothweiler Pulverfabriken, Nobel Industries had almost a complete monopoly in Britain. The in-fighting over the political and commercial aspects of nitrocellulose production has been excellently described by W. J. Reader in his history of I.C.I. [5].

Nitrocellulose is manufactured by nitrating cotton linters (short fibrous hairs on the cotton seed) or wood pulp with a mixture of nitric and sulphuric acids. By using a range of nitrocelluloses of different degrees of nitration and different viscosities, and ringing the changes on the resins added to improve hardness and adhesion, the plasticizers, the solvents, the diluents (cheaper solvents for the resins), and the dyes and pigments, a vast array of nitrocellulose lacquers were made available. Because of their ease of application and quick-drying properties, they were ideal for numerous industrial uses such as finishes for cars, furniture, leather, and paper. For instance, in the early 1900s it took 7–10 days completely to finish painting a car; the installation of stoving equipment reduced this time to 2–3 days. The introduction of nitrocellulose lacquer enabled the car to be re-coated after no more than about 30 minutes. Lacquers found a secure niche for themselves in the paint industry despite the subsequent introduction of newer types of non-cellulosic compositions. Nitrocellulose was not the only cellulose drivative to be used in finishes; a few others such as the ester cellulose acetate (and later cellulose acetobutyrate) and the ether ethyl cellulose, found special applications. Generally speaking, however, compared with nitrocellulose, the acetate has somewhat inferior solubility characteristics and ethyl cellulose is less resistant to water.

Water-borne paints. Formulators of paints could not ignore the attractions of a paint which would be miscible with water but capable, subsequently, of drying to a water-resistant coating, with the added bonus that the painter could wash out the brush with water instead of spirit.

Before 1900, water paints were mainly limited to limewash—a product of lime, tallow, and water—but at the turn of the century distempers made their appearance. For many years milk had been used, mixed with quicklime to provide a medium for whitewash. By isolating casein, the protein present in milk to the extent of about 3 per cent, and then mixing it with lime, as well as with pigments and a preservative, paint manufacturers were able to create a distemper which for some years was popular, particularly in America. British manufacturers favoured the use of glue rather than casein, and later marketed an improved distemper using glue in the aqueous phase and polymerized drying oil in the oil phase.

The history of distemper manufacture is the story of a long battle to overcome such defects as poor washability and a tendency to bacterial and fungal attack in the can. This ultimately led to the development of emulsion paints. Early emulsion paints were obtained by dispersing the pigment in an emulsion of a drying oil in water. In place of the drying oil, oleoresinous blends, and later oil-modified alkyds, were sometimes employed.

In the search for alternatives to paints based on drying oils Germany turned during the Second World War to polyvinyl acetate (PVA), and was the first country to produce PVA emulsion paints. Although German synthetic rubber (Buna) based on styrene–butadiene copolymer was first introduced in 1935 by I.G. Farben, it was the United States, with their enormous stocks of styrene and butadiene left over from its war-time effort, who were able to forge ahead in 1948 with styrene–butadiene latex paint. The styrene–butadiene copolymer ratio employed in synthetic rubber manufacture was altered to make it more suitable for emulsion paints. Nevertheless, PVA emulsion paints were soon to stage a come-back not only in Germany but in Britain, France, and Switzerland, and also in the U.S.A.

Miscellaneous finishes. Until the second half of the twentieth century there were only feeble attempts in the paint industry to introduce two-pack compositions; one part containing the resin vehicle and the second, smaller part, the catalyst. Mixing of the two portions would enable the overall composition to cold-harden by catalytic means. It was known, for instance, that phenolic, urea-formaldehyde, and melamine-formaldehyde resins were all curable (hardenable) by means of acid catalysts.

The outbreak of the Second World War resulted in the development, for use in aircraft construction, of new types of adhesives known as epoxide or epoxy resins. They were capable of being hardened by means of a great variety of agents, including some which were neither strong acids nor strong alkalis. Because of their good adhesion, hardness, and speed of curing, these epoxides were soon adopted by the paint and varnish industry, which however had still to deal successfully with the mechanical problems—for example, packaging, labelling, adequate mixing—associated with two-pack compositions.

A further wartime development, this time in Germany, concerned the use of di-isocyanates. These could be reacted with drying oil diglycerides to yield urethane oils, and with hydroxy polyesters to form polyurethane resins. Isocyanate resins could be used in single-pack as well as two-pack compositions, and could be cold-cured or formulated into air-drying paints. Despite their tendency to yellow on ageing, polyurethanes had come to stay, an important property being their capacity to yield very hard paint and varnish films.

For many centuries, up to the Second World War, bituminous paints were in great demand, partly because of their availability and partly because of their low cost. Some bituminous paints were merely solutions of asphalt, one of the most important of the latter being Gilsonite of American origin. Other finishes were obtained by dissolving vegetable pitches (for example stearine) or mineral pitches (for example coal-tar) in drying oils, the products providing useful stoving enamels. The bitumens were, however, useless in the preparation of white and light-coloured paints, and eventually, but not completely, gave way to the more superior alkyds.

The production of marine finishes increased with increasing world-wide mercantile tonnage. Paints for the superstructures of ships generally had to have improved resistance to the corrosive action of salt spray, whereas paints for surfaces which were continuously immersed in sea-water were formulated for antifouling properties, to resist the growth of marine fauna. The earlier types of antifouling paints were made by incorporating toxic substances, such as the oxides and other derivatives of copper and mercury, in a quick-drying medium. These paints performed their function by gradually releasing their toxic component.

Ancillary products. No special mention has so far been made of the many other products that were evolved for use with decorative finishes or for use by the furniture, car, shipbuilding, and other industries. Usually they were formulated for a special purpose so that they could be used as part of a paint system of which the finish was the final coating. Hence there came into being primers, undercoats, groundcoats, stains, sealers, putties, and fillers. Some of these products soon became major products of the paint industry.

No less important to the paint industry was the array of metal pretreatment products which were brought into being for the car and other industries. Some of these were formulated merely to help remove rust from ferrous metal surfaces; they mostly contained phosphoric acid as an important ingredient. Some functioned in other ways, to assist the adhesion of paint to metal surfaces (both ferrous and non-ferrous).

III. METHODS OF PAINT APPLICATION

From time immemorial paint has been applied by brush, and this method of painting is likely to continue indefinitely, particularly for home use, and for a number of reasons, not least of which is cheapness.

Probably the greatest advance made during the early 1900s in the field of paint technology centred on the introduction of the spray-gun [6], which first appeared in 1907. It heralded a minor industrial revolution, and spray application was rapidly adopted by the furniture, motor, and other industries which made use of coating compositions. By 1922 the Oakland motor-car was in production in the U.S.A. with a lacquer-sprayed body. A year previously, in Britain, a Cranco cellulose spray enamel, supplied by the Frederick Crane Co. had been applied to an Austin Seven.

In order to enable British industry to compete with the U.S.A. in the expertise of spray application, Nobel Chemical Finishes Ltd. in 1926 opened a demonstration centre in the Chiswick High Road, London (Fig. 23.4). It was a building fitted out with spray-booths, air conditioning, and the most modern spraying equipment then available. Free courses were given by a team of trained spray painters. It was estimated that some 80 per cent of the paint-shop staff of British car manufacturers received their first training at the Chiswick Centre, which was closed in 1928. The spray finishing of wooden furniture was also taught.

The greatest virtue of the spray-gun resided in the speed of application; its advent also helped the introduction of nitrocellulose lacquers, which are not ideally suited for brush application because of the volatility of the solvents used. Despite some loss of paint, arising from overspray because of the huge volume of air used, the speedy spray application of cellulose lacquers resulted in a savings of over $100 per car body and enabled the car industry to cut

FIG. 23.4. The first refinish training centre, Chiswick 1926. An everyday scene in the motor-car section showing work preparatory to spray painting and, in the foreground, an operator polishing the edge of a mudguard.

down enormously on the storage space set aside for the painting process. An additional feature of spray-application was the production of a finer finish, free from brush-marks. By the end of the 1940s, it was claimed for the U.S.A. that almost half of all paint was applied by the spray-gun (mainly for industrial use, but to some extent for decorative paints) and it was estimated that some 85–95 per cent of all industrial top-coats (final finishes) were spray applied.

Other methods for applying industrial finishes followed spraying. The dip tank, often of large capacity, was introduced. Paint losses and labour costs were minimized, but careful attention had to be paid to such technical points as paint viscosity and solids content and to temperature. Also introduced was flow-coating, the paint being held in a tank fitted with a circulating system and hosed on to the articles to be painted. Flow-coating was employed extensively during the Second World War for the painting of jerricans; the alternative use of a dip tank could lead to air locks.

The early 1940s heralded another method of paint application, namely electrostatic spraying. Known as the American Ransburg electrostatic painting process, it was introduced a few years later into Britain by H. W. Peabody

(Industrial) Ltd., the concessionaires. In this method, finely atomized paint particles, carrying an electric charge (high potential) are directed at the articles to be coated, which are on a conveyor at earth potential. A reduction in the volume of paint and of compressed air needed, as well as lower labour costs, were claimed for the process.

REFERENCES

[1] MATTIELLO, J. J. *Protective and decorative coatings*, Vol. III, p. 271. Chapman and Hall, London (1943).
[2] LINSTEAD, R. P. *Journal of the Chemical Society*, 1016 (1934).
[3] SCHULTE, E. In W. von Fischer (ed.), *Paint and varnish technology*, p. 7. Reinhold, New York (1948).
[4] PARKES, A. British Patent No. 2359 (1855).
[5] READER, W. J. *Imperial Chemical Industries—a history*, Vol. I. Oxford University Press, London (1970).
[6] O'REILLY, J. T. *Transactions of the Institute of Metal Finishing*, **31**, 314, (1954.)

BIBLIOGRAPHY

CHATFIELD, H. W. *Paint trade manual of raw materials and plant*. Croydon (1956).
DRUMMOND, A. A. *Introduction to paint technology*. Oil and Colour Chemists' Association, London (1967).
ELLIS, C. *The chemistry of synthetic resins*. Rheinhold, New York (1935).
FISCHER, W. VON *Paint and varnish technology*. Rheinhold, New York (1948).
HEATON, N. *Outlines of paint technology*. Griffin, London (1947).
Industrial nitrocellulose. Imperial Chemical Industries Ltd. (Nobel Division), Glasgow (1952).
Journal of the Oil and Colour Chemists' Association, **1–32** (1918–49).
KHRUMBAAR, W. *The chemistry of synthetic surface coating*. Rheinhold, New York (1937).
MATTIELLO, J. J. *Protective and decorative coatings*. United States Government Printing Office, Washington (1945).
MORGANS, W. M. *Outlines of paint technology*. Griffin, London (1969).
Official Digest of the Federation of Paint and Varnish Clubs. Philadelphia (1920-49).
Paint Manufacture, **1–19** (1931–49).
Paint, Oil and Colour Journal, **116** (1949).
Paint Technology, **1–14** (1936–49).
READER, W. J. *Imperial Chemical Industries—a history*. Two Vols. Oxford University Press, London (1970 and 1975).
Review of current literature relating to the paint, colour, varnish, and allied industries. Paint Research Station (1928–49).

24

PAPER

E. HAYLOCK

EARLIER volumes of this work have taken the history of paper-making in detail only up to 1800 (Vol. III, pp. 411–16), and for the sake of completeness alone it would be desirable to say something of nineteenth-century developments, which have been only very briefly alluded to (Vol. V, p. 73). In fact, this is an essential prelude to consideration of the history of the industry in the twentieth century, for it was in the nineteenth century that modern technological processes, in the manufacture of both raw materials and paper and board, were broadly established. Since then the equipment used has become much more sophisticated and much larger, but the basic principles are unchanged.

I. RAW MATERIALS

From late in the fifteenth century when the first paper mill in England was established, until the middle of the nineteenth, the principal raw material for making paper—by hand—was rags. With steadily increasing demand for paper products, pressure increased not only for alternative raw materials, but for quicker and more efficient methods of production. Indigenous sources of raw material supply were generally quite inadequate, and by the 1860s Britain, for example, was importing rags from a large number of countries. The process for pulp production is illustrated in Fig. 24.1. To some extent the problem was alleviated by developments in the use of esparto grass, found in North Africa and Spain. Indeed, esparto grass, converted into pulp, was still being used, especially by fine paper mills, as an ingredient of their furnish until well into the 1950s (Fig. 24.2). At the outset, esparto was found to satisfy many technical requirements; it was strong and quickly amenable to cleaning and bleaching. Its fibre length was comparatively uniform and it was used with complete success for the manufacture of high-quality printing paper, which absorbed ink quickly and was compatible with the type and blocks of that time. Some indication of the total use of this new material can be seen from the fact that in the late 1880s imports into Britain

FIG. 24.1. Pulp manufacture
from rags, around 1900
(Wolvercote, Oxford).
(a). First the rags are sorted,
and extraneous objects
such as buttons removed.

(b). The next step is to chop
the rags into small pieces,
which assists the digestion or
cooking process.

(c). When chopped into small
pieces the rags are fed into
boilers, often rotating. In
these boilers, in the presence
of pressurized steam and
chemicals, the rags are
'cooked', subsequently emerging as pulp.

Fig. 24.2. Pulp manufacture from Esparto grass. After a preliminary dusting treatment, which takes out grit and dirt, the grass is fed into digesters where, like rags, it is cooked in the presence of chemicals and steam under pressure, emerging from the digesters as pulp. The picture shows the top floor of a typical Esparto digestion 'house', c. 1950. The tops of three digesters can be seen securely fastened, thus containing an Esparto 'cook'.

reached nearly 250000 tons, compared with about 40000 tons of rags. By 1937, esparto imports had reached more than 360000 tons.

In terms of raw materials, there were two major developments in the nineteenth century. One was the realization that wood could be used as a raw material, leading to the development of the manufacture of wood pulp. With the advent of wood pulp altogether different methods had to be employed to prepare the 'stock' for the paper machine (Fig. 24.3). The other development, towards the end of the century, was the development of bleaching processes which enabled a much whiter pulp to be produced. Studies of how to use the vast timber resources of Scandinavia, Finland, and North America, in particular, were intensified after the development of mechanical wood pulp in Germany in the 1840s. In this process, logs were ground into fibre by a grinder (a large rotating grindstone) in the presence of water. The end-product was, however, of poor quality, especially in terms of strength and durability.

In 1854, Watt and Burgess took out a patent to manufacture wood pulp by direct digestion with caustic soda in the presence of steam under pressure. This soda process produced a pulp of much better quality and was widely used well into the 1860s. The equipment used can be likened to a huge domestic pressure cooker. This principle—the cooking of wood chips under pressure with chemicals—remains the chief system for the production of chemical pulp worldwide. A further important development was the discovery of the sulphite process, which used acidic liquor to produce a clean, fairly white pulp. The inventor was B. C. Tilghmann, of Philadelphia, who in 1866 set up a plant to produce sulphite pulp, but because of the need for acid-resistant equipment, which was not then available, this venture was not a success. His work, however, was carried on by C. D. Ekman who overcame the difficulties and set up a sulphite pulp mill in Sweden in 1872. Sulphite pulp remains one of the principal raw materials used in the paper industry.

The other principal chemical pulp still widely used in the manufacture of paper is sulphate, or kraft, pulp, invented in the early 1880s by C. F. Dahl. Although the method used is similar to the soda process, different chemicals are used to make the cooking liquor, and the resultant pulp is very different. Papers made from kraft pulp, bleached and unbleached, are used for a variety of packaging papers.

Almost concurrently with the development of wood pulps, methods of bleaching emerged, but there are no records to identify any one man with its invention, and it seems likely that various methods and processes were

FIG. 24.3. Wood pulp preparation for paper manufacture. The two principal machines used for this purpose are the breaker and beater. (a). Sheets of partly prepared wood pulp are seen being fed into a breaker which, with water, reconverts the sheets into pulp.

(b). The beater, not dissimilar in appearance, is used to produce certain changes in the fibres, so that on the paper machine wire they intermesh together, producing the characteristics of paper.

evolved by both pulp and paper manufacturers. Probably the simplest and oldest method is single-stage hypochlorite bleaching, but this has been little used for many years, except for non-wood fibres such as esparto and straw. Bleaching liquor is made by passing chlorine gas into lime in water, forming calcium hypochlorite.

The most common process in use during the first half of this century was bleaching in stages, in which the pulp is treated with a limited amount of chlorine gas, usually in a tower. Chlorine combines rapidly with lignin to form chlorolignin and hydrochloric acid, but the acid must at once be washed out before it attacks the cellulose. Chlorolignin left in the pulp does not dissolve in water, but is very soluble in alkali; caustic soda is consequently added. The pulp is then again filtered and washed. The pulp then contains very little lignin and other impurities, and it can readily be bleached to a good white colour with a small quantity of bleach liquor, as in the single-stage method.

This stage bleaching, described here in its simplest form, can be multiplied or repeated in sequence, bleaching sequences being denoted by initial letters. For example, CEHEDA denotes chlorination, extraction, hypochlorite, extraction, chlorine dioxide, acidification. The last stage is usually achieved by sulphur dioxide, which removes the last traces of chlorine and brightens the pulp. By these methods, sulphite pulps and kraft pulps can be bleached to a good white colour.

Concurrently with these developments in bleaching, the original mechanical (or groundwood) pulping process was improved in Europe and North America. By the end of the 1950s mechanical pulp was the principal ingredient in the pulp used for the manufacture of newsprint.

By 1950 methods of pulp manufacture could be classified into two main groups, mechanical and chemical, but the development of both systems produced pulps—and continue to do so—in batches. However, a new concept was introduced in the 1930s by a Swedish company, Kamyr AB, who developed a continuous process for chemical pulps. The problem of continuous digestion of fibrous materials had attracted inventors for years, but although a number of patents were taken out, few had any practical consequences.

Kamyr started working on the problem in 1938 and it took six years to produce normal kraft pulp in a continuous pilot plant, and another two years to ensure regular operation of the plant. During this long period of development, the company collaborated with the Swedish Royal Forest Industries, who owned the Karlsborg kraft mill where the pilot plant was installed. The mill provided wood chips, cooking liquor, steam, power, etc., as well as operators. Pulp and used liquor were returned to the mill in exchange. The original pilot plant had to be totally rebuilt twice before the company finally reached a point at which it was possible to embark on the construction of full-size plants.

An indication of the importance of this continuous pulping system can be seen from the relevant figures. At that time a batch digester produced about 10 tons per 24 hours per 1000 ft³ of wood. Under the same conditions, a continuous digester made 15–20 tons per 24 hours per 1000 ft³ of wood.

Although the development of mechanical, sulphate, and sulphite pulps are the base on which the modern paper industry is founded, there are other fibrous materials that are used for special purposes, or in areas where the material is abundant. Linen, straw, jute, hemp, manilla, and sugar-cane bagasse are some examples. Finally, there is waste paper, which has been used in the principal paper-making countries as an important source of raw material for a great many years. Indeed, the first recorded re-use of printed paper was in Denmark as long ago as 1695. In Britain, Matthias Koops was granted a patent in 1800 covering the re-use of printed papers, and since the middle of the nineteenth century waste paper has been widely used in commercial quantities in the production of paper and board. There are no reliable statistics of waste paper consumption in Britain before 1939, when the total amount used was 665600 tons against a total paper and board production in that year of 2631000 tons. By 1950, waste paper consumption had risen to 887900 tons, although paper and board output in that year had fallen slightly to 2617000 tons.

The principle involved in re-pulping waste paper is simply to reduce the paper to its individual fibres, and all the machinery used over the years has been designed first to clean the material and then to process the fibre into a condition in which it can be used on the paper machine. Early waste paper process systems used a *Kollergang* to mascerate the waste paper into pulp; this machine consisted of a pan in which two stones rolls revolved, and was very similar to the equipment used for milling flour. Since then a wide variety of machinery has been developed to treat clean waste grades, bitumenized papers, newsprint, domestic waste, and other papers and boards. By the 1950s there were twenty or so different items of equipment that could be used in various combinations to produce waste paper pulp for various end uses.

Although the importance of waste paper is that it is an indigenous raw material, the chief problem for more than fifty years has been in achieving a rate of collection corresponding with demand. In most countries using this material, there is a shortage when demand for paper and board is high, and a surplus when mill requirements fall. Another problem is that each time paper and board is re-pulped, it deteriorates in quality as a raw material, mainly because of the continual shortening of the fibre, and in this sense it is an asset of declining value.

In contrast to the great changes which occurred in the development of raw materials in the second half of the nineteenth century, there were no fundamental changes in the raw materials used, or in their methods of manufacture in the first half of the twentieth. Nevertheless, advances in the chemical industry and the application of its various products to paper-making were very important in the preparation of pulp from the raw materials available. Considerable strides were made by the chemical industry in the development of additives for loading and coating purposes; the more efficient treatment of water; and improved methods of treating waste water and fibre (effluent). Resin and gelatine, with alum added, continued in use as the principal sizing agents; machine sizing became common.

II. PAPER-MAKING MACHINES

The vast increase in the manufacture of paper and board from the raw materials described above took place almost entirely on paper machines, and the number of vats used declined compared with the nineteenth century: in Britain to 104 in 1900 from 130 in 1861. Again using Britain as an example, total production of paper and board went up from 96 000 tons in 1861 to 648 000 tons in 1900. By 1939 total output had gone up to the 2 631 000 tons already quoted above, but because of the Second World War, and its after-effects, total output in 1950 had still not reached the 1939 level.

The invention, in 1798, of the first continuous paper-making machine, is generally attributed to a Frenchman, Louis-Nicolas Robert, but it is fairly certain that his somewhat crude unit had a comparatively short life. It was not until 1816 that a continuous paper-making machine was operating in France, but this machine also had a short life. The first commercially successful paper-making machine was built in England by Bryan Donkin. The first English patent was obtained by John Gamble in 1801. His invention was essentially the same as that of Robert in France. Gamble became associated with the Fourdrinier brothers, and through this connection the Fourdriniers became interested in the paper-making machine. Subsequently, Henry and Sealy Fourdrinier and John Gamble were in 1807 granted a joint patent, which embodied numerous improvements made by Donkin and others. This machine, as far as the wet-end and press section are concerned, is virtually the same as the Fourdrinier machine of today.

The Fourdrinier family—whose name is synonymous with paper making throughout the world—were Huguenots who left France after the revocation of the Edict of Nantes, arriving in England via Holland. The two brothers were associated in a stationery business in London, but it was Henry who had the major interest in paper machines, and who spent a great deal of money in their development. Indeed, the Fourdriniers lost about £60000 during he first few years. Although Henry Fourdrinier did not personally invent the paper machine, there is little doubt that his determination was responsible for the development of the commercially successful continuous paper-making machine as it is known today. In this, he was fortunate in being assisted by Donkin, who was in charge of the Fourdrinier works at Dartford. Donkin was a brilliant engineer, subsequently (1838) becoming a Fellow of the Royal Society. He was, for example, associated with R. M. Bacon, printer of the *Norwich Mercury*, in developing printing machinery (Vol. V, p. 698). He was also connected with many other engineering projects. Over the years, particularly in the last quarter of the twentieth century, the Fourdrinier machine was refined to become a large, sophisticated piece of engineering.

In spite of the refinements that took place almost from the beginning— and continue to be made—the purpose and operation of the paper machine is relatively simple. It is designed to receive a mixture of prepared raw material (wood pulp, waste paper, or agricultural fibre, often mixed) with water in the proportion of 99 parts of water to one of fibre; to intermesh the fibres to form a flat sheet; and to remove the water in various stages so that the intermeshed fibres are left dry and compact in a continuous web.

At the wet end of the machine, the diluted pulp is projected on to an endless travelling belt of wire mesh under controlled conditions of speed and uniformity. As the belt moves forward it intermeshes the fibres and at the same time allows much of the water to drain through the wire mesh. Towards the end of the wet end, the wire passes over suction boxes which take out more water. At the end of the wire mesh the wet mass of pulp (the web) is lifted off by way of a roll, known as the couch, on to endless travelling felts which carry it through pressure rolls, known as the press part. This section simply consolidates the web and presses out more water so that when the web leaves the press section it consists of about 33 per cent pulp, possibly with chemical additives, and about 66 per cent water. In the final stage, the web is carried

FIG. 24.5. A much more sophisticated machine than the one shown in Fig. 24.4, installed c. 1900. The machine wire, or wet end, at the bottom of the picture is followed by the press section and the dry end, covered by a hood. Note the greatly increased number of heated rotating cylinders at the dry end.

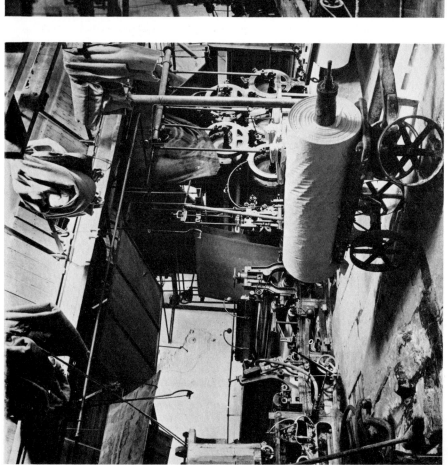

FIG. 24.4. An early paper machine built in 1856. Felts used on the machine can be seen hanging up to dry on the rafters of the machine house.

FIG. 24.6. Among the first paper machines to make newsprint in western Canada were these two units at the Powell River Division of MacMillan Bloedel, seen here with their crews, 1914. The machines were still in operation in the early 1970s.

to the dry end, where the remaining water is removed by a number of rotating heated metal cylinders, and wound into a reel. At this point the moisture content is 3 to 6 per cent.

From the base established by the Fourdriniers and Bryan Donkin, development of the continuous paper-making machine went ahead relatively fast. By the 1880s the machine had assumed a well-defined pattern, which was to last, with few changes, well into the 1950s. Of course, machines made by different manufacturers differed in detail, but the design was, for all practical purposes, the same, whether the machine was built in Britain, the U.S.A., or Europe. Wider and longer machines were made, and speeds of operation made much faster, as paper makers began to specialize in such products as newsprint, writing papers, or wrapping papers. By this time, they were able to acquire raw materials which were of good quality and readily available, and this led to the vast increase in paper-making in the latter part of the nineteenth century.

Another type of machine in use towards the end of the last century was the cylinder-mould machine, originally developed by John Dickinson in 1812. This type of machine, virtually unchanged in principle, is still widely used for the manufacture of boards. In this system, there was a vat, in which a cylinder-mould revolved. The pulp, diluted with water, was pumped into the vat with the fibres adhering to the wire mesh covering the cylinder-mould, building up to form a web of paper. A couch roll pressed further water out of the paper, and as soon as the web came into contact with a wet felt under the couch roll it left the wire cover of the mould and was carried on the felt through press rolls and then to the dryers, as on a Fourdrinier machine. Because the diameter, and consequently the wire surface, of the making-cylinder was very limited, heavy sheets could not be made on this machine. However, if the paper from two or more cylinders was united on one wet felt, it would, after having passed through the presses and dryers together, form a solid web.

Paper having two sides of different colour or quality could be made by assigning separate vats to each cylinder and supplying them with different pulps. Up to six cylinders are combined in modern machines, for the manufacture of heavy boards. The two earliest multi-cylinder board machines erected in England at the end of the nineteenth century were both imported from America, and were made by the Black-Clawson Company.

James Harper, paper-maker, of New Haven, Connecticut, patented and constructed a machine which was a combination of the Fourdrinier and the cylinder-mould machine. Paper was made in the ordinary way on the Four-drinier wire, which carried it to the couch-rolls. The wet felt was passed round a top couch-roll and in passing it picked up the sheet from the wire as it passed round the bottom couch-roll. This latter roll was constructed in the same way as the making-cylinder of a cylinder-mould machine. Although this first 'lick-up' machine was not successful, the idea persisted and today machines making thin papers use a felt which passes round the top couch-roll and removes the web from the wire in the process.

Towards the end of the nineteenth century the Fourdrinier machine and the cylinder-mould machine had become firmly established in design in most countries. In fact, many of the machines made about this time were still working up to 1950, although most had been improved and modernized, not so much in design, as by the much improved engineering skill available to paper-makers.

III. FASTER MACHINERY

In the first half of this century machines were not only faster, but were in many instances very much wider than the 100-inch which was usual at the end of the nineteenth century. The tendency was also for machines to be designed and made for specific purposes and papers, rather than for general use, which had been the practice. One example was the machine which produced wet crepe tissue, as described by the Lysle patent. The machine of 1890 was a lightly framed, slow-moving combination of rolls and dryers mounted on sleeve bearings. It was driven at a speed of 100–200 ft/min through a series of belts and pulleys, either by a water wheel or steam engine. There was no suction equipment to help de-water the sheet and the creping operation was, at best, very crude. Very little crepe was retained in the finished product by virtue of the position of the creping 'doctor' on top of the press; the vertical draw from the creping doctor; and the non-driven paper rolls used to carry the sheet into the first dryer. Drying cylinders were limited in steam pressure owing to inferior shell and head material and were not bored to obtain uniform heat transfer through the shell. Yankee dryers, or MG cylinders as they became known in Europe, had not yet been applied to the production of creped tissues. However, as early as 1880, the Yankee had been used to produce highly glazed paper. These machines, running at perhaps 200 ft/min maximum speed, were tremendous improvements over the Robert machine of 1800, which was turned manually by means of a hand crank, and achieved a speed of about 15 ft/min.

Electric motors were introduced to the tissue industry in the early 1900s and it was hoped that this more efficient source of motive power would increase machine speeds. The high friction drag of sleeve bearings, however, still imposed severe width and speed limitations and absorbed a great deal of power, so that wire widths generally remained in the 100–132-inch range, and maximum operating speeds were usually limited to about 700 ft/min. This was a severe obstacle to further progress and the industry was unable to solve the problem until 1922. In that year, the efforts of a very intensive research programme bore fruit in the form of the anti-friction bearing. The first paper machine with anti-friction bearings on its rotating equipment started up in 1922 and was an immediate success; it was a tissue machine. The power required to drive the machine, since the drag of the anti-friction bearings was only a fraction of that of sleeve bearings, was greatly reduced.

Very soon other machines were built using anti-friction bearings or existing ones were converted. As a result, both wire width and speeds began to go up. During the period between 1922 and 1942, operating speeds for tissue machines increased from 700 to 1500 ft/min, while wire widths increased from 132 inches to 190 inches. After the Second World War, the tissue industry resumed its quest for still higher speeds and increased production. Research had not been idle during the war years, and the results began to show shortly after the war. Paper companies came out with new concepts in forming equipment. Machinery builders improved casting techniques and were able to build rolls of higher nip-loading capacity and drying cylinders capable of withstanding higher steam pressures.

New, or improved, materials of construction were also made available along with new fabricating techniques. Machine speeds quickly climbed to around 1800–1900 ft/per min, where—with the exception of two companies holding exclusive patent rights to formation equipment which enabled them to attain operating speeds approaching 2500 ft/min—sheet formation became a limitation for more than a decade.

Evolution from an open headbox of steadily increasing height as speeds increased, to the development of an air-cushioned inlet around 1950, allowed minor quality improvements in basis-weight, profile, and uniformity of fibre distribution. However, the acute water-removal problems on the Fourdrinier wire caused by the large volumes of water, which were vital to the formation of a high-quality tissue sheet, continued to be a limiting factor. In the decade after 1950, these bottlenecks in formation and water handling, as well as in drying, were eliminated.

BIBLIOGRAPHY

CLAPPERTON, R. H. *The paper machine, its invention, evolution and development*. Pergamon Press, London (1967).

HAYLOCK, E. W. (ed.). *Paper—Its making, merchanting and usage* (3rd end.). Longman, London, for the National Assocation of Paper Merchants (1974).

HUNTER, D. *Paper-making through eighteen centuries* (1930).

SHORTER, A. H. *Paper making in the British Isles. An historic and geographical study*. David and Charles, Newton Abbott (1971).

WYATT, J. W. The art of making paper by machine. *Proceedings of the Institution of Civil Engineers*, **79**, 251 (1884–5).

25

CERAMICS

W. F. FORD

SINCE this chapter is concerned with primary production, the applications of ceramics are not completely covered, being necessarily restricted to consideration of the effects of industrial demand. In everyday life, however, the use of ceramics is widespread, among the more obvious being pottery, sanitary ware, building bricks for houses, and decorative tiles for walls and floors. In the electrical field the large ceramic insulators supporting the high-voltage power lines from pylons probably make most impact but there are other, less obvious, uses. The domestic electric fire needs a refractory insulating former on which to wind the resistance wire; today, silica glass meets this need and, having the great advantage of a very low coefficient of expansion, seldom fails by cracking due to rapid heating and cooling. All car ignition systems are dependent on ceramic insulators for the sparking-plugs and it is of interest that the alumina now used for this purpose was developed just in time for the Rolls-Royce Merlin engines that powered British fighters in the Battle of Britain. Radio and television sets have many ceramic components in the form of resistors, insulators, and capacitor dielectrics which must exhibit low losses at the very high frequencies used. The printed circuit, the heart of modern TV and radio sets, depends on ceramic substrates. Many gramophone pick-ups incorporate barium titanate crystals which convert mechanical movement into electrical voltage. The same transducer principle is also used for the modern gas-igniter. The most outstanding development in the Second World War was the pioneering work on ceramic magnets by the Dutch. In the form of 'soft' magnets they appear in 'ferrite rod aerials' in modern radio sets. As 'hard' magnets they function as tapes for audio-frequency recording, for the fields of small electric motors for kitchen mixers, toys, etc., and as magnetic catches.

Ceramics, one of the oldest technologies (Vol. I, Ch. 15; Vol. II, Ch. 8; Vol. IV, Ch. 11; Vol. V, Ch. 27), can now be said to have come into its own as a coherent body of technological and scientific knowledge. That coherence was being developed over the first half of this century, partly by the establishment of the institutions necessary for it. In the English-speaking world the period saw the birth of the American Ceramic Society in 1899 and the (now) British Ceramic Society in 1901, their publications serving to disseminate information on research and development in ceramics. From the beginning, such information emanated largely from industry but was amplified in America by the outputs of the geological institutes, the Bureau of Standards, and the university departments (that of Ohio State founded 1894), and in Britain by J. W. Mellor's school in Stoke-on-Trent. The end of the First World War saw the establishment in Britain of the British Refractories Research Association.

In the inter-war years the number of insitutes grew throughout the world; in the U.S.A., for instance, there were 16 ceramics schools in existence in 1939. In Britain, although there was a small amount of postgraduate activity in universities, the chief contribution to the ceramic literature came from industry and the research associations, amalgamation in the latter field giving rise to the British Ceramic Research Association in 1948. The Institute of Ceramics, the only qualifying body for the industry, was founded in 1955.

I. CERAMIC SCIENCE

Modern techniques for the study of materials were not, of course, available in 1900. J. W. Mellor, who may be said to have laid the foundations of ceramic science, had the petrological microscope, but the rest of his apparatus had to be home-made. Mellor carried out work on many aspects of ceramics and was particularly interested in the effect of heat on clay, which he began to study in 1906. His theory that the clay mineral kaolinite breaks down at 500 °C to give a mixture of alumina and silica received some confirmation when the then new technique of X-ray diffraction was applied to the problem in 1922, but subsequent work modified the conclusions. It is of interest that quartz, the naturally occurring form of silica was one of the first minerals subjected to X-ray examination by W. H. Bragg in 1914, one year after C. N. Fenner had formulated the basic nature of the polymorphism of silica. The optical microscope and X-ray diffraction are still two vital tools in ceramic science, but it is to the latter that we owe the development of our understanding of silicate chemistry, culminating at the end of our period in a definitive classification of clay mineral and allied structures. The electron microscope was first used as late as 1950 to provide the proof of clay particle shape that had been deduced from the structure and properties of clays (Fig. 25.1).

Many workers carried out fundamental research on clay–water systems, which are transformed from plastic masses to 'slips' as the water content

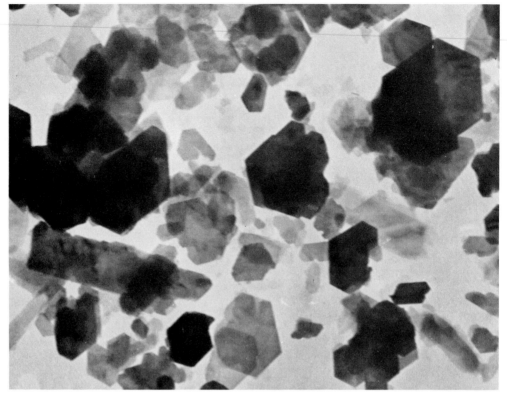

FIG. 25.1. An electron micrograph of china clay. ($\times$ 25 000)

rises. Study of the former state was hindered by the lack of an acceptable definition of plasticity, but progress was made by considering the large surface areas exhibited by clay particles and the concept of 'thick' water-films surrounding the clay particles. The concepts of the 'electrical double-layer' and the 'zeta-potential' were introduced in an attempt to explain the behaviour of colloidal suspensions. The mechanisms involved in the drying of clay were established.

The science of non-clay ceramics, which grew rapidly in importance during the period, was not neglected, but it is more convenient to deal with such aspects under the appropriate headings.

Although kinetics, the study of the rate of approach to equilibrium, had received little scientific treatment in the ceramics field, it was recognized that —particularly in silicates—equilibrium was not often established during the firing process and Mellor coined the term 'the chemistry of arrested reactions'. This approach can be said to have produced neglect of phase equilibrium principles, the practical importance of which seems first to have been realized by H. W. B. Roozeboom in 1899. Towards the end of the period, however, this neglect was being remedied, and important technological consequences arose from the application of phase equilibrium diagrams which the American geophysical school started to establish from the middle of the period.

II. CERAMIC WHITEWARES

The traditional recipe for bone china is 50 per cent bone ash, 25 per cent china clay, and 25 per cent Cornish stone, the British substitute for the feldspathic flux. Production of translucent ware of this composition continued and expanded during the period and was facilitated by the use of not more than 2 per cent of highly plastic clay, such as bentonite, to permit fabrication by machine. 'Triaxial' mixtures of clay, feldspar, and silica continued to be used for earthenware, vitreous china, and hard (continental) porcelain; an auxiliary flux was added in the case of American hotel china. The density and translucency of the first three bodies, varying little in composition, are determined by firing temperature, the attractive slightly bluish hue of porcelain being obtained by using reducing conditions in firing. In Britain the cessation of supplies of German hard porcelain for laboratory ware on the outbreak of the First World War led to the formation of the Worcester Royal Porcelain Co., which, in the early 1930s, began the production of oven-to-table ware and hotel ware in hard porcelain. Ceramic tiles continued to be generally based on triaxial mixtures, although in the U.S.A. tiles of higher resistance to crazing were made from pyrophyllite and talc. Natural, self-vitrifying, stoneware clays were still used for domestic pottery.

In Britain there was no departure from the wet method of mixing the three body constituents (which avoids a dust hazard), but control of slip compositon by weighing large volumes of it grew progressively and pumping of the individual slips to yield the correct proportions for the body began. The dry-mix process, used for instance in the U.S.A., led to the use of air-flotation and drying to produce standard clays; only after the middle of the century were such standard china- and ball-clays introduced in Britain. The control of calcined flint, used in place of ground silica, was improved by research which showed it to consist of cryptocrystalline quartz, water, and organic matter. The substitution of other feldspathic minerals for Cornish stone, the fluorine in which had to be removed, was effected after 1950.

In the 1940s remarkable advances in mechanizing the shaping of clay ware were facilitated by the introduction of the de-airing pug-mill to increase the

FIG. 25.2. The social fruits of technological progress; Longton in the Potteries, above 1910, below 1970.

FIG. 25.3. An electrically heated tunnel-oven for firing pottery.

workability of plastic masses. It became possible to produce earthenware plates, cups, and saucers at a rate exceeding 1000 an hour per machine operative. Slip-casting was mechanized, having been more adequately controlled by the earlier adoption of alkaline casting-slip.

Slip-cast ware was normally dried by natural convection but the much higher rates of machine production inevitably led to demands for accelerated drying. An important feature was that ware was formed on plaster moulds and careful preliminary drying to avoid cracking was essential before the ware was released from the moulds. Subsequently, the thin sections of most ware enabled rapid drying to be used, and it was found possible to employ radiant heating or fast hot-air streams.

The first half of this century brought changes in the method and technique of firing which greatly exceed any previous developments in the pottery industry. By 1950 the intermittent bottle-oven (Vol. V, p. 667), fired very inefficiently with coal and thus greatly contributing to atmospheric pollution (Fig. 25.2), had been virtually superseded by the much more efficient tunnel kiln, fired with gas or electricity (Fig. 25.3). Electricity, although expensive, made it possible to build twin-channel kilns for biscuit and glost (glazing) fires; to dispense with the saggars (fire-clay boxes) necessary to protect the ware against flame; and to reduce the firing time to cope with the increased rate of fabrication. To prevent sticking of the ware to the 'kiln furniture' on which it rested during firing, it had always been the custom in Britain to use ground flint as a bedding material. In 1928 practical tests were begun on the substitution of alumina for flint, so as to eliminate the risk of silicosis (pneumoconiosis) from exposure to silica dust. The change was confirmed in the Pottery (Health) Special Regulations, 1947.

In 1949, the year in which British regulations forbade the use of raw lead in glaze compositions, only one case of lead poisoning was reported, a situation in sharp contrast with 400 cases a year at the end of the nineteenth century. The risk to the worker applying glaze to the pottery body had been progressively reduced by the development of leadless glazes and lead bisilicate frits of low solubility. However, even today some countries permit the use of raw lead compounds under safeguards. Control of glazes developed from considerations of economy in glaze consumption and the prevention of glaze faults; it was obtained by means of grain-size and viscosity measurements. A great deal of work was carried out on the problem of the 'crazing' of glazes—that is, the appearance of surface cracks with time—and it was eventually realized that the relative contractions of body and glaze initially determined the residual stress in the glaze. When the body contracted more than the glaze, the latter was in compression and would not crack unless the stress was excessive, when 'peeling' resulted. An important discovery was that subsequent absorption of atmospheric moisture by the body could take place with time, the resulting 'moisture-expansion' producing a tensile stress in the glaze, relieved by crazing.

By 1950 detergents and dishwashing machines were showing their effects on on-glaze decoration and research work to improve the resistance of ceramic colours to alkali attack and abrasion began. Owing to a shortage of skilled painters for the larger scale production of under-glaze decorated bodies, ceramic transfers printed from lithographic plates were introduced.

III. HEAVY CLAY WARES

Heavy clay wares, the source of building bricks, roofing tiles, and sanitary pipes, have always (in Britain and elsewhere) been beset by the problem of widely fluctuating demand, partly brought about by the intervention of two world wars. After 1918 the output of building bricks rose rapidly, to satisfy housing demands, to 8000m. in 1938. From 1945 to 1949 output recovered only partially from 1200m. to 5200m. bricks. In the 1940s the industry faced progressively increasing competition from non-clay products and it was necessary to pay particular attention to reducing labour and fuel costs.

Building materials are made from common clays that would be unsuitable for fine ceramics and are so complex as to make them difficult subjects for scientific study. The solution of practical problems and the testing of products therefore tended to predominate in the earlier decades of the century. In Europe, many of the necessary tests had achieved the status of international standards as early as 1886, but systematic work on testing in Britain and the U.S.A. did not begin until the 1920s, aided in Britain by the establishment of the Building Research Station in 1922.

In spite of the large tonnages of raw material required, the winning of clay by hand had not been entirely eliminated by 1950, partly because of the necessary stripping and blending of separate strata when there were large variations in composition. Mechanical winning dates back to 1904 but was dependent for its full application on the great impetus given by war to the development of the internal combustion engine. It became possible to choose the machine—bucket excavator, dragline or shale-planer—to suit the physical characteristics of the clays. Modern transport facilities solved the problems of transporting clay over broken ground to the brickworks.

Machinery for crushing and grinding clays (Vol. V, p. 670), a necessary preliminary to the production of mouldable brick batches, was introduced in the nineteenth century; the period saw many improvements in the design of such machines and the adoption of electric power for driving them. The de-airing pug-mill, first introduced in the U.S.A. in 1926, was widely adopted for the extrusion of sanitary pipes but this vacuum technique was not applied to pug-mills used for the large-scale production of bricks by the wire-cut and stiff-plastic processes. The semi-dry process, in which high-pressure moulding offsets underdeveloped clay plasticity, continued as one of the three major brick-making processes, in all of which the emphasis was on designing machines to give ever greater rates of output.

The tunnel-dryer (Vol. V, p. 667) was not an invention of this century, but was progressively adopted and improved in its capacity to handle larger outputs, and attention was paid to the control of temperature and humidity and the effective use of waste heat. There was an increasing tendency to derive such waste heat from the firing operation, the efficiency of which was improved by using continuous kilns. Much effort was put into re-designing the original Hoffmann kiln, top-fired with coal; many large kilns with 80 chambers, holding 33 000 bricks, were built. Mechanical stoking was developed in Europe, and in the U.S.A. it was found economic to use oil and natural gas instead of coal and to fire heavy clay goods in tunnel kilns. Progress was considerable and resulted in large reductions in fuel consumption, diminution in smoke emission, and smaller variations in the quality of the product, owing to extensive use of instrumentation. Mechanical handling of the goods later made an important contribution to productivity, fork-lift trucks being brought into use after 1945.

The use of building bricks differs from that of pottery in that the former are built into structures, the properties of which are not entirely dependent on the properties of the individual bricks, which themselves have adequate compressive strength. Interest in the mechanical properties of 'load-bearing brickwork' always existed during the period, but it was not until the late 1950s that intensive study of the subject began with a view to determining the most economical use of brickwork structures.

Most countries now have standards for testing the many relevant properties of bricks. It is possible here to refer only to a few problems to which research contributed in the period under review. It was found, for instance, from X-ray studies that hydration of lime (causing 'lime-blowing') can be avoided by choosing a firing temperature high enough to ensure combination by chemical reaction but low enough to avoid excessive liquid formation. The control of fired colour was advanced by determining the effects of impurities, and particularly the role of iron and atmospheric conditions. A most important aspect is the existence in bricks of pore systems, through which soluble salts may be carried and deposited at the surface to produce unsightly 'efflorescence', or in which water may be frozen in winter, with resultant damage from its increase in volume. A vast amount of work on these mechanisms, and others affecting durability, was carried out from the beginning of the century, but complete solutions had not been found by 1950.

IV. REFRACTORIES

Few countries have sufficient indigenous resources of raw materials to cover the whole range of refractories and importation to the refractory-producing centres is therefore common. Melting and casting of the raw material to yield 'electro-cast' low-porosity products had begun in the U.S.A. and France towards the middle of the century, but the bulk of production was still by the traditional processes. Since porosity in the brick is an important factor in reducing durability in service, control of grain-size distribution to give optimum packing of particles—first enunciated by A. E. R. Westman and H. R. Hugill in 1930—was an essential part of research and development to reduce porosity. Except for the fabrication of complex shapes, the period saw the progressive adoption of the semi-dry process, using presses of ever-increasing capacity, to facilitate the reduction of porosity and to conform to specifications of much greater accuracy, so far as dimensions are concerned, than in heavy clay wares. Flexibility of production to some extent necessitated the continued use of intermittent kilns, the efficiency of which was improved by the use of 'hot-face' insulation, but fuel economy dictated the progressive introduction of continuous kilns. It is probable that up to 1950 the maximum firing temperatures used did not exceed 1500 °C, and it is important to note that great improvements—particularly in basic refractories—took place when much higher firing temperatures were adopted in the following decade.

Silica has been used as a refractory for over 250 years. The purer form known as ganister was used in England from 1859 until mining the rock became uneconomic; brick production was then based on quartzite rocks, the normal source in other countries. Although ganister had the advantages of responding better to wet grinding, this did not eliminate silicosis in the brick-works, but the Workmen's Compensation (Silicosis) Act of 1919 did much to reduce the risk by requiring rigid operational standards.

The first specifications for silica bricks (for the carbonizing industry) appeared in Germany as early as 1906, a notable achievement in view of the difficulties of testing at high temperatures in those days. Systematic work was carried out in England in the 1920s on the control of firing, which generally requires elimination of the original quartz so as to give the brick dimensional stability in service. Although the lime–alumina–silica phase diagram had been published in 1926, it was not until the 1940s that the effect of alumina content on the properties of lime-bonded silica bricks was appreciated and it was realized that a 10 °C increase in maximum working temperature could be obtained for each 0·1 per cent fall in alumina content. A U.S. patent appearing at this time introduced the 'super-duty' brick and defined it as one containing not more than 0·5 per cent of alumina, alkalis, and titania; subsequent experience and research showed that the latter oxide need not be so limited. Such considerations led to remarkable improvements in silica bricks, and the use of super-duty types persisted in basic open-hearth furnace roofs until the use of oxygen in melting raised the working temperature above the unfortunately low melting-point of silica (1725 °C).

In 1900 indigeneous firebricks, containing 30–44 per cent alumina, were the sole representatives of alumino-silicate refractories. By 1950 the range of alumina contents available had been greatly increased and refractories were being produced additionally from highly siliceous clays; calcined china clay; diaspore and flint clays; the sillimanite minerals; bauxite; and alumina. In Germany the practice of using only a small proportion of clay to bond *Schamotte* (pre-calcined fireclay) was adopted so as to secure efficient packing of the particles, which is not possible with water-slakeable grains of plastic clay. This technique, which promotes dimensional accuracy because of small drying and firing shrinkages, became generally used in Europe but elsewhere was confined to production based on materials exhibiting little or no plasticity. In Britain, fireclay batches contained only a small proportion of non-plastic 'grog', derived from product 'wasters'.

Sillimanite and kyanite from India, and andalusite from California, were introduced during the First World War, after a search for natural sources of the mineral which had been seen under the microscope in slagged firebricks and wrongly identified as sillimanite ($Al_2O_3.SiO_2$, the common formula of the three minerals). It was not until 1924 that the first version of the alumina–silica phase equilibrium diagram was published by N. L. Bowen and J. W. Greig in the U.S.A. This established that the effect of increasing alumina content above 5·5 per cent would be to increase refractoriness and that the only compound formed was not sillimanite but mullite ($3Al_2O_3.2SiO_2$) named previously from its rare occurrence in the Island of Mull, off the coast of Scotland. By the end of the period it was appreciated that although equilibrium was rarely established in practice the phase diagram provided a reliable guide to quality, especially if attention were paid to the effects of natural impurities in the materials. During the Second World War china clay, with a much lower iron impurity content than fireclay, was calcined and used as a grog for the production of bricks which had a high-temperature performance superior to that of firebricks and could be used instead of scarce sillimanite.

Dolomite, magnesite, and chrome ore, the raw materials for the production of basic refractories, were all in use in 1900 but continuous demand, particularly from steelmakers, for refractories that would withstand the increasing severity of operations led to remarkable improvements in the period. The discovery by P. C. Gilchrist in 1878 that phosphorus could be removed from steel melted in a dolomite-lined Bessemer converter (and subsequently on a dolomite hearth) was particularly important to Britain, which had dolomite as its only major source of basic refractories. Since dolomite is a double-carbonate, it has to be calcined at high temperatures to yield fully shrunk lumps containing magnesia and lime. The reaction of the lime with atmospheric moisture, giving 'hydration-expansion' and crumbling, cannot be eliminated, but it was found possible by careful control of grain-size to use freshly calcined dolomite for 'fettling' the hearth of steelmaking furnaces. The hydration problem was much more serious in the installation of a monolithic hearth or when dolomite bricks were required for furnace linings, but was solved by the use of tar to seal the grains against hydration; the first such bricks were made in England in the Second World War. Since the presence of carbon appreciably increased the resistance of dolomite to slag attack, economic service performance was achieved and was superior to that given by 'stabilized dolomite', a more expensive refractory made by means of a high-temperature reaction to remove the free lime.

The production of magnesite, a more slag-resistant refractory than dolomite, began in 1880; it was derived from Austrian rock with a high iron content. World industry became highly dependent on this source, which could be easily 'dead-burned' to eliminate hydration-expansion of the oxide. Alternative purer sources, to which Britain and the U.S.A. were restricted during the two world wars, resulted in serious hydration problems because higher calcination temperatures were needed. In 1937, however, there began in Britain the most striking development of the period: the successful extraction of magnesia from dolomite and sea-water. By 1948 one plant was producing 40 000 tons a year; its capacity in 1976 was 300 000 tons. Fig. 25.4 shows a flow diagram of the modern process for the production of sea-water magnesia of controlled composition; the pelletization process is used for the purer grades now required by the steel industry. The introduction of oil for firing the rotary kilns—thus raising the dead-burning temperature progressively to about 1900 °C, and completely eliminating hydration problems— is a post-1950 development.

By 1930 there had been little development in the use of chrome bricks, their main purpose being to serve as a neutral course between acid silica and basic magnesite. The use of magnesite, however, had increased considerably for furnace linings in the steel and cement industries, but poor resistance to temperature change, and their propensity to lose their working faces by 'slabbing-off', were continual sources of trouble. This mechanical loss was to some extent curbed by the use of chemically bonded bricks which were encased in sheet steel and were consequently not pre-fired. After the discovery in 1931 that the tensile strength of mixtures of magnesite and chrome ore was higher than that of either material alone, the first chrome-magnesite bricks appeared almost simultaneously in Britain, Germany, and the U.S.A. By 1935 the production of fired and chemically bonded bricks had been established, generally on the basis of 70 per cent coarse chrome ore and 30 per cent fine magnesite. Such bricks had a high resistance to thermal shock and were less liable to change size at high temperatures than magnesite, which they replaced in open-hearth furnaces. The 'all-basic open-hearth furnace', using the new bricks instead of silica in the roof, then became a target to be achieved; much work was carried out to improve the high-temperature strength and to reduce mechanical loss by 'slabbing' (Fig. 25.5). By 1939 the problem had been partly solved in Austria by extensive use of steel in a form of roof construction in which bricks were suspended and roof movement was controlled; elsewhere the high cost of the basic compared to

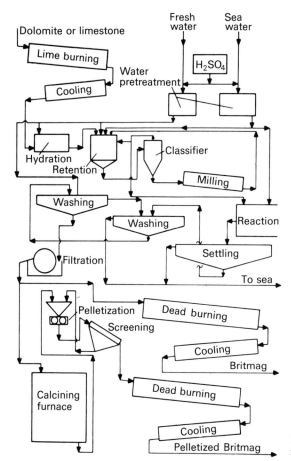

FIG. 25.4. Flow-sheet for the production of magnesia from the sea at Hartlepool.

FIG. 25.5. The 'slabbing' of basic bricks in the roof of an open-hearth steelmaking furnace.

the 'super-duty' silica brick had made it impossible to provide an economic solution by 1950. Success was, however, achieved in the next decade.

The great progress in the basic refractories field was partly due to the provision of new scientific data in the 1940s. Study of phase equilibrium diagrams relevant to the physical chemistry of steelmaking, together with the examination of used furnace linings, made possible the production of 'phase assemblages', from which the proportions of the minerals present in the solid state could be calculated from the chemical composition. The use of X-rays and the optical microscope uncovered the mysteries surrounding the constitutions of chrome ores and the effect of heat upon them.

Finally, not the least important development in refractories was the introduction of carbon blocks to replace firebricks in the hearths of blast-furnaces making pig-iron (Fig. 25.6). This practice, which started in Germany, had as its main objective the elimination of the very dangerous 'break-outs' of molten iron. Clay-bonded carbon, known as plumbago, had for many years been used for the production of crucibles, but the product intended for hearth construction could contain no clay and new fabrication techniques had to be evolved to make blocks from coke. Oxidation of carbon was prevented by the molten iron. Early experience was so successful that the 'all-carbon blast-furnace' seemed a possibility. These hopes were not realized because later experience showed that there was sufficient oxygen in the upper regions of the furnace to oxidize the carbon and preclude its use there.

FIG. 25.6. The corrugated design of carbon blocks used to construct the hearths of blast-furnaces melting iron.

V. THERMAL INSULATING MATERIALS

The rapid expansion during the first half of this century of the metallurgical, glass, carbonizing, and ceramic industries brought problems of fuel economy, which were partly solved by the adoption of thermal insulation. In 1900 the reduction of heat loss by the application of an insulating layer known (now) as 'backing insulation', to the furnace structure had just started. The original source for this purpose was German diatomite, a natural material consisting mainly of the fossil remains of plants. The grains of this mineral have a very high closed porosity, which considerably reduces the flow of heat through it in the form of 'loose-fill' or when made into bricks. Up to about 1920, diatomite from this and other purer sources was the standard insulating material. Rising furnace temperatures then made it obvious that insulating bricks stable at temperatures exceeding 800 °C (the limit for diatomite) were required, so that the thickness of the 'solid' furnace brick work could be reduced. In the 1940s natural vermiculite was rapidly heated to make it 'exfoliate' in order to produce highly porous grains; this yielded an insulating material suitable for a service temperature of 1100 °C. From the 1920s onwards, clays and other aluminosilicates were made into highly porous bricks, generally by the introduction of combustibles into the brick batch. Extensive use of the available range of alumina content made it possible to produce bricks classified in terms of the service temperature at which they exhibited minimum shrinkage. In the U.S.A., where most of the activity began, this classification was expressed in terms of the hundred digits of the Fahrenheit scale; for example, a type-23 brick could be exposed to a temperature of 2300 °F (1260 °C).

The use of backing insulation in continuous kilns saved much fuel. The additional cost of the insulation was quickly offset because many years of satisfactory operation were obtained, provided that the lining/insulation thickness ratio was chosen to avoid higher shrinkage rates of the lining bricks induced by this higher average temperature. In contrast, the insulation of continuous furnaces having 'dirty' atmospheres had to be approached with great care, since there is a balance between fuel saving and the rate of refractory wear; this is because liquid fluxes penetrate more deeply into insulated linings, owing to the reduced temperature gradient. A particular case was that of the silica roof of the open-hearth furnace where insulation could not be used; in fact, accumulating dust had to be regularly blown off the 'cold' surfaces.

In the case of intermittent kilns and furnaces, calculation showed that the heat saved when operating at the maximum temperature could be offset by the greater amount of heat required to bring the structure up to the required temperature, and that the cooling period could be disadvantageously lengthened. The development of the better qualities of product to serve as 'hot face insulation', thus replacing the composite lining, did much to remedy this situation, and was adopted for intermittent kilns and heat treatment furnaces.

BIBLIOGRAPHY

CHANDLER, M. *Ceramics in the modern world.* Aldus Books, London (1967).
GREEN, A. T., and STEWART, G. H. *Ceramics, a symposium.* British Ceramic Society, Stoke-on-Trent (1953).

26

THE TEXTILE INDUSTRIES: GENERAL SURVEY

D. T. JENKINS

MUCH of the history of the textile industries in the first half of the twentieth century stems from the changing volume and pattern of world trade, resulting from the increasing self-sufficiency of the newly industrializing countries and the emerging competition of these countries in world markets. Just as textiles had been at the forefront of the industrialization of the already developed economies—inspiring high levels of technological investigation and innovation, active entrepreneurship, and fundamental changes in the organization of production—so textiles were to emerge as important growth-areas in the subsequent development of other countries. The trend was appearing by the middle of the nineteenth century and became more apparent by the end; the developed countries were then experiencing a decline in the role of textiles in their economies, although, in general, their textile industries were continuing to grow. And then war, by disrupting trade and production in Europe, provided a respite from competition to other economies, allowing them to expand their home industries, to obtain self-sufficiency at home, and to capture neutral markets. The result, in the inter-war years, was a substantially lower volume of world trade, aggravated by high levels of protectionism, with increased competition for the smaller markets and the emergence of yet more countries as textile manufacturers. The Second World War again brought disruption of production and trade but competitive pressure emerged again with recovery from the dislocation of war.

Technological factors, which had been so vital to the growth of the textile industries in the early nineteenth century, were of less importance to the traditional industries by the early twentieth century. In two major areas however—namely artificial fibres and synthetic dyestuffs manufacture—they brought about fundamental changes, the impact of which were experienced, directly or indirectly, in all areas of textile manufacture.

I. THE ADVENT OF WORLD COMPETITION

The fiercest competition developed in the world cotton industry. In 1900 Britain still dominated cotton manufacture and trade, accounting for almost half of the world spindleage and almost three-quarters of the world trade in cotton goods. Her industry had been growing significantly, more than trebling its consumption of raw cotton between the early 1850s and 1913. But for Britain problems had begun to emerge by the 1860s; the cotton famine, resulting from the American Civil War, starved the industry of its raw material. Subsequent cyclical trade fluctuations and periods of active international competition created uncertainty and lack of confidence. The industry was being forced to adapt its production towards finer counts and cloths with the increased production of lower counts in the United States and elsewhere.

The United States was Britain's main competitor before the First World War. The British industry still relied on machinery originally developed early in the nineteenth century, but the U.S.A. was making important innovations in both spinning and weaving. Shortages of labour had encouraged the search for improved labour-saving machines, leading to the widespread introduction of the ring spindle (Table 26.1) and the Northrop automatic loom (Vol. V, p. 585). In 1909 the United States had 200 000 of the latter to Britain's 8000. Although British cotton manufacturers have been accused of being technologically antiquated, there appear to have been sound technical and business reasons why British entrepreneurs made little use of the new machinery. The mule (Vol. V, p. 577) was more satisfactory for medium and fine-count yarns and the short staple cotton used in Britain for low-count yarns was more susceptible to breakages with ring spinning. Likewise the automatic loom was more appropriate for long runs of coarser cloth made from better-quality yarn. Both machines required less skill to operate, but they were less versatile and more expensive to purchase [2].

TABLE 26.1

Mule and ring spindles in use in the cotton industry in 1913 [1]

	Mule spindles	(millions) Ring spindles	Total	% of ring spindles
United Kingdom	45·2	10·4	55·7	18·7
United States	4·1	27·4	31·5	87·0
Germany	5·1	6·1	11·2	54·5
France	4·0	3·4	7·4	45·9
Japan	0·1	2·2	2·3	95·7
World	71·3	72·2	143·5	50·3

Cotton-manufacturing industries were also expanding in France, Germany, India, and Brazil, but the most significant development, in terms of the later history of the world cotton trade, was in Japan. There the traditional peasant cotton industry began to be modernized in the 1890s and, within two decades, the country changed from being a large cotton yarn and piece importer to self-sufficiency in the home market and a net export position. The bulk of Japanese output consisted of very coarse yarns, spun from fibres imported from China, the United States, and India. And it was in China and India that Japan found her first major export markets. However, by 1913 Japan still accounted for only a tiny proportion of world cotton manufacture: 80 per cent of output still came from Europe, although the figure had been as high as 95 per cent three decades earlier.

The development of the world cotton industry had a substantial effect on other textile industries, particularly wool textiles. The wool textile industries of the traditionally exporting countries, Britain and France, had to adjust to increasing self-sufficiency in many of their markets. In Britain, the woollen and worsted sections fared rather differently. The cotton famine of the 1860s increased demand for worsted fabrics but also initiated a change in fashion to all-wool worsteds. The British industry had concentrated production on mixed worsted stuffs, with cotton warps, leaving the French industry to specialize in all-wool worsteds where it gained superiority in spinning, dyeing, and design, and in the flexibility of its response to fashion changes. British exports of mixed worsted stuffs fell by half in the last two decades of the nineteenth century and the impact on the trade in all-wool worsteds was even more severe. Exports decreased from about 43m. metres a year in the mid-1860s to about 12·7m. metres in the late 1890s, by which time the French industry was exporting about 68m. metres a year to Britain [3].

In spite of falling prices and profits, changing fashion, and increasing competition both at home and abroad, the British worsted industry was still expanding before the First World War. The number of spindles for spinning and doubling increased, employment rose, exports of yarn, tops, noils, and waste and of worsted coatings all showed improvement. At the same time, with falling prices and rising real incomes the home market was expanding. The woollen branch of the British wool textile industry faced competition rather more successfully, nearly doubling its export of woollen goods before the First World War. It was less affected by the uncertainties of changing fashion and found its home market buoyant. It reduced some of its costs by making more use of recovered wool (mungo and shoddy) and imported raw wool.

Both branches of the European wool textile industry were having to contend with increasingly complex and severe trade restrictions which varied from outright prohibition to various levels of duty by weight or value. These restrictions affected most of the overseas markets as these strove to develop their own industries. In Japan, for example, a wool textile industry was being established only in the late nineteenth century, but with the aid of tariff protection the value of its output then rose rapidly from an annual average of 4·36m. yen between 1899 and 1903 to 21·66m. yen between 1909 and 1913 [4].

II. THE FIRST WORLD WAR AND ITS AFTERMATH

War in 1914 brought disruption of production and trade for the countries directly involved and gave protection to other manufacturing countries. Japan, in particular, benefited. Not directly engaged in hostilities, and without the problems of physical damage and labour shortages, her textile industries were able to concentrate on supplying markets in India, China, and neighbouring countries, formerly to a great extent the preserve of British manufacturers. In 1913 Britain supplied 97 per cent of Indian imports of cotton piece goods, over a third of total British exports of them. By the early 1930s Japan and Britain were almost equally sharing a greatly decreased market. Indian home production had nearly trebled during the same period. During the war years the capacity of the British cotton industry remained static. Output decreased as a result of shortages of raw cotton and labour. Lack of shipping space and other export difficulties prevented the industry from continuing to supply all its foreign markets. Peace in 1918 brought a return to prosperity for a few months. Demand at home and abroad soared after wartime shortages. Prices and profits rose and a false sense of prosperity encouraged investment in new plant. But late in 1920 the boom disintegrated, prices fell rapidly, and the industry suddenly awoke to the changed world trading situation and entered two decades of overcapacity and attempts to adjust to reduced trade and a greater dependence on the home market. Immediately before the war Britain accounted for almost two-thirds of world trade in cotton yarns and piece goods and this trade amounted to about a quarter, by value, of all British exports. Three-quarters of the output of the British cotton industry was exported in 1913. International trade in cotton goods fell by over a third between 1913 and 1937 and the British share of that reduced total had fallen to almost a quarter immediately before the Second World War. The increased self-sufficiency of former markets and new competition, particularly from Japan, were further aggravated by trade restrictions, which became particularly severe after the impact of world depression in 1929.

The British cotton industry blamed some of its problems on cheap labour in India and Japan, but other European countries and the United States showed that in spite of high wages it was possible to compete, using good machinery and well-organized production. France, Germany, Italy, and Switzerland all managed to surpass their pre-war output by the late 1920s. Britain, however, failed to reorganize and reduce her surplus capacity. In spite of a few amalgamations and some attempts to improve the efficiency of the industry, the bulk of British production remained in the control of very small firms. Productivity, although increasing, failed to keep up with developments in the United States and Japan.

The problem was slightly alleviated by growing domestic demand for cotton goods, less severe competition for the finer section of the industry, and some protection within the home market after 1932. The British wool textile industry also gained advantage from rising home demand. Before the First World War it had increased its dependence on the home market and during the depressed years of the inter-war period it was therefore less affected by the vagaries of world trade. Production was somewhat disrupted during the First World War by shortages of labour and of supplies of raw wool from the Southern Hemisphere. Peace brought a sharp upsurge in demand but over-investment as a result was avoided. The breaking of the boom led to an awareness of the diminished export market, a situation that was to be aggravated in various ways during the inter-war period. The severest competition in foreign markets came from other European countries, notably Germany, France, and Czechoslovakia. The value of British exports of yarn and tissues declined from a post-war peak of £64·6m. in 1924 to a trough of £21·7m. in 1932, recovering only slowly to £30·7m. in 1937. Home demand recovered better and increased by about one-third during the two decades. Some sections of the industry were more affected than others. The expansion of the hosiery industry and fashion changes in favour of knitted goods adversely affected the weaving branch of the industry, whereas worsted spinning benefited for the same reason. The increasing demand for lighter fabrics worked to the disadvantage of manufacturers specializing in coarser and heavier yarns and cloths.

III. ARTIFICIAL FIBRES

By the 1930s the traditional staple textile industries were becoming aware of the growth of the manufacture of artificial fibres. The period between the initial development of rayon (originally known as artificial silk) in France in the late nineteenth century and the First World War was mainly one of experimentation. France remained the world's major rayon producer during the first decade of the century, but by 1913 Britain, through the activities of Courtaulds, had emerged as the largest producer, ahead of the United States, France, and Germany. In 1913, however, artificial fibres formed only a minute proportion of world textile production. Development continued during the war with the United States becoming the leading producer and other countries, including Japan, creating their own industries. Early production was concentrated on continuous filaments, using the nitrocellulose process which accounted for about half the world production in 1909. A further third of the production at that date was by the cuprammonium process but viscose production was rapidly replacing both immediately before the war. By 1927, 80 per cent of British rayon output was by means of the viscose process.

It was not until the 1920s and 1930s that the great boom in rayon production occurred (Table 26.2). High profits encouraged high levels of research and investment in the 1920s and staple fibre production rapidly gained in importance from the end of that decade. Rayon manufacture in Europe was dominated by Germany and Italy, producing 27 per cent and 14 per cent of world output respectively in 1939. At the same period Japan was producing 25 per cent of world output and the United States 17 per cent.

Before the First World War rayon was being used on a very limited scale for the production of knitted goods. By the 1920s it was firmly established

TABLE 26.2
World rayon production (000 tonnes) [5]

	United Kingdom	France	Italy	Germany	United States	Japan	World
1913	5·2	2·9	0·2	2·1	0·9	—	14
1924	10·0	6·0	10·5	10·5	16·3	0·5	62
1929	21·4	19·0	32·3	28·1	55·4	12·3	200
1934	40·5	27·9	48·7	46·2	95·4	71·4	374
1939	77·0	32·5	140·0	273·0	172·7	245·4	1022
1951	174·0	106·8	133·0	185·0*	588·1	167·1	1833

* West Germany.

as a stocking yarn. Its price remained higher than cotton and wool throughout the inter-war period, restricting its demand from the mass market, but
by the 1920s it was being used in conjunction with cotton for a variety of
clothing and furnishing fabrics. The demand for lighter fabrics was an advantage to the industry and by 1939 60 per cent of the British output of continuous filament was being used for weaving in the Lancashire cotton industry,
another 27 per cent went for knitting in the hosiery trade, and small quantities
were used for ribbons and braids and were exported. The wool textile industry made little use of the artificial fibre before the Second World War.

Besides changing fashions and the demand for lighter fabrics, the rapid
growth in world rayon production resulted from a variety of factors. Paramount were technological advances that improved production methods and
quality, and reduced costs. But also of importance was governmental encouragement. In a number of countries, including Germany, Italy, and Japan,
assistance was given in various forms of financial aid or restrictions on other
textile manufacture. In Japan, the use of raw cotton in textiles for home
consumption was banned in 1938, thus helping to create an upsurge in
artificial fibre output.

IV. DYESTUFFS MANUFACTURE

Another area of industry that was affected by changes in the textile industries and benefited immensely from technological improvements was the
manufacture of dyestuffs. Just as artificial fibres resulted from developments
in chemical research, so the production of artificial dyestuffs was closely
related to the growth of the chemical industry (Ch. 21). Before the First
World War artificial dyestuffs' production was dominated by Germany.
Indeed, Germany had almost total command over their supply, accounting
for 85 per cent of world output in 1913. German pre-eminence in this area
had developed rapidly, following the series of discoveries of synthetic dyestuffs between the 1850s and the 1870s, as a result of a number of factors.
Her industry was well organized into cartels with a high level of scientific
training and research and a dynamic sales policy. Strong German patent
restrictions and an unwillingness to issue licences prevented other countries
from benefiting from German chemical developments although the German
industry was often able to make use of foreign discoveries.

The British synthetic dyestuffs industry was almost totally overwhelmed.
It suffered from shortages of certain basic raw materials; high Government-
imposed duties on essential industrial alcohol; and a highly fragmented

organization, with the industry based almost entirely on small firms unable to provide the facilities for the necessary research. Nor could these firms develop economies in production and obtain the skilled technical labour required for production design and control.

The first challenge to German supremacy came during the First World War. The inability of textile manufacturers in many countries to obtain dyestuffs formerly supplied by Germany encouraged Switzerland to develop a synthetic dyestuffs industry and by 1920 that country was manufacturing 10 per cent of world production. Other countries, increasingly aware of the growing role of chemicals in industry and in warfare and of their dependence on German industry for dyestuffs, were encouraged to develop their own dyestuffs manufacture, leading in due course to severe world overcapacity in production. Most industrialized economies imposed import tariffs and restrictions and successfully encouraged chemical research. But German domination continued in the 1920s in those markets where there was no import prohibition. In Britain the Government responded in an unprecedented way. It assisted the formation of British Dyes Ltd. in 1915; by 1918, under the name of the British Dyestuffs Corporation, this accounted for 75 per cent of total British output. Import restrictions allowed the industry rapidly to provide the bulk of home artificial dyestuffs needs. By 1922, almost 80 per cent of consumption was met by the home industry, the proportion rising to over 90 per cent by 1928, by which time imports had been reduced to one-tenth of their pre-war level. Import restrictions and scientific advances, including the discovery of dyes for acetate rayon, allowed the British dyestuffs industry to create an efficient and substantial organization. It increased its share of world output from 3 per cent in 1913 to 12 per cent in 1937, although total world output increased by only 50 per cent over the same period. Germany and the United States remained the major manufacturing countries, with Switzerland also an important exporter. Synthetic dyestuffs industries were also developed successfully in Holland and Canada.

V. THE SECOND WORLD WAR AND ITS AFTERMATH

The dislocation of production and trade in textiles in Europe and Japan during the Second World War allowed the United States to increase capacity and exports. Production was less affected in Britain than in most other European countries but in European textile manufacture generally shortages of raw materials, restrictions on productive capacity, and inability to service overseas markets led to contraction. Output of artificial fibres failed to keep pace with demand but some significant developments occurred. After the discovery of nylon in the mid-1930s, commercial production was under way in 1940, although initial growth was slow with technical problems still remaining. The production of high-tenacity yarn, particularly for various war purposes including tyres and parachutes, was another major innovation.

The end of hostilities was followed by high world demand for textiles and considerable production problems in meeting this demand. In Britain, a labour shortage, caused particularly by the loss of women workers to other occupations, was aggravated by the uncertainty of the future world market. Experience within the textile industries varied. Wool textiles managed to achieve a steady recovery; worsted production in particular was continuing to benefit from the increasing demand for lighter fabrics and hosiery. Exports were assisted by rationing at home. British cotton manufacture recovered less well. Production problems held back exports. World trade had declined and many more countries were meeting their own requirements and competing on the international market.

Expansion continued in dyestuffs and in artificial fibres, including nylon and Terylene, which by 1955 made up nearly 10 per cent of world output of man-made fibres. There was also by this time a minute output of two protein fibres, Ardil and Fibrolane. Since the war the United States had become the major manufacturer of artificial fibres, accounting for almost one-third of world production. As a result of wartime changes of ownership most of her production was now under her own control. Artificial fibres were being used more widely in the other textile industries, including wool textile manufacture, and were becoming important for a variety of industrial uses. By the 1950s a quarter of the fibre used in cotton weaving in Great Britain was man-made.

The development of artificial dyestuffs and man-made fibres continued into the 1950s with both industries maintaining growth. The post-war recovery of textile manufacture from natural fibres came, however, to an abrupt halt in 1951. By that date demand was being met and surpluses in production were occurring. Overcapacity returned to the world industry and protectionist barriers were re-erected. World trade in both cotton and woollen manufactures diminished and a new era of tight competition and adjustment to changing markets was entered.

REFERENCES

[1] ROBSON, R. *The cotton industry in Great Britain*, p. 355. Macmillan, London (1957).

[2] SANDBERG, L. G. *Lancashire in decline: a study of entrepreneurship, technology and international trade.* Ohio State University Press, Columbus (1974).

[3] SIGSWORTH, E. M. *Black Dyke Mills: a history; with introductory chapters on the development of the worsted industry in the nineteenth century*, pp. 72–134. Liverpool University Press (1958).

[4] ALLEN, G. C. *A short economic history of modern Japan* (2nd edn.), pp. 71–8. Allen and Unwin, London (1962).

[5] MITCHELL, B. R. *European historical statistics 1750–1970*, pp. 454–5. Macmillan, London (1975).

ROBSON, R. *The man-made fibre industry.* Macmillan, London (1958).

BIBLIOGRAPHY

ALDCROFT, D. H. (ed.) *The development of British industry and foreign competition, 1875–1914: Studies in industrial enterprise.* Allen and Unwin, London (1968).

ALLEN, G. C. *British industries and their organization* (4th edn.). Longmans, London (1959).

COLEMAN, D. C. *Courtaulds: an economic and social history.* Oxford University Press, London (1969).

EWING, A. F. *Planning and policies in the textile finishing industry.* Bradford University Press (1972).

FABRICANT, S. *The output of manufacturing industries, 1899–1937.* National Bureau of Economic Research, New York (1940).

HAGUE, D. C. *The economics of man made fibres.* Duckworth, London (1957).

RAINNIE, G. F. (ed.) *The woollen and worsted industry: an economic analysis.* Clarendon Press, Oxford (1965).

RICHARDSON, H. W. The development of the British dyestuffs industry before 1939. *S.J.P.E.*, **9**, 110–29 (1962).

ROBSON, R. *The cotton industry in Britain.* Macmillan, London (1957).

——. *The man-made fibre industry.* Macmillan, London (1958).

SANDBERG, L. G. *Lancashire in decline: a study in entrepreneurship, technology and international trade.* Ohio State University Press, Columbus (1974).

27

TEXTILE MANUFACTURE

C. S. WHEWELL

I. INTRODUCTION

BY the end of the nineteenth century the general principles of converting wool, cotton, silk, flax, and jute into a wide variety of fabrics by mechanical means were firmly established and flourishing industries had grown up in many countries (Vol. V, Ch. 24). Since then textile processing has continued to attract the attention of ingenious men and there is now a vast patent and technical literature concerned with all aspects of textile fibre production and manipulation. Comparatively few of these patented processes have, however, been taken up by the industry, and perhaps more unexpectedly, there has often been a long interval, sometimes as long as forty years, between the patenting of an idea and its commercial adoption. This is not

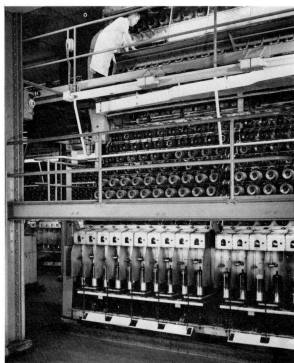

FIG. 27.1. A section of a continuous-type rayon spinning machine producing textile yarn.

FIG. 27.2. Production of rayon by the Industrial Rayon Corporation, continuous-spinning process introduced in 1938. The viscose is extruded and passed over godets at the top of the machine. Each of the subsequent stages of processing is carried out on separate thread advancing reels, located at successively lower levels, and the yarn is finally dried and collected.

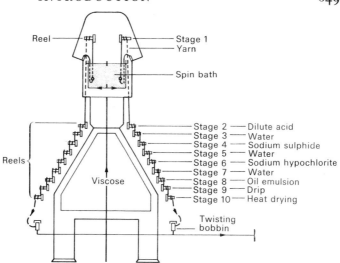

entirely due to the conservative nature of the textile industry: economic conditions may not be favourable or the need for the 'improvement' not evident.

The most profound change in the twentieth century has been the emergence and growth of the man-made fibres industry. The earliest man-made fibres were made from cellulose; viscose rayon in staple and continuous form being the oldest and most important. It was the outcome of research by C. F. Cross and E. J. Bevan, who in 1892 produced fibres by treating cellulose first with caustic soda and then with carbon disulphide to make cellulose xanthate; this was then dissolved in dilute caustic soda solution to produce a viscous liquid known as viscose. After being aged, this was extruded through fine holes in a spinneret and coagulated in a dilute acid solution to yield continuous filaments of regenerated cellulose. In 1904, the British rights for this process were bought by Courtaulds Ltd., who have been responsible for much of the research and development on viscose rayon. Of special importance was the invention in 1900 of the Topham box, a device for collecting extruded filaments on the inside of a cylindrical container rotating at high speed. Until 1914, viscose rayon was produced mainly in the form of continuous filaments, but the production of staple fibres by cutting tow (bundles of continuous filaments) into appropriate short lengths became increasingly important during the 1930s and later during the Second World War. Early batch methods of producing viscose rayon gave way to continuous methods such as those developed in 1934 by S. W. Barker and J. Nelson in Britain and in 1938 by the Industrial Rayon Corporation in the U.S.A. (Fig. 27.1). These techniques were based on the use of pairs of thread-advancing reels or skewed glass rollers whereby individual filaments could be treated with chemicals, thus obviating the need for processing the rayon 'cakes' formed in the Topham box. The properties of regenerated cellulose are improved when the wet filaments are stretched by being passed over two glass godet wheels some 25 cm in diameter, one rotating faster than the other. This continuous stretching process increases the orientation of the cellulose molecules and thus increases the strength of the filaments. Intensive research on the coagulation and stretching of viscose filaments led to a high tenacity viscose (marketed as Tenasco by Courtaulds in 1935) and to polynosic rayons of high strength and high wet modulus and with properties very similar to those of cotton. To produce Tenasco, the viscose filaments are coagulated in a bath with a high concentration of zinc sulphate and then stretched in hot water or hot acid solution. In the production of polynosic fibres, degradation of the cellulose is kept to a minimum, and regeneration and coagulation are effected slowly to facilitate gentle stretching in stages. Secondary cellulose acetate, made by hydrolysing cellulose triacetate, was discovered in 1904 by G. W. Miles (British Patent No. 19 330; 1905) and the production of filaments by extruding solutions of secondary acetate in acetone was the outcome of investigations by the Dreyfus Brothers who marketed the product under the name 'Celanese' in 1921. The production of cellulose triacetate fibres was delayed because of the lack of a suitable solvent, but the availability of cheap methylene dichloride after 1930 made commercial processes possible. Since that time triacetate fibres, for example Tricel (Courtaulds) and Arnel (Celanese Corp. U.S.A.), have firmly established themselves as textile raw materials.

With the discovery of nylon 66, a polyamide made from adipic acid and hexamethylene diamine by W. H. Carothers in 1937, a new synthetic fibre industry was born. The molten polymer was extruded through spinnerets and the cooled filaments then stretched by some 400 per cent to orient the molecules. This stretching develops the extraordinarily high strength of the filaments (4·6–5·8 g/denier). An alternative method of making a polyamide (nylon 6) from caprolactam was the topic of research carried out by P. Schlak in the laboratories of the I.G. Farbenindustrie in Germany, but commercial production of fibres from this polymer, although possible in 1929, did not, because of the war, become significant until after 1948. The next advance was due to British investigators, J. T. Dickinson and J. R. Whinfield, who in 1941 made a synthetic fibre from the polyester derived from terephthalic acid and ethylene glycol. Owing to wartime difficulties, this discovery was initially developed by the Du Pont Company in the U.S.A., and the fibres were marketed first by that Company as Dacron: later I.C.I. introduced their polyester under the name 'Terylene'. At about the same time a new class of fibres based on polyacrylonitrile appeared on the market. Probably the first was 'Fibre A', later named 'Orlon', from Du Pont in 1945. The Chemstrand Corporation introduced 'Acrilan' in 1952. During the period 1955–60 plants for producing acrylic fibres were erected in most countries with substantial chemical industries, the German installations bringing to fruition research and experimental production commenced before the Second World War.

These categories of fibre are today still the main sources of man-made fibres in staple and continuous filament form. When first introduced, man-made fibres were processed on traditional textile machinery to which relatively minor adjustments had been made; indeed, even today large weights are processed in this way. However, those properties of man-made fibres which are fundamentally different from those of natural fibres have been effectively exploited, resulting in processes such as yarn texturizing, heat setting, and tow-to-yarn processing which require completely new machinery.

The twentieth century has also been characterized by the growth of the chemical industry, which apart from making the raw materials for the manufacture of man-made fibres has also made available entirely new reagents which have found application in the processing and enhancement of textile fibres and fabrics, leading to improved performance and greater customer satisfaction. Moreover, this period has seen the start of systematic research for the textile industry in industrial laboratories, in Government-sponsored and other research institutions, and in universities. Of special significance was the setting up, after the First World War, of the British research associations (p. 130), financed partly by industry and partly by Government, and the generous funding of laboratories attached to the U.S. Department of Agriculture.

II. THE CONVERSION OF FIBRES INTO YARNS

Man-made fibres can be processed without any preliminary treatment, but natural fibres must first be cleaned. Wool has to be freed from wool grease, suint (sweat), and dirt; seeds, dirt, and other extraneous matter have to be removed from raw cotton. Until comparatively recently raw wool was purified by washing in warm soap and soda ash solution. Excessive matting of the wool must be avoided but on modern machines such as the Wool Industry

Research Association–Petri improved scourer, the product of research carried out in the early 1950s, the adjustment of the flow of liquor and the motion of the rakes minimizes this. Wool grease is recovered from the dirty liquor by acidification and separation or by centrifugation, and becomes the raw material from which lanolin and an increasing number of useful compounds such as cholesterol are obtained. Since the early 1940s soap, as the almost universal detergent for wool scouring, has in part been replaced by synthetic detergents, which were the outcome of German research to meet the shortage of soap during the First World War. The use of organic solvents for purifying wool has attracted considerable attention. The Maerton system, announced in 1898, involved extracting large batches (2270 kg) of wool with solvent naphtha and was used by the Arlington Mill, U.S.A., for some 35 years. After the end of the Second World War, Australian workers developed a method based on two extractions in white spirit and two treatments in water. An alternative process developed in Sweden involved extraction with kerosene. Since 1900, the firm Solvent Belge has been concerned with solvent extraction, and has operated a batch process of degreasing wool with hexane. More recently, the Sova process, associated with this company, has received much publicity. It consists in spraying wool successively with water, a mixture of water and isopropyl alcohol, and hexane, and a commercial plant has been in operation since 1967.

The purification of raw cotton is entirely mechanical. Cotton was for centuries 'picked' by hand, but most of it is now harvested by machinery. Mechanically harvested cotton is dirtier than the hand-picked product and necessitates more cleaning at the ginnery. Modern installations include considerable ancillary equipment such as drying machines. It was found that the optimum moisture content of cotton for ginning was about 5–10 per cent, and the availability of driers enables ginning to be done at any time, avoiding that loss of quality which results from ginning wet material.

The first process in traditional cotton yarn manufacture is to pull clumps of fibre from the bale and throw them on to a feed sheet or conveyor belt which delivers them into an opening machine. Cotton from different bales of the same quality or from bales of different qualities is blended to yield a mixture of fibres which will ensure good processing performance and give a uniform product at an acceptable cost. Older blending practice depended mainly on the skill of the operator, but over the years greater reliance has been placed on the results of laboratory measurements of fibre length and length distribution (for example, by the Suter–Webb comb sorter introduced in the early 1930s), of diameter (by the Micronaire method based on measuring the air porosity of a plug of fibres, introduced in 1946), and of strength (the Pressley bundle test has been used since 1940). The possibility of reducing the labour involved in bale-opening and blending was recognized in the 1920s and a patent specification by J. T. Tice in 1924 describes a machine in which the bales are placed into fireproof boxes and the cotton pulled off from below. Although there were some fundamental objections to these units, the delay in adopting them was mainly a consequence of a lack of incentive because of the availability of relatively cheap labour, and it was not until 40 years later that really satisfactory systems were introduced by Trützchler and by Rieter [1].

Opening and cleaning are carried out by various types of mechanical beating machines, the cotton now being transferred from one machine to another pneumatically. In the more traditional systems, mechanically cleaned cotton is delivered in the form of a lap, a thin layer of fibre which is rolled upon itself. To ensure reproducibility it is important that the lap weight is carefully controlled. The lap is conveyed to the card either manually or by a special conveyor such as that used on the S.M.C.A. picking and lapping machines introduced by Trützchler & Co. Many of the opening and cleaning machines are linked together, requiring comparatively little labour for operation, and are fitted with devices such as photoelectric units to control the height of cotton in a machine. The most important development has been direct transference of cotton from the opening machine to the card. For many years this was considered to be impossible, as the lap forming machine delivers fibre some ten times faster than the card takes up the lap. The difficulty has, however, now been resolved in systems such as the Rieter 'aerofeed system' in which one set of opening machines feeds several carding engines [1].

To obtain a mixture of wool fibres which is uniform in colour and of appropriate price, various lots of wool (dyed if a coloured product is required) are blended together. Older methods of pile, stack, or layer blending have over the years been replaced by less labour-intensive systems, like the Spenstead, introduced in the 1950s.

Carding. This key operation in wool yarn manufacture further opens and mixes tufts of fibres, converting them into a web which can be condensed into a soft rope or sliver. During this century carding machines have greatly improved, for serious research since 1930 has led to a clearer understanding of the behaviour of fibres in the machines. Delivery of material to the card is usually from hoppers to weighing devices of various degrees of reliability, but man-made fibres and some coarser wools are fed by chute directly on to a feed sheet. The central part of the card, the swift, with its attendant workers and strippers or floats, differs according to the system of yarn manufacture, and considerable attention has been given, particularly since 1940, to the card clothing. Working parts of the card covered with conventional clothing must be 'fettled' from time to time, that is, the fibre and grease embedded between the wires must be removed. The introduction, in 1959, of 'metallic clothing', that is, toothed wire or ribbon wound in a spiral groove cut in the cylinder has, however, greatly reduced the need for fettling.

Among the devices designed to improve the quality of the carded webs are the Peralta roller, introduced into Britain by the firm of Duesberg Bosson in 1938, and the Crosrol web purifier. These units, composed of two accurately turned metal rollers to which high pressure (approx. 4 tons) is applied hydraulically, crush any vegetable matter (burrs, etc.) or pieces of skin in the web as it is taken off the swift of the scribbler. The rollers are made slightly barrel-shaped so that an even pressure is obtained across their width. In the Crosrol device, the upper roll of a pair is set slightly askew, the angle of skew being adjusted so that the pressure between the rolls is uniform. These 15 cm-diameter rollers are smaller than the older Peralta rollers, and this reduction in size greatly facilitates the inclusion of the device in a conventional cotton card. Greater width and speed of carding machines have increased their productivity and there has also been a tendency towards simpler units, which is reflected in the growing acceptability of the Mackie system, introduced in 1948, for the manufacture of carpet yarns. Good results are obtained on a unit with only one carding cylinder, which is covered with metallic clothing.

High-production cotton carding is a consequence of improved engineering design and of the use of such devices as the Rieter patented roller clearing unit which replaces the fly comb [1]. In the Crosrol–Varga unit, which is incorporated in many new cards, but which can also be added to older models, the cotton is removed from the doffer (Vol. V, p. 572) and presented to the Crosrol unit by means of wire-clothed rollers.

Drawing and spinning. In cotton processing it was until about 1920 customary to pass the sliver through three pairs of rollers placed some distance apart. Bottom metal rollers with fine flutes running lengthways across them were positively driven but the top rollers, covered with leather or synthetic rubber, were driven by frictional contact with the sliver. Several methods of applying pressure to these rollers have since been devised, but in all cases the pressure is adjusted so that the fibres can be drawn past each other to effect attenuation. The most important advance resulted from the investigations of Fernando Casablancas, which in 1912 culminated in a series of patents which describe the two-apron system for drawing. When applied to a three-roll unit, the front and back pairs of rollers are the same as on a conventional unit but the middle pair is modified so that each carries a short endless 'apron' extending almost to the front roller. These aprons, the bottom one of which is positively driven, are kept in position by bars which support the forward end, and the fibres passing through the unit are therefore gripped not only by the back, middle, and front pairs of rollers, but also supported between the gentle nips of the two aprons. This gives greater fibre control and enables higher drafts to be obtained than formerly. Although apron drafting was first adopted on cotton drawing and spinning machinery, it also finds application in worsted yarn manufacture. Today, millions of cotton spindles are equipped with drafting aprons. High-draft spinning was envisaged by Casablancas, for the patent for the Casablancas compound drafting system, now used to obtain drafts of several hundreds, was filed in 1920. A system using a single bottom apron on which rested two small slip rollers was patented by J. L. Rushton in 1924, and is adopted in present-day Rieter, Platt, and Saco-Lowell machines. Automatic control of sliver regularity did not emerge until the 1950s when the Raper autoleveller, a mechanical device, became available for worsted drawing. In 1953, however, work commenced on incorporating the principles of the Uster electronic evenness tester for measuring the regularity of yarns and slivers in a Saco-Lowell drawframe. The outcome of this collaboration between the Zelleweger Company and the Saco-Lowell workshops was the Saco-Lowell–Uster Versamatic A.D.C. drawframe, the letters A.D.C. signifying 'automatic draft control'.

In the worsted industry three systems of drawing had been established by the end of the nineteenth century (Vol. V, p. 576): (1) the English, open, or Bradford system; (2) the cone system; and (3) the Continental, French, or Porcupine system. In 1929, the Anglo-Continental system, intended to incorporate the desirable features of the English and the Continental systems was introduced, but it failed to develop. After the Second World War, there was considerable incentive to reduce labour and power costs by reducing the number of operations in conventional processing, and the most important development in the English system was the introduction of the Raper auto-leveller, which made possible fewer operations between combing and spinning. This machine consists essentially of a wheel resting in a groove through which the sliver runs, the wheel 'feeling' the thickness of the sliver. When attached to the feed-end of a gill box or draw box this sensing unit measures the thickness of the ingoing sliver, and by means of an entirely mechanical control system the draft given to the sliver is automatically altered according to the thickness. The inventor of this process (Raper) was also responsible for the pressure drafter, a long metal box with an adjustable top which is altered in position according to the weight of sliver being processed. An alternative method of drafting was invented at about the same time by G. H. Ambler and incorporated in the Ambler Superdraft (ASD) spinning unit. Drafts of 100–150 were obtained with consequent savings in labour, machinery, and power. The unit consists of a slubbing intake guide; a pair of small diameter positively driven rollers; and a flume, which is a channel-shaped guide of rectangular cross-section converging slightly towards the forward end. The placing of the unit with respect to the front roller nip of the spinning frame is very important. The superdraft spinning system offered many attractions and in 1958 was incorporated in what became known as the 'New Bradford System' of drawing and spinning, by which great economies could be effected.

'American' drawing systems owe much to developments in cotton machinery, and were built to process wool tops as quickly as possible. They were at first used only for spinning short wools to fairly coarse counts, but their scope was greatly increased by the introduction of well-designed and precision-built intersecting gill boxes for high-speed operation such as the Warner and Swasey Pin drafter.

Traditional spinning techniques were well established by the end of the nineteenth century and mules, as well as flyer, cap-, and ring-frames were highly developed (Vol. V, p. 577). Over the years, however, mule spinning has declined in popularity in spite of desirable characteristics claimed for mule-spun yarns. Cotton and worsted mules have been replaced by ring spinning machines which originated in America in 1832, and even woollen mules have given way to woollen ring-frames, particularly since 1960.

Since the late 1950s, however, attention has been directed to completely new methods of spinning, of which break spinning, also known as open-end (O.E.) spinning, is probably the most important. Although the first Czechoslovakian O.E. spinning machine was not in commercial operation until 1960,

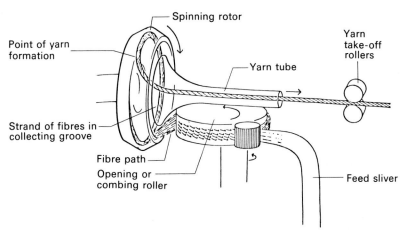

FIG. 27.3. Yarn formation in the rotor of an open-end spinning machine.

the idea behind this type of spinning was not completely new. A British patent by A. W. Metcalf was filed in 1901 and the Danish engineer S. Berthelsen described in his patent of 1937 a method very similar to that used today. The most important O.E. spinning system and the one used in practice is rotor spinning (Fig. 27.3). A sliver of fibres is fed to an opening device, usually a spiked roller, and opened so that the fibres move forward more or less individually. The fibres are carried by an air stream to be collected on the inner surface of a cup-shaped rotor which revolves at 30000 to 60000 rev/min. During each revolution of the rotor a thin layer of fibres is deposited in the collecting groove on the inner surface of the cup. With successive rotations layers of fibres are laid one upon the other and a strand is built up. This is subsequently peeled from the groove, twist being simultaneously introduced by the rotation of the rotor to form a yarn. Other methods of break spinning use an air vortex or an electrostatic field to manipulate the fibres.

In conventional spinning the fibres are held together by twist but in twist-less spinning they are temporarily held together by an adhesive which can readily be removed subsequently. The first machine was based on the 1954 patent of B. Lawrence. Roving is drafted by conventional means and then pressed on to a roller carrying a thin film of adhesive (starch solution). While in contact with the film the fibres are condensed by rubbing rollers, and the yarn is then dried before being wound on a bobbin. The production rate is high but the yarn is suitable for weft only. In the more recent process from the Fibre Research Institute, T.N.O., in Delft [2], false twist is inserted to give temporary strength to the wet yarn before it is wound on a package which is then steamed and dried. A suspension of powdered potato starch is used as the adhesive and the false twister is of the air-vortex type.

The Repco self-twist spinning machine embodies a new concept of spinning. The process originated in the C.S.I.R.O. laboratories in Australia and is based on patents granted in 1964 to D. E. Henshaw and C.S.I.R.O. A typical machine handles eight roving ends at a time. A strand is produced from each of the rovings by a modified apron drafting unit followed by a pair of high-speed reciprocating rollers which introduce alternating twist along the length of the strand. Pairs of these strands are allowed to twist about each other in a controlled way, the tendency of each to untwist creating the self-twist when two strands are brought together. The yarn is then wound on a cheese package by high-speed uptwisting, and is steamed before winding to eliminate snarling. The machine is capable of exceptionally high production, one with eight drafting lines and four take-up heads producing as much medium to fine yarn as 100 conventional spindles.

Tow-to-top processes. These processes were devised to convert tow directly into top, sliver, or yarn, and several ingenious machines were introduced. Two methods of tow conversion have been particularly successful, namely the cutting method and stretch breaking.

In the Pacific converter, a unit of the first type, invented in 1950 by the Pacific Mills Corporation of Laurence, Massachusetts, and developed and built by the Warner and Swasey Company of Cleveland, Ohio, tow, which contains between 100 000–200 000 continuous filaments, is fed from large spools to feed rolls from which it emerges as a flat web. The web is cut into oblique strips by a helical cutter and these are fed forward through two fluted rollers which separate those ends of the fibres adhering as a result of being crushed. It then passes between fluted rollers and a leather apron so that the top of the web moves forward relatively faster than the bottom. This causes a shuffling action and the web is drafted at the same time. It is then formed into a coherent sliver by a scroll roller placed diagonally across the carrying apron. The top so produced can be processed on the Bradford System or on a Warner-Swasey pin drafter.

The Greenfield top was first made in Courtaulds' Greenfield Factory in 1939. Tow is cut with a helical cutter within an angle of 5° to 15° to the tow length. The cut fibres are carried forward by the cutting rollers on to an endless apron which feeds them into a conventional gill-box set to give the required draft and pin action to convert them into a top.

In the Seydel system of stretch-breaking, dating from the 1940s, twistless tow—for example, 100000 denier 1·5 d.p.f. nylon tow—is stretched by successive stages to an extension just short of its breaking extension. The stretched tow then passes into a breaking section where filaments are stretched between two pairs of rollers to breaking point. The resultant fibres, which vary in length from 4 to 20 cm, are passed through a twist-tube and may be crimped before being collected.

Another stretch-break tow-to-top machine was a development of an idea patented by J. L. Lohrke in 1929 and which became the basis of the Perlok process. The modern form of the machine, now known as the turbostapler, was introduced in 1954 and found to be particularly suitable for acrylic tows. Tow is fed into the machine and passed under tension between heater plates carefully maintained at the appropriate temperature. It is then cooled by a blast of air and passes, still under tension, to the breaking zone, which consists of two pairs of rollers, the delivery rollers rotating faster than the receiving pair. Breaker bars made of carborundum or other hard material are placed between these pairs of rollers, the blades of the bars intermeshing so as to apply a breaking stress to the filaments in the tow. The final section of the machine is a crimping device of the stuffer box type.

An important feature of this machine is the possibility it offers of making what are called 'high bulk' yarns. The product from the turbostapler is in a strained state, but can be relaxed and set by a steaming and vacuum treatment. When a yarn made from a blend of this material with unset strained material is subsequently relaxed by treatment in hot water, the unset fibres shrink but the set fibres do not. The shrinkage causes the set fibres to buckle or form loops, thereby increasing the bulk of the yarn.

Textured yarns. Researches on increasing the bulk of the continuous filament yarns so that they have some of the characteristics of staple yarns have led to the introduction of a completely new range of textile yarns, now referred to as textured yarns. The main texturing processes are as follows.

(i) Yarn is highly twisted, heat set, and untwisted either continuously or in three stages. This highly successful sequence was, in fact, the subject of a patent granted in 1932 to the Swiss firm of Heberlein and Co., A.G., but it fell into disuse because of the limitations of the man-made fibres available at that time. It was not until after the introduction of nylon and polyester that the full potential of the process was realized, for these materials enabled wash-fast effects to be obtained. Further patents filed in 1953 and 1954 describe the procedure for bulking man-made filaments, the yarns being referred to as 'Helenca' yarns. Continuous production of these yarns is based on false twist crimping and depends for its success on the sophisticated engineering of the texturizing machines for the false twist spindle inserts twist at the rate of up to 1·5m. turns per minute. The yarns are also highly elastic and find use in such products as swimwear, ski-pants, and upholstery. The elasticity of the yarns can be reduced by heating and stretching after texturizing, and typical products are associated with trade marks such as 'Saaba', 'Crimplene' etc. These yarns were first produced by bulking fully drawn yarn, but in the late 1960s and early 1970s interest focused on 'producer bulking', by which the extruded filaments are first partially drawn before being fully drawn and textured in the final texturing operation.

(ii) Filament yarn is continuously crammed into a heated wedge-shaped stuffer box where it folds up in a concertina-like fashion. It is set either in the box or subsequently by a separate operation. The first patents appeared in 1951, but the one of greatest commercial significance is that assigned to Joseph Bancroft and Sons of Wilmington, U.S.A. in 1953. The trade marks 'Banlon', 'Textralised', 'Spunised', and 'Trycola', refer to products made on the stuffer-box principle.

(iii) Heated yarn is passed over a knife-edge, the process being known as edge crimping. The way in which this procedure increases yarn bulk is illustrated by drawing a hair over the thumb-nail, when the hair forms itself into coils or spirals. When filaments in yarn form are subjected to this knife-edge treatment, the formation of these loops or spirals in the individual filaments increases the bulk of the yarn as a whole. Yarns made in this way are described by the trade mark 'Agilon'.

(iv) Heated yarn is passed two or three times between a pair of gear wheels, being slightly longitudinally displaced on each passage.

(v) Yarn is knitted into a fabric which is then heat-set and the yarn, which retains the distortions imposed by the knitting, is then unravelled. This is referred to as the knit-de-knit process.

(vi) Loops are formed in individual filaments by overfeeding them into a turbulent air-stream. Patents covering this process date from 1952, and are held by the Du Pont Company of Wilmington, U.S.A. 'Taslan' is the registered trade mark for the yarns produced by this method, which can be applied to most continuous fine filament yarns.

III. CONVERSION OF YARNS INTO FABRICS

Although the ancient operations of knitting and weaving still dominate the scene, there are indications that some unconventional methods of making fabrics, usually classified as 'non-wovens', are sufficiently attractive economically to justify their further development. The relative significance of knitting and weaving has altered during this century, knitting having become progressively more important, although some of the predictions of the decline of the weaving industry have been very much exaggerated.

Weaving. The nineteenth century was a time of extraordinary inventiveness in weaving devices (Vol. V, p. 579), and by 1900 power looms had replaced hand looms for most commercial purposes. The automatic loom, in which the weft supply is maintained automatically, was the subject of patents by J. H. Northrop during the years 1889 to 1894, industrial machines being made by the Draper Corporation in 1895. From this time onward it has grown in popularity, replacing the non-automatic models except for special purposes.

Appropriate yarn preparation is essential for all successful weaving, especially automatic weaving. Packages of yarn from the spinning frame are generally not of the most appropriate form for use as weft or warp or for knitting and must be rewound. Winding also provides an opportunity to remove from the yarn such irregularities as slubs, thick ends, snarls, loose waste, and doubler's knots by passing it through a clearing device placed between the supply package and the winding drum or spindle. Some of the long-established clearers have since the end of the Second World War, been replaced by electronic clearers, notably the photo-electric clearer and the capacitance clearer.

Cotton warps and many types of warps made from man-made fibres are usually sized to reduce breakages in weaving caused by abrasion. Even as late as 1940, sizing held an element of mystery but this is gradually being dispelled, largely because of the need to develop new techniques for sizing man-made fibres, but also because of fundamental and painstaking investigations made at the British Cotton, Silk, and Man-Made Fibres Research Association (the Shirley Institute). The principle of the operation remains the same but greater reproducibility of results has followed the introduction of new sizing materials, such as chemically modified starches, carboxymethyl-cellulose, and water-soluble polymers, as well as the control of the amount of size applied using such devices as the Shirley automatic size box (introduced

FIG. 27.4. Saurer '100 W' multicolour automatic pirn changing loom for weaving woollen and worsted fabrics.

in 1942), and the control of the moisture content and the tension of dried yarn. Although the conventional arrangement is still used for sizing cotton yarns, machines for sizing rayon usually incorporate seven or nine drying cylinders of small diameters. Cylinders are now made from stainless steel instead of copper and the yarn may be prevented from sticking to the surface of the cylinders by coating them with Teflon, a fluorinated hydrocarbon polymer.

The principles on which looms are based have altered little, but the gradual advance in weaving technology has resulted not only in greatly improved productivity but also in better working conditions. The mass of belts and pulleys which characterized weaving sheds of the nineteenth century were, during the years 1920 to 1940, replaced by electric motors driving individual looms, while first tungsten electric lamp bulbs and then fluorescent tubes replaced the unsatisfactory gas mantles used for lighting. As to the looms themselves, they are wider and their appearance and design have markedly improved. Picks of weft are inserted at ever-increasing speeds, and where some of the more recent machines have been installed the characteristic clatter of the traditional weaving shed has given way to a busy hiss of high-speed water- or air-jet looms. While the delivery of warp threads and the take-up of woven cloth on the cloth beam have been the subject of much research leading to electrically controlled, rather than mechanical, warp-stop motions, more attention has been given to devising new methods of inserting weft, the aim being to eliminate the shuttle.

The modern conventional automatic loom is a sound piece of equipment, attaining an efficiency of 90–95 per cent, that is, the loom is productive for 90–95 per cent of its running time. The Northrop type of automatic pirn (reel) change retains its popularity and the performance of these looms has been improved by using hydraulic or hydropneumatic shock-absorbers on the shuttle brake, which automatically adjusts itself to variations in the impact of the shuttle. A further important development was the marketing in 1930, by the Swiss firm of Maschinenfabrik Saurer, of a new model (Fig. 27.4) which operates from one major shaft rather than from two, as on other automatic looms. Moreover the shafts which lift the warp threads are raised from below, thereby eliminating the superstructure associated with most types of loom, and greatly improving the appearance of the loom and of a shed housing a battery of such looms. The 50 000th machine of this type was installed in 1965.

The building of automatic looms suitable for weaving multicoloured fabrics necessitated a great deal of research to ensure that a pirn of yarn of the right colour was inserted in the shuttle to replace a used-up pirn, and major loom makers like Maschinenfabrik Ruti, and Crompton and Knowles, Inc. now market efficient machines [3].

Labour demands can be reduced still further by the use of the 'ALV' bobbin loader, developed by the firm George Fischer of Schaffhausen. The apparatus is the subject of a 1912 patent but was not introduced to the trade until the Milan Exhibition of 1959. The usual 24-pirn battery is replaced by a container in which as many as 180 pirns can be stacked. The base of the container has a moveable shutter which controls the delivery of a pirn on to the pirn guide and then by an ingenious pneumatically controlled system into the shuttle. The Leesona Corporation's 'Unifil' winder, introduced to the European market in 1957 but used in the U.S.A. for some time before, is another very successful device for increasing productivity in a weaving establishment. Here a pirn winder is incorporated in the loom. Weft yarn is wound on to the pirn from large packages and these full pirns are fed to the shuttle in the usual way. Each pirn is ejected from the shuttle, guided to a stripper, and returned to the winding beam. Some 12 pirns are in circulation at one time, only six or seven being full of yarn. This is in marked contrast to the 200–300 pirns per loom required in conventional automatic weaving. While the Northrop battery system has remained in favour, the alternative method of automatic replacement of whole shuttles rather than of pirns has fallen into disuse. It was popular for a time because any type of shuttle can be used and also as it was particularly suitable for weaving silk and silk-type fabrics.

In spite of the sophistication of many of the modern automatic looms, there were obvious advantages in developing shuttleless looms. The most widely used of these is the Sulzer weaving machine which uses a small gripper or 'bullet' (9 cm × 1·4 cm and weighing 38 g) to carry yarn from a package and insert it into the shed made by the warp threads (Fig. 27.5). The first patent for a weaving machine incorporating a gripper and a fixed supply of yarn appears to be that of Carl Pastor of Krefeld in 1911, but application for the important patents was not made until 1929, by R. H. Rossman of Munich. The first were granted in 1930 and later research and development work by members of the Sulzer Organization was covered by patents over the period 1933 to 1950. Commercial production of a machine weaving one colour of weft commenced in 1950, and this was followed by the introduction of two-colour machines (the 'ZS' range) in 1955, and four-colour machines (the 'U.S.' range) in 1959.

Although the Sulzer machine is relatively expensive, its productivity, reliability, high efficiency, and versatility ensured its commercial success, and by 1969 no fewer than 20 000 had been sold. The principle of the machine is as follows. Weft is taken from a large package and is fed to the gripper shuttle which is then driven through the shed. The yarn is tensioned and held in position by two clamps. It is cut off outside the clamps and beaten up into the cloth. The free ends of each pick are turned into the cloth to produce a tucked-in selvedge. Throughout the operation the weft is completely under control.

An alternative mechanical method for inserting weft without using a shuttle is the rapier system. The idea is not new, but commercial success came in 1945, with a machine developed by the Draper Corporation. Two carriers or rapiers enter the shed of warp threads simultaneously, one from each side. One of these carries the yarn and they meet half way across the warp, reversing direction at this point, where the weft yarn is transferred from the one

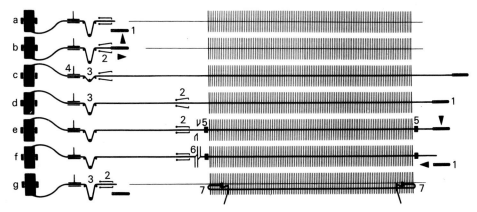

FIG. 27.5. Principles of the Sulzer weft insertion system.
(a). Projectile 1 moves into the picking position.
(b). Projectile feeder 2 opens after the projectile has gripped the end of the weft thread presented to it.
(c). Projectile has drawn the thread through the shed, during which time the weft tensioner (3) and the adjustable weft brake (4) act to minimize the stress on the thread at the moment of picking.
(d). Projectile 1 is stopped and pushed back inside the receiving unit housing, while the weft tensioner (3) holds the thread lightly stretched. At the same time feeder 2 moves close to the edge of the cloth.
(e). Feeder 2 grips the thread, while the selvedge grippers (5) hold the weft at both sides of the cloth.
(f). Thread is severed by the scissors (6) on the picking side and released by projectile (1) on the receiving side. The ejected projectile (1) is then placed on the conveyor which carries it outside of the shed back to the picking position.
(g). The thread has now been beaten up by the reed. The needles (7) tuck the thread ends into the next shed (tucked-in selvedge). The length of thread slackened by the return of projectile feeder (2) is taken up by the weft tensioner (3). The next projectile is brought to the picking position.

rapier to the other. Since weft yarn is supplied from one side of the loom only, the selvedge at one side is a series of 'hairpin loops' and at the other cut ends. In other looms, for example the Iwer machine, only one rapier is used.

The earliest patent for using a jet of air to propel a shuttle across warp threads was granted to J. C. Brooks in 1914. A different method of pneumatic picking is described in the patent of E. H. Ballou taken out in 1929, but the most important patent was not widely publicized until 1954. This machine was called the Maxbo loom after its inventor Max Paabo. Efficient control of the air-stream is the main reason for its success. A measured amount of weft from a supply package is projected through the shed formed by the warp threads, and after insertion the pick is cut at both edges of the warp. Although the loom has some limitations, the picks are inserted silently and quickly. Related to this machine is the water-jet loom, particularly suited to synthetic fibre yarns. It was the outcome of investigations by the Czech engineer, Vladimir Svaty, and was introduced to Western Europe at the Brussels exhibition in 1955, although a complete installation of 150 water-jet looms had then been in operation for some time in Czechoslovakia.

All the looms discussed so far produce cloth by projecting weft across the warps by a reciprocating mechanism. The limitations of this were recognized early in the history of loom making and some type of circular weaving was envisaged. The design of such a machine presented many obstacles but some success was achieved by the Saint Freres, the Dautricourt, and other circular looms, ingenious pieces of equipment which have found special application in weaving jute and asbestos, but which can, it is claimed, be used to make relatively plain fabrics from most other conventional textile yarns. In these machines the shuttle is kept in continuous circular motion through a shed formed by vertical warp threads arranged around the circumference of a circle. The main difficulty lies in the beating up of the weft.

Carpet weaving forms a special branch of the weaving industry. During the 1930s the production of tapestry and Chenille carpets declined rapidly, the former being discontinued in favour of the products of the Gripper Axminster loom. The new tufted carpets originated in the U.S.A., the first being made in 1949. The principle of the machine used is similar to that of a sewing-machine, but it is fitted with a row of closely-spaced needles instead of with one needle only. The length of the loop stitch can be controlled and the loop can be cut or uncut. Tufts are held in position by a coating of rubber or other adhesive, and tufted carpets are commonly backed with polyurethane foam. The tufted carpet industry has grown rapidly, largely because of the high speed of production and low labour costs, so that by 1963 it accounted for 70 per cent of the total carpet production in the U.S.A. Its main drawback was the limitation of design, but this has now been overcome by improvements in machinery and techniques for printing tufted carpets. Warp knitted structures made on Raschel machines are used both as loop pile and cut pile carpets.

Knitting. This century has seen a striking increase in the scope and importance of both warp- and weft-knitting, due in part to the acceptance of man-made fibres [4]. The knitting industry was originally mainly concerned with making hosiery and underwear (Vol. V, p. 595), but by 1900, significant amounts of knitted outerwear were beginning to appear. This branch of the industry has continued to grow and there has been a parallel increase in the production of knitted piece goods from which garments can be made by 'cut-and-sew' methods (Ch. 28). Silk hose were first replaced partially by cellulose acetate, but with the availability of nylon in 1940 the industry entered a new era. Lightweight shear stockings of unusual strength could be made on conventional fully fashioned or Cotton's machines, the operation being simplified by the introduction in 1950 of the 'loopless toe', which made unnecessary the usual subsequent operation of linking.

The setting properties of nylon enables satisfactory stockings to be made more simply on circular machines. The yarn is knitted on narrow circular machines in the form of a long tube which is then permanently set in the required shape by being drawn over a 'former' shaped like a human leg and then heated or steamed at a high temperature. The commercial acceptance of this procedure owes much to the development since the 1940s of methods for producing a series of complete stockings on one machine and for separating them. These seamless stockings are of good fit and find a ready market. With the introduction of highly elastic textured nylon yarns, the setting and shaping could be dispensed with, as the elasticity ensured a good fit, and fewer sizes were necessary to meet customer demands. Since the early 1960s stockings have been increasingly replaced by tights, and machine developments have now enabled them to be produced in one piece, although many are still made by joining together two 'long stockings'.

In 1900 the XL machine, a revolving cam system, was built by Stretham and Johnson, for making half-hose, socks, etc. It was converted to a revolving needle cylinder by Spiers in 1912, but the most important advance was the introduction of the Komet double-cylinder machine in 1921, by the British Wildt Company, now a member of the Bentley Engineering Group.

Many more fabrics are now made on larger circular knitting-machines [5]. The trend towards these machines was due partly to the influence of the American industry, which concentrated on the 'cut-and-sew' trade, in contrast to the British industry which emphasized 'knitting-to-fit', and had consequently built machines in several diameters. The first movement towards greater production was to increase the number of feeders, a decision which represented a change in attitude, for there are technical advantages in using a small number of feeders; for example, spirality is less. The commercial acceptance around 1950 of this principle demonstrated the great potential of multi-feed machines.

Tension control of yarns being fed to the knitting area is a significant factor in ensuring good knitting and a great deal of attention has, over the years, been devoted to this aspect of knitting-machine design. Other considerations must however, be taken into account. Observations in 1953 on the importance of stitch length led research workers at the Hosiery and Allied Trades Research Association in Britain (HATRA), to devise methods for controlling this parameter. Their studies on the development of a yarn-speed meter and a yarn-length counter resulted in the concept of feeding a known length of yarn to the knitting area rather than relying, as previously, on tension control. Several of these 'positive feed' devices are now available. In the first, due to HATRA and demonstrated to the public in 1959, yarn from the cone comes into contact with the high-friction non-slip surface of a roller driven at a speed which bears an appropriate relationship to that of the knitting circle. It was later in 1961, that I. K. Rosen introduced the simpler 'trip-tape' system which is now widely adopted.

Advances in machine design resulted in the manufacture of new knitted structures. For example, in 1908–9 machines for making interlock fabric were introduced by Scott and Williams, and in 1936 patterned interlock fabric appeared on the market. Serious inroads were made into the woven fabric trade with the production of single-jersey and double-jersey fabrics, the development by the U.K. Bentley group of the disc-type selection machine known as the 9RJ being a major advance [5].

A wide range of fabric structures, including the relatively new ones such as Punto-di-Roma and double piqué, can be made on modern machines. Considerable design scope is offered by many purely mechanical machines introduced during the years 1930–50 by such firms as Wildt and Mellor Bromley, but the availability of electronic equipment has increased the design potential of knitting-machines still further. Electronic devices were formerly mainly used as stop motions to detect thread or needle-butt breakages as soon as they occurred, but the invention by F. Morat of an important machine shown first in 1965 and known as the 'Moratronik' is based on electromagnetic selection of needles on a jacquard head circular knitting-machine. Flat knitting-machines similarly have improved in performance and scope, and produce an extraordinarily wide range of patterns and constructions.

The growth in warp knitting followed the introduction of nylon and other man-made fibres, and tricot and raschel machines, such as those produced by Liba and by Karl Mayer are good examples of modern engineering, with greatly increased production speeds. The use of light alloys, the redesigning of the drive through the thread guide bars, and a reduction in the distances travelled by moving parts have all contributed to this. Of special interest is the use, around 1940, of the compound or tubular needle on the 'fly needle frame' (F.N.F.) machine (Fig. 27.6). Tubular needles were in fact invented in 1858, but were not produced commercially at that time because of insurmountable engineering difficulties. The needle consists of a tongue which

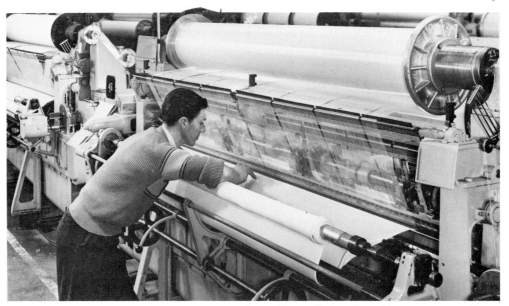

FIG. 27.6. The F.N.F. warp knitting (fly needle frame) machine.

rises and falls in a cylinder carrying a hook. It makes the knitting cycle less complex than that on the conventional bearded-needle machine, for the movement required from the needle is reduced by making both tongue and cylinder move in appropriate and opposite directions. The F.N.F. machine was introduced in 1946, although designed in 1938, and gave great impetus to the tricot trade which had started in 1930 mainly for making lingerie from cellulosic rayon yarns. It was essentially a high-speed machine operating at more than twice the speed of its competitors, and had several interesting features; it was provided with an eccentric drive instead of cams, and thread was fed positively to the knitting area at a constant rate irrespective of the beam diameter. The F.N.F. machine, has, however, subsequently declined in popularity mainly because the needles were cast in type-metal busses in groups of four and could not be replaced singly. Several of the ideas incorporated in this machine have, however, been used in the more common and less expensive bearded-needle machines, in which each needle moves in a groove in a needle bar, the groove supporting the stern of the needles. In more recent machines which use compound needles the difficulty inherent in the F.N.F. machine has been overcome.

Other methods of producing fabric. New possibilities were opened up when the idea was put forward of strengthening webs of any type of fibre by mechanical entanglement [6]. This was effected mechanically by punching the web with barbed needles, thereby drawing fibres through it. This process was first applied in 1900 to make paddings, rugs, and carpet underlays from webs of waste fibres and such materials as coconut fibres, jute, and ramie. The introduction of man-made fibres and improvements in the design of these needle looms widened the variety of textiles that could be made.

In a recent machine built by Hunter (U.S.A.) since 1964, the needles are arranged at an acute angle to the baseboard and this machine has been developed by the Chatham Manufacturing Co. (U.S.A.) to produce, by what is known as the Fiberwoven system, a wide range of blankets and other 'woollen-type' fabrics. In many cases the web is supported on, or interleaved with, a woven or knitted gauze-like fabric or a matrix of yarns.

Webs of fibres can also be given cohesion and strength by the incorporation of thermoplastic fibres or by application of an adhesive. Lack of suitable adhesives limited advances in this technique until the end of the Second World War, when developments in polymer chemistry led to new adhesives which found ready acceptance in the production of non-woven fabrics. Of special interest are synthetic rubber latexes based on butadiene or acrylic derivatives, which are sprayed on, or applied from a lick roll, to the web supported on a mesh screen. In all the methods listed above good preparation of the web is all important. A card or Garnett machine is the obvious piece of equipment for making webs, several layers of carded web being laid on top of one another until an appropriate thickness is reached. Because of the operating action of a card, the web tends to be stronger lengthways than widthways, and to minimize this effect webs can be cross laid, that is, fed on to a lattice which moves backwards and forwards across a second lattice moving at right-angles to it.

Probably the most uniform web is produced by laying down the fibres in a stream of air. The most successful method of doing this was patented by F. M. Buresh in 1948 and from this patent the popular machine known as the Rando-Webber has been developed by the Curlator Corporation (U.S.A.).

Webs can be given cohesion by stitching, usually in lines about 1 cm apart. This process was originally used to make waddings and similar materials but the idea has been developed, particularly in the Soviet Union and Eastern Europe, so that more sophisticated 'fabrics' can be made. Of particular significance is the Arachne process, demonstrated at the 1961 Prague symposium, but based on ideas developed over a period of some 20 years. A web is fed into the back of a warp knitting-machine, and the closely set warp yarns being delivered to the needles are knitted through it. From East Germany come three processes known respectively as Malipol, Maliwatt, and Malimo. The first can be regarded as a tufting process. The second entails stitching through a sheet of yarns. In the Malimo process a web of fibres forms the base for the stitching [6].

Although the production of woven velvets and plushes has been firmly established for many years the introduction of man-made fibres, particularly acrylics, opened up new possibilities. The lustrous appearance of these materials made them useful alternatives to mohair and cotton but more interesting was the utilization of the contractile properties of fibres such as those based on polyvinyl chloride. The fact that these fibres shrink when heated above a certain temperature was for long regarded as a serious limitation on their use, but this defect has been turned to advantage by a process for making pile fabrics patented in 1939 by the Bradford firm of Lister and Co. Ltd. The cut pile in these fabrics consists of a mixture of contractile and non-contractile fibres. When the cloth is placed in hot water or steamed during finishing the contractile fibres shrink, thereby producing a 'two-height' pile, which resembles the surface of a natural fur, for animal furs are usually made up of a fine short ground coat and longer coarse guard hairs. Fabrics made by the two-height pile technique have found ready acceptance as simulated furs, largely because of their relatively low cost. They can be produced by weaving, or by knitting on machines such as the Wildman-Jaquard in which a miniature carding machine producing a sliver is attached to a knitting machine [4]. The sliver is fed by V-shaped claw devices into the machine adjusted to knit a plain stitch fabric, and ingenious finishing techniques convert the sliver-knit fabric into a surprising range of simulated furs and related products.

IV. DYEING AND FINISHING

During this century the outlook of the dyeing and finishing industry has markedly changed, both as a result of the rapid growth of the chemical and dyestuffs industry, and of research on the chemical reactivity of textile fibres. Natural dyes have been replaced almost completely by synthetic dyes and entirely new chemicals have found extensive use in making textile fabrics more suitable for the purposes for which they are intended.

Advances in the chemistry of dyes during the twentieth century have made available a vast range of dyes which are listed and classified in that well-known reference work, the *Colour Index*, first issued by the Society of Dyers and Colourists (Bradford) in 1924, subsequent editions appearing in 1956 and 1971. Man-made fibres presented many challenges to the dyemaking industry, but these have been successfully met and are discussed in another section of the present work. Some notable advances should, however, be listed here [6]: (i) synthesis of indigo (1894); (ii) introduction of more extensive range of anthraquinone dyes (1901 onwards); (iii) advances in azoic dyeing of cotton due to discovery of naphthol AS (1912–13); (iv) discovery of disperse dyes (1923); marketing of copper phthalocyanine as Monastral Blue BS (1934); (v) production of Procion reactive dyes capable of being attached chemically to the fibre (1956); (vi) development of sublistatic (transfer) printing (1960s).

The processes used for finishing wool, cotton, and flax fabrics at the end of the last century, were labour-intensive and often time-comsuming. Cotton was bleached by boiling in lime or caustic soda solution followed by treatment with bleaching powder or sodium hypochlorite. Over the years, owing largely to the introduction of stainless steel and to the fall in price of hydrogen peroxide, hypochlorites have been replaced by hydrogen peroxide. In addition the bleaching of cloth in rope form has given way to open-width processing, and batch processing in kiers and storage 'boxes' has been replaced by continuous operations in which the cloth is successively impregnated with caustic soda, heated in a J-box, impregnated with alkaline hydrogen peroxide, heated in a second J-box, and finally washed free from reagents in an openwidth washer. Ranges of this type were introduced by Du Pont as early as 1941 and are gradually replacing the older equipment. The introduction of 'optical bleaching agents' after the Second World War set completely new standards of whiteness.

Man-made fibre fabrics are dyed after being washed, but nylon and polyester fabrics require an additional process to prevent them from shrinking irregularly and unpredictably when treated in hot or boiling water. Before dyeing therefore they must be 'heat set', that is, heated at 180–220 °C while held at the appropriate width on a stenter. The success of heat setting greatly contributed to the commercial acceptance of truly synthetic fibres in 1939.

Cotton cloths crease readily, necessitating frequent washing and ironing. With the object of producing crease-resistant cotton cloths the firm of Tootal Broadhurst Lee in Manchester commenced a research programme in 1918, and in 1928 filed a patent for a process which added a new dimension to textile finishing. The fabric is impregnated with the components of a synthetic resin, for example urea formaldehyde and a catalyst such as dihydrogen ammonium phosphate, and is then dried and baked at a temperature of 140–150 °C, when a condensation polymer is formed inside the fibre. Subsequent washing completes the process and the finished fabric has greatly increased resistance to creasing. The effectiveness of the process was originally attributed to the deposition of polymer in the fibres, but later investigations have shown that new cross-links are formed between the cellulose chains. The original Tootal process has been followed by many others based on cross-linking by the use of other reagents.

Polymer or resin finishing and cross-linking treatments have led to other new developments in cotton and rayon processing, for they can also be used for permanent finishes and what are known as durable press effects. The underlying principle is to impregnate the fabric with a cross-linking agent, then flatten or calender it to give good appearance, and finally apply heat to bring about the cross-linking and so fix the finish. Similarly, the pre-sensitized material can be made into a garment, such as, trousers, the creases inserted, and the garment heated to fix the creases permanently in the desired position.

The Second World War stimulated the development of finishes designed to make cotton fabric flame-resistant, water-repellent, and immune to attack by micro-organisms (mildew). During the war, attention was concentrated on performance and rigorous standards had to be met. Water-repellency was achieved by wax-aluminium acetate type reagents, even though they were removed in dry cleaning. To prevent this, reagents such as stearoamido-methyl pyridinium chloride were introduced in 1947 under the trade names of Velan PF (I.C.I.) and Zelan (Du Pont), and later metal complexes and silicones (1958) were used for the same purpose.

The production of flame-resistant cotton was the concern of scientists immediately after the First World War, and temporary finishes based on such reagents as ammonium phosphate and mixtures of borax and boric acid were popular. A treatment based on deposit of stannic oxide, which was not so readily removed by washing, was invented by A. G. Perkin in 1913, but it was not until the Second World War that permanent finishes based on THPC (tetrakis hydroxymethyl phosphonium chloride), and on mixtures of antimony oxide and chlorinated hydrocarbons were evolved. The growing concern about the large numbers of fatal accidents due to inflammable fabrics has, since the last war, led to the introduction of regulations demanding that cotton and viscose rayon be made flame-resistant.

Resistance to mildew is achieved by the use of such compounds as salicyl-anilide and copper salts, the latter being particularly effective.

During the Second World War special aspects of wool finishing were also important. To prevent wool socks and fabrics from shrinking during washing they were given chlorination treatments. A particularly successful one entailed exposing the fabric to the action of chlorine gas (1933). Since the war a great deal of work has been carried out on making wool resistant to shrinking and this has led to many successful processes based on permonosulphuric acid (1956), on DCCA (dichloroisocyanuric acid) (1955), and on deposition of polymers on the surface of the wool fibres (1949 onwards) [8]. As a result of intensive research carried out since 1930, wool can now be made resistant to attack by clothes moths and other insects, chlorine-containing compounds such as the Eulans (1928 onwards), DDT (1943), and Mitin FF (1939) being widely used.

REFERENCES

[1] HONEGGER, E. *Ciba Review*, No. 3, 3 (1965).
[2] SELLING, H. J. *Twistless yarns*. Merrow, Watford (1971).
[3] HONEGGER, E. *Ciba Review*, No. 2, 3 (1966).
[4] *Ciba Review*, No. 6, 5 (1964).
[5] HURD, J. C. H., *Proceedings of the Thirteenth Canadian seminar*, 29 (1972).
[6] *Ciba Review*, No. 1, 10 (1965).
[7] ROWE, F. M. *The development of the chemistry of commercial synthetic dyes (1856–1938)*. Institute of Chemistry, London (1938).
[8] *Wool Science Review*, No. 36, 2 (1969).

BIBLIOGRAPHY

VON BERGEN, W. *American wool handbook* (3rd end.). Wiley-Interscience, New York (1970).
DUXBURY, V., and WRAY, G. R. (eds.) *Modern developments in weaving machinery*. Columbine Press, Manchester (1962).
HAMBY, D. (ed.) *American cotton handbook* (3rd edn.), Vols. 1 and 2. Interscience, New York (1966).
LORD, P. R. (ed.) *Spinning in the '70s*. Merrow, Watford (1970).
MARSH, J. T. *Introduction to textile finishing* (2nd edn.). Chapman and Hall, London (1966).
MONCRIEFF, R. W. *Man-made fibres* (6th edn.). Heywood Books, London (1975).
PRESS, J. J. (ed.) *Man-made fibres encyclopedia*. Interscience, New York (1950).
Review of Textile Progress, **1–18**, Textile Institute, Manchester, and Society of Dyers and Colourists, Bradford (1949–67).
SPIBEY, H. (ed.) *British wool manual* (2nd edn.). Columbine Press, Buxton (1969).
Textile Progress, **1–7**. Textile Institute, Manchester (1969–75).
Textile terms and definitions. Textile Institute, Manchester (1975).
TROTMAN, E. R. *Dyeing and chemical technology of textile fibres* (5th edn.). Griffin, London (1975)
WRAY, G. R. (ed.) *Modern yarn production from man-made fibres*. Columbine Press, Manchester (1960).

28

CLOTHING MANUFACTURE

H. C. CARR

IN general the period 1900 to 1950 saw very little which could be identified as conceptually new in the clothing industry, although continuous refinement and adaptation took place. Such developments as did occur were concentrated at the beginning or end of this period: in pressing and cutting equipment and the evolution of new stitch types from 1900 to 1920 and in large-scale sizing surveys, faster machines, and production engineering developments in the 1940s and early 1950s.

I. THE APPLICATION OF STATISTICS TO THE DEVELOPMENT OF SIZE SCALES

The objectives of the systematic study of human measurements are to improve the fit of ready-to-wear garments cut from flat patterns. H. Wampen in 1900 claimed to have measured vast numbers of human figures, but the statistical basis of his work is unclear, and he shows a constant tendency to relate his work to artists' models. He did, however, elucidate the concept of proportions, an idea which formed the basis of much pattern-cutting practice.

H. Simons in 1933 acknowledged much more statistical information, in particular from the Medical Department of the U.S. Army, gathered in 1917 and 1919. He reports on surveys conducted by the U.S. Government in the First World War, in which a thousand recruits in each of the given chest and waist measurements were measured to establish a number of mean average body dimensions as a guide for making patterns for uniforms.

The techniques used in later, large-scale surveys attempted to overcome three problems. First, how to choose the key items on which to base the sizing systems. Secondly, how to decide what particular values of the key items should constitute each separate size. Thirdly, how to avoid the distortion of average measurements arising from the presence of individuals with unusual measurements in the size categories. The first problem was solved in terms of correlation, or the degree of association between pairs of measurements, a high multiple coefficient implying that the key measurements account for most of the variation in particular measurements. The second problem demanded that when size groups were constructed from the key measurements the individuals within the group had to be as similar as possible; or, when the variation attributable to the key measurements had been removed, the residual standard deviation had to be very small. The third problem was overcome by the use of regression coefficients to smooth out random fluctuations in the sample.

The first large-scale survey of women's sizes was undertaken in the U.S.A. in 1939–40 and published in 1941; it covered 10000 women. In Holland, 5000 women were measured between 1947–50. Since this was the era of the 'new look', and the waist was the critical area of fit, this led the survey to make waist circumference and height the key measurements. The survey was not accepted by the Dutch clothing industry. In Britain, 5000 women were measured in 1951, and the results of the survey published in 1957. The methods used, and the results obtained, were broadly similar to those of the American survey. The British survey discovered that in two obvious respects most existing size charts were based on false assumptions. The mean average height of women was assumed as 5 ft 6 in (168 cm) instead of the correct 5 ft 3 in (160 cm) and the relationship between hip girth and bust girth was found to be wrong in many size charts. The American survey resulted in the publication of a proposed commercial standard in 1952; but the British survey led to a British Standard only in 1963.

II. CUTTING ROOM METHODS

Marking operations. Improvements in marking operations aimed to reduce the work content involved in duplicating markers and at the same time to impose control over the amount of material used, since material constituted approximately half the total cost of garment manufacture in the earlier years of the twentieth century.

In 1897 the Marsden lay marker was developed. This was made of lightweight pattern card. The outlines of the patterns were perforated by a hole-punching machine, similar to a sewing-machine. The marker was laid on top of the spread lay and powdered with french chalk. When the marker was removed, a line of chalk dots remained embedded in the material. This device found its chief use in men's tailoring factories, where standard markers could be used repeatedly, in particular on trimmings and trousers. It was quicker than chalking directly on to material, but the powdering process could be fairly lengthy.

In the 1930s the dress industry in the U.S.A. developed the use of paper markers. Seven copies could be obtained simultaneously by the use of three double-sided carbons. The marker was then attached to the top of the spread. In 1929, M. E. Popkin described a method of photographic duplication, using a camera carrying large rolls of paper film [1]. Popkin claimed that only three different widths of marker need be made to cover all fabrics, but today this would not be considered to give acceptable fabric utilization. In 1947 the Heavy Clothing Working Party Delegation from Britain to the U.S.A. saw only one factory using blueprinting photographic equipment [2]. Spirit duplicating and the wider use of photographic methods of duplicating markers were not established until after 1950. Similarly, miniature lay planning, using one-fifth-scale patterns in order to provide the overall view of markers necessary in a conceptualizing rather than an analysing process, was not developed until the early 1950s. Only then could serious attempts be made to improve the efficiency of markers and utilization of material.

Spreading operations. The origin of the spreading machine is not known. In 1920 B. W. Poole described a spreading machine fitted with a set of edge guides [3]. The machine had a carriage running on rails fitted to the edge of the table. This carried a piece of fabric which unwound as the machine was pushed along the spread. The number of plies was counted automatically. By the 1940s the machines were more sophisticated, with a motorized drive and a platform for the operator to ride on.

Cutting operations. In 1888 Jacob Bloch perfected a circular-knife portable cutting machine operated on a d.c. electricity supply. One problem with the round knife was that cutting curves was slow because the cloth had to be raised on the curves so that it would reach the furthest part of the circular blade; the cut would otherwise not be the same on the bottom as on the top of the spread. Hence, G. P. Eastman, in the U.S.A., in the same year developed a portable straight-knife cutting machine, which reciprocated vertically (Figs. 28.1(a) and (b)). In 1938 the basic machine was improved significantly by the incorporation of mechanical sharpening devices to replace the hand methods formerly used [4]. One further development was the transfer of die-cutting techniques for small parts, such as collars, from the shoe industry

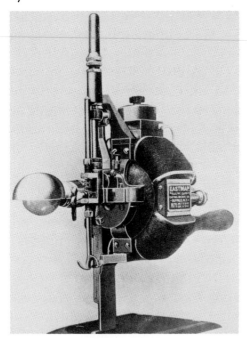

FIG. 28.1(a). The Eastman portable straight-knife cutting machine, *c.* 1900, an example of a series which began with the original invention in 1888.

(b). Later models of Eastman straight knives in use in a cutting room in Britain, *c.* 1925.

into a few U.S.A. clothing factories in the 1940s, although die cutting had been known for many years before that.

III. NEEDLE POINT ENGINEERING

The basic two-thread lock-stitch sewing-machine (Vol. V, p. 588) is designed to drive, in correct timing, the elements of the sewing mechanism. The needle passes through the fabric plies to its lowest position, with the eye carrying the thread below the throat plate hole, and the presser-foot holding the fabric stationary on the feed dogs. As the needle begins to rise, the thread loops away from the needle, to be picked up by the point of the hook. The hook makes the loop big enough to pass completely round the bobbin case. As the needle continues to rise the take-up lever also rises, drawing up the 'intraloop' of top thread and underthread. As the needle reaches its highest point the thread is locked in the centre of the fabric, and the next cycle begins. During the first part of the descent of the needle, the feed dogs carry the fabric forward the distance of one stitch, and the presser-foot rises sufficiently to allow the fabric plies to move forward.

All the above elements of the sewing mechanism were available to sewing-machine manufacturers in 1900. The pointed needle, with a straight grooved eye, set in a vertically reciprocating needle bar, was used by Isaac Singer in 1850, as was the take-up lever. A. B. Wilson developed the four-motion feed dog in 1854, although Singer had used a vertical, spring-loaded presser-foot in conjunction with less efficient feeding devices. The combined use of these two elements allowed continuous sewing in either a straight or curved line. Two methods of carrying the underthread were in use in 1900: the shuttle which reciprocated in a horizontal plane, first used by B. Thimmonier in 1832 and developed as a double-pointed shuttle by Wilson in 1849; and a rotary hook, with stationary circular bobbin (Figs. 28.2(a) and (b)), developed by Wilson in 1851.

It was Singer who first used a foot treadle to drive his machines in 1850; a battery-driven electric motor was used to drive a domestic sewing-machine in 1885; and in 1891 the first industrial sewing-machines (in fact for sewing carpets) were fitted with individual electric motors. The 1907 Census of Production in Britain counted over 90000 power-driven machines, but over 60000 treadle machines were then still in use in the clothing industry [5]; treadle machines continued to be used in significant numbers until well into the 1920s. From about 1870 the manufacturing of sewing-machine mechanisms was precise enough to make it possible to drive a sewing-machine at a

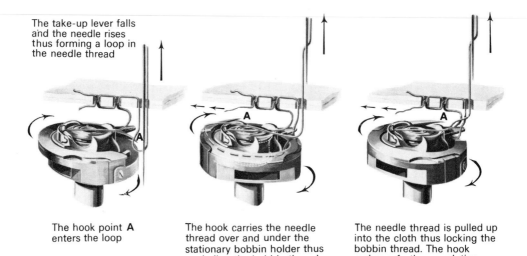

The take-up lever falls and the needle rises thus forming a loop in the needle thread

The hook point **A** enters the loop

The hook carries the needle thread over and under the stationary bobbin holder thus encircling the bobbin thread

The needle thread is pulled up into the cloth thus locking the bobbin thread. The hook makes a further revolution while the next loop is forming

FIG. 28.2.(a). The Singer oscillating-hook machine, 1908, showing the stitch-forming action.

(b). The same machine partially sectioned for exhibition.

higher speed than could be achieved by a foot treadle. By 1900 the typical power-driven clothing factory used line shafting with steam, gas, or later electricity as the motive power. A common arrangement in a Lancashire shirt factory of the 1920s was a long line of double benching, driven from an overhead shaft with flat leather belting, and using a 5-hp d.c. motor (Fig. 28.3) [6].

B. Frank records the claim that 4000 stitches per minute could be achieved by 1900 [7], but elsewhere R. Heywood claims only that the early individual electric motors, rated at $\frac{1}{4}$ hp ran at 1425 stitches per minute [8]. Later motors introduced in the 1930s, rated at $\frac{1}{3}$ hp or $\frac{1}{2}$ hp, enabled about 3000 stitches per minute to be attained. In the U.S.A. in the 1940s, machines produced about 5000 stitches per minute.

The history of the two-thread lockstitch sewing-machine from 1900 to 1950 is basically the story of refining and adapting an already invented mechanism in order to achieve greater productivity. In 1938, W. Riches still felt it important to describe what he called the 'vibrating shuttle' machine as well as the 'central bobbin' machine. Later generations of machines, however, all used the principle of the rotary hook with stationary bobbin, since this could be adapted to ever higher sewing speeds. In addition machines introduced in

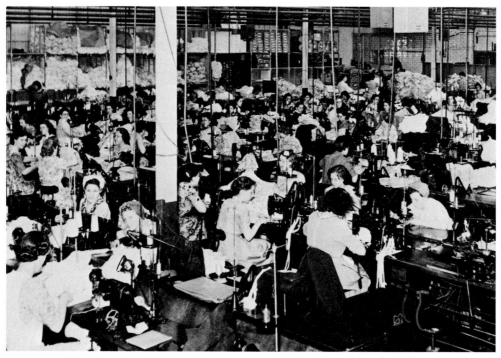

FIG. 28.3. A typical pre-1930s sewing room showing line and bench shafting.

the U.S.A. in the 1940s, but not in Britain until the 1950s, incorporated automatic or semi-automatic lubrication.

IV. MACHINES EMPLOYING OTHER STITCH TYPES

British Standard 3870: 1965 (*Schedule of stitches, seams and stitchings*), lists 72 stitch types in eight classes: chain-stitch, hand-stitch, lock-stitch, multi-thread chain-stitch, overedge-stitch, flat-seam stitch, single-thread lock-stitch, and combination stitch.

Single-thread chain-stitch (101) became a practical possibility when J. E. A. Gibbs developed the first rotary chain-stitch hook in the 1850s. Two-thread chain-stitch (401) was developed by W. O. Grover a few years earlier, in an attempt to do away with reloading the bobbin of the basic two-thread lock-stitch. More complex types became available by the 1890s. Single-thread and two-thread overedging began in the late 1850s and 1860s, but three-thread overedging stitches, classes 504 and 505, were not developed until 1897 and 1903 respectively by J. Merrow. These machines used a curved, eye-pointed needle in conjunction with an underthread and cover thread looper, making the edge covering much firmer and more suitable for securing the edges of easily frayed woven fabrics. The development of this machine encouraged factory production, since it provided an economical way of finishing edges in bulk for the large manufacturer, but it was too costly for the homeworker. In 1914 J. P. Weis and C. E. Johnson brought out an over-edging machine with four needles, one looper, and one combined looper and spreader; its stitch was similar to the modern 506. The development which made these machines possible was the refining of what was at first no more than an extra needle underneath the machine into an open-ended looper.

In 1891 H. H. Fefel, of the Union Special Machine Company, produced the first flat-seam stitch (601) using two needles, a spreader and a looper. In the same year G. Munsing developed stitch type 603, using two needles, two spreaders on top and one looper underneath, and type 604 with an additional needle. In 1906 S. Borton, of the Willcox and Gibbs Sewing-machine Company, was granted a patent for the first of the four-needle flat-seam stitches, using four loopers and one spreader. This became available commercially bearing the brand name 'Flatlock' in 1912. These complex stitch types had two advantages; firstly, suitability for use with knitted fabric structures, since the stitch type would elongate at least as much as the fabric; secondly, the ability to cover both raw edges when the seam was sewn with the edges of the fabric components overlapping instead of being face to face, as was more usual. This machine resulted in large savings because formerly the production of flat seams entailed separate seaming, trimming, and seam-covering operations. Flatlock machines were developed later than multithread chain-stitch, because in most cases the stitch formation of flat seam stitches was similar to that of chain-stitch, with the addition of one or more covering threads. The potential speed of the first Flatlock machines was 2800 stitches per minute but the later, automatically lubricated, versions achieved 3800 stitches per minute [9].

The blind-stitch principle, in which a curved, eye-pointed needle penetrates the surface of the fabric and emerges on the same side as it enters, produces a stitch which does not appear on the other side of the fabric. This principle made its appearance in the U.S.A. in 1878, but the first commercially practical machine was developed by C. A. Dearborn in 1900 for the blind felling of the bottom of boys' short trousers [10]. An improved machine was introduced in 1902 for the padding of jacket collars and lapels.

In general the sewing-machine does not simply mechanize hand sewing, but produces stitches of an entirely different type. In hand sewing the needle passes right through the fabric and re-enters it from the other side. Problems arise in mechanizing this process because only a short length of thread can be used and therefore only a short seam sewn. However, a machine was patented in 1933 which overcame the problem by using a 'floating' two-pointed needle [11]. The commercial possibilities of this principle were exploited from the late 1940s in a machine for 'hand stitching' the edges of men's jackets.

V. THE EVOLUTION OF AUTOMATIC MACHINERY

The basic sewing-machine does no more than produce stitches continuously. Every other feature of the sewn seam is determined by the actions of the operator: chiefly the shape of the sewing line, and the matching and fitting of the parts being assembled. The first machines which sewed a predetermined pattern of stitches were for buttonholing and button sewing.

The patent for the first automatic buttonhole machine was granted to J. Reece in 1881. The machine cut and sewed an eyed buttonhole. The increasing number of clothing factories prompted the design of a buttonhole machine specially for use on men's tailored suits in 1908, with stitching more closely resembling the effect produced by hand. Indeed the machine was known as the 'Hand Hole' machine. In 1940 a new fast machine was produced for shirt and dress buttonholes, but the Second World War curtailed its development until the late 1940s.

Button-sewing machines were developed in the late 1880s and bartacking machines, first used for finishing the ends of buttonholes, followed. These machines were based on cam-controlled mechanisms. With hindsight, it is perhaps surprising that this line of development was not pursued more assiduously before the mid-1950s, when a spate of cam-based developments of machines for small-profile sewing operations took place.

Instead, the coupling of auxiliary devices to sewing-machines engaged the attention of a number of engineers working in sewing factories. There were ingenious ideas for the continuous application of trimmings such as ruffled borders on curtains. In 1916 appeared the 'Weis Design Cutting and Sewing-machine' incorporating cutting, feeding, and sewing in order to cut and sew sleeve and leg sections for knitted garments continuously. One of the larger overall manufacturers in the U.S.A. developed, about 1930, a line-shaft table equipped with 14 twin-needle chain-stitch machines for making shoulder straps for overalls. One operator supervised all the machines, rectifying thread breaks and replenishing rolls of material. A separate machine cut banding into strap lengths. The claimed capacity of this arrangement was 3000 dozen overalls per day [12]. Unfortunately, most of these ingenious operations represented a blind alley, because they were suitable only for operations involving the simplest forms of handling. Not until further work was done on clamps, clamp technology, hydraulic and pneumatic operation, and electronic controls, all coming after 1950, could automatic sewing-machines play a more significant part in the manufacture of garments.

VI. PRESSING OPERATIONS

In 1900 pressing was carried out using two devices. One was the traditional hand iron. The second was the 'jumper' iron, fastened to a lever mounted on a table and operated by pressure on a foot pedal. The irons were heated by a mixture of compressed air and gas, and the temperature was controlled by the size of the flame. The moisture necessary for pressing came from a damp cloth spread over the garment.

The steam press was invented between 1898 and 1905 by Adon J. Hoffman, a New York tailor, who, on breaking his arm, found himself unemployed, and conceived a machine for pressing clothing. Hoffman's machine consisted of a cast-iron box or head, which moved up and down on to a wooden table by the operation of a foot pedal. A small boiler at the right of the table produced steam, which was sprayed from the head on to the garment. By the time steam presses were introduced into Britain in 1909 they incorporated a heated work-holder or 'buck'. Until 1919 all steam presses had a flat head and buck which sandwiched the garment, with steam coming from the head. The crucial addition to the machine came at this time in the form of a vacuum device, sucking air through the buck. This enabled heat and moisture to be removed from the garment and prevented the bubbled effect so easily obtained on the early presses. In addition the linkages between the pedal and head were improved to make locking easier, resulting in greater pressure between the head and the buck. These improvements influenced clothing design because they made it possible to press sharp creases in trousers and pleats in skirts. A comparison of modern clothing with nineteenth-century photographs brings home the significance of steam pressing.

As soon as the basic steam press was fully developed, steam and vacuum were applied to the 'jumper' iron (Fig. 28.4). In addition both this iron with its buck, and the steampress itself, were produced in many different shapes to suit the particular shapes of garment parts in both final pressing and under-pressing operations. This development entailed the use of much larger boilers and vacuum motors with piped distribution systems throughout the clothing factory.

VII. PRODUCTION ENGINEERING IN THE SEWING ROOM

It is very difficult to generalize about production engineering (Ch. 43) in the sewing room because throughout its history the clothing industry has exhibited great diversity. The range of garment types has meant a large range of work-content, from a few minutes to several hours per garment. The factories which manufactured garments between 1918 and the Second World War varied in size from a few workers to several thousand. In 1942 establishments in Britain employing over 100 formed 14 per cent of the total, but they employed 57 per cent of the industry's labour force [13]. This situation reflected some concentration into larger establishments owing to wartime conditions. From the time the sewing-machine industry was well established, the basic tool of the clothing industry has been cheap and readily available. Even in 1898 the output of the Singer factory at Kilbowie, Scotland, was 10 000 sewing-machines per week [14]. From the beginning, a ready trade in second-hand sewing-machines developed. This enabled the small entrepreneur to enter the industry with very little capital outlay. His main needs were the availability of markets and design flair. Since the industry was labour-intensive, and labour costs comprised about half the total conversion

FIG. 28.4(a). Bespoke tailoring section. The largest number of expert men tailors under one roof are here occupied in the production of Montague Burton bespoke tailored garments. Note the 'jumper' irons on the right.

FIG. 28.4(b). A scene in the making up room; 3500 girls are employed in this room alone. Again, note the 'jumper' irons in the foreground, and the line shaft bench arrangement of sewing-machines.

costs, there has always been pressure to keep labour costs low in order to remain competitive in a highly fragmented industry. Hence women and immigrants have always formed a very high proportion of the total labour force and the earliest Trade Boards included several for sections of the clothing industry, evidence of a relatively high incidence of low wages in these sections. The larger factories tended to concentrate in those sectors of the industry in which the work content of the garment was bigger, such as mens wear, or where the fashion element was less important, such as shirts. It was in those factories that the opportunities for production engineering were greatest.

One characteristic of the clothing industry was the long time-lag between the invention of new machines and their wide commercial application. A number of reasons have been suggested for this: they include the effect of fashion in increasing the variability of assembly processes; the availability between the wars of a low-paid labour force; the small size of many clothing businesses, which could not use sophisticated equipment economically; and the traditional outlook of many factory managements. In addition, two more radical explanations may be put forward. First, the sewing-machine, although a flexible and long-lasting power tool, produced only very low added value compared to machines in other industries, because the typical production unit was one machine with one operator, and because the average sewing operation contained a very high proportion of handling time. This meant that little money was available to employ specialist staff to investigate and improve the method of manufacture, or to replace obsolescent production machinery. Secondly, the level of sophistication required to advance from basic sewing equipment was very high, owing to the nature of the raw material. Fabric webs are limp: the kernel of the difficulty lies in the fact that they bend in all directions. It was, therefore, much more difficult to invent jigs and automatic equipment for performing sewing operations.

However, in 1920 Poole was able to describe a system for producing coats, used in a New York menswear factory, which divided the manufacturing process into 45 operations. This system employed 11 different sewing-machines, but over half the operations were performed by hand [15]. In 1929, Popkin described a method for producing coats, involving 117 operations, using 15 different sewing-machines, but with less than a quarter of the work performed by hand [16]. In the 1930s and 1940s production practice in the U.S.A. and Britain tended to diverge. In the U.S.A. the best practice attacked the problems of handling within the operation, developing many ingenious home-made guides and attachments, using self-contained sewing units with individual motors instead of a shaft-driven line of machines, and encouraging the maintenance of buffer stocks between operations so that, with the operator in command of her own output, individual piecework payment methods might have their full impact. Inter-operation handling was attacked by the use of overhead electrical distribution systems, which facilitated, with single-unit sewing-machines, complete mobility of layout [17]. In Britain, some larger factories adopted the same general approach, but many adopted a conveyor-based production system. The system attacked only inter-operation handling, which in most factories represented the smaller component of the total handling time. It reduced the tasks of supervision in many respects, but, in reducing work in progress to a minimum, it reduced the opportunity of applying piecework payment methods. It tended to divert attention away from improving individual operations, and created in a highly variable work situation a production line geared to the pace of the slowest operator or the longest operation, thus inducing lower productivity of labour [18].

REFERENCES

[1] POPKIN, M. E. *Organisation, management and technology in the manufacture of men's clothing*. Pitman, London (1929).
[2] *Board of Trade Working Party Reports: Heavy clothing*. H.M.S.O., London (1947).
[3] POOLE, B. W. *The clothing trades industry*. Pitman, London (1947).
[4] PRANSKY, A. I. The history of the cloth cutting machine. *Bobbin*, **7**, 23 (1965). and *Eastman 50th anniversary*. Eastman, London (1971).
[5] WRAY, M. *The women's outerwear industry*. Duckworth, London (1957).
[6] HEYWOOD, R. *Clothing Institute Journal*, **12**, 431 (1964).
[7] FRANK, B. *The progressive sewing room*. Fairchild Publications, New York (1958).
[8] HEYWOOD, R. *Clothing Institute Journal*, **12**, 433 (1964).
[9] SCHEINES, J. (ed.) *Apparel engineering and needle trades handbook*. Kogos, New York (1960)
[10] *Tailor and Cutter*, 21 June 1935.
[11] GILBERT, K. R. *Sewing-machines*. H.M.S.O., London (1970).
[12] *Automatic equipment and work aids in the sewing room*. American Apparel Manufacturers' Association, New York (1963).
[13] *Board of Trade Journal*, **151**, 478 (1945).
[14] *Tailor and Cutter*. 31 March 1898.
[15] POOLE, *op. cit.* [3].
[16] POPKIN, *op. cit.* [1].
[17] *Board of Trade Working Party Reports: Heavy clothing* (Appendix III). H.M.S.O., London (1947).
[18] Ibid.

BIBLIOGRAPHY

BARKER, H. A. F. *The economics of the wholesale clothing industry of South Africa 1907–1957.* Pallas, Johannesburg (1962).

Board of Trade Working Party Reports: Heavy clothing. H.M.S.O., London (1947).

Board of Trade Working Party Reports: Light clothing. H.M.S.O., London (1947).

Board of Trade Working Party Reports: Rubber proofed clothing. H.M.S.O., London (1947).

BOARD OF TRADE (KEMSLEY, W. F. F.) *Women's measurements and sizes.* H.M.S.O., London. 1957.

DOBBS, S. P. *The clothing workers of Great Britain.* Routledge, London (1928).

FINLAY, E. Trends in the men's tailoring industry of Great Britain. Thesis (unpublished) presented for the degree of Ph.D. in the University of London (1947).

GILBERT, K. R. *Sewing-machines.* H.M.S.O., London (1970).

KALECKI, E. Historical outline of developments in ironing and pressing. *Clothing Institute Journal*, **20**, 387–94 (1972).

LYONS, L., ALLEN, T. W., and VINCENT, W. D. F. *The sewing-machine: An historical and practical exposition of the sewing-machine from its inception to the present time.* John Williamson, London (*circa* 1930).

O'BRIEN, R., and SHELTON, W. C. *Women's measurements for garment and pattern construction.* U.S. Department of Agriculture, Washington (1941).

POOLE, B. W. *The clothing trades industry.* Pitman, London (1920).

POPKIN, M. E. *Organisation, management and technology in the manufacture of men's clothing.* Pitman, London (1929).

RICHES, W. *Lock-stitch and chain-stitch sewing-machines.* Longmans, London (1938).

SEIDMAN, J. *Labour in twentieth century America: the needle trades.* Farrar and Rinehart, New York (1942).

SIMONS, H. *The science of human proportions.* Clothing Designer Co. Inc., New York (1933).

SITTIG, J., and FREUDENTHAL, H. *De juiste maat.* L. Stafleu, Leyden (1951).

WAMPEN, H. *Anthropometry; or the geometry of the human figure.* John Williamson, London (1900).

WRAY, M. *The women's outerwear industry.* Duckworth, London (1957).